Mechanics of Hydraulic Fracturing

Mechanics of Hydraulic Fracturing

Experiment, Model, and Monitoring

Edited by

Xi Zhang
China University of Geosciences
Wuhan, China

Bisheng Wu
Tsinghua University
Beijing, China

Diansen Yang
Wuhan University
Wuhan, China

Andrew Bunger
University of Pittsburgh
Pittsburgh, PA, USA

WILEY

This edition first published 2023

Registered Office(s)
John Wiley & Sons, Inc., 111 River Street, Hoboken, NJ 07030, USA

Editorial Office
Boschstr. 12, 69469 Weinheim, Germany

For details of our global editorial offices, customer services, and more information about Wiley products visit us at www.wiley.com.

Wiley also publishes its books in a variety of electronic formats and by print-on-demand. Some content that appears in standard print versions of this book may not be available in other formats.

Library of Congress Cataloging-in-Publication Data
Names: Zhang, Xi (Geologist), editor.
Title: Mechanics of hydraulic fracturing : experiment, model, and monitoring / edited by Prof Xi Zhang, China University of Geosciences, Prof. Bisheng Wu, Tsinghua University, Prof. Diansen Yang, Wuhan university, Prof. Andrew Bunger.
Description: First edition. | Hoboken, NJ, USA : Wiley, 2023. | Includes bibliographical references and index.
Identifiers: LCCN 2022030857 (print) | LCCN 2022030858 (ebook) | ISBN 9781119742340 (hardback) | ISBN 9781119742418 (adobe pdf) | ISBN 9781119742456 (epub) | ISBN 9781119742487 (oBook)
Subjects: LCSH: Hydraulic fracturing.
Classification: LCC TN871.255 .M43 2023 (print) | LCC TN871.255 (ebook) | DDC 622/.3381–dc23/eng/20220907
LC record available at https://lccn.loc.gov/2022030857
LC ebook record available at https://lccn.loc.gov/2022030858

Cover Design: Wiley
Cover Image: © Studio One-One/Getty Images

Set in 9.5/12.5pt STIXTwoText by Straive, Pondicherry, India

Contents

List of Contributors

Andrew P. Bunger
Department of Civil and Environmental Engineering, University of Pittsburgh, Pittsburgh, PA, USA
and
Department of Chemical and Petroleum Engineering, University of Pittsburgh, Pittsburgh, PA, USA

Z. R. Chen
CSIRO Energy, Clayton South, Victoria, Australia

A. R. Cruden
School of Earth Atmosphere & Environment, Monash University, Clayton, Victoria, Australia

B. Damjanac
Itasca Consulting Group, Minneapolis, MN, USA

Emmanuel Detournay
Department of Civil, Environmental, and Geo-Engineering, University of Minnesota, Minneapolis, MN, USA

C. Detournay
Itasca Consulting Group, Minneapolis, MN, USA

E. V. Dontsov
ResFrac Corporation, 555 Bryant St., #185 Palo Alto, CA, USA

D. I. Garagash
Department of Civil and Resource Engineering, Dalhousie University, Halifax, Nova Scotia, Canada

D. Gunaydin
Department of Civil and Environmental Engineering, University of Pittsburgh, Pittsburgh, PA, USA

Y. Han
Aramco Services Company: Aramco Research Center, Houston, TX, USA

R. G. Jeffrey
SCT Operations Pty. Ltd., Wollongong, New South Wales, Australia

E. A. Kanin
Project Center for Energy Transition and ESG, Skolkovo Institute of Science and Technology (Skoltech), Moscow, Russian Federation

Olga Kresse
Schlumberger, Houston, TX, USA

Brice Lecampion
Geo-Energy Laboratory, Gaznat Chair on Geo-Energy, Ecole Polytechique Fédérale de Lausanne, EPFL-ENAC-IIC-GEL, Lausanne, Switzerland

Z. Lei
Department of Hydraulic Engineering, Tsinghua University, Haidian, Beijing, China

Mengli Li
Key Laboratory of Geotechnical and Underground Engineering of Ministry of Education, Tongji University, Shanghai, China
and
Department of Geotechnical Engineering, College of Civil Engineering, Tongji University, Shanghai, China

Dong Liu
Geo-Energy Laboratory, Gaznat Chair on Geo-Energy, Ecole Polytechique Fédérale de Lausanne, EPFL-ENAC-IIC-GEL, Lausanne, Switzerland

John Napier
Department of Mining Engineering, University of Pretoria, Hatfield, Pretoria South Africa

A. A. Osiptsov
Project Center for Energy Transition and ESG, Skolkovo Institute of Science and Technology (Skoltech), Moscow, Russian Federation

Panos Papanastasiou
Department of Civil and Environmental Engineering, University of Cyprus, Nicosia, Cyprus

A. Peirce
Department of Mathematics, University of British Columbia, Vancouver, British Columbia, Canada

Lei Tao
State Key Laboratory of Oil and Gas Reservoir Geology and Exploitation, Chengdu University of Technology, Chengdu, China

S. T. Thiele
Helmholtz Institute Freiberg for Resource Technology, Helmholtz-Zentrum Dresden-Rossendorf, Freiberg, Germany

Shouceng Tian
State Key Laboratory of Petroleum Resources and Prospecting, University of Petroleum (Beijing), Beijing, China

M. Torres
Itasca Consulting Group, Minneapolis, MN, USA

Haizhu Wang
State Key Laboratory of Petroleum Resources and Prospecting, University of Petroleum (Beijing), Beijing, China

Xiaowei Weng
Retired, Schlumberger, Houston, TX, USA

Bisheng Wu
Department of Hydraulic Engineering, Tsinghua University, Beijing, China

Diansen Yang
School of Civil Engineering, Wuhan University, Wuhan, China

J. Yang
Institute of Rock and Soil Mechanics, Chinese Academy of Sciences, Wuhan, China

Xi Zhang
Faculty of Engineering, China University of Geosciences, Wuhan, China
and
SCT Operations Pty Ltd, Wollongong, Australia

Fengshou Zhang
Key Laboratory of Geotechnical and Underground Engineering of Ministry of Education, Tongji University, Shanghai, China
and
Department of Geotechnical Engineering, College of Civil Engineering, Tongji University, Shanghai, China

Guangqing Zhang
College of Petroleum Engineering, China University of Petroleum-Beijing, Beijing, China

Yun Zhou
State Key Laboratory of Geomechanics and Geotechnical Engineering, Institute of Rock and Soil Mechanics, Chinese Academy of Sciences, Wuhan, Hubei, China

Dawei Zhou
College of Petroleum Engineering, China University of Petroleum-Beijing, Beijing, China

Haiyan Zhu
State Key Laboratory of Oil and Gas Reservoir Geology and Exploitation, Chengdu University of Technology, Chengdu, China

Foreword

For about 15 years until his retirement from the Commonwealth Scientific and Industrial Research Organization (CSIRO) in 2015, Rob Jeffrey led a group of scientists, technical officers, students, and academic visitors, all engaged in hydraulic fracturing research. Under Rob's direction, the CSIRO Hydraulic Fracturing (CSIRO HF) group, located in Melbourne, was a vibrant center, unique in the diversity of its activities that involved field testing, laboratory experiments, theoretical modeling, and numerical simulations. Many authors of this book have been associated with the CSIRO HF group, and their contributions are tributes to Rob's inspired leadership.

A permanent focus of the CSIRO HF group was to support the development of hydraulic fracturing as a means to precondition a rock mass for cave inducement of ore bodies and coal mine goafs. This application of hydraulic fracturing to mining was Rob's brainchild, an invention foreshadowed by the experience and expertise Rob gained in the United States while working, first, for the mining industry and later for the petroleum industry in research and development. His patented technique is now being used routinely in mines in several countries, including Australia, Chile, and Indonesia. Preconditioning by hydraulic fracturing has a profound economic impact on hard rock mines relying on block caving and on longwall coal mines, in some cases ensuring the economic viability of these operations.

One of Rob's outstanding achievements at CSIRO was the establishment of a world-class laboratory dedicated to hydraulic fracturing research with a large polyaxial cell as its centerpiece. A particular point of pride for Rob was the heavy-duty equipment used to conduct field tests. The lab was the birthplace of innovative experiments; some were conducted in glass or polymethyl methacrylate (PMMA) to enable visualization of the fracture evolution and photometric measurements of the aperture field to test computational models, while others explored the interaction of a hydraulic fracture with discontinuities or with a free surface, to cite a few. Although experiments in the field or in the lab are Rob's passion, he was always enthusiastically supportive of theoretical and computational work.

Some members of the former CSIRO HF group have moved to academic positions in China, Europe, and North America, but they remain engaged in hydraulic fracturing research. With this migration, the theoretical expertise and know-how for innovative experimental work gained in the CSIRO lab has been kept alive and further flourished. This is one of Rob's enduring legacies.

On a personal note, I am indebted to Rob for giving me the wonderful opportunity to work in his group. It was initiated at the ARMA Vail Symposium in 1999, when Rob asked me "would you be interested to spend some of your sabbatical in Melbourne ...." This led to a long-term collaboration and a chance to discover Australia on multiple occasions.

Emmanuel Detournay, University of Minnesota,
Minneapolis, USA

Preface

This volume of contributed chapters comes from the network of researchers working on mechanics of hydraulic fracturing in theory, experiments, and applications. Many of these contributors comprised the participants in a series of Hydraulic Fracturing Summits held from 2001 to 2015. These workshops provided a venue for lively debates and effective exchanges of ideas. The success of these Summits is attributed to Rob Jeffrey and Emmanuel Detournay, who initiated this series of annual meetings and fostered the community in their collaborations. They recognized the scientific significance of this fast-growing field prior to the surge of unconventional oil and gas reservoir stimulation, and through this workshop mentored a generation of hydraulic fracturing researchers.

This volume is devoted to celebrating the 70th birthday of Rob Jeffrey, who was the program leader at Commonwealth Scientific and Industrial Research Organisation (CSIRO), Australia, when three of the editors worked there. Rob is one of the pioneers for the application of hydraulic fracturing in coal seam gas reservoir stimulation and the foundation of the knowledge on hydraulic fracture growth in naturally fractured rocks. During his forty years working in hydraulic fracturing field, he launched many important research directions, being among the first to work in areas such as fracture network formation, novel fracture monitoring methods, mine-through mapping of full-scale hydraulic fractures, and use of hydraulic fracturing to modify rock mass properties for both block caving and longwall coal mining. All of his work brought together theoretical and experimental methods in order to address new issues with rigor and pragmatism. In recognition of Rob's outstanding contributions both to the field and to their careers, his Hydraulic Fracturing Summit colleagues and others are happy to contribute this book that is in his honor.

This volume contains 17 chapters, covering a wide range of relevant research topics and important application areas. The editors wish to thank every contributor for their diligence and careful choice of topics. The chapters will provide the first comprehensive story on these experimental and field monitoring results, including comparisons of several numerical methods with industrial applications. This book will benefit a wide spectrum of readers, ranging from newcomers seeking an efficient orientation to the field to seasoned experts looking for a unique set of references on some of the most important issues within the hydraulic fracturing community over the past two decades.

The editors are grateful to those anonymous reviewers of the book proposal for their positive comments and constructive suggestions and to the many volunteer reviewers who spent tremendous amount of time and effort in the evaluation of each chapter. We also owe much to Dr. Rituparna Bose, Layla Harden, and Stefanie Volk at Wiley, who are aware of the rapid-growing knowledge and social impact of hydraulic fracturing, and along with other editorial team members supporting this book. Without their vision, enthusiasm, and diligent efforts, this book would not have been possible.

Xi Zhang, China University of Geosciences, Wuhan, China
Bisheng Wu, Tsinghua University, Beijing, China
Diansen Yang, Wuhan University, Wuhan, China
Andrew Bunger, University of Pittsburgh, Pittsburgh, PA, USA

1

Hydraulic Fracture Geometry from Mineback Mapping

R. G. Jeffrey

SCT Operations Pty. Ltd., Wollongong, New South Wales, Australia

KEY POINTS

- Details of full-size hydraulic fracture geometry are only available from mapping of mined fractures.
- Hydraulic fractures mapped in naturally fractured coal, sandstone, and in crystalline rock show many similarities that result from interactions with natural fractures and shear structures.

1.1 Introduction

Hydraulic fractures have been mapped during mining in a range of rock types and in a variety of geologic and in situ stress settings. Mapping has occurred for fractures placed into clay and soil [1], welded tuff [2], coal ([3–5]), andesite [6], and other crystalline and metamorphic rocks in porphyry ore bodies [7–9].

Coal seams are typically fractured for the purpose of stimulating production of seam gas (coalbed methane), either for commercial use of the gas or to improve drainage of gas from the coal before mining. Approximately 50 hydraulic fractures have been mined and mapped in detail in coal seams in the United States and Australia ([3, 4, 10, 11]; Jeffrey et al. 1993). In comparison, fractures mapped in other materials (soil and rock) total less than 20, with 10 of these located in porphyry copper and gold orebodies ([6, 7, 9]; containing intrusive monzonites and metamorphosed volcanic sediments. These fractures were placed as part of investigations into fracture geometry expected to be produced by hydraulic fractures used to precondition the orebody in advance of mining. The fractures placed into welded tuff were part of early research into hydraulic fracture growth [2] and the soil fractures were part of a study of fracturing for waste remediation [1].

By comparing the fracture geometry mapped in these different natural materials, common and disparate features of the fractures are highlighted. To help with the comparison, dimensionless groups that have been shown to be important in hydraulic fracture growth are calculated or estimated for each mapped fracture. There is an extensive body of work using experimental, analytical, and numerical methods to investigate interactions of hydraulic fractures with bedding and natural fractures and faults. This paper limits itself to the mapped geometry exposed by mining of full-size hydraulic fractures and presents a comparison of features found in coal, sandstone, and stronger metamorphic or igneous rocks.

1.2 Summary of Mapped Fracture Geometries

Selected hydraulic fractures that were placed in coal, sandstone, and hard rock are described in this section. For each fracture the treatment parameters, rock properties, and in situ properties are listed.

1.2.1 Fractures in Coal

Two fractures placed into coal seams will be described, one located in Australia as described by Jeffrey et al. (1993) and one located in the United States [10]. The fracture descriptions include details of the treatment and site characterization, including rock mechanical properties and in situ stress data.

Mechanics of Hydraulic Fracturing: Experiment, Model, and Monitoring, First Edition.
Edited by Xi Zhang, Bisheng Wu, Diansen Yang, and Andrew Bunger.

1.2.1.1 DHM-7 Fracture

As part of a program to better understand hydraulic fracture stimulation of coal, well DHM-7 was fractured using linear gel with sand proppant [10]. DHM-7 was drilled and completed open hole through the Blue Creek coal seam in Alabama. The well was located over the Oak Grove Mine in the Warrior Basin. Table 1.1 summarizes the parameters of the site and the treatment parameters. It is difficult to measure σ_H in coal because the cleat and natural fractures make overcoring impractical and the determination of fracture initiation is difficult when using a microfrac stress test. A value for σ_H that is larger than the vertical stress has been selected because the mine back mapping revealed no development of propped fractures at right angles to the direction of the main vertical fracture branch.

Room and pillar mining exposed the fracture in the coal rib around the sides of two pillars and along the coal rib nearest the well (Figure 1.1). A vertical and horizontal propped fracture (T-shaped) was found with the vertical part consisting of a number of parallel strands. The vertical fracture extended for more than 30 m to the north of the well. A horizontal propped fracture, extending over the vertical fracture, was located at the coal-roof rock interface and was mapped in detail (see Figure 1.2). The fracture was not mined to the south, but based on the area and fracture widths mapped, the propped fracture mapped on the north side of well DHM-7 was estimated to contain approximately 75% of the sand proppant injected [12].

The treatment pumped into DHM-7 used a non-crosslinked guar-based gel fluid that was injected at an average rate of 8.3 barrels/min (0.022 m^3/s). A thick resin-coated sand system was pushed into the near-well part of the fracture on the morning following the main treatment. This resin coated sand was used to test its ability to stabilize the wellbore region. It is designed to retain 80% of the sand's permeability after curing. Much of the vertical fracture exposed nearest the wellbore, at location A in Figure 1.1, was filled with this resin-coated sand. Samples of this propped fracture were excavated and taken from the mine for later analysis and display. No resin-coated sand was found at the next exposure at location B. Mapping of the horizontal fracture was done at the detail level represented in Figure 1.2 along all propped exposures.

Table 1.1 DHM-7. Coal, well, and treatment parameters.

Parameter	Value	Units	Description
σ_H	>8.1	MPa	Maximum horizontal stress
σ_h	6.2	MPa	Minimum horizontal stress
σ_v	8.1	MPa	Vertical stress
P_o	<3	MPa	Pore pressure (estimated)
k	1.2	md	Permeability, millidarcy
E	4000	MPa	Young's modulus of coal
ν	0.3		Poisson's ratio of coal
Q	0.022	m^3/s	Injection rate
μ	25×10^{-9}	MPa s	Apparent fluid viscosity, at 170 s^{-1}
Depth	331.3	m	Depth to top of Blue Creek seam
r	0.108	m	Wellbore radius

The mapped geometry in DHM-7 and other cases presented below rely on the proppant to mark the fracture path. The hydraulic fracture typically extends beyond the proppant, especially when less viscous fluids are used. Hydraulic fractures cannot in general be found or mapped if they do not contain proppant. The distribution of the proppant, especially in a horizontal fracture, depends on both the fracture width and the fluid velocity field. Proppant transport in horizontal fractures is an area of study that has received little attention, primarily because horizontal fracture growth is thought to be a rare occurrence at depths greater than approximately 300 m. The hydraulic fractures described below that were placed into the orebody at Northparkes at a depth of 580 m were horizontal. T-shaped fractures are common in stimulation of coal and better models that can deal with horizontal fracture growth and the associated proppant transport problem would be welcome.

1.2.1.2 DDH 190 Fracture

An uncased HQ-size cored borehole (DDH 190) was drilled through the German Creek coal at Central Colliery in Queensland and was hydraulically fractured using a borate crosslinked hydroxypropyl guar gel fluid. The site was characterized by undertaking well testing, stress measurement, core testing, and fracture testing before the main fracture treatment. Table 1.2 summarizes the site parameters relevant to the treatment as given by Jeffrey et al. [5].

Mapping of the fracture during and after development of the roadways in this area of the mine revealed a vertical fracture in the coal that extended into the roof rock (Figure 1.3). The fracture trace in the roof rock (Figure 1.3a) was primarily a single fracture, but interactions with natural fractures resulted in the formation of some offsets and short parallel branches. The vertical fracture trace in the coal at the north side of 13 cut-through (Figure 1.3b) was typical of other vertical sections mapped at this site, consisting of a single fracture that interacted with bedded and sheared coal. The 150 mm-thick mid-seam shear zone (mssz) runs through much of the German Creek seam and is composed of sheared coal, with particles

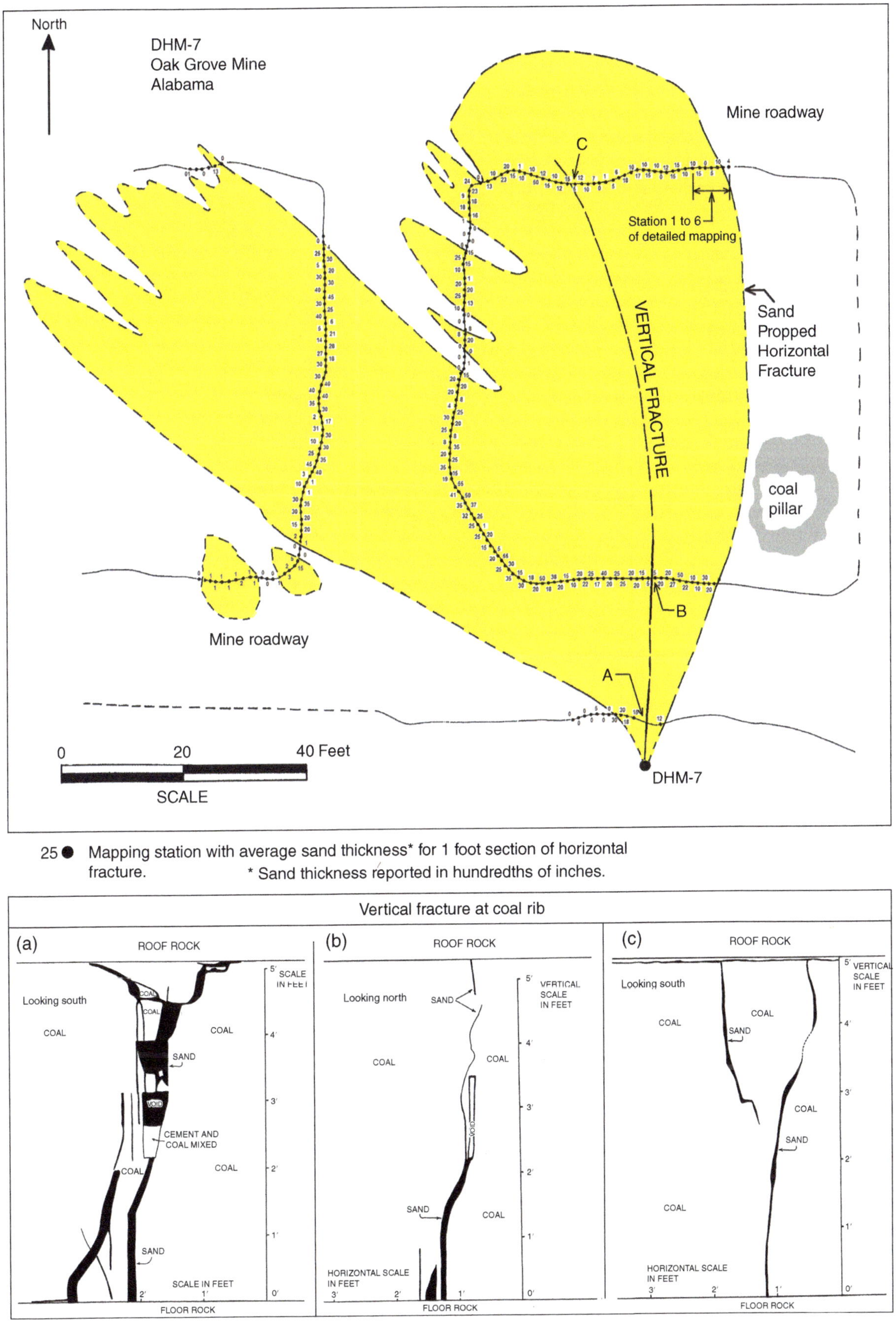

Figure 1.1 A plan view of the propped hydraulic fracture mapped at the DHM-7 site is shown in the top drawing while three mapped vertical fracture sections exposed at the coal rib are shown in the lower drawing. Mapping of the horizontal fracture occurred along the ribs of the roadways where the fracture was exposed near the roof (*Source:* Boyer et al. [12]/Gas Research Institute).

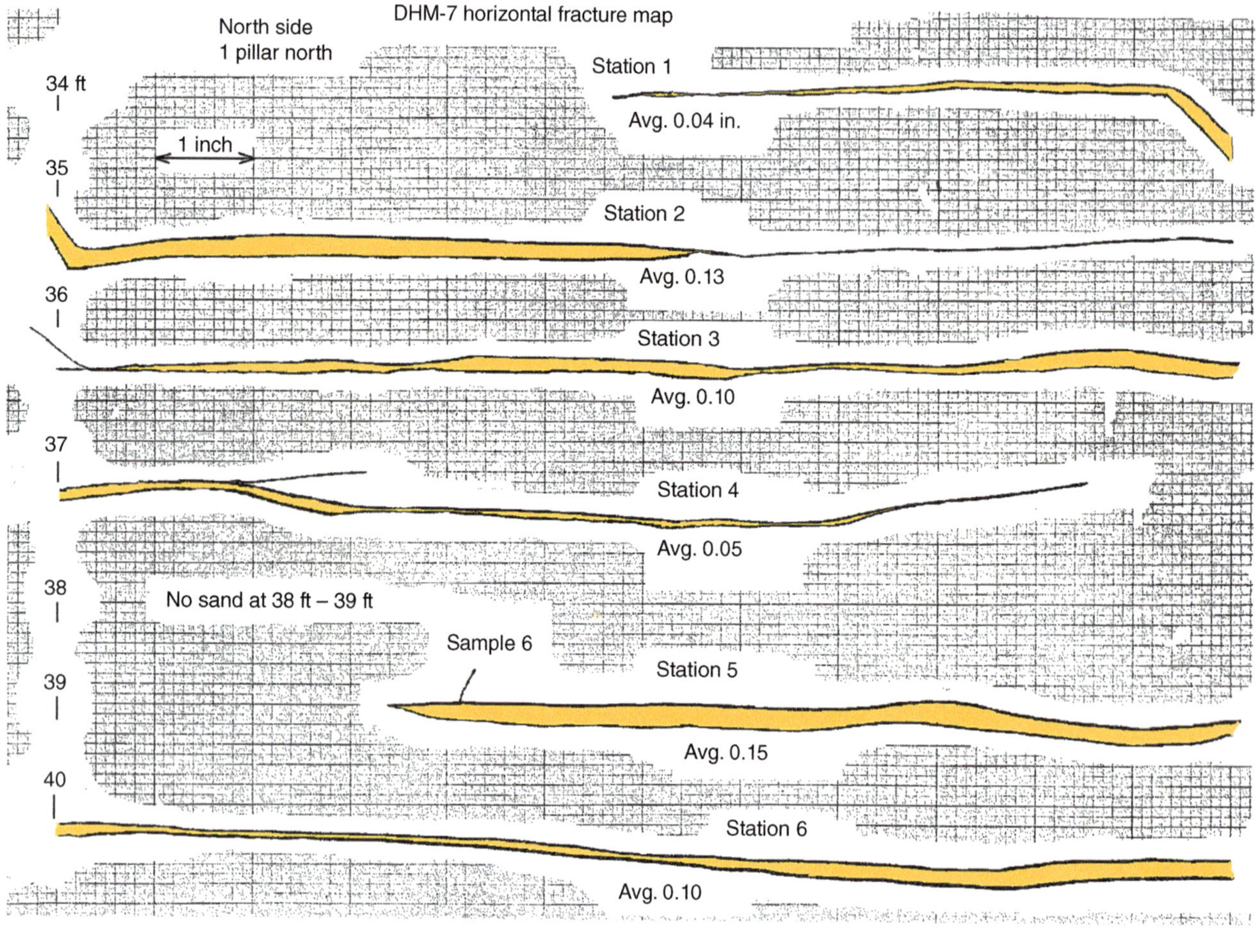

Figure 1.2 Seven linear feet of horizontal fracture mapped along the north side of the coal pillar (*Source:* Boyer et al. [12]/Gas Research Institute). The location of this portion of the horizontal fracture is indicated in Figure 1.1 and labeled as station 1 through 6. The entire horizontal fracture was mapped at this level of detail and the numbers in Figure 1.1 along the pillar boundary indicate the average propped width in hundredths of inches (e.g. 30 represents 0.30 in.).

Table 1.2 DDH 190 site parameters.

Parameter	Value	Units	Description
σ_H	>4	MPa	Max horizontal stress, in roof
σ_h	2.5	MPa	Min horizontal stress in coal
	1.9	MPa	Min horizontal stress in roof
σ_v	4.5	MPa	Vertical stress
P_o	1.08	MPa	Pore pressure
k	4.2	md	Permeability, millidarcy
E	2 000	MPa	Young's modulus, coal
	25 000	MPa	Young's modulus, roof rock
ν	0.35		Poisson's ratio, coal
	0.13		Poisson's ratio, roof rock
Q	0.002 5	m^3/s	Injection rate
μ	610×10^{-9}	MPa s	Apparent fluid viscosity at 170 s^{-1}
Depth	193.5	m	Depth to top of German Creek seam
r	0.048	m	Wellbore radius

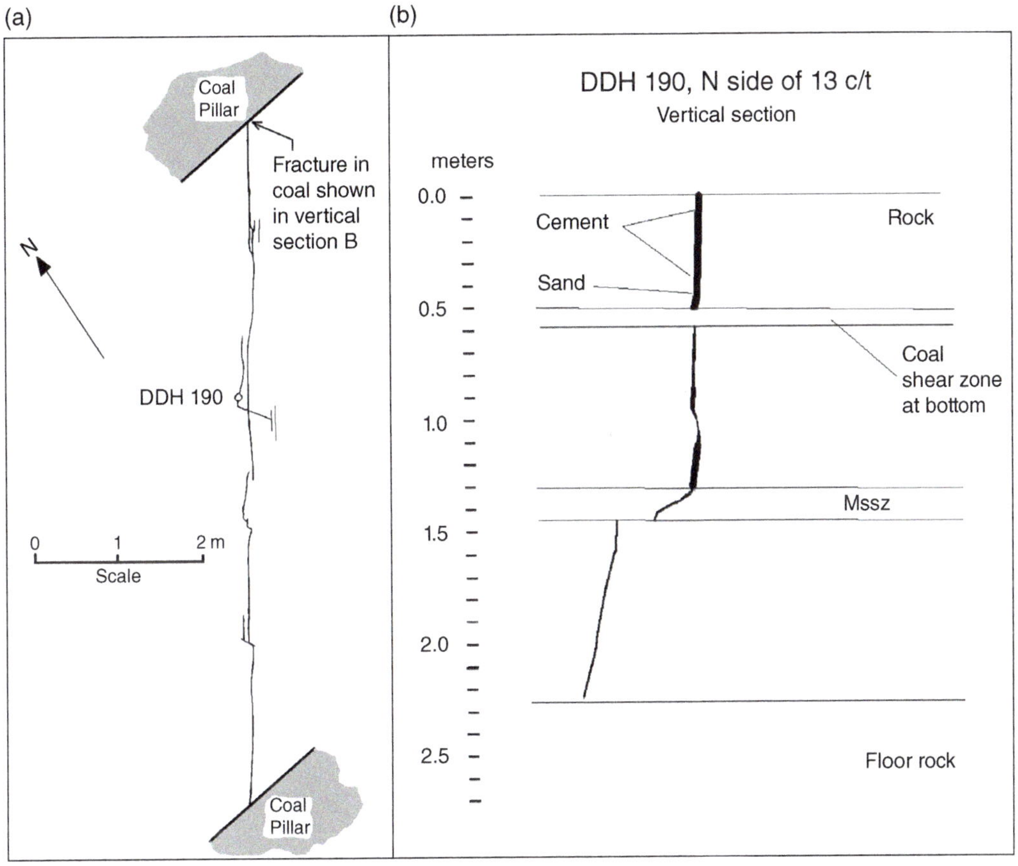

Figure 1.3 Looking down on propped fracture trace (a) in sandstone roof rock at 13 cut-through. Vertical section (b) showing propped fracture exposed on north coal rib of 13 cut-through near borehole DDH 190 (*Source:* Jeffrey et al. [5]/Coalbed Methane Association of Alabama).

ranging from clay size to a few centimeters in size. The mssz is softer and weaker than the coal above and below it. This hydraulic fracture and others mapped in this coal seam often developed an offset across the mssz.

The fracture at DDH 190 extended into the lower stress roof rock with only 12% of the proppant injected estimated to be accounted for by the propped fracture in the coal seam. The mapping clearly shows the trace of the propped hydraulic fracture in the roof and in the coal, but does not reveal if the fracture was growing primarily laterally or vertically at the sections mapped. Modeling of this fracture suggests upward growth of 7 m into the roof rock at the borehole [13].

1.2.2 Fractures in Hard Rock

Hydraulic fracturing is used in mining to induce caving and to precondition rock for caving [7]. More recently, preconditioning has been used in areas of high stress as a means of reducing the potential for the occurrence of large, damaging seismic events [14]. A total of nine fractures have been mined and mapped at four metalliferous mine sites in Australia [7, 9, 15] and Chile [6]. The fractures described by Jeffrey et al. [9] will be compared to the fractures placed into coal seams.

1.2.2.1 Northparkes E48 Mapped Fractures

Six hydraulic fractures were placed ahead of a tunnel at 580 m depth at the E48 Northparkes mine as part of a mine-through experiment [9]. The fractures were propped with colored plastic and sand and were monitored by microseismic and tiltmeter arrays. The mapped hydraulic fractures at Northparkes consist of nearly horizontal segments with offsets at intervals along them produced as the fracture grew into and along dipping veins, natural fractures, and shear zones. Fracture branches and sub-parallel propped sections were also mapped, making up 10–15% of the total fracture extent. The rock mass at the site is naturally fractured, containing approximately five natural fractures per meter. Additional details of the site, fracturing, and mine-back can be found in Jeffrey et al. [9]. Table 1.3 lists site and rock parameters. The stress directions given in Table 1.3 are based on overcoring data measured near the fracture site.

Table 1.3 Site and rock parameters for Northparkes E48.

Parameter	Value	Units	Description
σ_H	40	MPa	Max horizontal stress, 290^0, 8^0 dip
σ_h	22	MPa	Min horizontal stress, 22^0, 11^0 dip
σ_v	15	MPa	Vertical stress, 165^0, 76^0 dip
P_o	<1	MPa	Pore pressure
k	0.005	md	Permeability, millidarcy
E	50 000	MPa	Young's modulus
ν	0.2		Poisson's ratio
Q	0.007 5	m^3/s	Injection rate
μ	610×10^{-9}	MPa s	Apparent crosslinked gel viscosity at 170 s^{-1}
	1×10^{-9}	MPa s	Viscosity of water
Depth	580	m	Depth below surface
r	0.048	m	Wellbore radius

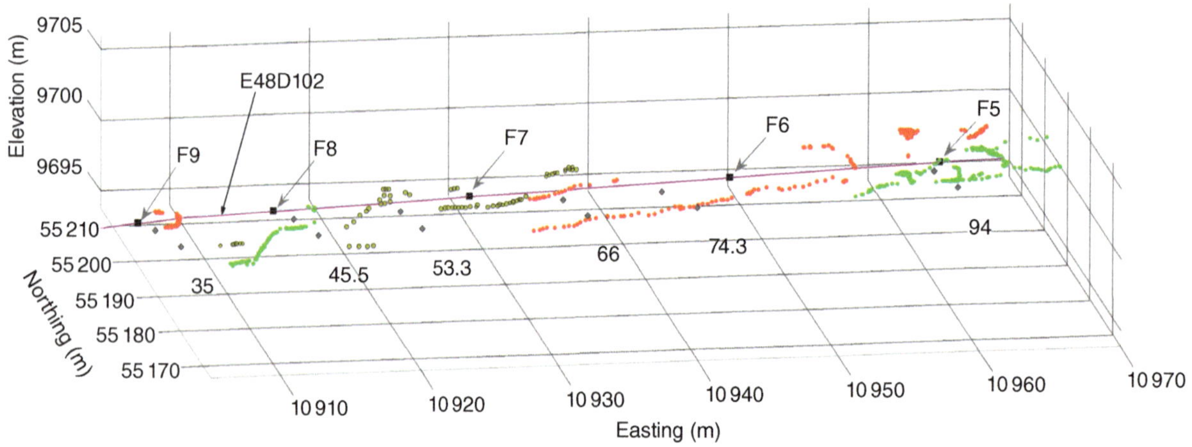

Figure 1.4 Hydraulic fractures mapped along a tunnel at the Northparkes E48 mine (*Source:* Jeffrey et al. [9]/Society of Petroleum Engineers). The initiation point of each fracture is indicated along the borehole which is drawn with a purple line.

The minimum stress was nearly vertical with the maximum stress oriented nearly horizontal and directed approximately east–west.

The hydraulic fractures were mapped along the E48D102 tunnel as it was driven. The fractures were horizontal with steps along their path often forming where they interacted with natural fractures and shear structures. These offsets were large enough to increase the average dip because the fracture trace following a stair-step pattern. Figure 1.4 shows an overview of the five fractures that were propped with colored plastic and mapped along the sides of the tunnel.

Fractures 6 and 7 were placed using crosslinked guar gel while fractures 5, 8, and 9 were placed using water. The colors of the points mapped along each fracture shown in Figure 1.4 correspond to the colors of the plastic proppant used. The grid lines shown are the mine coordinates in meters. The initiation point of each fracture is indicated by a black square symbol and the borehole is shown by the line connecting these symbols. Figure 1.5 contains a more detailed sketch of Fracture 8, which contained the largest offset or step mapped in any of the fractures at this site. The injection interval, which was in the borehole E48D102, is shown. This interval is located approximately 2.5 m out of the plane of the fracture trace shown because the borehole was drilled along the centerline of the tunnel. Fracture growth is likely to have been semi-radial from the injection interval and should not be visualized as occurring purely along the fracture trace. Fracture 7 contained an offset similar in size to the one shown in Figure 1.5 and a series of smaller offsets (but consisting of 200–400 mm steps) were

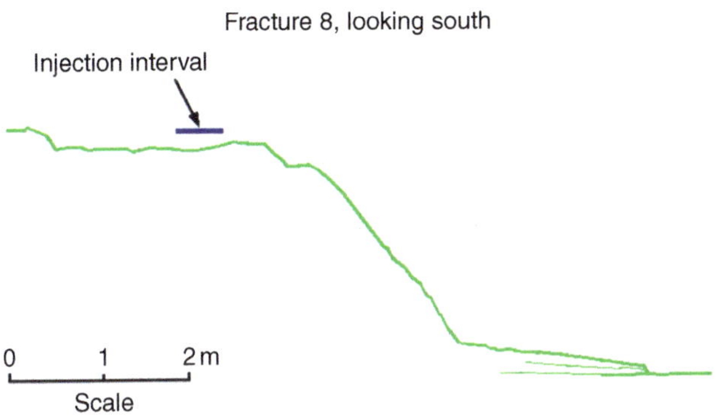

Figure 1.5 The mapped trace of fracture 8, exposed along the south side of the tunnel, at the Northparkes E48 mine-back site (*Source:* Jeffrey et al. [15]).

mapped at one exposure of Fracture 5. Therefore, at this site the crosslinked gel did not produce more planar fractures compared to the water driven fractures.

Details of the fracture geometry were not collected except along the sides of the tunnel and occasionally across the tunnel face and back. This is a common difficulty experienced when mapping hydraulic fractures in a commercial mine. The tunnelling or coal extraction operations take precedence over mapping. In the case of the E48 site, the tunnel is extended in approximately four-meter intervals using a drill-blast-muck-support cycle. Mapping is restricted to occur between the muck and support or support and drill steps.

1.2.3 Other Mapped Fractures

Hydraulic fractures have been placed into a range of rock and soil materials followed by mining and mapping. Warpinski et al. [16] and Warpinski and Teufel [2] describe hydraulic fractures placed into volcanic tuff and Murdoch [1] describe fractures placed into clay and soil at shallow depth. Details of these fractures can be found in the papers cited.

1.3 Comparison of Mapped Fracture Geometries

Nondimensional parameters are useful in helping to determine the type of fracture growth to expect in different rocks and at different sites. For example, the dimensionless toughness or dimensionless viscosity can be used to determine if the treatment was carried out in the fracture toughness or fluid viscosity dominated regime [17]. Viscosity dominated growth leads to more planar fracture geometries [18].

The fracture width restrictions resulting from offsets and branches along the fracture path are potential sites for proppant bridging and the narrow width at these locations result in a higher injection pressure and slower fracture growth compared to more planar fracture conditions [19]. The growth of closely spaced fractures has been shown to be well described by several dimensionless parameters [8, 20]. Results from numerical calculations of the path of a new fracture placed next to an existing fracture were presented by Bunger et al. [20] and were compared to mine-back and laboratory experiments in Bunger et al. [8]. The fractures mapped and described above will be compared with each other and are further categorized by calculating values for the dimensionless viscosity and differential stress parameters.

1.3.1 Dimensionless Parameters

The nondimensional parameters used are listed in Table 1.4. The physical parameters that are used in defining the nondimensional parameters are listed in Table 1.1. The nondimensional viscosity is used to determine whether the fracture growth is dominated by rock fracture toughness or by frictional losses associated with viscous fluid flow in the fracture channel [17]. The dimensionless stress is commonly used in studies of hydraulic fractures interacting with other hydraulic fractures or crossing natural fractures [8, 18, 19]. Two values of dimensionless differential stress are defined because a hydraulic fracture growing with an

Table 1.4 Dimensionless groups used in comparisons.

Name	Parameter	Notes
Dimensionless viscosity	$M = \dfrac{3\pi^2 \mu Q_0 E'^3}{256 K_{Ic}^4}$	If $M < 0.25$, regime is toughness dominated. If $M > 1$, regime is viscosity dominated.
Dimensionless differential stress	$d_{13} = \dfrac{(\sigma_1 - \sigma_3)}{\sigma_3}$	d_{13} applies to fracture growth in the direction of σ_1.
Dimensionless differential stress	$d_{23} = \dfrac{(\sigma_2 - \sigma_3)}{\sigma_3}$	d_{23} applies to fracture growth in the direction of σ_2.

Table 1.5 Nondimensional parameters for three sites.

Site	M	d_{13}	d_{23}
Northparkes F8	7.7	1.7	0.5
DHM-7	4.5	0.4	0.3
DDH 190 coal	3.3	0.8	0.6
DDH 190 roof	2000	1.5	0.7

orientation such that it opens against σ_3 will be subject to a range of differential stresses, depending on the location of the point being considered with respect to the other two principal stresses. These parameters are therefore used here as reference values in the comparisons of overall fracture growth behavior as determined by mine-back mapping.

The values of the three dimensionless parameters at each of the three mapping sites are listed in Table 1.5.

1.4 Fracture Geometry Summary

The *M* value listed for the Northparkes site applies to fractures created using water, which was the case for Fracture 8, while the treatment in DHM-7 used linear gel and a crosslinked gel was used in DDH 190. The *M* values below 0.25 represent a fracture regime that is toughness dominated while an *M* value of one or greater is viscosity dominated. All of the fractures described here fall into the viscosity dominated regime. The fracture growth regime for the part of the fracture that extended into the roof rock at DDH 190 was strongly viscosity dominated.

Natural fractures and shear zones were crossed by the hydraulic fractures at all three sites. The downward growth of the fracture at the DDH 190 site was blunted by a soft clay at the coal-floor rock interface.

Fracture 8 at Northparkes entered a 45° dipping natural fracture and grew along it for approximately 2.5 m before exiting down dip. The exit point coincides with additional calcite mineralization in the natural fracture. This section of the structure would be stronger in shear, resulting in higher shear-generated tensile stress and we propose that this allowed tensile fractures or wing fractures [21] to form at that location, allowing the hydraulic fracture to escape the natural fracture.

Table 1.6 provides a summary of the features that are present in the fractures at the three sites for two different resolutions. The resolution is defined as the smallest feature on the hydraulic fracture that is included in describing the fracture geometry. In Table 1.6, nonplanar means the main fracture channel is not in a single plane while branching means the fracture treatment has propped several fractures over parts of its extent that may be parallel or at an angle to the main fracture.

The propped fracture at the DHM-7 site was T-shaped, with each segment planar when viewed with a resolution of 1 m. When viewed in more detail (with a resolution of 0.01 m) offsets and branches are apparent as shown in Figures 1.1 and 1.2. A T-shaped fracture is considered to be a branched fracture, because the horizontal and vertical segments are both considered as branches of the fracture channel. In Table 1.6, the DHM-7 fracture is considered to be nonplanar and branched at the coarser resolution. T-shaped fractures represent a case where the hydraulic fracture is diverted into the horizontal direction rather than crossing the horizontal feature and continuing to grow vertically. Horizontal fractures associated with T-shaped fractures have been mapped to extend 10–100 s of meters without diverting back into the coal or up into the roof. The mechanics of this observed feature of these fractures has not been completely explained.

The qualitative description in Table 1.6 shows that the classification regarding planarity, branches, and offsets depends on the scale of the observation. When a more detailed view is available, the fractures at all the sites show offset, nonplanar, and branched growth. The main factor that controls the development of these features in a hydraulic Efracture is the presence of natural fractures or bedding interfaces. The fracture growth regime and the deviatoric stress, at least in the range represented by the three sites reviewed here, do not seem to result in a strong change

Table 1.6 Fracture geometry versus resolution of observation.

	Resolution 1 m			Resolution 0.01 m		
Site	**Offsets**	**Nonplanar**	**Branching**	**Offsets**	**Nonplanar**	**Branching**
Northparkes	Yes	Yes	No	Yes	Yes	Yes
DHM-7	No	Yes	Yes	Yes	Yes	Yes
DDH 190 coal	No	No	No	Yes	Yes	Yes
DDH 190 roof	No	No	No	Yes	Yes	Yes

in the fracture geometry, at least as expressed by mapping of its propped extent. Network like fracture development would be expected as the deviatoric stress is reduced toward zero, but we do not have mapped fractures representing that stress condition.

1.5 Conclusions

Three sites at which hydraulic fractures have been mined and mapped are described. Measured parameters are given for each site that are then used to find the dimensionless viscosity and dimensionless deviatoric stress values for the sites.

The mapped fractures at all three sites are found to be similar in the sense that, when they are studied at a detailed scale, they all contained offsets, were nonplanar, and contained branches. The common factor between all sites is that the rock (or coal) was naturally fractured. The nondimensional parameters were not found to be useful in predicting the amount of offsetting, nonplanar growth, or branching. Mapping fractures at sites with lower deviatoric stress may show a sensitivity to that parameter in predicting network such as fracture development.

References

1 Murdoch, L.C. (1995). Forms of hydraulic fractures created during a field test in Overconsolidated glacial drift. *Quarterly Journal of Engineering Geology* 28: 23–35.

2 Warpinski, N.R. and Teufel, L.W. (1987). Influence of geologic discontinuities on hydraulic fracture propagation. *Journal of Petroleum Technology* 39 (2): 209–220.

3 Diamond, W.P. and Oyler, D.C. (1987). Effects of Stimulation Treatments on Coalbeds and Surrounding Strata. *USBM Report RI 9083.*

4 Elder, C.H. (1977). Effects of Hydraulic Stimulation on Coalbeds and Associated Strata. *USBM Report RI 8260.*

5 Jeffrey, R.G., Enever, J.R., Ferguson, T. et al. (1992). Small-scale hydraulic fracturing and mineback experiments in coal seams. *Proceedings of the 1992 International Coalbed Methane Symposium*, Tuscaloosa, (17–21 May). Paper 9330.

6 Chacon, E.; Barrera, V.; Jeffrey, R.; van As, A. Hydraulic fracturing used to precondition ore and reduce fragment size for block caving. In *MassMin* 2004; Santiago; pp 429–534.

7 van As, A. and Jeffrey, R.G. (2002). Hydraulic fracture growth in naturally fractured rock: mine through mapping and analysis. *5th North American Rock Mechanics Symposium and the 17th Tunnelling Association of Canada Conference*, 9. Toronto: ARMA.

8 Bunger, A.P., Jeffrey, R.G., Kear, J. et al. (2011). Experimental investigation of the interaction among closely spaced hydraulic fractures. *45th US Rock Mechanics/Geomechanics Symposium*. American Rock Mechanics Association. https://www.onepetro.org/conference-paper/ARMA-11-318. Accessed 28 January 2016.

9 Jeffrey, R.G., Bunger, A.P., Lecampion, B. et al. (2009). Measuring hydraulic fracture growth in naturally fractured rock. *Proceedings SPE Annual Meeting*, 18. New Orleans: SPE.

10 Boyer, C.M., Stubbs, P.B., Schwerer, F.C. et al. (1986). Measurement of coalbed properties for hydraulic fracture design and methane production. *Proceedings of the 1986 Unconventional Gas Technology Symposium*, Louisville, (18–21 May). Paper SPE 15258.

11 Steidl, P.F. (1991). Observations of Induced Fractures Intercepted by Mining in the Warrior Basin, Alabama. *GRI Topical Report No. GRI-91/0327.*

12 Boyer, C.M., Stubbs, P.B., Schwerer, F.C., et al. (1986b). Mapping of the DHM-7 Hydraulic Fracture. Appendix I in: Measurement of static coalbed reservoir conditions for hydraulic fracture design; final report, volume II, GRI-**87**/0082.2. Chicago, IL, Gas Research Institute, Dec. 1986.

13 Jeffrey, R.G., Settari, A., and Smith, N.P. (1995). A comparison of hydraulic fracture field experiments, including mineback geometry data, with numerical fracture model simulations. *Proceedings SPE Annual Meeting*, (591–606). Dallas: SPE.

14 Board, Mark, Tony Rorke, Gary Williams, and Nick Gay. "Fluid injection for rockburst control in deep mining." In *The 33rd US Symposium on Rock Mechanics (USRMS)*. 1992.

15 Jeffrey, R.G., Zhang, X. and Bunger, A.P. (2010). Hydraulic fracturing of naturally fractured reservoirs. *Thirty-Fifth Workshop on Geothermal Reservoir Engineering*, 9. Stanford: Stanford University.

16 Warpinski, N.R., Schmidt, R.A., and Northrop, D.A. (1982). In-situ stresses: the predominant influence on hydraulic fracture containment. *Journal of Petroleum Technology* 34 (3 (March)): 653–664.

17 Detournay, E. (2004). Propagation regimes of fluid-driven fractures in impermeable rocks. *International Journal of Geomechanics* 4 (1): 35–45.

18 Weng, X., Kresse, O., Chuprakov, D. et al. (2014). Applying complex fracture model and integrated workflow in

unconventional reservoirs. *Journal of Petroleum Science and Engineering* 124: 468–483. https://doi.org/10.1016/j.petrol.2014.09.021.

19 Zhang, X. and Jeffrey, R.G. (2006). The role of friction and secondary flaws on deflection and re-initiation of hydraulic fractures at orthogonal pre-existing fractures. *Geophysical Journal International* 166: 1454–1465.

20 Bunger, A.P., Zhang, X., and Jeffrey, R.G. (2012). Parameters affecting the interaction among closely spaced hydraulic fractures. *SPE Journal-Richardson* 17 (1): 292–306.

21 Fairhurst, C. and Cook, N.G.W. (1966). The phenomenon of rock splitting parallel to the direction of maximum compression in the neighborhood of a surface. *Proceedings First Congress International Society for Rocks Mechanics*. Lisbon.

2

Measurements of the Evolution of the Fluid Lag in Laboratory Hydraulic Fracture Experiments in Rocks

Dong Liu and Brice Lecampion

Geo-Energy Laboratory, Gaznat Chair on Geo-Energy, Ecole Polytechique Fédérale de Lausanne, EPFL-ENAC-IIC-GEL, Lausanne, Switzerland

2.1 Introduction

The strong nonlinear coupling between elastic deformation and lubrication flow yields a strong suction in the near-tip of a propagating hydraulic fracture. It results in a steep fluid pressure gradient and a negative fluid pressure singularity near the tip. As a fluid cannot sustain traction, a fluid-less cavity nucleates such that the fluid front lags behind the propagating fracture tip [1, 2]. Depending on the permeability of the surrounding medium, this cavity (commonly referred to as a fluid lag) can be filled either by the vaporized fracturing fluid or by the medium pore fluid which leaks into this lag region [2–4].

The existence of such a fluid lag region near the fracture tip had been postulated since the first modeling efforts on hydraulic fracture growth [5, 6]. It was also observed in a number of experiments in the laboratory [7–9] as well as in the field [10]. Early attempts to estimate the extent of such a fluid lag [11] did not account for the complete elastohydrodynamics coupling. The first quantitative estimation of the fluid-lag size as a function of the fracture velocity V, material properties (elastic plane-strain modulus E' and fracture toughness K_{Ic}), fluid viscosity μ and in situ confining stress σ_3 was obtained for the steadily moving semi-infinite hydraulic fracture [2, 3]. Garagash and Detournay [2] demonstrated that for an impermeable material the lag size increases with the decrease of a tip dimensionless toughness $\kappa \propto K_{Ic}\sigma_3^{1/2}\left(E'V^{1/2}\mu^{1/2}\right)$. Extensions to the case of a fluid-infiltrated permeable material was investigated in [4, 12], and the effect of a shear-thinning fluid rheology on the lag size is reported in [13] for the same semi-infinite hydraulic fracture problem.

For a finite plane-strain hydraulic fracture, the spatio-temporal evolution of the fluid lag was first clarified theoretically in the impermeable medium case [14, 15]. The fluid lag is largest at early time where its extent is initially governed by the plane-strain dimensionless toughness [14] (with larger lag size for smaller dimensionless toughness). Under a constant injection rate, the fluid front is predicted to coalesce with the fracture front over a timescale $t_{om} = 12\mu E'^2/\sigma_3^3$ – thus indicating a strong dependence of the fluid-lag size on the minimum in situ confining stress. The same timescale of coalescence is found for a radial hydraulic fracture [16]. The evolution of both the fluid and fracture fronts have been subsequently included in a number of simulators for finite hydraulic fractures [17–22].

Using transparent brittle materials has allowed to experimentally measure the evolution of the fluid lag under low confining stress [23]. A good match between experiments and numerical predictions has been reported [24]. However, rocks are nontransparent and typically exhibit a quasi-brittle behavior that may result in non-negligible process zone size. The critical importance of the coexistence of a fluid lag and a process zone has been recently highlighted theoretically for both the steadily moving semi-infinite [25] as well as finite plane-strain HFs [26–28] clarifying previous qualitative estimations [29]. The role of the fluid lag for a fluid-driven fracture is analogous to that of the process zone in fracture mechanics. The fluid lag ensures the finiteness of the fluid pressure behind the fracture front while the process zone ensures the finiteness of the stresses immediately in front of the fracture front.

Precise experimental measurements of the evolution of the fluid lag in quasi-brittle rocks are particularly difficult. Some measurements (along a single given direction) have been reported in nontransparent materials using acoustic methods, notably in sandstone [8, 30], cement and plaster [30, 31]. In this chapter, using extensive acoustic imaging [32], we report the quantitative spatio-temporal evolution of the complete two-dimensional fluid and fracture fronts

Mechanics of Hydraulic Fracturing: Experiment, Model, and Monitoring, First Edition.
Edited by Xi Zhang, Bisheng Wu, Diansen Yang, and Andrew Bunger.

(during both the injection and shut-in phases) on three experiments performed in tight quasi-britle rocks (Carrara marble and Zimbabwei gabbro).

2.2 Materials and Methods

2.2.1 Materials and Experimental Set-up

Carrara marble and Zimbabwei gabbro have been used. They both exhibit very low porosity and permeability such that any fluid-leak-off remains negligible over the timescale of these laboratory experiments. These two rocks differ in their average grain sizes, elastic and fracture properties (see Table 2.1).

The rock samples are cut and rectified into a 250 mm cubic sample prior to their loading into a true triaxial frame. A wellbore is drilled in the middle of the block and a radial notch is created to control precisely the initiation of the fracture. The confinement is applied by symmetric pairs of flat jacks in the three axis of the poly-axial reacting frame. A compressive stress up to 20 MPa can be applied on each axis separately. The fracturing fluid is injected in a central wellbore by a syringe pump (ISCO D160) at a constant flow rate (in the range of 0.001–107 ml/min). An interface vessel is used to avoid filling the pump with the very viscous fracturing fluid. Such an interface vessel increases the compliance of the injection system which is directly proportional to the volume of fluid in the injection line from the pump. Due to the compliance of the injection system, upon fracture initiation, the flow rate entering the fracture is not equal to the pump injection rate Q_o during a transient phase that may last over the whole experiment duration [15, 35]. A needle valve is placed in the injection line close to the well-head to control the release of fluid compressed during the pressurization phase. Using volume conservation within the injection system (pump to fracture inlet), the flow rate $Q_{in}(t)$ entering the fracture can be estimated by taking the derivatives of the fluid pressure measurements [32].

An active acoustic imaging system is integrated in the loading frame to monitor the fracture geometry as shown in Figure 2.1. We have 64 piezoelectric transducers with 32 as sources and 32 as receivers, respectively. We use a source function generator connected to a high-power amplifier to send a Ricker excitation signal at a central frequency of 750 kHz. The source signal is routed to one of the source transducers via a multiplexer such that the signals received by the 32 receivers are recorded with a sampling frequency of 50 MHz. As the switch between sources is limited by the multiplexer, the excitation of a given source is repeated 50 times, and the data stacked to improve the signal to noise ratio. Spanning of the 32 sources defines an acoustic scan and takes about 2.5 s. In addition to acoustic data, we record the fluid pressure upstream and downstream of the valve located near the well-head, volume, and pressure of each flat-jack pairs or displacement of the tip surface near the wellbore at 1 Hz.

2.2.2 Methods

We reconstruct the fracture extent based on the time-arrival of the waves diffracted by the fracture front [32]. We pick the arrival time of these diffracted waves manually and then reconstruct the fracture front with the knowledge of different source–receiver locations and wave velocities. We refer to [32] for the complete details of the solution of this inverse problem which is performed independently for every active survey conducted during the experiment.

The fracture width can be estimated from the attenuation of the acoustic transmission [32, 36]. However, the existence of the fluid lag makes it inappropriate to use this method, which is based on the assumption of a single fluid layer between two smooth surfaces [37]. We therefore compute the evolution of the transmitted energy for given pairs of source–receiver transducers. We combine it with the diffracted fracture front to qualitatively estimate the growth of the fluid lag and process zone. Following [32, 34], we focus on the time evolution of the attenuation ratio of transmitted waves for facing source–receiver pairs. We first compute the signal strength $E_i^{1/2}$ of a given transmitted wave arrival (P wave or S wave) for the acoustic scan i as

$$E_i^{1/2} = \sqrt{\sum_{j=j_{min}}^{j_{max}} u_i^2(t_j)} \tag{2.1}$$

Table 2.1 Rock properties.

Rock	E (GPa)	ν	K_{Ic} (MPa m$^{1/2}$)	Grain size (mm)	Porosity (in %)
Carrara marble	60.6	0.30	1.55 ± 0.11	0.1–0.2	0.6
Zimbabwe gabbro	99.7	0.29	2.79 ± 0.11	1–3	0.32

Porosity represents the connected volume porosity measured using a gas pycnometer. Elastic properties have been measured from static triaxial tests, while the mode-I fracture toughness K_{Ic} estimated from semicircular bending test following the recommended ISRM methods (*Source:* Adapted from [33]) (see more details in [34]).

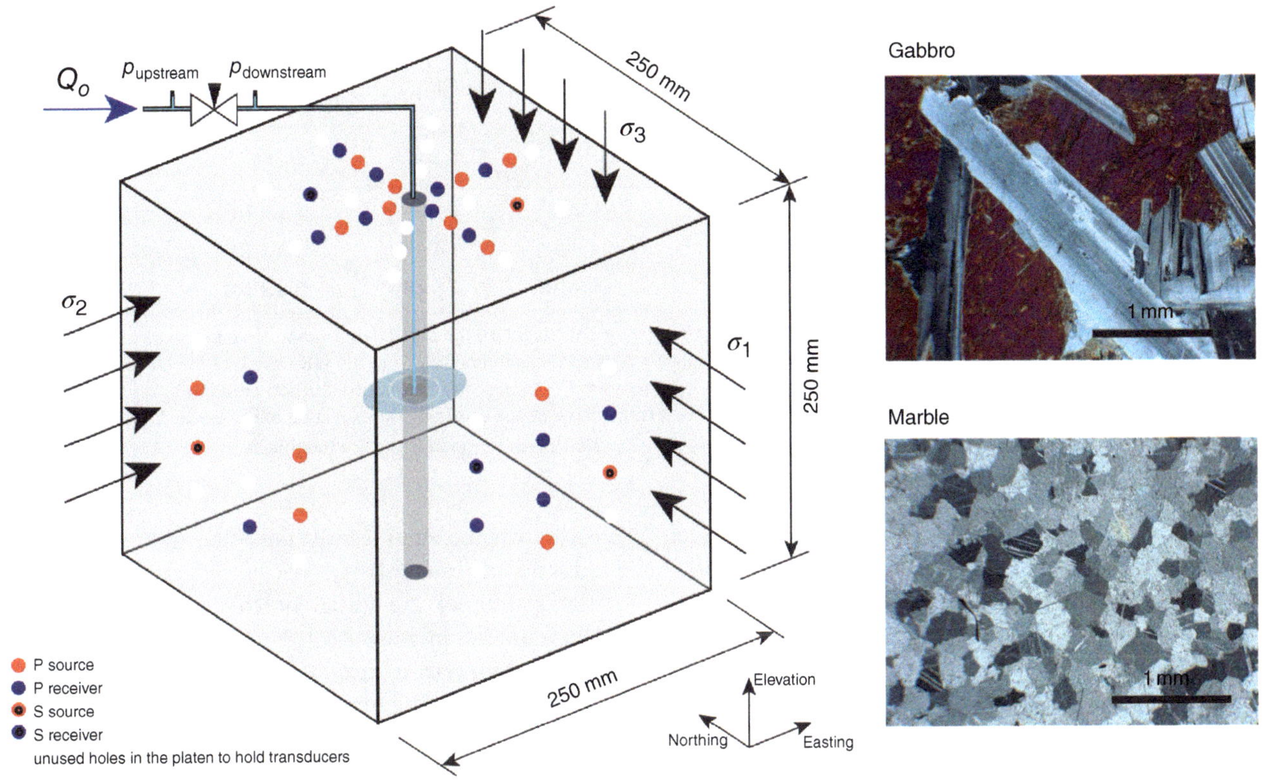

Figure 2.1 Schematic illustration of the cubic sample showing the transducers array disposition (left) and thin sections of the tested Zimbabwe gabbro and Carrara marble (right). Additional holes are available in the platens allowing the use of various transducer dispositions. Two facing platens share the same transducers disposition and source/receivers transducers are alternatively located on opposite platens for robustness.

where $u_i(t)$ is a low-pass filtered (at 2 MHz) waveform centered on the interest arrival with a size of $\left(t_{j_{max}} - t_{j_{min}}\right) = 14\,\mu s$ using a tapered Hamming window. We choose a reference signal obtained before fracture initiation and define the attenuation ratio as $(E_i/E_{ref})^{1/2}$.

2.2.3 Experimental Design

We impose a bi-axial state of stress on the blocks $\sigma_1 = \sigma_2$, and set the minimum stress σ_3 parallel to the wellbore to favor a planar horizontal fracture. We keep the same deviatoric stress $\sigma_1 - \sigma_3 = 10$ MPa for all the experiments reported here. The details of the different injection parameters (fluid, rate) for these experiments are reported in Table 2.2. These experiments were designed in order to probe the so-called viscosity dominated regime where a fluid lag exists at early time. This is achieved by using two different highly viscous silicone oils (see Table 2.2).

We estimate the dominant propagation regime based on the theory of linear hydraulic fracture mechanics (LHFM) [38]. Owing to our experimental configuration where the fracture initiates from a cylindrical notch perpendicular to the minimum stress, the fracture can be assumed to propagate as a radial fracture. We thus use the scaling analysis described in [15] for such a configuration in order to estimate the different characteristic time scales governing the propagation. Several competitive mechanisms are indeed at play such as the effect of the competition between

Table 2.2 Sample configuration and experimental parameters for HF injections.

Rock sample	Fracturing fluid	μ (Pa.s)	σ_3 (MPa)	$\sigma_1 = \sigma_2$ (MPa)	Injection rate Q_o (ml/min)	System compliance U (ml/GPa)
MARB-005	Silicone oil A	100	10	20	0.2	282.5
MARB-007	Silicone oil B	1000	0	10	0.02	319.5
GABB-002	Silicone oil A	100	0	10	0.05	255.1

The system compliance U is estimated from the pressurization phase prior to fracture initiation (see more details in [15, 32, 34]).

Table 2.3 Characteristic timescales for a radial hydraulic fracture: viscosity–toughness transition $t_{mk} = \mu'^{5/2} Q_o^{3/2} E'^{13/2}/K'^9$, fluid-lag disappearance timescale $t_{om} = E'^2\mu'/\sigma_3^3$, viscosity–toughness transition when the system compliance dominates the fracture growth $t_{\hat{m}\hat{k}} = E'^{5/2}\,U^{1/2}\mu'/K'^3$, compressibility–viscosity transition $t_{\hat{m}m} = t_{mk}\chi^{-3/2}$ and compressibility–toughness transition $t_{\hat{k}k} = t_{mk}\chi^{-5/3}$.

Rock sample	t_{prop} (s)	t_{mk} (s)	t_{om} (s)	$t_{\hat{m}\hat{k}}$ (s)	$t_{\hat{m}m}$ (s)	$t_{\hat{k}k}$ (s)	$\chi = t_{mk}/t_{\hat{m}\hat{k}}$ (s)
MARB-005	≈ 582	1.2×10^5	5.3×10^3	6.0×10^3	1.3×10^3	8.1×10^2	20.2
MARB-007	≈ 2711	1.2×10^7	2.5×10^{14}	6.4×10^4	1.4×10^4	9.0×10^3	19.0
GABB-002	≈ 8199	1.9×10^3	5.5×10^{13}	3.4×10^3	4.5×10^3	5.0×10^3	5.6×10^{-1}

For experiments where the applied minimum confining stress equals zero, we put weight on the top platen in some cases to ensure a good contact between acoustic sensors and the top surface of the rock. We calculate t_{om} using the estimated σ_3 from the self-weight of the block and the added weight on the top platen. We estimate t_{mk} using the pump injection rate Q_o during the fracture propagation duration t_{prop}, the elastic properties and toughness values in Table 2.1. Note that $E' = E/(1-\nu^2)$ is the plane-strain elastic modulus related to the elastic Young's modulus E and Poisson's ratio ν, $\mu' = 12\mu$ the effective viscosity, and $K' = \sqrt{32/\pi}K_{Ic}$ is an effective mode I fracture toughness.

viscous dissipation and fracture energy but also the impact of the injection system compressibility on fracture initiation. We report in Table 2.3 the estimation of all these different timescales as well as the measured total fracture propagation time t_{prop}. We used the known material properties, injection rate, and measured injection system compliance for these estimations.

Comparing these different timescales with the propagation duration allows to grasp the propagation regime of these different experiments. Notably, the fracture grows strictly in the toughness dominated regime with no fluid lag when the propagation duration t_{prop} is much larger than the viscous to toughness transition timescale t_{mk} (see Table 2.3 for definition). This is never the case for these three experiments where viscous effects are prominent due to the large viscosity of the fluid used, and $t_{prop} \lesssim t_{mk}$ as a result. The timescale of coalescence of the fluid and fracture fronts t_{om} is much larger than the overall reported experiment duration. A fluid lag is thus always present and the fracture growth is in the transition between lag and viscosity dominated regimes for these three experiments.

On the account of the system compliance and the finite size of the wellbore and notch, we also evaluate corresponding timescales related to the system compliance following scalings in [15]. As shown in Table 2.3, $t_{om} \gg t_{prop}$ and $t_{\hat{m}m}, t_{\hat{k}k}$ are both very large and cannot be neglected compared with the propagation duration. This indicates a very important effect of the system compliance for all experiments. Moreover, the two dimensionless parameters $\mathcal{L} = \ell_o/(E'U)^{1/3}$~$10^{-3}$ and $\mathcal{A} = a/(E'U)^{1/3} \sim 10^{-2}$, which characterize, respectively the effect of the notch size ℓ_o and the wellbore radius a, indicate a negligible effect of these two early-time initiation factors when comparing to the system compliance effect.

It is important to recall that for such type of viscosity/lag dominated experiments, the fracture initiation pressure can be significantly impacted by the rate and fluid viscosity used. Large viscous effect impedes the pressurization of the initial notch and thus increases fracture initiation pressure [39]. In addition, the fluid pressure continues to rise after nucleation, as the flow rate entering the fracture increases very slowly – such as the fluid volume in the wellbore continues to be pressurized. As a result, the maximum recorded wellbore pressure (also inappropriately referred to as the breakdown pressure) can be much larger than the initiation pressure [15]. As a result of these strong viscous effects, the fluid lag is large and the flux entering the fracture $Q_{in}(t)$ is initially much smaller than the injection pump rate. The evolution of the inlet flux $Q_{in}(t)$ reflects directly the fluid front growth. Following [32], we estimate the entering flow rate by taking the derivatives of the fluid pressure measurements and using the volume conservation within the injection system (pump to fracture inlet). However, the accuracy of this estimation of the entering flow rate is somehow limited. It depends significantly on the accuracy of the measurement of the fluid pressure and the system compliance upstream and downstream the valve. Oscillations in the pressure measurement sometimes appears when the valve position is not strictly fixed during the injection most likely due to poor cleaning of very viscous fluids used in a previous test. This ultimately results in oscillations of the inlet flux in some cases (see Figure 2.4 in particular).

2.3 Results

2.3.1 MARB-005 – A HF Growth with a Fluid Lag

MARB-005 presents a hydraulic fracture growth characterized by an important fluid lag. As shown in Figure 2.2, the

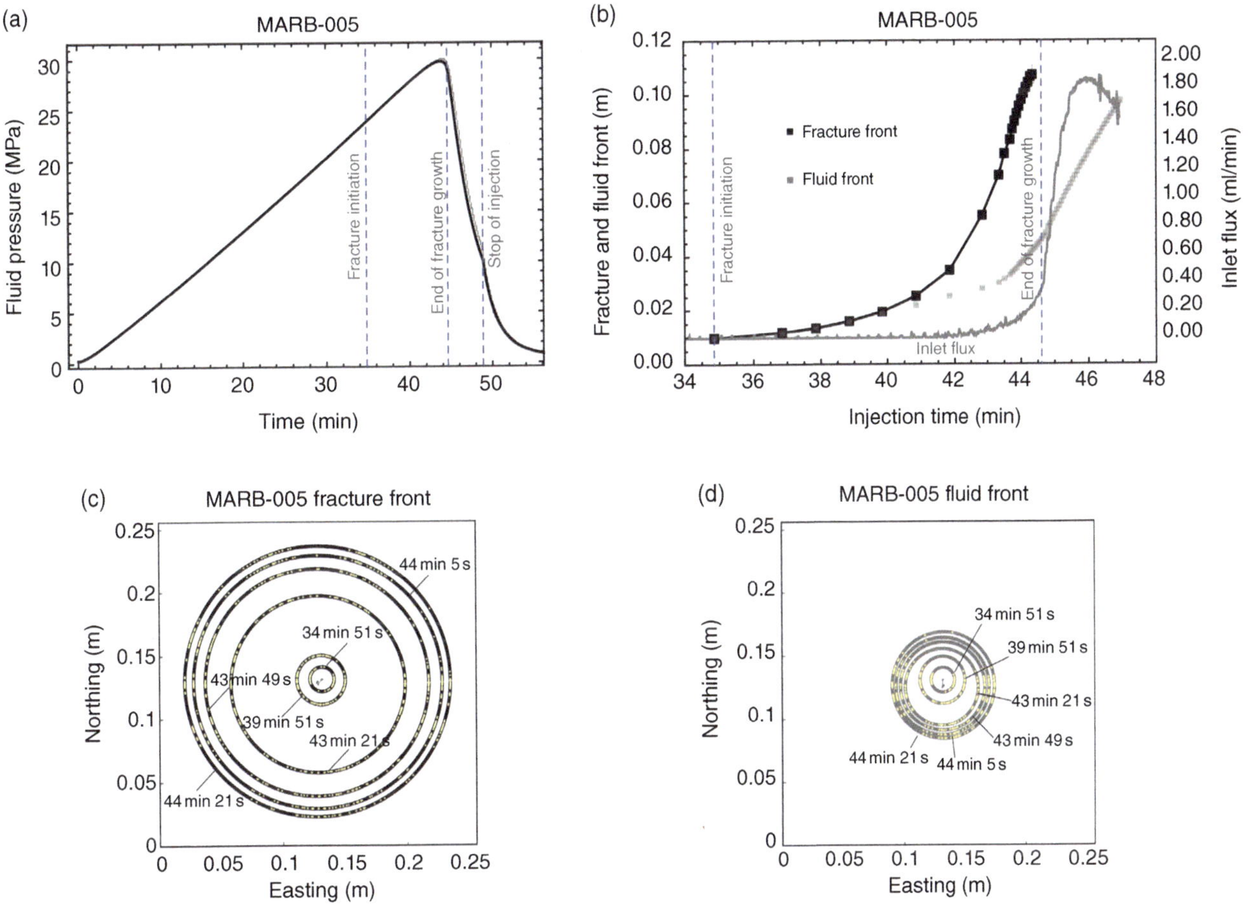

Figure 2.2 MARB-005: time evolution of (a) the injection pressure, (b) radius of the fracture and fluid fronts, estimated inlet flux, and the footprint of the (c) fracture and (d) fluid fronts. The dots on the reconstructed fronts indicate the diffracting locations for different source-receiver pairs. The vertical lines in (a) and (b) indicate respectively the time of fracture initiation (determined by the volume change of the top-bottom flat-jacks), the time where the fracture front reaches the edge of the sample (determined by the downstream pressure evolution) and the stop of the injection.

downstream pressure continues to increase after the fracture initiation has been detected via the acoustic monitoring system. This results from the large pressure gradient in the viscous fluid flow: the increase of the fracture volume upon the fracture initiation does not lead to the instant drop of the fluid pressure at the inlet. This is completely different from a toughness-dominated hydraulic fracture where the pressure distribution inside the fracture is almost uniform and the inlet pressure drops immediately when the fracture starts to grow: as a result, the maximum downstream pressure at the break down corresponds to the fracture initiation when the fracture toughness dominates the HF growth. Another difference of the lag dominated experiments from the toughness dominated experiments is that the growth of the fluid front does not coalesce with that of the fracture front and is characterized by the increase of inlet flux. An obvious fluid front growth often appears much later after the fracture initiation.

In parallel to the diffracted fracture and fluid front, the transmission attenuation also confirms the presence of a fluid lag. As shown in Figure 2.3, the signal strength of transmitted waves decreases upon the arrival of the fracture front and regains its amplitude following the arrival of the fluid front. It then subsequently decreases as the fluid layer thickness increases which results in more loss of the transmitted energy.

2.3.2 MARB-007 – A HF Growth during and after the Injection

MARB-007 presents also an important fluid lag but with a much larger initiation net pressure than MARB-005 (see Table 2.4). Such a large initiation pressure is possibly due to the 10 times more viscous fluid adopted, which represents a much more important effect of the viscous fluid flow. More interestingly, MARB-007 is characterized by a continuous fracture growth after the shut-in of the valve

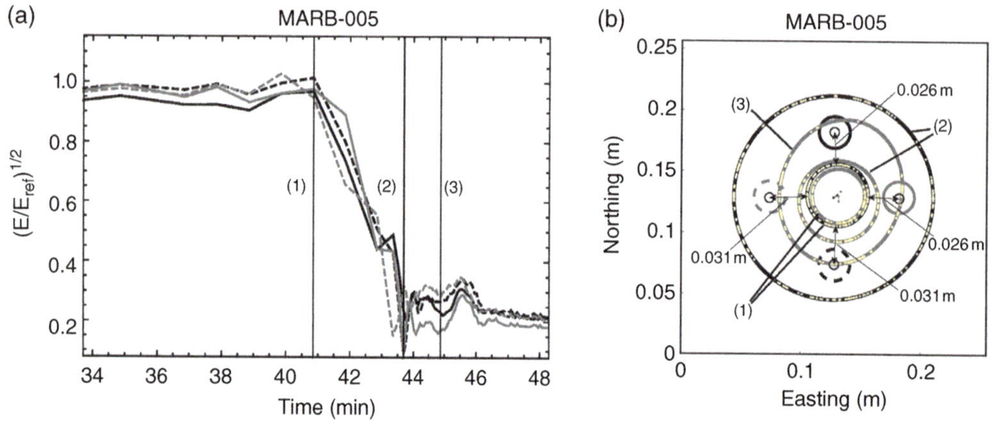

Figure 2.3 MARB-005: time evolution of (a) the signal strength of transmitted compressional waves, and (b) the fracture geometry at different times of the fracture growth. The dots on the reconstructed fronts indicate the diffracting locations for different source-receiver pairs. Locations of the transducers whose acoustic energy attenuation ratio is displayed in figure (a) are indicated by circles in figure (b) using the same line style. The fracture (in black) and fluid (in gray) extent associated with the numbered vertical lines in figure (a), is also shown in figure (b) using the same number (note that at snapshot 3, the fracture front already reached the sample boundaries).

Table 2.4 Wave velocities in the block and net pressures ($p_f - \sigma_3$) at fracture initiation and at break-down (maximum encountered pressure).

Rock	V_p (m/s)	V_s (m/s)	Initiation net pressure (MPa)	Peak net pressure (MPa)
MARB-005	6249.8 ± 54.0	3229.9 ± 176.2	13.84	19.76
MARB-007	6220.3 ± 45.4	3305.8 ± 36.9	45.45	52.79
GABB-002	6835.3 ± 18.5	3754.7 ± 23.2	11.61	35.34

V_p and V_s are measured directly with the cubic rock samples during the pressurization stage prior to fracture growth. The mean and standard deviation values are calculated based on the vertical and horizontal ray paths for the first 10 acquisitions before the fracture initiation (16 vertical ray paths for the calculation of V_p and 5 horizontal ray paths [in the north–south and east–west directions] for that of V_s).

in the injection line as shown in Figure 2.4. Due to the strong viscous effect, excess elastic energy is stored in the medium and the propagation continues after shut-in until the fluid pressure gradient vanishes inside the fracture and the energy release rate falls below the critical fracture energy. Such fracture growth after shut-in has been investigated theoretically in the case of a negligible fluid lag [40].

2.3.3 GABB-002 – A Point-Load Like HF Growth

For GABB-002, the fluid front stays within the wellbore/notch during the whole duration of the fracture propagation as shown in Figure 2.5. No clear diffraction associated with the growth of a fluid front can be distinguished on the acoustic data before the fracture front reaches the end of the sample (see Figure 2.5). As a result, this experiment is a typical illustration of the early-time/large lag limit [16] with the additional complexity of the importance of the injection system compliance. In that limit where the fluid remains close to the injection point, the injected fluid acts as a point load P in the fracture center. For a radial fracture and such a point-like loading, the stress intensity factor is given as [41]

$$K_I = P(\pi R)^{3/2} \tag{2.2}$$

Assuming a simplified pressure distribution over a given area (for example bounded by the notch size), we can estimate the net loading as

$$P \approx (p_{\text{downstream}} - \sigma_3)\frac{\pi}{4}d_{\text{notch}}^2 \tag{2.3}$$

with $d_{notch} = 0.022$ m. From the time-evolution of both the downstream pressure and fracture radius, we can thus estimate an apparent fracture toughness. Such an apparent toughness as function of fracture radius is displayed in Figure 2.6. Such an apparent toughness is much smaller than the reported fracture toughness for gabbro (see Table 2.1). Uncertainties of up to 1 cm for the fracture front location have been estimated during the inversion process for this lag dominated experiment (which we have used to

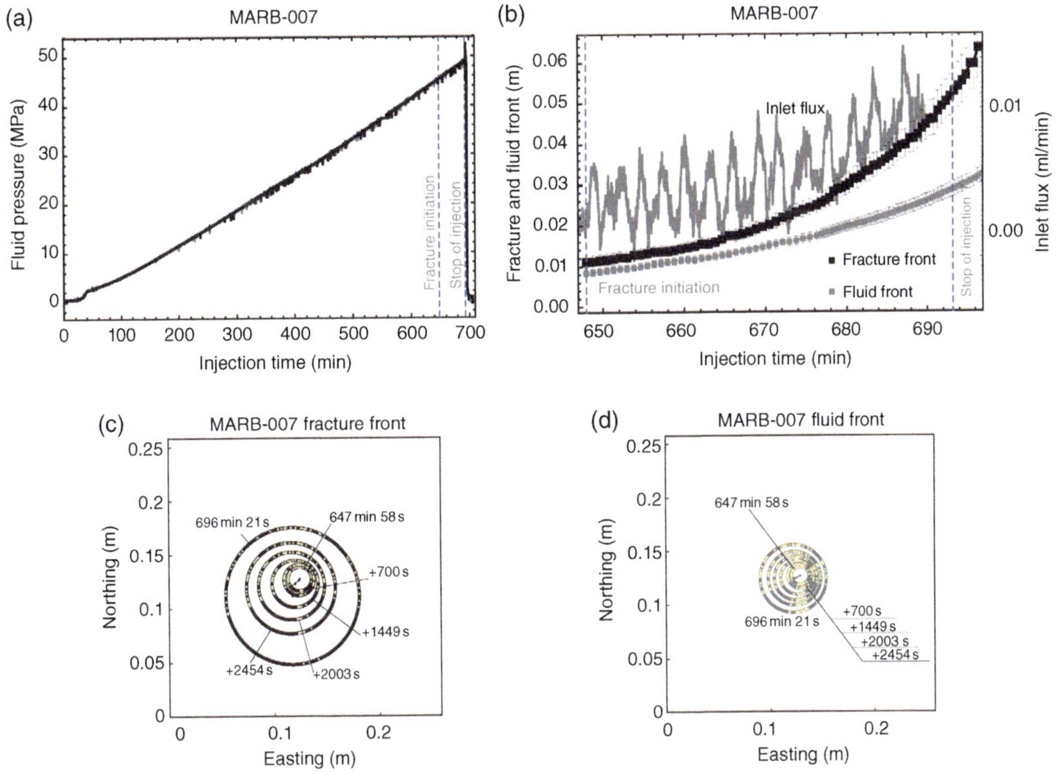

Figure 2.4 MARB-007: time evolution of (a) the injection pressure, (b) radius of the fracture and fluid front, and estimated inlet flux, and the footprint of the (c) fracture and (d) fluid front. The dots on the reconstructed fronts indicate the diffracting locations for different source-receiver pairs. The vertical lines indicate respectively the time of fracture initiation (determined by the sudden vertical displacement near the wellbore on the top surface), and the stop of the injection. The oscillations observed in the estimated flux results from the oscillations of the pressure measurement, which are related to the oscillating position of the needle valve during this particular test.

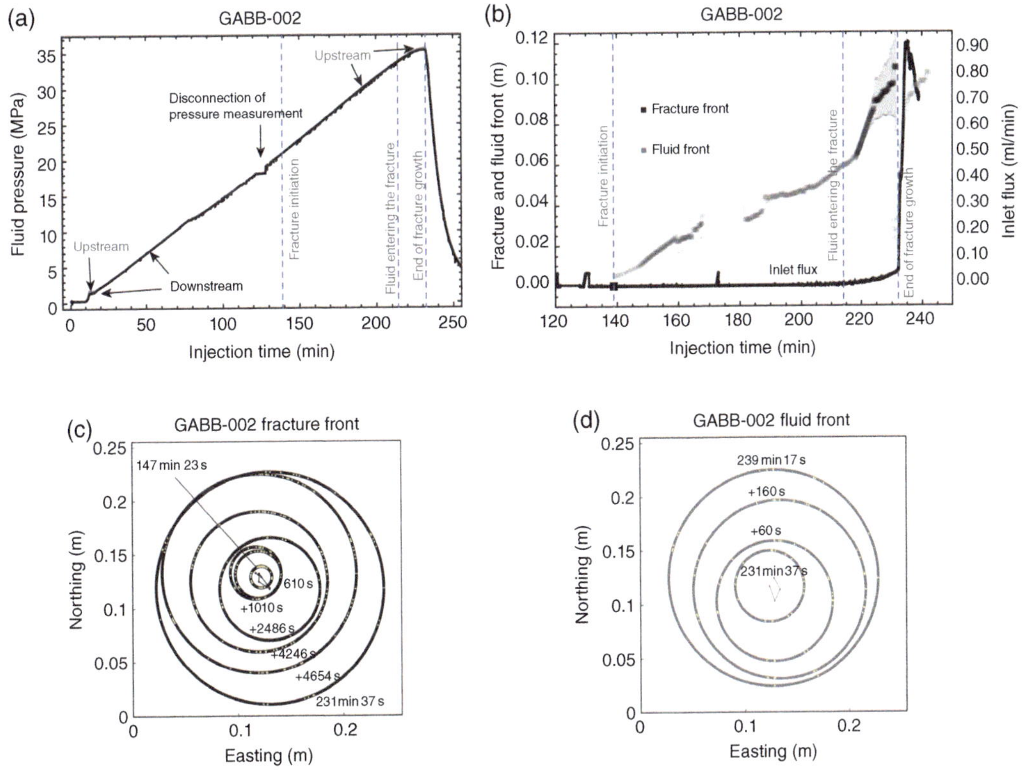

Figure 2.5 GABB-002: time evolution of the (a) injection pressure, (b) radius of the fracture and fluid fronts, and estimated inlet flux, and footprint of the (c) fracture and (d) fluid fronts. The dots on the reconstructed fronts indicate the diffracting locations for different source-receiver pairs. The vertical lines indicate respectively the time of fracture initiation (determined by the start of the visible diffraction), the start of the fluid front growth (determined by the increase of the inlet flux), and the time where the fracture front reaches the sample boundaries (determined by the very fast increase of the inlet flux). No stress is applied in the vertical direction in that experiment.

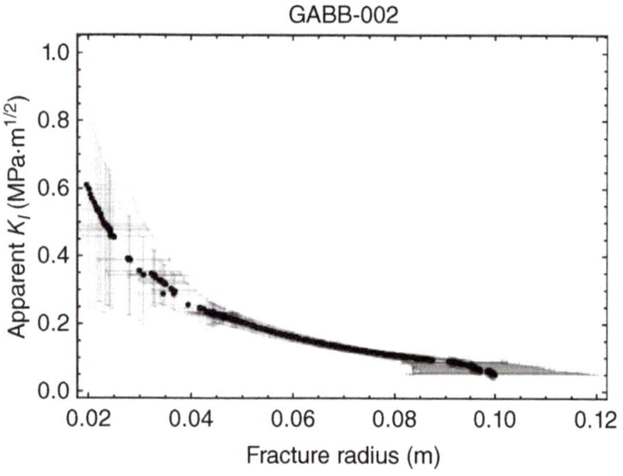

Figure 2.6 GABB-002: the evolution of the estimated apparent toughness. The error bars are estimated based on the uncertainties of the fracture front measurement.

estimate the error bar on such apparent toughness estimate in Figure 2.6). Such a large uncertainty on the fracture radius is intrinsically related to the low amplitude of the diffraction associated with the rather small fracture width near the fracture front when a large lag is present. Such an apparent toughness estimate has – of course – to be taken with caution. First, the area and exact profile of the fluid pressure around the injection point is strongly approximated. One should probably exclude the wellbore diameter from the loaded area, but as already mentioned, nothing indicates that the fluid penetrates inside the fracture during the experiment prior to the fracture front reaching the specimen boundaries – such that the area over which the net loading occurs is probably overestimated. This cannot explain the discrepancy with the gabbro fracture toughness measurement. One possible reason is that the size of the process zone is large and interacts strongly with the specimen boundary thus resulting in such an apparent "softening." It is worth noting that this apparent toughness evolution is different from that of toughness-dominated HF experiments in gabbro [34], where an apparent toughening occurs before the drop of the apparent toughness. This discrepancy needs to be further investigated in relation to the quasi-brittle behavior of this material and its interplay with the fluid–solid coupling.

2.4 Discussions and Conclusions

2.4.1 Resolution of the Fluid Front Location

The resolution of the reconstructed fracture and fluid front depends on the number of the picked diffraction arrivals. The fracture and fluid front presents large uncertainties at early and late time due to the limited number of picked source-receiver pairs. We indicate the corresponding diffracting locations as the dots along the reconstructed fracture/fluid front for all three experiments in Figures 2.2–2.5, 2.7, and 2.8. MARB-005 has the best coverage of the diffraction points, while GABB-002 presents the least coverage with a fracture/fluid radius uncertainty of centimeters. As a result, the fracture footprints in Figures 2.2–2.5, 2.7, and 2.8 should be taken with caution – one should combine it with the coverage condition and the corresponding estimation accuracy for further analysis.

Lag-dominated experiments are characterized by a specific evolution of the transmitted energy for facing source–receiver pairs perpendicular to the fracture plane. It first attenuates upon the arrival of the fracture front, then reincreases following the arrival of the fluid front, and finally decreases again due to the increase of the fluid layer thickness. This re-increase of the transmitted energy occurs prior to the arrival of the diffracted fluid front at the given transducer pairs location in the MARB-007 and GABB-002 experiments (Figures 2.7 and 2.8). Such an inconsistency may possibly result from the fact that the location of the diffraction fluid front actually corresponds to the maximum change of the width profile gradient rather than the actual fluid front position – owing to the non-locality of the fracture elastic deformation. As shown in Figure 2.8, the re-increase associated with the fluid front arrival does not always occur for shear waves due to the lack of shear stiffness of the fracturing fluid [32].

2.4.2 Quasi-Brittle Effects

As shown in Figures 2.3, 2.7, and 2.8, the attenuation of the transmitted energy occurs prior to the fracture front arrival. Such an attenuation ahead of the estimated fracture front from the diffracted waves is likely related to the presence of a fracture process zone ahead of the tip where micro-cracking occurs. Shear waves are more sensitive than compressional waves to the acoustic attenuation and sometimes indicates a larger process zone size (see Figure 2.8). Moreover, postmortem observations of the fracture path also indicate a strong quasi-brittle effect with the presence of crack bridges as shown in Figure 2.10. We design here the experiments using the LHFM theory, but we are well aware that these quasi-brittle effects would enter. As pointed out in [26], the presence of a process zone would bring into play another time scale $t_{cm} = E'^2 \mu'/\sigma_c^3$ related to the nucleation of a cohesive zone and a characteristic length scale $L_{coh} = E' w_c/\sigma_c^3$ for this process zone (σ_c and w_c indicate respectively the maximum traction in the cohesive zone model and the critical opening beyond which no cohesive forces would apply). When the process zone size is not negligible compared to the fracture extent,

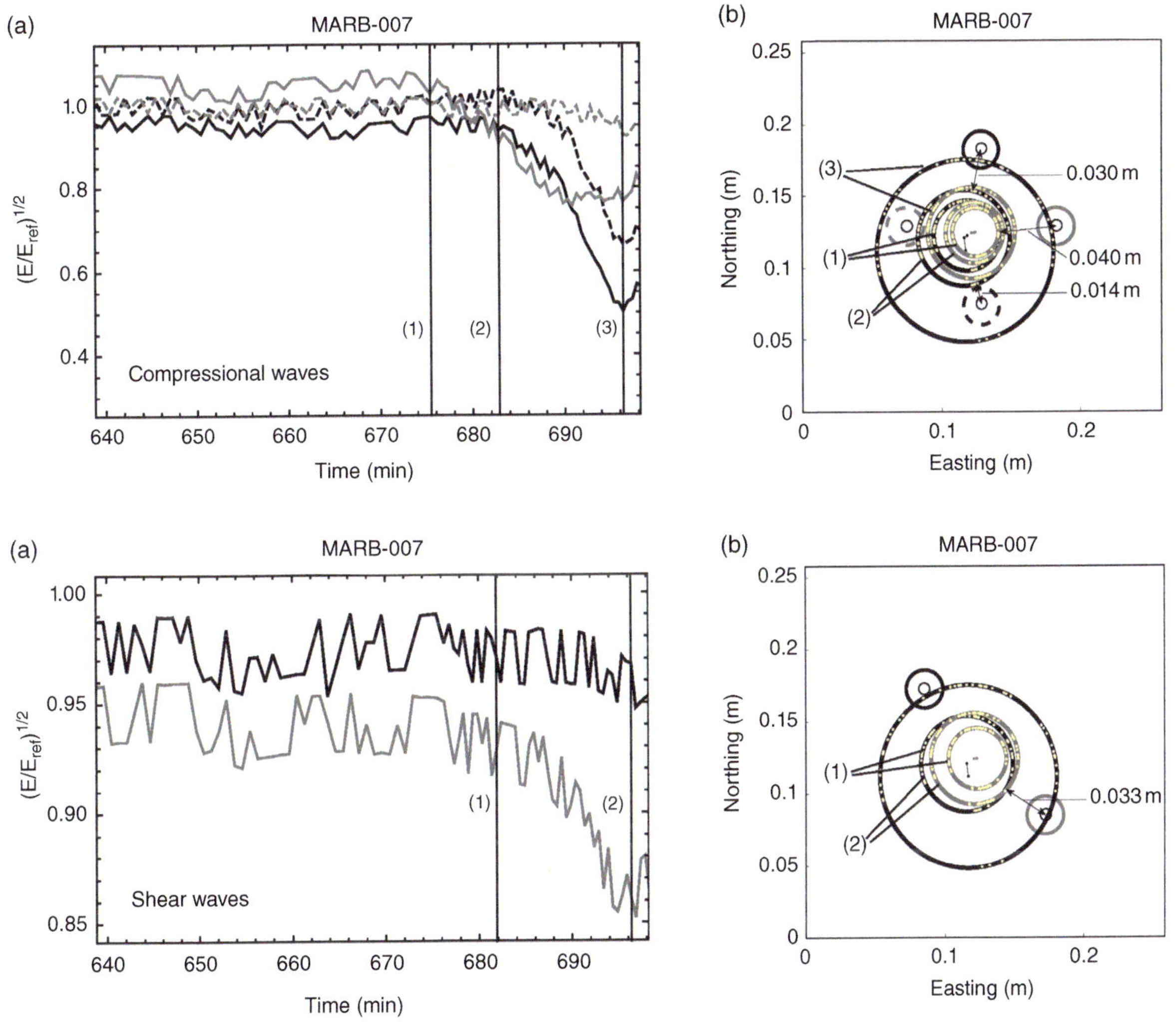

Figure 2.7 MARB-007: time evolution of the signal strength of transmitted waves, and (b) the fracture geometry at different times of the fracture growth. The dots on the reconstructed fronts indicate the diffracting locations for different source-receiver pairs. Locations of the transducers whose acoustic energy attenuation ratio is displayed in figure (a) are indicated by circles in figure (b) using the same line style. The fracture (in black) and fluid (in gray) extent associated with the numbered vertical lines in figure (a), is also shown in figure (b) using the same number.

the fracture growth deviates from the LHFM solutions as a result of the interplay between the fluid front and the rough cohesive zone. During this process, a larger confining stress σ_3/σ_c would result in more energy dissipation in the viscous flow inside the rough cohesive zone and therefore more significant deviations from LHFM – a larger fracture width, fluid lag and net pressure, and a shorter fracture length. A better understanding of these quasi-brittle effects requires a more quantitative measurement of the fracture geometry and process zone during the fracture growth.

2.4.3 Hydraulic Fracture Surfaces

The silicone oil left on the fracture surfaces makes it difficult to measure quantitatively the roughness profile of the created fractures. We therefore only show some photos of the fractured surfaces in Figure 2.9. A grooving pattern is clearly visible in MARB-007, while this cannot be clearly observed in the other two experiments.

The fracture roughness seems to depend both on the grain size of the material as well as the loading conditions during the fracture growth. From visual inspection, the roughness in gabbro is much larger than that in marble. For the two marble experiments, MARB-007 presents a rougher fractured surface. This is probably related to a more important effect of the lag/viscous flow in MARB-007, where the fluid pressure distribution is less uniform.

The roughness of GABB-002 is much larger than that of toughness dominated experiments performed in the same rock as shown in Figure 2.10 (note that we need a hammer

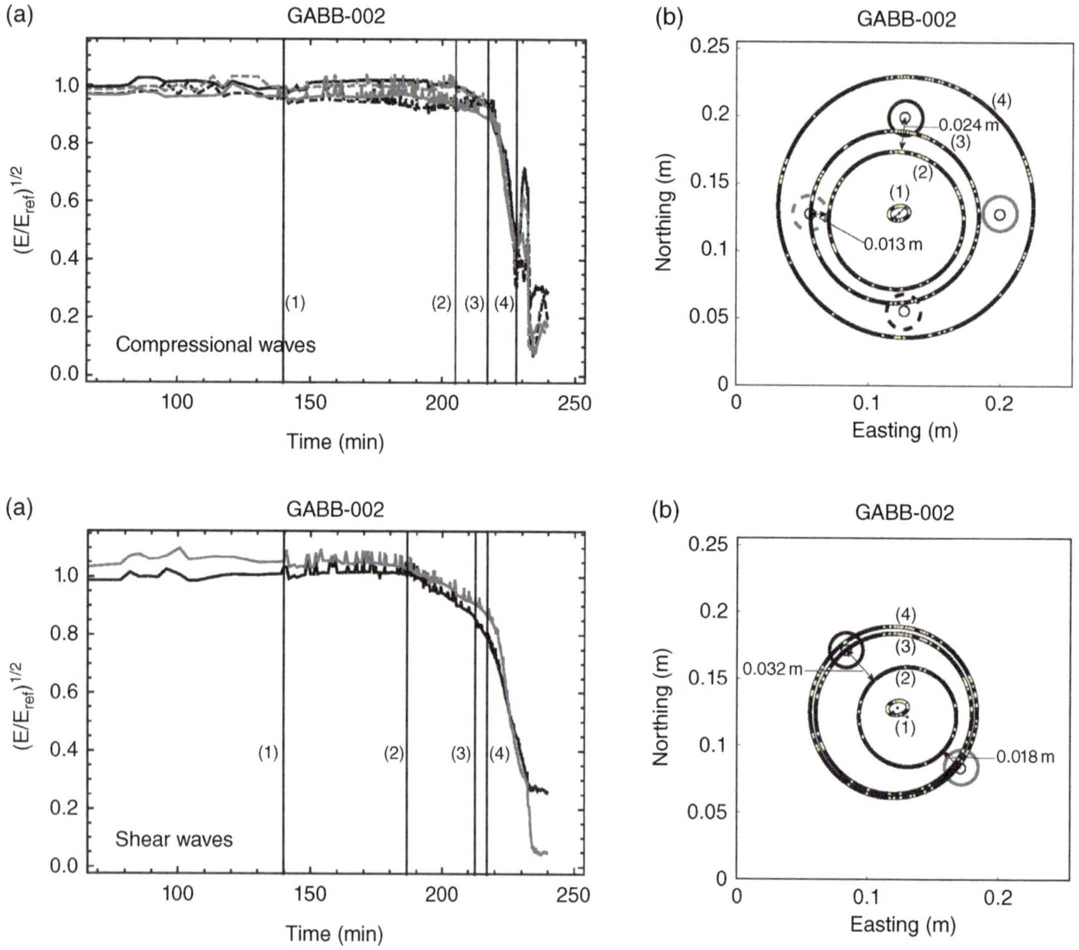

Figure 2.8 GABB-002: time evolution of (a) the signal strength of transmitted waves and (b) the fracture geometry at different times of the fracture growth. The dots on the reconstructed fronts indicate the diffracting locations for different source-receiver pairs. Locations of the transducers whose acoustic energy attenuation ratio is displayed in figure (a) are indicated by circles in figure (b) using the same line style. The fracture extent associated with the numbered vertical lines in figure (a), is also shown in figure (b) using the same number. The fluid front enters into the fracture nearly at the end of the fracture propagation and is therefore not shown here.

Figure 2.9 Fracture path of the three lag-related experiments and fractured surface of the southwestern lower part of (a) MARB-005 and the southwestern part of (b) MARB-007 and the southwestern lower part of (c) GABB-002. Note that the MARB-007 is created by two injections with the second one performed at a much larger rate one day after the first injection reported in Figure 2.4.

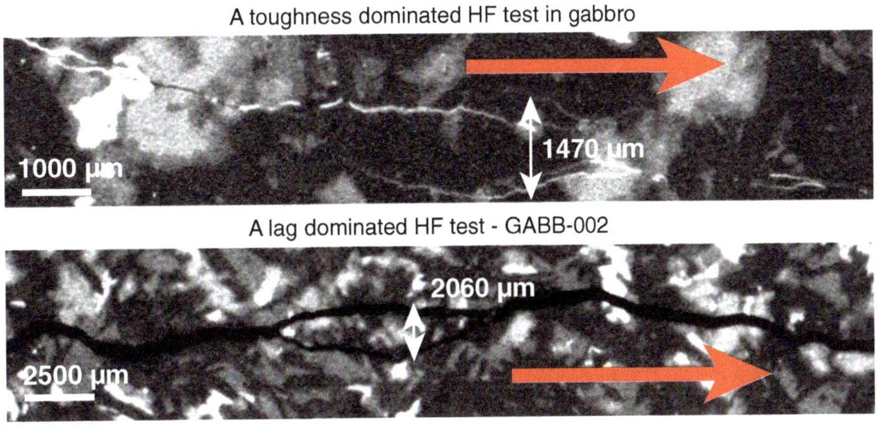

Figure 2.10 Postmortem CT images of the hydraulic fracture path of a toughness dominated HF test (GABB-001, see more experimental details in [33]) and the lag-dominated HF test - GABB-002. Note that the scales in these two images are different and that the crack-bridges observed in both tests testify an important quasi-brittle effect. The red arrow indicates the direction of fracture propagation. (*Source:* Liu [33], EPFL).

to separate the two fractured surfaces of GABB-002 even though the fracture width is much larger than that of the toughness dominated experiments). However, this rougher surface does not lead to the increase of the fracture apparent toughness (one basic assumption of the fracture length-dependent apparent toughness model [42]: a rougher profile results in a larger real fractured area and therefore a larger fracture energy/apparent toughness within the same nominal fractured area under the assumption of the same surface energy). For GABB-002, the apparent toughness instead presents a softening behavior with a smaller value compared to the toughness estimated from dry fracture mechanics tests (Figure 2.6). This mismatch may be due to the quasi-brittle characteristic of gabbro: many of the crack bridges as shown in Figure 2.10 may still be bounded during the fracture growth. The observed fracture roughness profile of GABB-002 may result from the latter splitting of crack-bridges after the fracture front reaches the specimen edge. The splitting of these crack bridges could be done either by the passage of the fluid front or by the physical splitting with a hammer after the experiments.

2.4.4 Conclusions

We have reported three hydraulic fracture experiments performed in quasi-brittle rocks characterized by an obvious fluid lag: a typical HF growth with a fluid lag (MARB-005), a HF growth before and after the shut-in (MARB-007) and a point-load like HF growth (GABB-002). In these lag related HFs, the pressure gradient is large in the viscous fluid flow and the inlet fluid pressure does not respond instantly to the increase of the fracture volume due to fracture initiation. As a result, the initiation pressure is large and the maximum pressure is much larger than the initiation pressure. This difference is more significant when the viscous flow plays a more important role as expected from theoretical predictions [15]. The fracture surface roughness seems to correlate with the grain size. A larger viscous effect and a larger fluid lag creates rougher surfaces than those reported in toughness dominated experiments [33] – an observation that needs further quantitative experimental investigations.

In the case of a large fluid lag, the small width near the fracture front results in a relatively weak diffraction of the fracture front. Despite this uncertainty in the fracture front measurement, we find that the evolution of the acoustic transmission agrees well with that of the fracture and fluid fronts in general. We observe that the transmission attenuation occurs prior to the fracture front arrival, most likely due to the presence of a fracture process zone in such quasi-brittle rocks. Further improvement of the resolution of the fracture and fluid fronts from acoustic measurements would benefit from full waveform forward and inverse simulations. Moreover, in order to better link quantitatively these experiments to theoretical predictions, direct measurement or better estimation of the time-evolution of the entering flux is needed.

Data Availability

The complete dataset of these experiments are made publicly available at https://doi.org/10.5281/zenodo.3901193 (MARB-005) and https://doi.org/10.5281/zenodo.5525773 (MARB-007 and GABB-002) in the hope of fostering future theoretical and experimental efforts.

Appendix A Determination of the Time of Fracture Initiation

We determine the time of fracture initiation by combining the following observable.

1) Decrease of the volume in the top-bottom pair of flat-jacks; or increase of the displacement of the top block

surface near the wellbore measured by the linear variable differential transformer (LVDT) when there is no confining stress applied on the top platen.

2) Increase of the inlet flux or change of the pressure difference between the upstream and downstream. This is only valid for toughness-dominated experiments, otherwise it represents the moment when the fluid starts entering into the fracture.
3) Deviation of the downstream pressure from the linear pressurization. This is valid for toughness-dominated experiments, otherwise the deviation might not occur or occurs with an important delay after initiation.
4) Start of a visible diffraction due to the growing fracture front. This depends strongly on the quality of the active acoustic measurement. It possibly leads to a delay in the estimation compared with the exact initiation time.

The fracture initiation time in MARB-005, MARB-007, and GABB-002 is determined, respectively by the evolution of top-bottom flat jacks and the appearance of the acoustic diffraction, the evolution of the LVDT, and the combination of the evolution of the LVDT, the deviation of the downstream pressure, and the appearance of the acoustic diffraction.

References

1 Desroches, J., Detournay, E., Lenoach, B. et al. (1994). The crack tip region in hydraulic fracturing. *Proceedings of the Royal Society of London. Series A: Mathematical and Physical Sciences* 447 (1929): 39–48.

2 Garagash, D.I. and Detournay, E. (2000). The tip region of a fluid-driven fracture in an elastic medium. *Journal of Applied Mechanics* 67: 183–192.

3 Lister, J.R. (1990). Buoyancy-driven fluid fracture: the effects of material toughness and of low-viscosity precursors. *Journal of Fluid Mechanics* 210: 263–280.

4 Detournay, E. and Garagash, D. (2003). The near-tip region of a fluid-driven fracture propagating in a permeable elastic solid. *Journal of Fluid Mechanics* 494: 1–32.

5 Khristianovic, S. and Zheltov, Y. (1955). Formation of vertical fractures by means of highly viscous fluids. *Proceedings 4th World Petroleum Congress*, Vol. II, Rome, pp. 579–586.

6 Geertsma, J. and De Klerk, F. (1969). A rapid method of predicting width and extent of hydraulically induced fractures. *Journal of Petroleum Technology* 21 (12): 1571–1581.

7 Biot, M., Medlin, W., and Masse, L. (1981). Laboratory experiments in fracture propagation. *Eastern regional meeting*, no. SPE-10377, Society of Petroleum Engineers, Columbus, OH.

8 Medlin, W. and Masse, L. (1984). Laboratory experiments in fracture propagation. *Society of Petroleum Engineers Journal* 24 (3): 256–268.

9 Johnson, E. and Cleary, M.P. (1991). Implications of recent laboratory experimental results for hydraulic fractures. *Low Permeability Reservoirs Symposium*, no. SPE-21846, Society of Petroleum Engineers.

10 Warpinski, N. (1985). Measurement of width and pressure in a propagating hydraulic fracture. *Society of Petroleum Engineers Journal* 25 (1): 46–54.

11 Jeffrey, R.G. (1989). The combined effect of fluid lag and fracture toughness on hydraulic fracture propagation. *Low Permeability Reservoirs Symposium*, no. SPE 18957, Society of Petroleum Engineers.

12 Kanin, E., Garagash, D., and Osiptsov, A.A. (2020). The near-tip region of a hydraulic fracture with pressure-dependent leak-off and leak-in. *Journal of Fluid Mechanics* 892: A31.

13 Moukhtari, F.E. and Lecampion, B. (2018). A semi-infinite hydraulic fracture driven by a shear thinning fluid. *Journal of Fluid Mechanics* 838: 573–605. https://doi.org/10.1017/jfm.2017.900.

14 Garagash, D.I. (2006). Propagation of a plane-strain hydraulic fracture with a fluid lag: early-time solution. *International Journal of Solids and Structures* 43 (18–19): 5811–5835. https://doi.org/10.1016/j.ijsolstr.2005.10.009.

15 Lecampion, B., Desroches, J., Jeffrey, R.G., and Bunger, A.P. Experiments versus theory for the initiation and propagation of radial hydraulic fractures in low permeability materials. *Journal of Geophysical Research - Solid Earth* 122: 1239–1263.

16 Bunger, A.P. and Detournay, E. (2007). Early-time solution for a radial hydraulic fracture. *Journal of Engineering Mechanics* 133 (5): 534–540.

17 Lecampion, B. and Detournay, E. (2007). An implicit algorithm for the propagation of a hydraulic fracture with a fluid lag. *Computer Methods in Applied Mechanics and Engineering* 196 (49–52): 4863–4880.

18 Bunger, A. and Cruden, A. Modeling the growth of laccoliths and large mafic sills: Role of magma body forces. *Journal of Geophysical Research Solid Earth* 116 (B2): B02203.

19 Gordeliy, E. and Detournay, E. (2011). A fixed grid algorithm for simulating the propagation of a shallow hydraulic fracture with a fluid lag. *International Journal for Numerical and Analytical Methods in Geomechanics* 35 (5): 602–629.

20 Gordeliy, E., Abbas, S., and Peirce, A. (2019). Modeling nonplanar hydraulic fracture propagation using the XFEM:

an implicit level-set algorithm and fracture tip asymptotics. *International Journal of Solids and Structures* 159: 135–155.

21 Shen, Y. (2014). A variational inequality formulation to incorporate the fluid lag in fluid-driven fracture propagation. *Computer Methods in Applied Mechanics and Engineering* 272: 17–33.

22 Mollaali, M. and Shen, Y. (2018). An Elrod–Adams-model-based method to account for the fluid lag in hydraulic fracturing in 2D and 3D. *International Journal of Fracture* 211 (1–2): 183–202.

23 Bunger, A.P. (2005). Near-surface hydraulic fracture. Ph.D. thesis. University of Minnesota.

24 Bunger, A.P., Gordeliy, E., and Detournay, E. (2013). Comparison between laboratory experiments and coupled simulations of saucer-shaped hydraulic fractures in homogeneous brittle-elastic solids. *Journal of the Mechanics and Physics of Solids* 61 (7): 1636–1654.

25 Garagash, D.I. (2019). Cohesive-zone effects in hydraulic fracture propagation. *Journal of the Mechanics and Physics of Solids* 133: 103727.

26 Liu, D. and Lecampion, B. (2021). Propagation of a plane-strain hydraulic fracture accounting for a rough cohesive zone. *Journal of the Mechanics and Physics of Solids* 149: 104322.

27 Liu, D. and Lecampion, B. (2019). Propagation of a plane-strain hydraulic fracture accounting for the presence of a cohesive zone and a fluid lag. *53rd US Rock Mechanics/Geomechanics Symposium*, ARMA.

28 Liu, D. and Lecampion, B. (2019). Growth of a radial hydraulic fracture accounting for the viscous fluid flow in a rough cohesive zone. *ARMA-CUPB Geothermal International Conference*, American Rock Mechanics Association.

29 Rubin, A.M. (1993). Tensile fracture of rock at high confining pressure: implications for dike propagation. *Journal of Geophysical Research Solid Earth* 98 (B9): 15919–15935.

30 De Pater, C., Desroches, J., Groenenboom, J., and Weijers, L. (1996). Physical and numerical modeling of hydraulic fracture closure. *SPE Production & Facilities* 11 (2): 122–128.

31 Groenenboom, J. and van Dam, D.B. (2000). Monitoring hydraulic fracture growth: laboratory experiments. *Geophysics* 65 (2): 603–611.

32 Liu, D., Lecampion, B., and Blum, T. (2020). Time-lapse reconstruction of the fracture front from diffracted waves arrivals in laboratory hydraulic fracture experiments. *Geophysical Journal International* 223 (1): 180–196.

33 Liu, D. (2021). Hydraulic fracture growth in quasi-brittle materials. Ph.D. thesis. EPFL.

34 Kuruppu, M., Obara, Y., Ayatollahi, M. et al. (2014). ISRM-suggested method for determining the mode I static fracture toughness using semi-circular bend specimen. *Rock Mechanics and Rock Engineering* 47 (1): 267–274.

35 Lhomme, T.P.Y. (2005). Initiation of hydraulic fractures in natural sandstones. TU Delft, Delft University of Technology.

36 Kovalyshen, Y., Bunger, A.P., Kear, J., and Kasperczyk, D. (2014). Comparison between ultrasonic and photometric methods for hydraulic fracture laboratory monitoring. *International Journal of Rock Mechanics and Mining Sciences* 70: 368–374.

37 Groenenboom, J. and Fokkema, J.T. (1998). Monitoring the width of hydraulic fractures with acoustic waves. *Geophysics* 63 (1): 139–148.

38 Detournay, E. (2016). Mechanics of hydraulic fractures. *Annual Review of Fluid Mechanics* 48: 311–339.

39 Garagash, D. and Detournay, E. (1997). An analysis of the influence of the pressurization rate on the borehole breakdown pressure. *International Journal of Solids and Structures* 34 (24): 3099–3118.

40 Möri, A. and Lecampion, B. (2021). Arrest of a radial hydraulic fracture upon shut-in of the injection. *International Journal of Solids and Structures* 219: 151–165.

41 Tada, H., Paris, P.C., and Irwin, G.R. The stress analysis of cracks, Del Research Corp, Hellertown PA.

42 Liu, D., Lecampion, B., and Garagash, D.I. (2019). Propagation of a fluid-driven fracture with fracture length dependent apparent toughness. *Engineering Fracture Mechanics* 220: 106616.

3

Mapping Hydraulic Fracture Growth Using Tiltmeter Monitoring Technique

Z. R. Chen[1] *and R. G. Jeffrey*[2]

[1] *CSIRO Energy, Clayton South, Victoria, Australia*
[2] *SCT Operations Pty. Ltd., Wollongong, New South Wales, Australia*

KEY POINTS

- The forward fracture model provides a lindear relationship between model parameters and the fracture-induced deformation.
- A linear inverse problem is solved to map fracture geometry by inversion of tilt observations.
- Application to field tilt data predicts a nearly horizontal, asymmetrical fracture, agreeing with offset monitoring borehole intersection data.

3.1 Introduction

As a proven highly effective technique for production enhancement in the petroleum industry, hydraulic fracturing has also found various successful applications: determination of in situ stress in rock mechanics, preconditioning rock for caving or inducing rock to cave in mining [1], mitigation of seismic and rockburst hazard [2], creation of geothermal energy reservoirs in renewable energy [3], and underground disposal of toxic or radioactive waste, underground sequestration of carbon dioxide, and remediation of contaminated ground in environmental engineering [4, 5]. Hydraulic fracture growth is a complicated process governed by the treatment parameters, mechanical property parameters of the rock and fracturing fluid, in situ stress and geological discontinuities within the rock formations. The resulting geometry of the hydraulic fracture has a critical effect on its success and is often proving to be more complex than initially designed/expected. Applying methods to monitor hydraulic fracture growth is of great importance to develop better understanding of fracturing mechanisms, to evaluate the effectiveness and efficiency of hydraulic fracturing, and to optimize hydraulic fracture design and treatment.

Due to the complexity of typical reservoirs, the absence of adequate hydraulic fracture models that can model a fracture comprehensively, the uncertainty in data collected and the lack of sufficient data, it is generally difficult to obtain a full knowledge about fracture geometry and growth or to verify the model-predicted fracture geometry [6–8]. Various fracture diagnostic techniques have been developed to provide valuable information about fracture growth, improving our understanding of complex hydraulic fracture behavior [6–8]. Among them, tiltmeter hydraulic fracture monitoring has proven to be able to provide reliable information on fracture azimuth, dip, volume, fracture center offset, or asymmetric fracture growth [9–18], and it can also provide useful information on fracture length, height, and width when using downhole tiltmeter mapping [19–21]. High-precision borehole tiltmeters, typically with a resolution better than five nrad, contain two orthogonal electrolytic level sensors that measure tilt angles (rotation) in two orthogonal vertical planes with reference to the vertical gravity vector. Surface tiltmeters are installed in boreholes and sand, placed around each instrument, couples them to surrounding rock mass deformation. When a hydraulic fracture is created in a rock formation, the opening of fracture faces deforms the surrounding rock mass and the rock deformation increases as the fracture grows. This fracture induced rock deformation can be detected by the tiltmeter array. The inversion of the fracture induced tilt observations can provide information about the hydraulic fracture growth. The analytical solution of a uniform dislocation (opening) across a rectangular fracture plane in a homogeneous elastic half-space has often been used in the inverse analysis as a forward fracture model to

Mechanics of Hydraulic Fracturing: Experiment, Model, and Monitoring, First Edition.
Edited by Xi Zhang, Bisheng Wu, Diansen Yang, and Andrew Bunger.

infer the fracture geometry parameters which best fit the measured tilt vectors at each tiltmeter station [10, 13, 14, 18, 21, 22]. The analytical solution of the fracture induced deformation is a nonlinear function of the fracture orientation, so the nonlinear inversion problem has commonly been approximated by a linearized inverse problem and solved using a nonlinear least squares method [10, 13] or a Bayesian inversion approach [23]. For a nonlinear inverse problem, the misfit function can have multiple local minima, and the minimization of the linearized problem may converge to a model estimate corresponding to a local minimum or may lead to a wrong solution for some strongly nonlinear problems. Inversion of fracture induced deformation measurements also provides information on fracture opening distribution and growth [4, 15, 16, 24]. The inversion method consists of two separate inversion steps using the same tilt data set; the fracture orientation is estimated using a simple forward fracture model in the first step, and the inferred fracture orientation is then used to define the fracture plane (the fracture orientation is fixed) and to determine the opening distribution over the defined fracture plane in the second step. Since the fracture-induced deformation is a function of all the model parameters and the estimates of model parameters are coupled in minimizing the misfit function, inversion for part of all the model parameters in separate steps may not find the optimum solution and may lead to unreliable model parameter estimates [25].

In this chapter, the concept of eigenstrain has been used to characterize fracture geometry and determine fracture induced deformation. The fracture model provides a linear relationship between the model parameters (eigenstrain) and the induced deformation. So, a linear inverse problem needs to be solved to find the fracture geometry using the deformation observations. The advantage of developing a linear forward model is that the fruitful methods for solving linear inverse problems, and the uniqueness, stability and resolution of the solutions have been well established. In this chapter, the Bayesian inversion approach has been adopted to solve the linear inverse problem in mapping fracture geometry. The inverse analysis method has been applied to field tilt data collected for monitoring hydraulic fracturing growth during preconditioning of a roof rock over a coal longwall panel. The predicted fracture geometry has been compared with the intersection data observed in offset monitoring boreholes.

3.2 Forward Problem Formulation

3.2.1 Forward Model Definition

As shown in Figure 3.1, consider a planar fracture in an isotropic homogeneous elastic half-space. The orientation of the fracture plane is described by the angles of strike (φ) and dip (θ). The strike direction represents the line of intersection of the fracture plane with a horizontal plane (the ground surface), given as the angle of strike ($0 \le \varphi < 2\pi$) measured clockwise from north. The dip direction is at an angle of $\pi/2$ to the right of the strike direction in the horizontal plane. The dip angle ($0 \le \theta \le \pi/2$) is the dihedral angle between the fracture and horizontal planes, representing the inclination of the fracture plane below the horizontal plane. The fracture is modeled as a Volterra dislocation (displacement discontinuity) vector $\boldsymbol{b}$ across the fracture plane. $\boldsymbol{n}$ is the unit vector normal to the fracture plane, $\boldsymbol{s}$ is the unit vector tangent to the fracture in the strike direction, and the unit vector $\boldsymbol{t}$ is orthogonal to $\boldsymbol{s}$ and is in the direction of projection of the dip direction onto the fracture plane. $\boldsymbol{\xi}$ is the reference point on the fracture plane. The local fracture coordinate system consists of the set of orthogonal vectors $\boldsymbol{t}$, $\boldsymbol{s}$, and $\boldsymbol{n}$. The components of the dislocation vector $\boldsymbol{b}$ in the local fracture coordinate system ξtsn represent the dip-slip (b_t), strike-slip (b_s), and opening (b_n), respectively. The projection of the dislocation vector $\boldsymbol{b}$ onto the normal $\boldsymbol{n}$ to the fracture plane, $(\boldsymbol{b} \cdot \boldsymbol{n})\boldsymbol{n}$, represents the opening of the fracture. The slip, $\boldsymbol{b} - (\boldsymbol{b} \cdot \boldsymbol{n})\boldsymbol{n}$, is the projection of the dislocation vector $\boldsymbol{b}$ onto the fracture plane, representing the relative shear motion between the fracture surfaces. The rake angle ($-\pi < \gamma \le \pi$) measures the direction of the slip vector relative to the strike direction on the fracture plane.

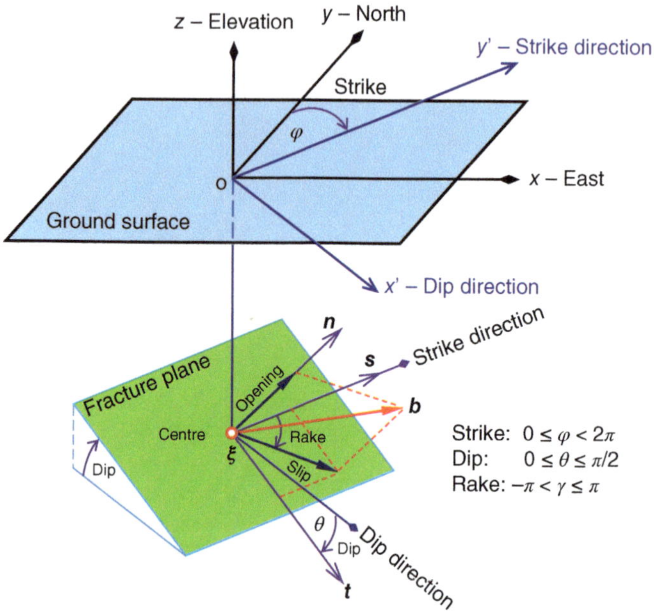

Figure 3.1 Definition of fracture plane orientation and displacement discontinuity vector.

For a fracture plane of given strike angle (φ) and dip angle (θ), the direction cosines of the unit vectors $\boldsymbol{t}$, $\boldsymbol{s}$, and $\boldsymbol{n}$ in the global coordinates system $oxyz$ can be expressed respectively as,

$$t = \begin{bmatrix} t_1 \\ t_2 \\ t_3 \end{bmatrix} = \begin{bmatrix} \cos\theta\cos\varphi \\ -\cos\theta\sin\varphi \\ -\sin\theta \end{bmatrix}, s = \begin{bmatrix} s_1 \\ s_2 \\ s_3 \end{bmatrix} = \begin{bmatrix} \sin\varphi \\ \cos\varphi \\ 0 \end{bmatrix},$$

$$n = \begin{bmatrix} n_1 \\ n_2 \\ n_3 \end{bmatrix} = \begin{bmatrix} \sin\theta\cos\varphi \\ -\sin\theta\sin\varphi \\ \cos\theta \end{bmatrix} \tag{3.1}$$

The components b_i ($i = 1, 2, 3$) of the dislocation vector $\boldsymbol{b}$ in the global coordinate system $oxyz$ can be expressed as,

$$\boldsymbol{b} = b\boldsymbol{l} = b\begin{bmatrix} l_1 \\ l_2 \\ l_3 \end{bmatrix} = \begin{bmatrix} b_1 \\ b_2 \\ b_3 \end{bmatrix} = b_t\boldsymbol{t} + b_s\boldsymbol{s} + b_n\boldsymbol{n} \tag{3.2}$$

where b and the unit vector $\boldsymbol{l}$ represent the magnitude and direction of the dislocation vector $\boldsymbol{b}$, respectively.

3.2.2 Forward Model

The displacement field u_i induced by a dislocation vector $\boldsymbol{b}$ across the fracture plane S can be expressed as [26, 27]

$$u_i(\boldsymbol{x}) = \frac{1}{8\pi\mu}\int\int_S b_j n_k \omega^i_{jk} dS = \frac{1}{16\pi(1-\nu)}\int\int_S b_j n_k \overline{\omega}^i_{jk} dS \tag{3.3}$$

where n_k are the direction cosines of the unit vector normal to the surface element dS, μ is the shear modulus, and ν is the Poisson's ratio. The summation convention is assumed unless otherwise stated in this work. The kernel function ω^i_{jk} and $\overline{\omega}^i_{jk}$ are obtained from the nuclei of strain in an isotropic homogeneous elastic half-space as

$$\omega^i_{jk} = \frac{\mu}{2(1-\nu)}\overline{\omega}^i_{jk} = \frac{\lambda+\mu}{\lambda+2\mu}\left[\lambda\delta_{jk}\frac{\partial u^l_i}{\partial \xi_l} + \mu\left(\frac{\partial u^j_i}{\partial \xi_k} + \frac{\partial u^k_i}{\partial \xi_j}\right)\right] \tag{3.4}$$

where λ is the lame's constant. u^j_i is the ith component of the displacement vector at (x_1, x_2, x_3) due to a single force with magnitude $16\pi\mu(1-\nu)$ at (ξ_1, ξ_2, ξ_3) in the jth direction, $\partial u^l_i/\partial\xi_l$ (summed over $l = 1, 2, 3$) correspond to the displacements from basic Galerkin vectors for a double force in the jth direction without moment when $j = k$ or a double force in the jth direction with moment in the direction normal to the ox_jx_k plane when $j \neq k$. Because of the symmetry $\omega^i_{jk} = \omega^i_{kj}$, there are only six independent functions ω^i_{jk} necessary for determination of the ith component of the displacement u_i, and each of them represents a characteristic deformation pattern. For example, ω^i_{11} represents the displacement field due to a displacement discontinuity b_1 across a surface element with its normal in the x_1 direction, i.e. representing a vertical fracture in opening mode in the x_1 direction.

The detailed expressions for various types of nuclei of strain have been given by Mindlin [28] for the case of full-space solutions, Mindlin and Cheng [29] for half-space solutions, Yu and Sanday [30] for joined half-spaces solutions, and Singh [31] and Pan [32, 33] for multilayered half-spaces solutions. Analytical expressions based on integration of Eq. (3.3) over rectangular dislocation surfaces in an elastic half-space have been obtained for determining the deformation and stress induced by various dislocations [10, 26, 34–39]. These analytical solutions together with numerical solutions based on Eq. (3.3) have been widely used as efficient forward fracture models in inverse analysis of quasi-static deformation measurements (displacements, strains, and tilts) for mapping the slip distributions over fault planes [40–44] and for mapping hydraulic fracture geometry [10–14, 23]. These analytical solutions are nonlinear functions of the orientation (strike and dip) of the fracture plane, so mapping the fracture geometry (orientation and size) generally needs to deal with nonlinear inverse problems.

Following the way of treating the dislocation as an equivalent inclusion [45–47], we define the eigenstrain ε^*_{jk} associated with the equivalent inclusion as

$$\begin{aligned} \varepsilon^*_{jk}dV &= b_{jk}dS, && \text{for a volume inclusion} \\ \varepsilon^*_{jk} &= b_{jk}\delta(\boldsymbol{S}-\boldsymbol{\xi}), && \text{for a plane inclusion} \end{aligned}, \tag{3.5}$$

with

$$b_{jk} = \frac{1}{2}(b_j n_k + b_k n_j), \tag{3.6}$$

where $\delta(\boldsymbol{S}-\boldsymbol{\xi})$ is the Dirac delta function, which is unbounded when $\boldsymbol{\xi}$ is on $\boldsymbol{S}$ and zero otherwise, and $\int_V \delta(\boldsymbol{S}-\boldsymbol{\xi})d\boldsymbol{\xi} = \int_S dS$. The tensor b_{jk} combines the dislocation vector $\boldsymbol{b}$ with the fracture plane unit normal vector $\boldsymbol{n}$ to form a second-order symmetric tensor, representing the intensity of the dislocation. The tensor b_{jk} is refered to as the potency density tensor [48].

Then the induced displacement by the equivalent inclusion with the eigenstrain ε^*_{jk} can be expressend as

$$u_i(\boldsymbol{x}) = \frac{1}{16\pi(1-\nu)}\int\int\int_V \varepsilon^*_{jk}\overline{\omega}^i_{jk}dV = \frac{1}{16\pi(1-\nu)}\int\int_S b_{jk}\overline{\omega}^i_{jk}dS \tag{3.7}$$

Since ε^*_{jk} and b_{jk} are symmetric second-order tensor, $\varepsilon^*_{jk} = \varepsilon^*_{kj}$ and $b_{jk} = b_{kj}$, only six components are necessary to represent the displacement discontinuity and the induced deformation field. Equation (3.7) shows that the induced displacement u_i is a linear function of the eigenstrain ε^*_{jk} and can be expressed as a linear combination of the six characteristic deformation patterns $\overline{\omega}^i_{jk}$ multiplied by the corresponding eigenstrain ε^*_{jk} or its intensity b_{jk}.

The volume integral form of Eq. (3.7) generally provides an approximated method for modeling the far-field deformation of a dislocation by its equivalent inclusion. Vasco et al. [43] have applied the volume integral form of Eq. (3.7) as a forward model in the inversion of surface displacement and strain observations to map the underground distribution of volume expansion due to magna intrusion. The advantage of this forward model is that it deals with a linear problem and no initial deformation source geometry needs to be assumed. Since the three-dimensional distribution of the eigenstrain is modeled in the forward problem, it significantly increases the number of unknown model parameters and their uncertainties, compared with the methods that consider the two-dimensional distribution on a known fault plane.

In tiltmeter monitoring techniques, the tiltmeter measures small angular movement of the rock mass along its two orthogonal axes at the tiltmeter station. The measured tilt angles are related to the skew symmetric part of the displacement gradient in the rock mass by the relation

$$T_i = \frac{\partial u_i}{\partial x_3} - \frac{\partial u_3}{\partial x_i}, \qquad (i = 1, 2) \tag{3.8}$$

Substituting Eq. (3.7) into Eq. (3.8), we have the tilt field

$$T_i(\boldsymbol{x}) = \frac{1}{16\pi(1-\nu)} \iint_S b_{jk} \tau^i_{jk} \, dS, \quad (i = 1, 2)$$
$$\text{where } \tau^i_{jk} = \frac{\partial \overline{\omega}^i_{jk}}{\partial x_3} - \frac{\partial \overline{\omega}^3_{jk}}{\partial x_i} \tag{3.9}$$

3.2.2.1 Point Source Dislocation Singularity Model

For an elementary infinitesimal dislocation loop dS of strength b_j and unit normal n_k, the induced displacement and tilt fields by a point source dislocation singularity are given by,

$$u_i(\boldsymbol{x}) = \frac{1}{16\pi(1-\nu)} b_{jk}\overline{\omega}^i_{jk}, \quad (i = 1, 2, 3)$$
$$T_i(\boldsymbol{x}) = \frac{1}{16\pi(1-\nu)} b_{jk}\tau^i_{jk}, \quad (i = 1, 2) \tag{3.10}$$

The matrix form of the second-order symmetric tensor b_{ij} can be expressed as

$$\begin{bmatrix} b_{11} & b_{12} & b_{13} \\ & b_{22} & b_{23} \\ Sym & & b_{33} \end{bmatrix} = b \begin{bmatrix} n_1 l_1 & (n_1 l_2 + n_2 l_1)/2 & (n_1 l_3 + n_3 l_1)/2 \\ & n_2 l_2 & (n_2 l_3 + n_3 l_2)/2 \\ Sym & & n_3 l_3 \end{bmatrix} \tag{3.11}$$

The contributions of the unit vector of the normal, $\boldsymbol{n}$, and the unit vector of the dislocation direction, $\boldsymbol{l}$, are symmetric and can be interchanged without affecting the eigenstrain tensor b_{ij}. So, for the point dislocation model, the displacement and tilt fields, Eq. (3.10), are not affected by interchanging the unit vectors of $\boldsymbol{n}$ and $\boldsymbol{l}$, i.e. the normal vector to the fracture plane could be the dislocation direction vector and vice versa. The eigenstrain and induced displacement and tilt fields are also unchanged by replacing $(\boldsymbol{n}, \boldsymbol{l})$ with $(-\boldsymbol{n}, -\boldsymbol{l})$. This means that the direction of the fracture plane and dislocation vector cannot be resolved by matching the point source dislocation model with the deformation measurements without applying additional constrains or information during the inverse modeling. This is the well-known issue of fault plane-auxiliary plane ambiguity in seismic moment tensor inversion of the point source dislocation model [49]. However, for a dislocation in opening model (e.g. tensile-opening dominated hydraulic fracture), the dislocation vector is normal to the fracture plane, i.e. $\boldsymbol{n} = \boldsymbol{l}$, so the fracture direction can be uniquely determined using the point source dislocation model with the constraint/convention $(0 \le \theta \le \pi/2)$ on the dip angle.

The induced displacement and tilt fields can be expressed in matrix form as

$$\begin{bmatrix} u_1(\boldsymbol{x}) \\ u_2(\boldsymbol{x}) \\ u_3(\boldsymbol{x}) \end{bmatrix} = \frac{1}{16\pi(1-\nu)} \begin{bmatrix} \overline{\omega}^1_{11} & \overline{\omega}^1_{22} & \overline{\omega}^1_{33} & 2\overline{\omega}^1_{12} & 2\overline{\omega}^1_{13} & 2\overline{\omega}^1_{23} \\ \overline{\omega}^2_{11} & \overline{\omega}^2_{22} & \overline{\omega}^2_{33} & 2\overline{\omega}^2_{12} & 2\overline{\omega}^2_{13} & 2\overline{\omega}^2_{23} \\ \overline{\omega}^3_{11} & \overline{\omega}^3_{22} & \overline{\omega}^3_{33} & 2\overline{\omega}^3_{12} & 2\overline{\omega}^3_{13} & 2\overline{\omega}^3_{23} \end{bmatrix} \begin{bmatrix} b_{11} \\ b_{22} \\ b_{33} \\ b_{12} \\ b_{13} \\ b_{23} \end{bmatrix}$$

$$\begin{bmatrix} T_1(\boldsymbol{x}) \\ T_2(\boldsymbol{x}) \end{bmatrix} = \frac{1}{16\pi(1-\nu)} \begin{bmatrix} \tau^1_{11} & \tau^1_{22} & \tau^1_{33} & 2\tau^1_{12} & 2\tau^1_{13} & 2\tau^1_{23} \\ \tau^2_{11} & \tau^2_{22} & \tau^2_{33} & 2\tau^2_{12} & 2\tau^2_{13} & 2\tau^2_{23} \end{bmatrix} \begin{bmatrix} b_{11} \\ b_{22} \\ b_{33} \\ b_{12} \\ b_{13} \\ b_{23} \end{bmatrix} \tag{3.12}$$

In far-field monitoring, due to the limitations of field conditions for instrument deployment the observation points are often far, compared to the characteristic length of the fracture, from the fracture. The point source dislocation model is generally used as an efficient forward model in mapping fracture volume and orientation using far-field observations. For an array of n tiltmeters located at different observation stations $\boldsymbol{x}_i$ $(i = 1, 2, ..., n)$, we have a set of $2n$ observations (2 tilt components for each tiltmeter). The point source singularity model predictions of the tilts at the observation points can be expressed in a general form as

$$\boldsymbol{d}_{2n} = \boldsymbol{G}_{2n\times 6}\boldsymbol{p}_6 \tag{3.13}$$

where $\boldsymbol{d}$ denotes the column vector consisting of $2n$ tilt observations, $\boldsymbol{p}$ denotes the model parameter vector consisting of six eigenstrain components, and the linear forward model operator $\boldsymbol{G}$ maps the model parameter vector $\boldsymbol{p}$ on the $2n$ observations $\boldsymbol{d}$ through Eq. (3.12). In practical monitoring, we normally have more observations than the number of unknown model parameters for the point source singularity model, i.e. $2n > 6$. So, we are often confronted with an overdetermined inverse problem in monitoring fracture using the point source singularity forward model.

3.2.2.2 A General Distributed Dislocation Model

A fracture of general shape can be modeled using distributed eigenstrains $b_{jk}(\xi)$ across the fracture plane S, and the induced displacement and tilt fields can be obtained by integrating Eq. (3.12) over the fracture plane as

$$\begin{gathered}
\begin{bmatrix} u_1(\boldsymbol{x}) \\ u_2(\boldsymbol{x}) \\ u_3(\boldsymbol{x}) \end{bmatrix} = \frac{1}{16\pi(1-\nu)} \iint\limits_S \begin{bmatrix} \overline{\omega}_{11}^1 & \overline{\omega}_{22}^1 & \overline{\omega}_{33}^1 & 2\overline{\omega}_{12}^1 & 2\overline{\omega}_{13}^1 & 2\overline{\omega}_{23}^1 \\ \overline{\omega}_{11}^2 & \overline{\omega}_{22}^2 & \overline{\omega}_{33}^2 & 2\overline{\omega}_{12}^2 & 2\overline{\omega}_{13}^2 & 2\overline{\omega}_{23}^2 \\ \overline{\omega}_{11}^3 & \overline{\omega}_{22}^3 & \overline{\omega}_{33}^3 & 2\overline{\omega}_{12}^3 & 2\overline{\omega}_{13}^3 & 2\overline{\omega}_{23}^3 \end{bmatrix} (\boldsymbol{x}, \boldsymbol{\xi}) \begin{bmatrix} b_{11}(\boldsymbol{\xi}) \\ b_{22}(\boldsymbol{\xi}) \\ b_{33}(\boldsymbol{\xi}) \\ b_{12}(\boldsymbol{\xi}) \\ b_{13}(\boldsymbol{\xi}) \\ b_{23}(\boldsymbol{\xi}) \end{bmatrix} d\boldsymbol{\xi} \\
\begin{bmatrix} T_1(\boldsymbol{x}) \\ T_2(\boldsymbol{x}) \end{bmatrix} = \frac{1}{16\pi(1-\nu)} \iint\limits_S \begin{bmatrix} \tau_{11}^1 & \tau_{22}^1 & \tau_{33}^1 & 2\tau_{12}^1 & 2\tau_{13}^1 & 2\tau_{23}^1 \\ \tau_{11}^2 & \tau_{22}^2 & \tau_{33}^2 & 2\tau_{12}^2 & 2\tau_{13}^2 & 2\tau_{23}^2 \end{bmatrix} (\boldsymbol{x}, \boldsymbol{\xi}) \begin{bmatrix} b_{11}(\boldsymbol{\xi}) \\ b_{22}(\boldsymbol{\xi}) \\ b_{33}(\boldsymbol{\xi}) \\ b_{12}(\boldsymbol{\xi}) \\ b_{13}(\boldsymbol{\xi}) \\ b_{23}(\boldsymbol{\xi}) \end{bmatrix} d\boldsymbol{\xi}
\end{gathered} \tag{3.14}$$

In practical applications of Eq. (3.14) in fracture monitoring, we need to estimate the continuous distribution of eigenstrains $b_{jk}(\xi)$ by inversion of a finite, discrete set of observations, which is a typical ill-posed Fredholm integral equation of first kind. A common way to deal with the continuous inverse problem is to convert it into a discrete inverse one by approximating the continuous distribution of model function with a finite number of discrete model parameters using collocation methods. The induced displacement and tilt are approximated by the discrete forward model as

$$\begin{gathered}
\begin{bmatrix} u_1(\boldsymbol{x}) \\ u_2(\boldsymbol{x}) \\ u_3(\boldsymbol{x}) \end{bmatrix} = \frac{1}{16\pi(1-\nu)} \sum_{j=1}^{m} \iint\limits_{s(j)} \begin{bmatrix} \overline{\omega}_{11}^1 & \overline{\omega}_{22}^1 & \overline{\omega}_{33}^1 & 2\overline{\omega}_{12}^1 & 2\overline{\omega}_{13}^1 & 2\overline{\omega}_{23}^1 \\ \overline{\omega}_{11}^2 & \overline{\omega}_{22}^2 & \overline{\omega}_{33}^2 & 2\overline{\omega}_{12}^2 & 2\overline{\omega}_{13}^2 & 2\overline{\omega}_{23}^2 \\ \overline{\omega}_{11}^3 & \overline{\omega}_{22}^3 & \overline{\omega}_{33}^3 & 2\overline{\omega}_{12}^3 & 2\overline{\omega}_{13}^3 & 2\overline{\omega}_{23}^3 \end{bmatrix} (\boldsymbol{x}, \boldsymbol{\xi}) d\boldsymbol{\xi} \begin{bmatrix} b_{11} \\ b_{22} \\ b_{33} \\ b_{12} \\ b_{13} \\ b_{23} \end{bmatrix}^{(j)} \\
\begin{bmatrix} T_1(\boldsymbol{x}) \\ T_2(\boldsymbol{x}) \end{bmatrix} = \frac{1}{16\pi(1-\nu)} \sum_{j=1}^{m} \iint\limits_{s(j)} \begin{bmatrix} \tau_{11}^1 & \tau_{22}^1 & \tau_{33}^1 & 2\tau_{12}^1 & 2\tau_{13}^1 & 2\tau_{23}^1 \\ \tau_{11}^2 & \tau_{22}^2 & \tau_{33}^2 & 2\tau_{12}^2 & 2\tau_{13}^2 & 2\tau_{23}^2 \end{bmatrix} (\boldsymbol{x}, \boldsymbol{\xi}) d\boldsymbol{\xi} \begin{bmatrix} b_{11} \\ b_{22} \\ b_{33} \\ b_{12} \\ b_{13} \\ b_{23} \end{bmatrix}^{(j)}
\end{gathered} \tag{3.15}$$

The fracture plane is discretized into m rectangular elements, each of which has a uniform eigenstrain with collocation at the element center. For an array of n tiltmeters located at different observation stations $\boldsymbol{x}_i$ ($i = 1, 2, ..., n$), we have a set of $2n$ observations (2 tilt components for each tiltmeter). The discrete fracture model predictions of the tilts at the observation points can be expressed in a general form as

$$\boldsymbol{d}_{2n} = \boldsymbol{G}_{2n \times 6m}\boldsymbol{p}_{6m} \tag{3.16}$$

The linear forward model operator $\boldsymbol{G}$ relates the $2n$ observations $\boldsymbol{d}$ to the $6m$ model parameters $\boldsymbol{p}$ through Eq. (3.15). In order to well approximate the fracture and map fracture geometry in near-field monitoring, the fracture is often discretized into a large number of small elements and the number of unknown model parameters is generally far greater than the number of available observations, i.e. $2n < 6m$. In this case, we are faced with an underdetermined inverse problem.

3.3 Bayesian Inversion Method

The forward models, Eqs. (3.12) and (3.15), provide the theoretical relationship between the induced deformation (displacement and tilt) and the fracture geometry expressed in terms of the eigenstrains. If we define the deformation observations as data denoted by a column vector $\boldsymbol{d}$ of length N, and the eigenstrains as fracture model parameters $\boldsymbol{p}$ denoted by a column vector of length M, respectively, then the measured data vector $\boldsymbol{d}$ can be related to the model parameter vector $\boldsymbol{p}$ in the linear form as

$$\boldsymbol{d} = \boldsymbol{G}\boldsymbol{p} + \boldsymbol{\epsilon} \tag{3.17}$$

where the error vector $\boldsymbol{\epsilon}$ contains the model and measurement error.

Compared to the linear fracture model with the eigenstrains being chosen as the model parameters, the conventional forward fracture model defines a nonlinear function of the fracture dip and strike angles. The nonlinear forward model leads to a nonlinear inverse problem in estimating the fracture geometry [11, 13, 14, 18, 21]. The common way of solving a nonlinear inverse problem is by iteration of an approximated, linearized problem, which may result in a wrong solution for some problems of strong nonlinearity.

The methods for solving linear inverse problems, and the uniqueness, stability, and resolution of the solutions have been well established. In this chapter, the Bayesian inversion approach is adopted to solve the linear inverse problem. The Bayesian approach treats the model parameters and observations as random variables. Assuming that the prior information on both the data and model parameters have Gaussian forms and are independent of each other, then the Gaussian probability density functions are defined, respectively, as [50]

$$\begin{aligned}\rho(\boldsymbol{d}) &= \left[(2\pi)^N \det(\boldsymbol{C}_d)\right]^{-1/2} \exp\left[-\frac{1}{2}(\boldsymbol{d}-\boldsymbol{d}_0)^T \boldsymbol{C}_d^{-1}(\boldsymbol{d}-\boldsymbol{d}_0)\right] \\ \rho(\boldsymbol{p}) &= \left[(2\pi)^M \det(\boldsymbol{C}_p)\right]^{-1/2} \exp\left[-\frac{1}{2}(\boldsymbol{p}-\boldsymbol{p}_0)^T \boldsymbol{C}_p^{-1}(\boldsymbol{p}-\boldsymbol{p}_0)\right]\end{aligned} \tag{3.18}$$

where $\boldsymbol{d}_0$ and $\boldsymbol{p}_0$ are the vectors of the a prior means for the data vector and model parameter vector, respectively, and $\boldsymbol{C}_d$ and $\boldsymbol{C}_p$ are the corresponding a prior covariance matrices. The subscript T denotes the transpose.

For the linear forward model, the posterior information on the model parameters is then given as

$$\sigma(\boldsymbol{p}) = \text{const} \cdot \exp\left[-S(\boldsymbol{p})\right] \tag{3.19}$$

where the misfit function $S(\boldsymbol{p})$ is given by

$$S(\boldsymbol{p}) = \frac{1}{2}\left[(\boldsymbol{G}\boldsymbol{p}-\boldsymbol{d})^T \boldsymbol{C}_d^{-1}(\boldsymbol{G}\boldsymbol{p}-\boldsymbol{d}) + (\boldsymbol{p}-\boldsymbol{p}_0)^T \boldsymbol{C}_p^{-1}(\boldsymbol{p}-\boldsymbol{p}_0)\right] \tag{3.20}$$

The solution of the model parameter $\boldsymbol{p}$ to the linear inverse problem is obtained by maximizing the posterior probability density $\sigma(\boldsymbol{p})$ or equivalently minimizing the misfit function $S(\boldsymbol{p})$. The posterior probability density $\sigma(\boldsymbol{p})$ for the Gaussian linear model is also Gaussian with the posterior mean $\tilde{\boldsymbol{p}}$ which minimizes the misfit function $S(\boldsymbol{p})$

$$\tilde{\boldsymbol{p}} = \boldsymbol{p}_0 + \boldsymbol{C}_p\boldsymbol{G}^T\left(\boldsymbol{G}\boldsymbol{C}_p\boldsymbol{G}^T + \boldsymbol{C}_d\right)^{-1}(\boldsymbol{d}-\boldsymbol{G}\boldsymbol{p}_0) \tag{3.21}$$

and the posterior covariance

$$\begin{aligned}\tilde{\boldsymbol{C}}_p &= \left(\boldsymbol{G}^T\boldsymbol{C}_d^{-1}\boldsymbol{G} + \boldsymbol{C}_p^{-1}\right)^{-1} \\ &= \boldsymbol{C}_p - \boldsymbol{C}_p\boldsymbol{G}^T\left(\boldsymbol{G}\boldsymbol{C}_p\boldsymbol{G}^T + \boldsymbol{C}_d\right)^{-1}\boldsymbol{G}\boldsymbol{C}_p\end{aligned} \tag{3.22}$$

In the generalized inverse analysis of the linear inverse problem (3.17), the model resolution matrix is commonly introduced to characterize the generalized inverse solution [51]. The model and data space resolution matrices are defined, respectively, as

$$\begin{aligned}\boldsymbol{R}_p &= \boldsymbol{G}^{-g}\boldsymbol{G} \\ \boldsymbol{R}_d &= \boldsymbol{G}\boldsymbol{G}^{-g}\end{aligned} \tag{3.23}$$

where $\boldsymbol{G}^{-g}$ is a generalized inverse of $\boldsymbol{G}$, called the Moore–Penrose pseudoinverse.

The model resolution $\boldsymbol{R}_p$ characterizes whether the model parameters can be independently predicted or resolved. If $\boldsymbol{R}_p$ is equal to the identity matrix $\boldsymbol{I}$, then each model parameter is perfectly resolved. On the other hand, if any of the diagonal elements of $\boldsymbol{R}_p$ are much less than one, then the corresponding model parameters are poorly resolved. The data resolution matrix $\boldsymbol{R}_d$ characterizes how the data can be fitted by the generalized solution. If $\boldsymbol{R}_d$ is equal to the identity matrix $\boldsymbol{I}$, then the data is fitted by the model exactly. The connection of the model resolution matrix to the Bayesian inversion is given by [50]

$$\boldsymbol{R}_p = \boldsymbol{C}_p\boldsymbol{G}^T\left(\boldsymbol{G}\boldsymbol{C}_p\boldsymbol{G}^T + \boldsymbol{C}_d\right)^{-1}\boldsymbol{G} = \boldsymbol{I} - \tilde{\boldsymbol{C}}_p\boldsymbol{C}_p^{-1} \tag{3.24}$$

The model resolution matrix $\boldsymbol{R}_p$ defined by Eq. (3.23) depends on the properties of the forward model operator $\boldsymbol{G}$ only, while the counterpart defined by Eq. (3.24) in the Bayesian inversion incorporates the prior information about the prior covariances of data and model parameters. Both of them do not depend on specific data or model parameters. So, they can be analyzed and assessed before conducting the monitoring. This provides a quantitative guidance for designing and optimizing a tiltmeter array, especially for far-field monitoring using the point source singularity model Eq. (3.13) for which no information (orientation or volume) about fracture geometry is needed in calculating the forward model operator $\boldsymbol{G}$.

3.4 Field Applications

Surface tiltmeter technique was applied to monitoring hydraulic fracturing growth during preconditioning of a roof rock over a coal longwall panel at the Longwall 103 site, Narrabri Coal Operations, near Narrabri, NSW, Australia [25]. Five fractures were placed at a depth of 160–180 m into vertical borehole 103AA, and were monitored by an array of 12 surface tiltmeters. In this chapter, the field tilt data collected for monitoring fracture F103AA03 has been analyzed using the proposed inversion method to map the fracture geometry. Fracture F103AA03 was placed at a depth of 168.1 m in borehole 103AA, and a total of 16 m^3 of water was injected over 40 minutes at an average rate of 0.4 m^3/min. For the details of the hydraulic fracturing treatment and surface tiltmeter monitoring, refer to [25, 52].

3.4.1 Inversion Results Using the Point Source Forward Model

The point source forward fracture model is generally used to map fracture orientation and volume in far-field monitoring. The estimation of fracture orientation using the point source fracture model is affected by near-field effects such as fracture size effect and asymmetric fracture growth because of the model error. To minimize the near-field effect, six tiltmeters at the outer boundary of the monitoring array have been used in the inverse analysis. The predicted fracture volume, dip and dip direction, together with comparison of the measured tilt vectors and model predictions at the end of the injection are shown in Figure 3.2. Fracture F103AA03 was predicted to dip N193° at a dip angle of 6.8° with a maximum volume of 8.4 m^3. The predicted tilt vectors are in good agreement with the measured ones and are consistent with a horizontal fracture deformation field. The predicted dip angle of 6.8° agrees well with a nearly horizontal fracture orientation confirmed by intersection data collected in neighboring monitoring boreholes [25]. Previous inversion of the tilt data using all the 12 tiltmeters consistently produced a dip of 20°, which was believed to be too large and was likely to be affected by some movement of the center of opening of the fracture with respect to the borehole, reflecting fracture asymmetric growth [25].

3.4.2 Inversion Results Using the General Planar Forward Model

The general planar forward fracture model, which is able to model the near-field effect, has been used to estimate the fracture growth using tilt data from all the 12 tiltmeters. To carry out the analysis, a square planar region of side 200 m is established and centered at the depth of 168.1 m in borehole 103AA. The square region within which the forward model fracture will grow is discretized into a uniform square mesh of size 5 m. Piecewise constant fracture opening over each square element with collocation at element centers has been used for computation of the forward model operator $\boldsymbol{G}$. The prior information on the fracture opening and tilt measurements may have a significant effect on the solution of this underdetermined inverse problem. To avoid overfitting the model and to get a physically reasonable prediction of the fracture opening, a forward modeling is first conducted to estimate the fracture opening using the measured injection pressure data, and the difference between the tilt measurements and the model pedictions is checked to estimate the error level of the tilt measurements. With the error level analysis, a prior distribution of fracture opening, varying linearly from a maximum of 4 mm at the borehole to zero at the boundary of the square region, is adopted in the Bayesian inversion, together with the positivity constraint applied on fracture opening. The prior standard deviation for the data and fracture opening are 0.06 μrad and 1.5 mm, respectively. The predicted fracture opening distribution and the comparison of the measured and model predicted tilt vectors at the end of the injection are shown in Figure 3.3. Fracture

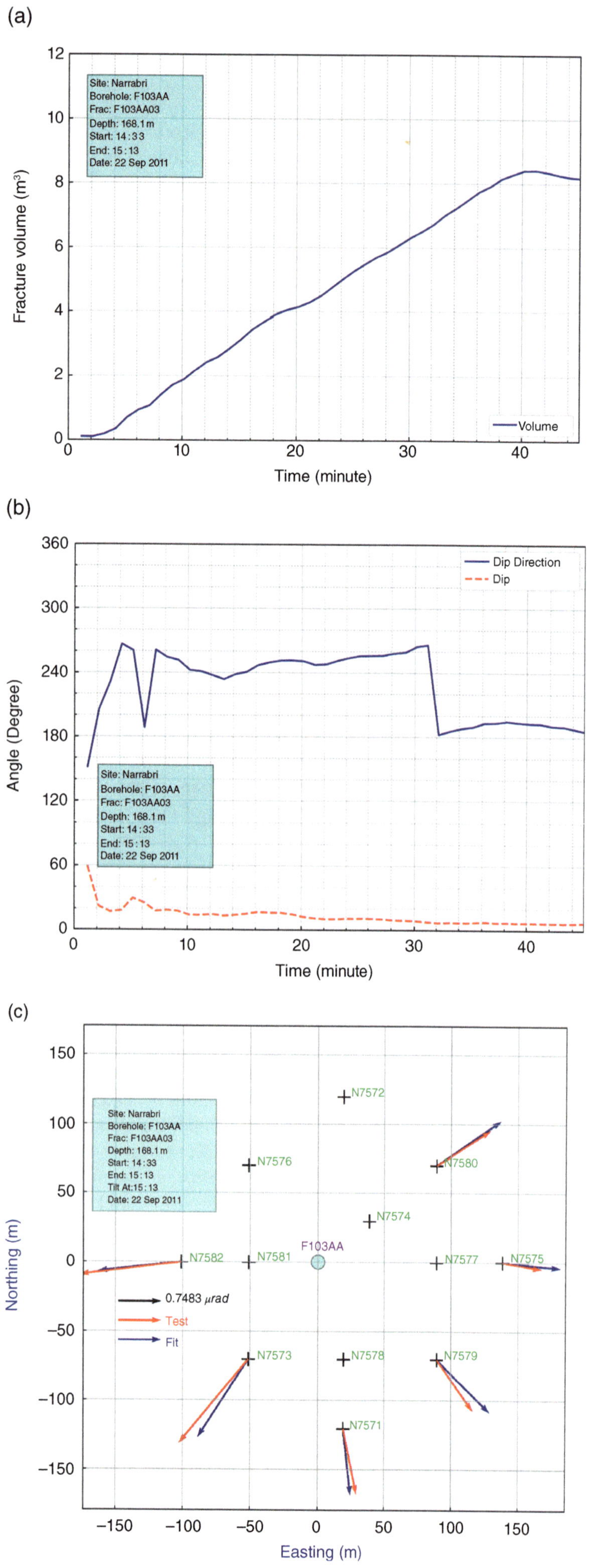

Figure 3.2 Inversion results for volume and orientation for F103AA03: (a) volume; (b) dip and dip direction; and (c) measured and predicted tilt vectors.

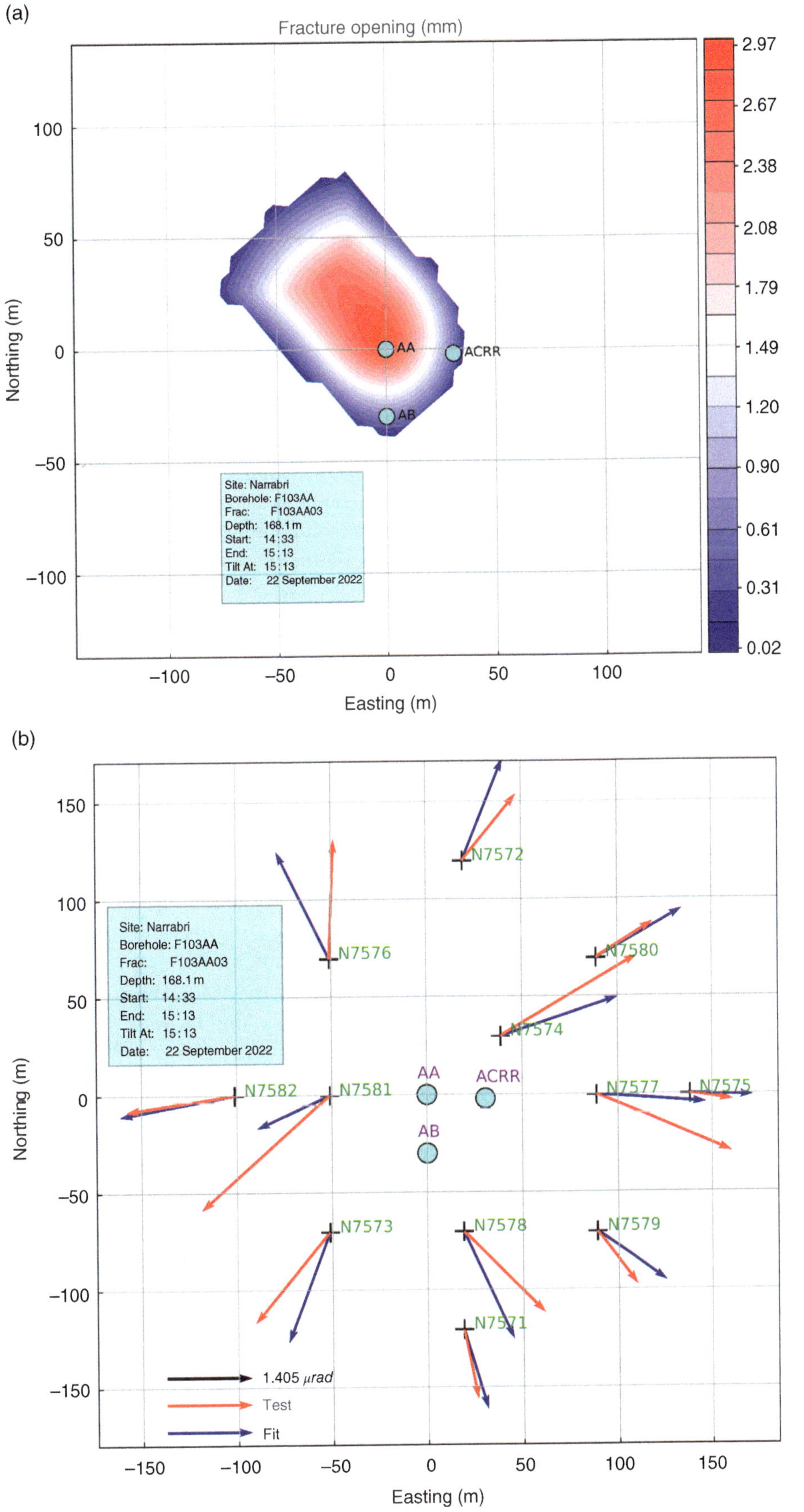

Figure 3.3 Inversion results for fracture opening distribution for F103AA03: (a) fracture size and opening distribution; (b) measured and predicted tilt vectors.

F103AA03 was predicted to dip N230° at a dip angle of 7° with a volume of 11 m^3. The predicted dip of 7° is much close to that of 6.8° inferred in the far-field inversion and agrees well with a nearly horizontal fracture orientation confirmed by intersection data collected in neighboring monitoring boreholes AB and ACRR [25]. The predicted fracture opening distribution also shows that fracture F103AA03 intersects boreholes AB and ACRR, and grows asymmetrically with respect to borehole 103AA, preferentially to the northwest roughly along the strike direction. Overall, the predicted tilt vectors fit the measured tilt data well in both direction and magnitude except that the model underestimates the tilt magnitude by 1.4 μrad for tiltmeter N7581.

3.5 Conclusions

The concept of eigenstrain defined by the unit vector normal to the fracture plane and the displacement discontinuity vector has been used to characterize the fracture geometry. The forward fracture model based on the definition of eigenstrain provides a linear relationship between the model parameters and the induced deformation, and allows use of a linear inverse problem in mapping fracture geometry using the tiltmeter monitoring technique. The proposed analysis approach has been applied to a set of field tilt data collected for monitoring a hydraulic fracturing growth. The predicted fracture geometry agrees well with a nearly horizontal fracture orientation confirmed by intersection data from measurements in neighboring monitoring boreholes. The inversion also provides information on the fracture opening distribution and asymmetric fracture growth. Overall, the predicted tilt vectors provided an accurate fit to the measured tilt data.

For the case of a nearly horizontal fracture growing asymmetrically with respect to the borehole, and when the near-field effect caused by the asymmetric fracture growth is detected by the surface tiltmeter array, then the inversion using the point source fracture model generally overestimates the fracture dip. In this case, including only data from far-field tilt measurements in the inverse analysis will improve the estimate of the fracture dip, but generally underestimate the fracture volume. Using the general planar forward fracture model in the inverse analysis may capture the asymmetric fracture growth effect, and generally result in improved estimates on both fracture dip and volume.

Acknowledgments

ZC thanks CSIRO for granting permission to publish this work.

References

1 Jeffrey, R.G. and Mills, K.W. (2000). Hydraulic fracturing applied to inducing longwall coal mine goaf falls. *The 4th North American Rock Mechanics Symposium*, pp. 423–430. Seattle, WA, America Rock Mechanics Association.

2 Board, M., Rorke, T., Williams, G., and Gay, N. (1992). Fluid injection for rockburst control in deep mining (ARMA-92-0111). *The 33rd U.S. Symposium on Rock Mechanics (USRMS)*, pp. 111–120. Sant Fe, NM.

3 Pine, R.J. and Batchelor, A.S. (1984). Downward migration of shearing in jointed rock during hydraulic injections. *International Journal of Rock Mechanics and Mining Sciences & Geomechanics Abstracts* 21: 249–263.

4 Du, Y.J., Aydin, A., and Murdoch, L. (1993). Incremental growth of a shallow hydraulic fracture at a waste remediation site, oakbrook, Illinois from inversion of elevation changes. *International Journal of Rock Mechanics and Mining Sciences & Geomechanics Abstracts* 30: 1273–1279.

5 Mendelsohn, D.A. (1984). A review of hydraulic fracture modeling – part 1: general concepts, 2D models, motivation for 3D modeling. *Journal of Energy Resources Technology-Transactions of the ASME* 106: 369–376.

6 Cipolla, C.L. and Wright, C.A. (2000). State-of-the-art in hydraulic fracture diagnostics. *The SPE Asia Pacific Oil and Gas Conference and Exhibition*, pp. 649–663. Brisbane, Queensland, SPE-64434-MS.

7 Cipolla, C.L. and Wright, C.A. (2002). Diagnostic techniques to understand hydraulic fracturing: What? Why? And how? *SPE Production & Facilities* 17: 23–35.

8 Warpinski, N.R. (1996). Hydraulic fracture diagnostics. *Journal of Petroleum Technology* 48: 907–910.

9 Castillo, D.A., Hunter, S.P., Harben, P. et al. (1997). Deep hydraulic fracture imaging: recent advances in tiltmeter technologies. *International Journal of Rock Mechanics and Mining Sciences* 34: 47.

10 Davis, P.M. (1983). Surface deformation associated with a dipping hydrofracture. *Journal of Geophysical Research* 88: 5826–5834.

11 Evans, K. (1983). On the development of shallow hydraulic fractures as viewed through the surface deformation field:

part 1-principles. *Journal of Petroleum Technology* 35: 406–410.
12 Evans, K. and Holzhausen, G. (1983). On the development of shallow hydraulic fractures as viewed through the surface deformation field: part 2-case histories. *Journal of Petroleum Technology* 35: 411–420.
13 Gazonas, G.A., Wright, C.A., and Wood, D.M. (1988). Tiltmeter mapping and monitoring of hydraulic fracture propagation in coal: a case study in the Warrior Basin, Alabama. In: *Geology and Coal-Bed Methane Resources of the Northern San Juan Basin, Colorado and New Mexico* (ed. J.E. Fassett), 159–168. Rocky Mountain Association of Geologists.
14 Lecampion, B. and Gunning, J. (2007). Model selection in fracture mapping from elastostatic data. *International Journal of Solids and Structures* 44: 1391–1408.
15 Olson, J.E., Du, Y., and Du, J. (1997). Tiltmeter data inversion with continuous non-uniform opening distributions: a new method for detecting hydraulic fracture geometry. *International Journal of Rock Mechanics and Mining Sciences & Geomechanics Abstracts* 34: 436.
16 Pandurangan, V., Peirce, A., Chen, Z.R., and Jeffrey, R.G. (2018). Tiltmeter mapping of measured nonsymmetric hydraulic-fracture growth in a conglomerate/sandstone formation using the implicit level-set algorithm and the extended kalman filter. *SPE Journal* 23: 172–185.
17 Smith, M.B., Ren, N.K., Sorrells, G.G., and Teufel, L.W. (1986). A comprehensive fracture diagnostics experiment: part 2 - comparison of fracture azimuth measuring procedures. *SPE Production Engineering* 423–431.
18 Wright, C.A., Davis, E.J., Minner, W.A. et al. (1998). Surface tiltmeter fracture mapping reaches new depths – 10,000 feet, and beyond? *Proceedings – SPE Rocky Mountain Regional/Low Permeability Reservoir Symposium and Exhibition*, pp. 135–146. Denver, Colorado, SPE paper 39919.
19 Griffin, L.G., Wright, C.A., Davis, E.J. et al. (2000). Surface and downhole tiltmeter mapping: An effective tool for monitoring downhole drill cutting disposal. *The 2000 SPE Annual Technical Conference and Exhibition*, Dallas, Texas, SPE 63032.
20 Warpinski, N.R., Branagan, P.T., Engler, B.P. et al. (1997). Evaluation of a downhole tiltmeter array for monitoring hydraulic fractures. *International Journal of Rock Mechanics and Mining Sciences* 34: 329–341.
21 Wright, C.A., Davis, E.J., Wang, G., and Weijers, L. (1999). Downhole tiltmeter fracture mapping: A new tool for direct measurement of hydraulic fracture growth. *Rock Mechanics for Industry* Vols 1 and 2: 1061–1068.
22 Palmer, I.D. and Moschovidis, Z.A. (1990). Fracture geometries consistent with tiltmeter data and pressure and tracer communication in the Glisp 1-1 stimulation. *Journal of Canadian Petroleum Technology* 29: 56–66.
23 Lecampion, B., Jeffrey, R., and Detournay, E. (2005). Resolving the geometry of hydraulic fractures from tilt measurements. *Pure and Applied Geophysics* 162: 2433–2452.
24 Peirce, A. and Rochinha, F. (2012). An integrated extended Kalman filter-implicit level set algorithm for monitoring planar hydraulic fractures. *Inverse Problems* 28: 015009.
25 Jeffrey, R.G., Chen, Z., Mills, K.W., and Pegg, S. (2013). 45 Monitoring and measuring hydraulic fracturing growth during preconditioning of a roof rock over a coal longwall panel. In: *Effective and Sustainable Hydraulic Fracturing* (ed. A.P. Bunger, J. McLennan and R. Jeffrey), 893–914. InTech.
26 Press, F. (1965). Displacements strains and tilts at teleseismic distances. *Journal of Geophysical Research* 70: 2395–2412.
27 Steketee, J.A. (1958). On volterras dislocations in a semi-infinite elastic medium. *Canadian Journal of Physics* 36: 192–205.
28 Mindlin, R.D. (1936). Force at a point in the interior of a semi-infinite solid. *Physics* 7: 195–202.
29 Mindlin, R.D. and Cheng, D.H. (1950). Nuclei of strain in the semi-infinite solid. *Journal of Applied Physics* 21: 926–930.
30 Yu, H.Y. and Sanday, S.C. (1991). Elastic fields in joined half-spaces due to nuclei of strain. *Proceedings of the Royal Society of London Series A-Mathematical Physical and Engineering Sciences* 434: 503–519.
31 Singh, S.J. (1970). Static deformation of a multilayered half-space by internal sources. *Journal of Geophysical Research* 75 (17): 3257–3263.
32 Pan, E. (1997). Static Green's functions in multilayered half spaces. *Applied Mathematical Modelling* 21: 509–521.
33 Pan, E. (2019). Green's functions for geophysics: a review. *Reports on Progress in Physics* 82 (106801): 52.
34 Chinnery, M.A. (1961). The deformation of the ground around surface faults. *Bulletin of the Seismological Society of America* 51: 355–372.
35 Mansinha, L. and Smylie, D.E. (1971). Displacement fields of inclined faults. *Bulletin of the Seismological Society of America* 61: 1433–1440.
36 Okada, Y. (1985). Surface deformation due to shear and tensile faults in a half-space. *Bulletin of the Seismological Society of America* 75: 1135–1154.
37 Okada, Y. (1992). Internal deformation due to shear and tensile faults in a half-space. *Bulletin of the Seismological Society of America* 82: 1018–1040.
38 Savage, J.C. and Hastie, L.M. (1966). Surface deformation associated with dip-slip faulting. *Journal of Geophysical Research* 71: 4897–4904.
39 Yang, X.M. and Davis, P.M. (1986). Deformation due to a rectangular tension crack in an elastic half-space. *Bulletin of the Seismological Society of America* 76: 865–881.

40 Du, Y.J., Aydin, A., and Segall, P. (1992). Comparison of various inversion techniques as applied to the determination of a geophysical deformation model for the 1983 borah peak earthquake. *Bulletin of the Seismological Society of America* 82: 1840–1866.

41 Harris, R.A. and Segall, P. (1987). Detection of a locked zone at depth on the parkfield, California, segment of the San-Andreas fault. *Journal of Geophysical Research-Solid Earth and Planets* 92: 7945–7962.

42 Segall, P. and Harris, R. (1987). Earthquake deformation cycle on the San-Andreas fault near parkfield, California. *Journal of Geophysical Research-Solid Earth and Planets* 92: 10511–10525.

43 Vasco, D.W., Johnson, L.R., and Goldstein, N.E. (1988). Using surface displacement and strain observations to determine deformation at depth, with an application to long valley caldera, California. *Journal of Geophysical Research-Solid Earth and Planets* 93: 3232–3242.

44 Yabuki, T. and Matsuura, M. (1992). Geodetic data inversion using a bayesian information criterion for spatial-distribution of fault slip. *Geophysical Journal International* 109: 363–375.

45 Backus, G. and Mulcahy, M. (1976). Moment tensors and other phenomenological descriptions of seismic sources .2. Discontinuous displacements. *Geophysical Journal of the Royal Astronomical Society* 47: 301–329.

46 Eshelby, J.D. (1961). III elastic inclusions and inhomogeneities. In: *Progress in Solid Mechanics* (ed. I.N. Sneddon and R. Hill), 89–140. North-Holland, Amsterdam.

47 Mura, T. (1987). *Micromechanics of Defects in Dolids*, 2nde. Kluwer Academic Publishers.

48 Ampuero, J.P. and Dahlen, F.A. (2005). Ambiguity of the moment tensor. *Bulletin of the Seismological Society of America* 95: 390–400.

49 Aki, K. and Richards, P.G. (2002). *Quantitative Seismology*, 2nde. Sansalito, CA: University Science Books.

50 Tarantola, A. (2004). *Inverse Problem Theory and Methods for Model Parameter Estimation*. Philadelphia, PA: Society for Industrial and Applied Mathematics.

51 Aster, R.C., Borchers, B., Thurber, C.H. et al. (2019). *Parameter Estimation and Inverse Problems*, 3rde. Amsterdam: Elsevier Science Bv.

52 Jeffrey, R.G., Chen, Z.R., Zhang, X. et al. (2015). Measurement and analysis of full-scale hydraulic fracture initiation and reorientation. *Rock Mechanics and Rock Engineering* 48: 2497–2512.

4

Experimental Observations of Hydraulic Fracturing

Guangqing Zhang and Dawei Zhou

College of Petroleum Engineering, China University of Petroleum-Beijing, Beijing, China

4.1 Introduction

During hydraulic fracturing, fluid is pumped at high pressure into the underground formation to create fluid-driven fractures, thus enhancing reservoir conductivity. Since hydraulic fracturing has been applied to substantially improve gas and oil production in unconventional reservoirs over the past two decades, a good understanding of hydraulic fracturing has been pursued by many researchers [1–5]. With the advance of our knowledge in hydraulic fracturing treatment, this technology has also been employed in environmental engineering and enhanced geothermal systems (EGS) [6–8].

The characteristics and mechanisms of fluid-driven fractures have been extensively investigated in the laboratory. Many experimental studies are conducted to investigate mechanisms that dominate the initiation and propagation of hydraulic fractures. Due to the challenge in measuring fracture sizes under in situ tectonic conditions, laboratory experiments are often the best means to study the propagation mechanisms of hydraulic fractures, especially for finding the factors controlling fracture growth in specific lithological conditions in order to maximize the oil and gas production. Laboratory results can also benchmark numerical simulation models, which are expected to be used for fracturing treatment design, and then have directly led to successful ideas for on-site hydraulic fracturing design. In this chapter, we will review some recent laboratory works related to hydraulic fracture growth in various rocks such as oil and gas reservoirs, shallow formation associated with environmental remediation, and EGS reservoirs at depth and will compare the operational conditions to optimize the stimulated reservoir volume under given conditions.

4.2 Experimental Setup on Laboratory-Scale

The experimental setup contains a loading system, injection system, and monitor system (Figure 4.1). Two types of loading facilities, including cylindrical biaxial and cubic loading frames, are commonly used to study fracture growth under both biaxial and true triaxial stress states [9]. The true triaxial compression tests controlled by three servo pumps can better replicate in situ stress states. However, due to the finite specimen size, the influence of end shear friction between specimen and loading plate should be carefully checked, because it disturbs the assumed uniform stress within the specimen and thus deflects fracture path, especially near the specimen boundary. To minimize the effect of induced frictional stress on fracture propagation, Teflon sheets between specimen and steel plates are used [10, 11]. Feng et al. [12] measured the coefficient of friction between granite and alloy steel platen surfaces subjected to true triaxial loading, ranging from 0.146 to 0.157. They claimed that a mixture of stearic acid and Vaseline (MSV) with a friction coefficient of 0.018 is optimally selected for reducing the friction effect. To better mimic the stress states underground, the correct triaxial stresses and nearly negligible end friction at the contact surface between specimen and loading frame should be satisfied at most for a loading system.

Mechanics of Hydraulic Fracturing: Experiment, Model, and Monitoring, First Edition.
Edited by Xi Zhang, Bisheng Wu, Diansen Yang, and Andrew Bunger.

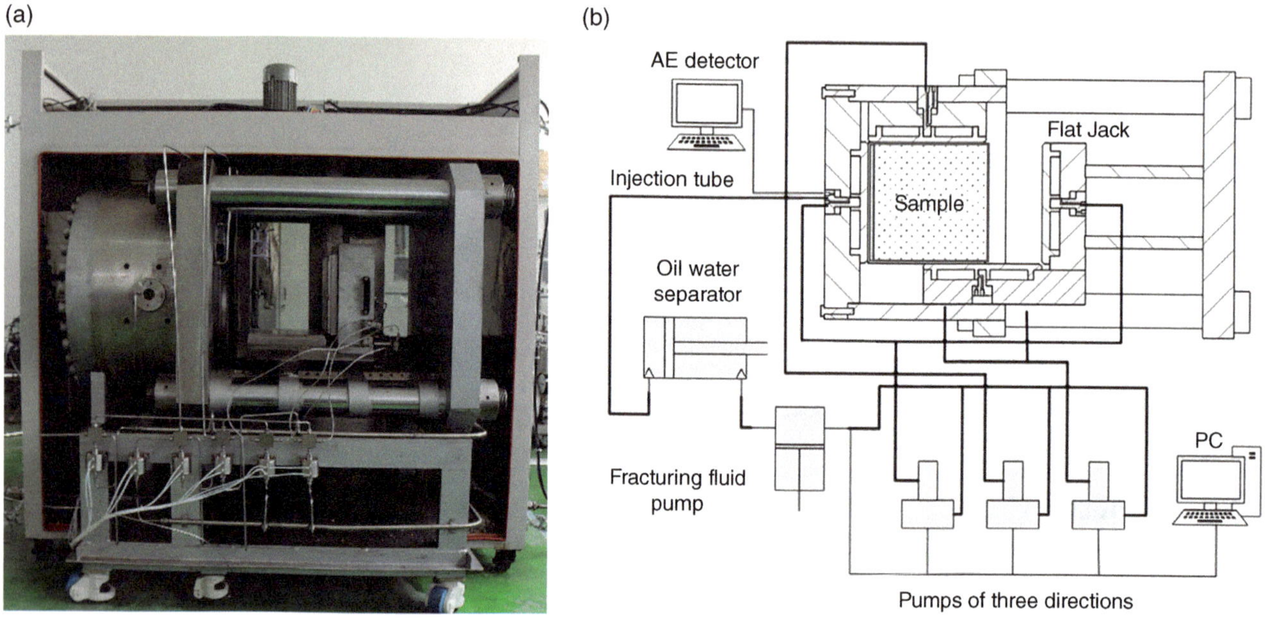

Figure 4.1 (a) Photo of the hydraulic fracturing test rig and (b) schematic of the experimental setup at China University of Petroleum at Beijing.

The second important issue is about the injection system, which comprises injection schemes and injection fluids adopted for the targeted tests. In contrast to the conventional hydraulic fracturing tests using monotonic injection scheme, cyclic, and pulse injection schemes, called fatigue hydraulic fracturing, are now used to optimize fracture network scopes and mitigate larger seismic events on sites [7, 8, 13, 14]. In unconventional reservoir stimulation, compared to the common water-based fracturing fluid, the nonaqueous fracturing fluid is now drawing broad attention because it can promote production, minimize formation damage, and conserve water resources [15–17]. To consider these new fracturing methods, laboratory experiment setup must be improved to be adaptable to new injection schemes and fracturing fluids.

Clearly, real-time monitoring methods are important to identify the fracture propagation behaviors in transparent and opaque solids. The commonly used methods include acoustic emission (AE) monitoring [18], active ultrasonic monitoring [10, 19], and optical fiber monitoring [20, 21] in opaque samples. AE is the most powerful method in the laboratory by far to investigate fluid-driven fracture by tracking fracture propagation. Transparent materials, such as polycarbonate and glass, as testing samples or loading plates, are often used for simulating fracture growth recorded by a high-resolution digital camera [1, 2]. The evolution of fracture shape and opening can be clearly visualized via camera and subsequent image processing. The above means of monitoring provide an insight into fracture mechanisms of hydraulic fracture.

4.3 Laboratory Investigation of Fluid-Driven Fractures in Various Applications

4.3.1 Hydraulic Fracturing in Oil and Gas Reservoirs

4.3.1.1 Basic Issues of Breakdown Pressure and Fracture Geometry

The analysis of injection pressure is critically important for understanding hydraulic fracture (HF) growth, as schematically shown in Figure 4.2. Fracture initiation pressure (Pi) occurs when a deviation in linearity is observed in the plot,

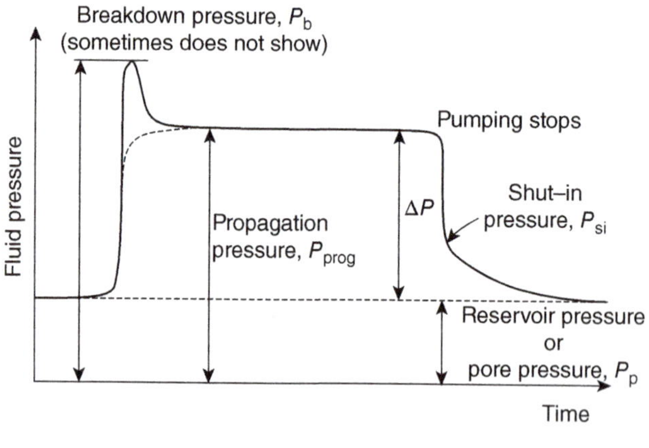

Figure 4.2 The relationship between injection pressure and time of hydraulic fracturing treatment (*Source:* Yew and Weng [22]/ With permission of Elsevier).

suggesting energy dissipation through fracturing. The fracturing fluid propagates a fracture stably with pressure increasing until the maximum pressure, namely breakdown pressure (P_b), is reached. At this point, the pressure usually drops due primarily to the unstable fracture propagation. Then a nearly constant propagation pressure is reached to propagate fractures in the formation. Lecampion et al. [23] performed laboratory tests to compare the difference between P_i and P_b on low permeability rock, stating that they significantly depend on the fracturing fluid pressurization rate and viscosity. Gunarathna and da Silva [24] analyzed the influence of effective vertical stress (σ_v') on the P_b based on several shale reservoirs and EGS hydraulic fracturing projects. It was observed that the ratio of P_b to σ_v' is usually larger than 2.0 in the shallow reservoir, while it is approaching 1.0 in deep formation. Recently, cyclic or fatigue (i.e. dynamic loading) hydraulic fracturing is proposed to reduce P_b by up to 10% in EGS and unconventional reservoirs [7, 14]. Compared to the dynamic loading, fatigue of fluid penetration is in a delayed manner that initiates a fracture when wellbore pressure is lower than P_b after some time [25].

It is crucial to understand the general HF geometries that occur in the laboratory to upscale indoor research to the field application. The various HF geometry features, such as length, height, aperture, spacing, and roughness, are strongly dependent on formation properties and treatment parameters. More specifically, the induced fracture geometry is mostly related to the applied stress state, borehole and perforation geometry, fracturing fluid viscosity and injection rate, and rock structure and scale [26]. It is accepted that HFs propagate along a plane perpendicular to the least stress. A combination of sample slicing and acoustic emission (AE) may be the best method to visualize fracture geometry on a laboratory scale. Based on observations of a series of experiments, [26] proposed three basic fracture patterns, including offsetting branches, traversing coalescence and smooth reorientation, which yields a complex HF network. At the micro-scale, optical microscope, and SEM are used to observe the fracture morphology for investigating its related fracture mechanism [27–29], showing tensile-dominated fractures along natural fractures (NFs) and grain boundary prevail. Bunger and Lecampion [30] reviewed fracture height growth, borehole tortuosity, fracture network, and simultaneous fracture growth that emerged over past decades with respect to laboratory study, field application and numerical modeling. They pointed that fluid viscosity, injection rate and in situ stresses are likely to control all these issues.

When we design the hydraulic fracturing experiments, two important issues, suitable scaling of physical phenomena and stable fracture propagation, should be necessarily addressed. Some problems, such as a large-sized wellbore, low applied stress, high injection rate, can release excessive energy to a propagating fracture upon the P_b occurs, resulting in a dynamic initiation. At this moment, the injection pressure drops sharply, and meanwhile, the fracture reaches the sample boundary. Some complex fractures observed from the experiments are attributed to the near-wellbore tortuosity other than that induced by fracture propagation. In the laboratory experiments, scaling laws related to fracture initiation [31] and fracture propagation [32] as an experimental design basis should be considered.

4.3.1.2 Multiple Hydraulic Fracture Growth

Multiple HFs consist of an array of multi-staged or multi-stranded fractures induced by fracturing fluid. It was first inferred from microseismic monitoring in the field by Fisher et al. [33]. Lately, a large number of parallel or sub-parallel HFs were directly observed in the drilled shale cores of hydraulic fracturing test sites (HFTS) [34, 35]. Several critical parameters, such as fracture net pressure, fracture spacing, in-situ stresses and pore pressure distribution, commonly affect multi-fracture initiation and propagation.

In the early stage, extensive experimental works have been done on horizontal fracturing to investigate multi-fractures [36–39]. A propagating fracture competition means that there exists a balance between attraction and repulsion forces. It is well known that the subsequent fracture can deviate away or curve toward the previous fractures due to suppression and minimum principal stress (σ_3) rotation induced by stress shadow, respectively [40–42]. A higher fracture net pressure is needed than a single fracture to counteract work done by nearby HFs. Bunger et al. [40] presented experimental investigations of sequential fracturing, i.e. injection into one cluster at a time, in gabbro blocks. It is observed that a subsequent fracture is likely to curve toward a previous fracture when the σ_3 perpendicular to the fracture is small (Figure 4.3a). In addition to the influence of elastic deformation induced by fluid flow, Zhou et al. [42] investigated the effects of poroelasticity on the multi-fracture propagation in the porous sandstone. It is observed that the previous fracture (HF1) propagation is likely to be asymmetric or off-balanced [3]. A subsequent fracture directly curves toward and merges into a previous fracture (Figure 4.3b), showing a strong attraction enhanced by the combined effects of stress shadow and pore pressure around the multiple fractures.

At laboratory scale, there are few published results about the simultaneous fracturing of multiple clusters in the laboratory scale. In light of the experiment's limited sample size and effective monitoring methods, the HF initiation sequence and fluid injection volume for each cluster are poorly understood in simultaneous fracturing.

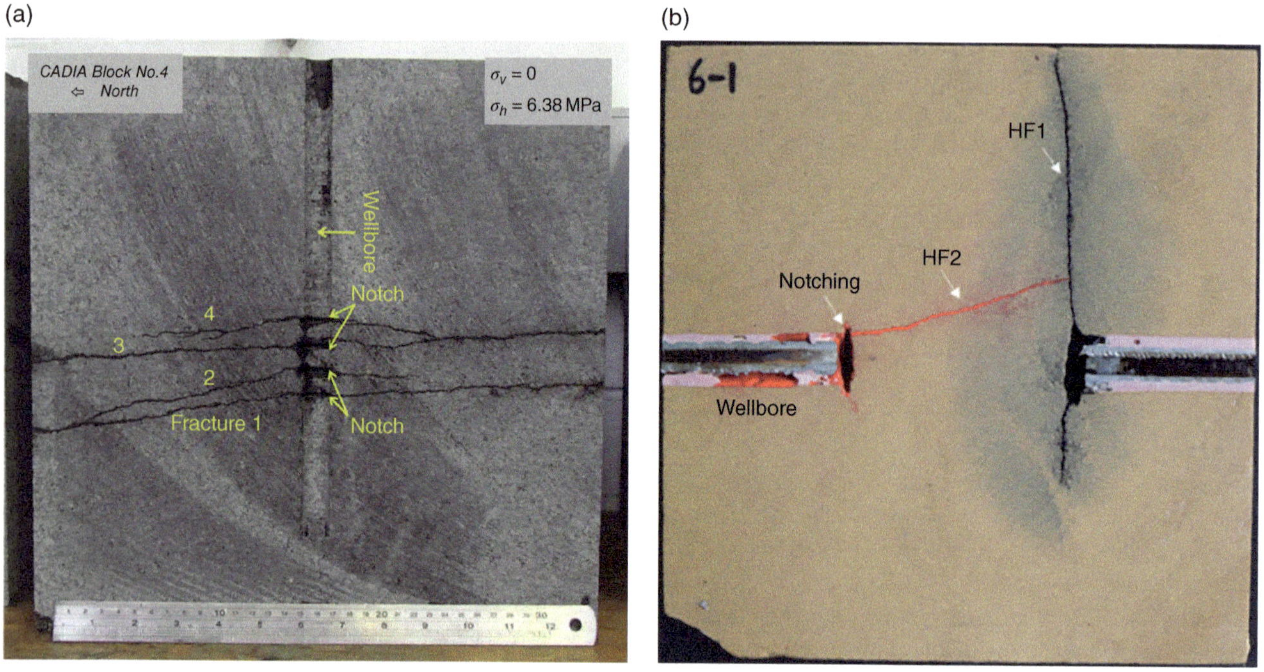

Figure 4.3 (a) Coalescing and subparallel fracture paths are observed in fracture paths on the middle section of the specimen subjected to σ_3 of 0 MPa (*Source:* Bunger et al. [40], American Rock Mechanics Association). (b) The subsequent hydraulic fracture 2 (HF2) curves toward the previous hydraulic fracture 1 (HF1) (*Source:* Zhou et al. [42], Society of Petroleum Engineers).

4.3.1.3 Interactions Between Hydraulic Fractures and Natural Fractures

The preexisting discontinuities may be in various natural fractures (NFs), bedding planes, faults, joints, etc. Interactions between HF and NF inevitably occur in naturally fractured formations. As an HF is approaching NFs, some possible interaction behaviors, including opening, offsetting, crossing and arresting, are expected, as illustrated in Figure 4.4.

The orientation of HF concerning NF, stress state, mechanical properties of NF (e.g. interface friction coefficient), fracturing fluid properties and fracture net pressure are responsible for the interaction behaviors. Blanton [43] proposed HF tends to cross NF when deviatoric stresses and approaching angles are high (Figure 4.4). Similarly, Renshaw and Pollard [44] stated a dry fracture would cross an NF if the normal stress applied on the NF surface is large enough to prevent its sliding so that a tensile fracture can initiate from the other side of the interface. Daneshy [45] conducted experimental work and concluded three controlling factors, including the strength of the weakness plane, orientation, and stress difference regarding the principal stress. He found that the influence of small NFs compared to the HF on a propagating fracture is local and insignificant.

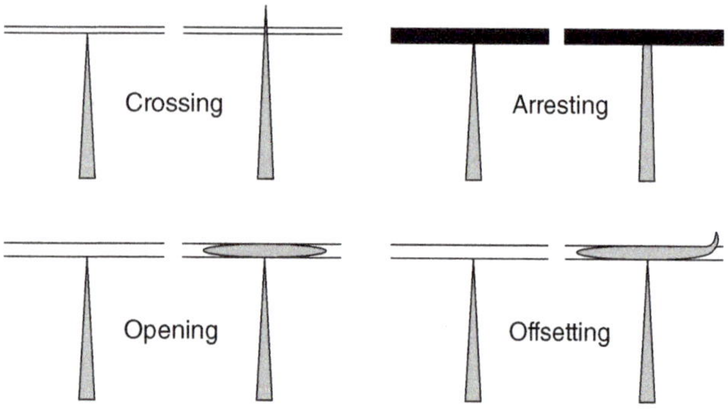

Figure 4.4 Sketch of various possible interactions between HF and NF (*Source:* Sarmadivaleh [9]).

Fracture arresting is that a propagating HF does not cross an NF and terminates at the NF interface. Olson et al. [5] concluded from a series of hydraulic fracturing tests on hydrostone blocks that oblique HF-NF intersections are likely to divert an HF, while the orthogonal intersections may result in arrest or bypassing of HF. This conclusion only represents the well-cemented NF, analogous to quartz cemented fractures in the formation. If the NF interface is poorly cemented or bonded, such as the calcite cement always weaker than the surrounding host rock in shale, it possibly arrests a propagating HF. Bunger et al. [19] investigated the effects of frictional strength of an unbonded machined interface of NFs in sandstone and crystalline rock on the HF growth. It is observed that the HF will always cross NF orthogonally without offsetting,

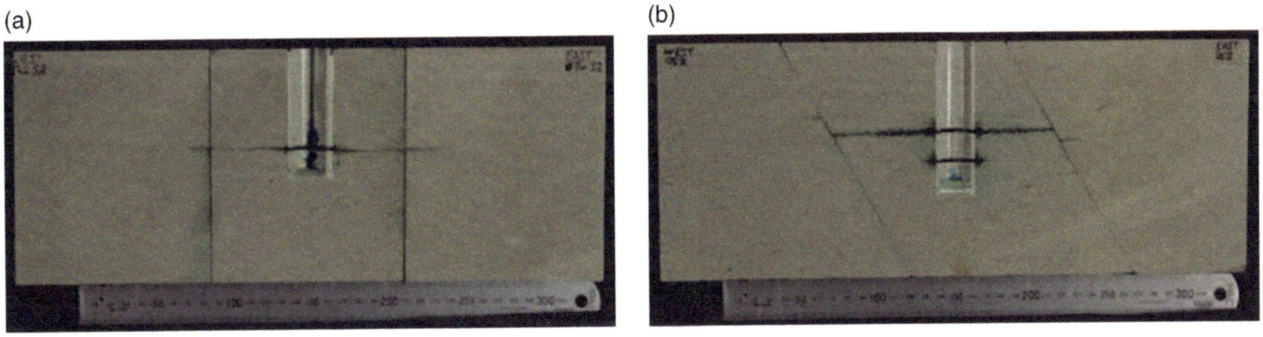

Figure 4.5 Cross-section of HF crossing interface (a) orthogonally without offset and (b) obliquely with offsetting (*Source:* Bunger et al. [19], American Rock Mechanics Association).

while the crossing with offset occurs in oblique NF (Figure 4.5). They considered the stress needed to promote crossing depending on the intersection angle between HF and NF. But at a low approach angle, the NF will possibly slide or shear rather than cross.

Besides the physical and mechanical properties of NF, the injection schemes, such as fracturing fluid viscosity and injection rate, also play an essential role in determining HF behaviors by various fracture propagating velocities and fracturing fluid leak-off. De Pater and Beugelsdijk [46] investigated the effects of flow rate and fluid viscosity on HF propagation in synthetic samples with randomly distributed NFs. He found that an increase in flow rate or viscosity can promote HF to cross the NF. Llanos et al. [11] provided experimental results for the effects of viscous fluid flow on the orthogonal fracture crossing, suggesting a fluid with higher viscosity can promote an HF cross the discontinuity. In other words, high-viscosity fluids are not likely to enter and pressurize the NF as easily as low-viscosity fluids.

Kolawole and Ispas [47] presented an up-to-date review of numerical, analytical, and experimental investigations on HF-NF interactions, documenting possible interaction behavior scenarios. Hydraulic fractures connect with numerous NFs in the form of shear slippage to form fracture networks, which can be categorized into orthogonal fracture networks, sub-orthogonal fracture networks and linear-to-sublinear fracture networks. Fan and Zhang [48] conducted laboratory experiments to investigate the features of fracture networks in naturally fractured blocks containing two groups of orthogonal cemented fractures. The results indicate that shear slippage along NFs is the dominant cause of fracture network in naturally fractured formation.

4.3.1.4 Fracture Propagation Through the Layered Formation

Layered formations can be considered as rocks characterized by varying physical or mechanical properties and in situ stress states. Fracture height is mostly dominated by mechanical properties, in-situ stress contrast and interface properties among layered rocks. More specifically, the least principal stress contrast between adjacent layers is the governing factor that controls whether a fracture can break through the layered rocks or not [49]. When the stress contrast between layers is large enough, the fracture growth tends to be contained and even terminated. Time-dependent or viscoelastic stress relaxation is proposed to explain the mechanism of variations in stress magnitudes from layer to layer [50]. Compared to the commonly assumed constancy of the least principal stress in brittle rocks (e.g. sandstone), it has a tremendous increase in relatively ductile rock (e.g. shale).

Both stress contrast induced by difference in layer stiffness and layered rock mechanical properties are the primary consideration for the laboratory-scale simulation of HF propagation. Daneshy [51] indicated that the difference of physical and mechanical properties of layered limestone and sandstone is insufficient to stop fracture growth at the interfaces, except for a highly weak interface (Figure 4.6a). Athavale and Miskimins [52] assessed the effects of rock properties and stress contrasts on the HF growth in small scale, laminated rocks under true triaxial stresses. The layered interface could impede or redirect fracture growth locally resulting in nonplanar and branching fractures (Figure 4.6b).

AlTammar et al. [1] also revealed the effects of flow properties on the 2D fracture propagation in layered materials using digital image correlation (DIC). The layered rock-like sample is placed between two transparent plates to allow the camera to take videos. It is observed that a single fracture goes through the interface from hard to soft layer. When a fracture propagates from a soft to a hard layer, a fracture tends to kink because the tensile stress induced by the approaching fracture initiates a new fracture on the other side (Figure 4.7). It is indicated that a tendency of fracture turning, reorientation and kinking increases

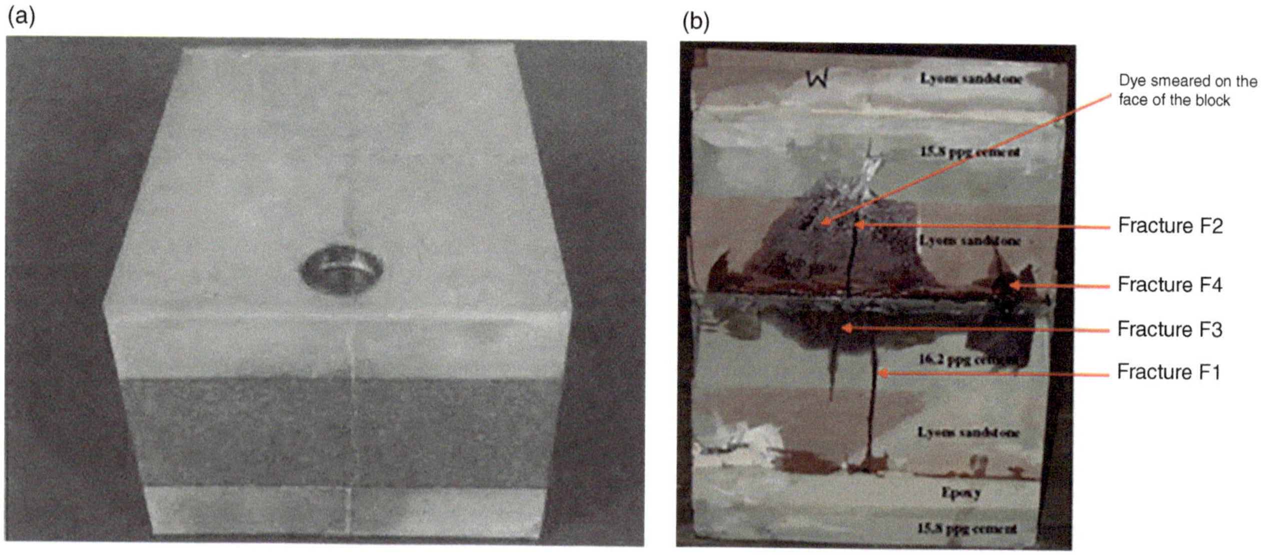

Figure 4.6 Photos of induced HF in layered rocks. (a) one fracture propagates through the interface (*Source:* Daneshy et al. [51], Society of Petroleum Engineers). (b) multiple fractures initiate from the interface to form complex fractures (*Source:* Athavale et al. [52], American Rock Mechanics Association).

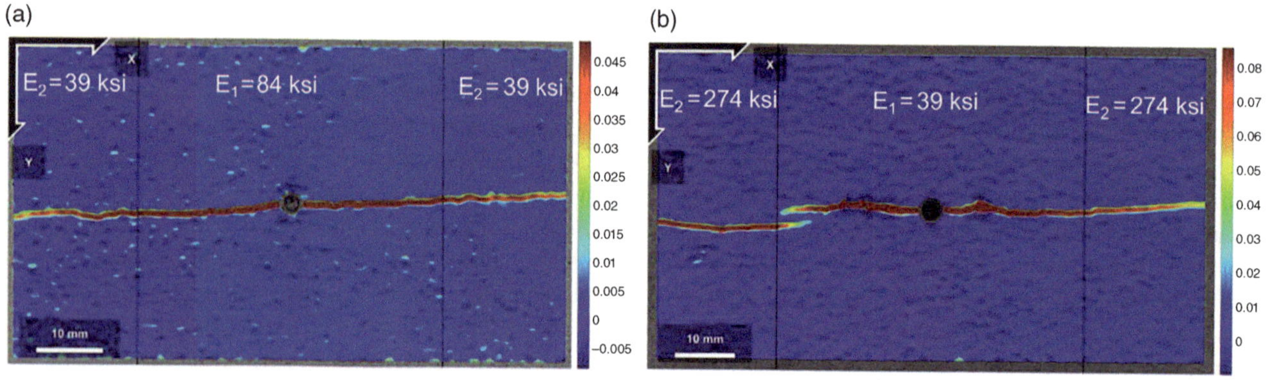

Figure 4.7 Incremental images of y-normal strain field of a fracture propagating from (a) soft to hard layer and (b) hard to soft layer (*Source:* AlTammar et al. [1], Society of Petroleum Engineers).

as the contrast in mechanical properties between layers increases. To date, limited laboratory works are performed on the influence of stress contrast on fracture propagation due to the difficulty in applying layered stress.

4.3.1.5 Nonlinear Fracturing in the Deep Reservoir

The deep reservoir is usually characterized by high stress, high temperature, and pressure. In such situations, the down-hole pressure estimated by the linear elastic fracture models is obviously lower than that measured in the field, suggesting that the traditional knowledge of linear elastic mechanics seems invalid for describing HF growth in the deep reservoir [53]. So, the effects of nonlinear deformation or plasticity should be considered in the prediction of fracture growth.

The dilation of fracture tip due to the effect of inelastic deformation during HF propagation is proposed to explain the observed high net pressure. Compared to non-confinement, effective fracture toughness (EFT) of sandstone increased by approximately 500% at confining pressure of 9 MPa [54]. The main reason why the EFT increases significantly is that the plasticity plays a shielding mechanism around the fracture tip, resulting in high net pressure and large fracture width [53, 55]. Xing et al. [56] summarized their published works related to the nonlinear fracturing characteristics in the deep reservoir. There are three main factors, including the fracture process zone (FPZ), microcrack band and plastic zone, dominating the fracture tip in the nonlinear fracturing. FPZ always occurs ahead of the

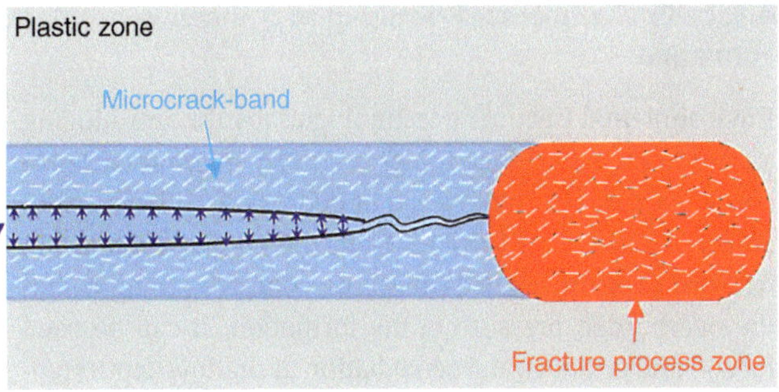

Figure 4.8 Schematic illustration of HF in nonlinear fracturing (*Source:* Xing et al. [56]/IOP Publishing/ CC BY-3.0).

fracture tip, due to microcrack generation and coalescence. When HF propagates into the FPZ, a microcrack band is subsequently distributed on both sides of HF (Figure 4.8). Meanwhile, HF opening will compress and yield the surrounding rock to create the plastic zone. Characterizing the thermoplastic behavior of rock is fundamental to understanding HF nonlinear fracturing in the deep reservoir. To describe the temperature-dependent mechanical behavior of shale in the deep reservoir with high temperatures, a thermoplastic constitutive model is proposed, which incorporates the temperature-dependent Drucker–Prager hardening rule, thermal loading/unloading criteria, and completely coupled stress–strain–temperature relation [57]. In short, the combined effects of high temperature and high confining pressure significantly dominate the fracture tip behavior of nonlinear fracturing in the deep reservoir.

4.3.1.6 Cyclic Fracturing

The behaviors of brittle rocks subjected to cyclic loading are significantly different from monotonic loading [58]. This difference exists not only in compressive stress but also in tensile stress. Erarslan and Williams [59] investigated mode-I fracture toughness using static and cyclic loading, showing that the static fracture toughness can be reduced by up to 46% when the testing is under cyclic loading. So, unlike monotonic or conventional fracturing, the cyclic fracturing in the form of various injection rates and scheduled pumping timetables is a potential for enhancing the production of unconventional oil and gas reservoirs.

Recently, a low cycle injection fracturing method was proposed by Zhou et al. [14, 60], suggesting fracturing with increasing cyclic pumping pressure and cyclic frequency during the process instead of focusing on reducing the pre-breakdown pressure can induce more complex fracture networks (Figure 4.9). Cyclic fracturing can improve the initiation rate of fractures from closely spaced perforation clusters in one treatment. The related factors, such as tri-axial stresses, cyclic times, open-hole length, cyclic frequency, and total times, dominate the fracture complexity. Besides, the frequency of cyclic fracturing will be a challenge for field application due to its limitation of fracturing equipment. Based on the preliminary works, the cyclic fracturing is likely to be an alternative for multistage fracturing treatments in field applications, especially for shale reservoirs.

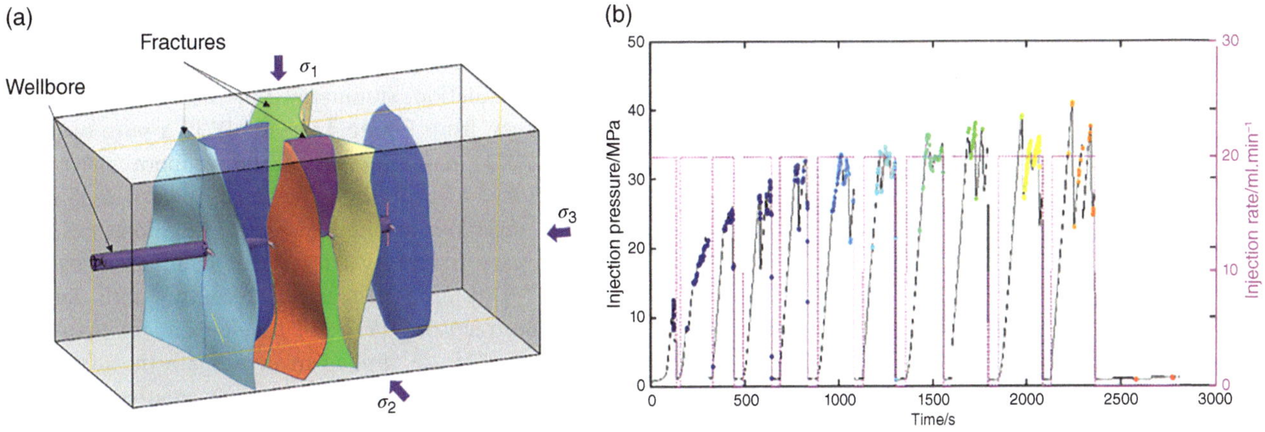

Figure 4.9 (a) 3D morphology of all fractures entirely initiates from four perforation clusters by using cyclic pumping and (b) its related injection pressure, injection rate, and AE events (*Source:* Zhou et al. [60]/Springer Nature).

4.3.2 Environmental Fracturing in a Shallow Formation

Environmental fracturing technologies involve introducing a fluid to the subsurface under relatively high pressures to create fractures in shallow soil or bedrock with low permeability. The fracture initiates from the borehole when the fracture pressure exceeds the host cohesive strength and the overburden pressure of the formation. It can be used for soil and groundwater remediation at shallow depth (typically less than 20 m) to increase the effective permeability, thus enhancing the mass transfer of contaminants and increasing the contact area between contaminants and amendments [61]. Now the environmental fracturing can be divided into three general categories, i.e. hydraulic fracturing, pneumatic fracturing, and blast fracturing [62].

Though there are many works on the in situ tests, few laboratory investigations on the fracture characteristics can be found. Consolidation plays a vital role in determining the fracture propagation direction without considering geological characteristics. The consolidation conditions can be determined by the ratio of horizontal to effective vertical stresses, i.e. $K_0 = \sigma_h/\sigma_v$. K_0 is evaluated by the over-consolidation ratios (OCR) ranging from 3 to 8, similar to K_0 of 0.8–1.4 [63]. In under-consolidated sediments ($K_0 < 1$), the steeply inclined HFs can be induced. In normally consolidated sediments ($K_0 \approx 1$), the fracture path is controlled by the geological structure, such as the NF and bedding. In over-consolidated sediments ($K_0 > 1$), the sub-horizontal or bowl-shaped fractures are expected to be created. As contaminants distribute in layered formation, the larger value of K_0 will be favored to form approximately horizontal fractures within contaminated layers. As shown in Figure 4.10, an idealized HF induced at shallow depths is slightly elongated and curved toward the free ground surface, which is verified by the detailed mapping in field excavations in over-consolidated sediments [64–66].

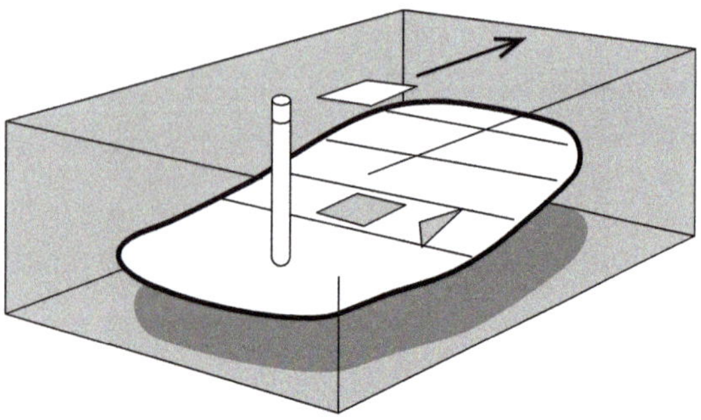

Figure 4.10 The perspective of idealized gently dipping HF at shallow depth (*Source:* Murdoch and Slack [64]/American Society of Civil Engineers).

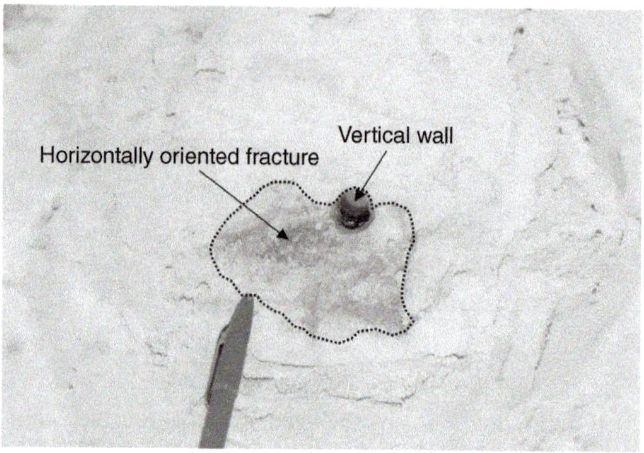

Figure 4.11 Photos of fracture surface in a vertical well with OCR = 8 and K_0 = 1.4 (*Source:* Alfaro et al. [68], American Rock Mechanics Association).

The mechanisms related to soil fracturing include shear failure, tensile failure, and fracture mechanics. Murdoch [67] conducted hydraulic fracturing using rectangular blocks of silty clay confined within a triaxial loading cell. The induced fractures are characterized by narrow, fluid-filled discontinuities normal to least principal stress. He stated that the influence of pore fluid infiltration close to fracture tip results in numerous steps and ridges on fracture surfaces. Alfaro and Wong [68] used a cylinder soil sample with a diameter of 37 cm and height of 40 cm to investigate the effect of in situ stresses on the fracture orientation and growth. To obtain various OCR, the fully saturated soil was firstly subjected to vertical stress and subsequently unloaded. A horizontally tensile fracture was created under the high OCR = 8 and K_0 = 8, as shown in Figure 4.11.

4.3.3 Hydraulic Stimulation in EGS

HF is vital to enhance permeability in the development of EGS. A great majority of EGS projects showed that bottomhole pressure has approached or even exceeded σ_3 during stimulation, summarized by McClure and Horne [6]. As seen from Figure 4.12, an HF in a pure opening mode (POM) may initiate from the wellbore and an NF has activated away from the wellbore. In early EGS stimulation, mechanisms of pure shear stimulation (PSS) [69] and primary fracturing with shear stimulation leak-off (PFSSL) [70] are most used. Recently, both fatigue hydraulic fracturing [71] and time-dependent hydraulic fracturing [72] are applied to investigate the feasibility of permeability enhancement by newly creating fractures. Mixed-mechanism stimulation (MMS) includes opening and sliding preexisting fractures to create new fractures forming fracture networks.

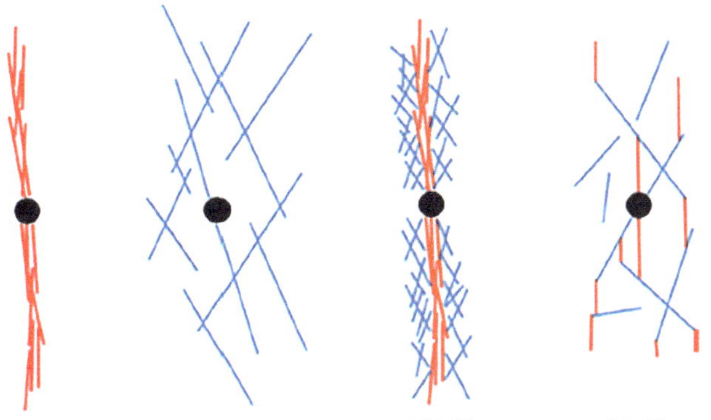

Figure 4.12 Schematic diagram of four conceptual models for mechanism in EGS stimulation. The cycle represents the wellbore. The full lines and dotted lines represent induced fractures and preexisting fractures (*Source:* McClure and Horne [6]/With permission of Elsevier).

In contrast to the monotonic pumping in conventional hydraulic fracturing, the fatigue hydraulic fracturing involves a combination of progressive cyclic and pulse hydraulic fracturing [7]. Two pump systems, one for cyclic injection with progressively increasing pressure and another for dynamic pressure pulse with various frequencies, are used simultaneously. Cyclic injections on laboratory cubic granite with a side size of 10 cm are conducted under true triaxial stresses [8]. The results show that stepwise pulse pressurization (SPP) creates the largest fracture length, resulting in the highest increase in the stimulated formation permeability. Compared to the continuous injection, cyclic pulse injection is prone to creating intergranular fractures induced by tensile or shear failure. So, the intergranular fractures complicate the fracture path leading to large tortuosity. A significant part of input hydraulic energy is converted into damage by pulse pressure cycles and fracture tip depressurization [73]. Patel et al. [74] used prebreakdown cyclic injection into Tennessee sandstone to decrease the P_b and increase the damage surrounding the fracture. The FPZ is characterized by microfracture or damage developed by cyclic injection and is connected to the main HF (Figure 4.13). The measured mean width of the process zone induced by cyclic injection is twice that generated by the conventional injection. Xing et al. [75] used acoustic emission (AE) to monitor the growth of microcrack bands surrounding HF in high-temperature granite. AE source analysis of the fracture mechanisms in the microcrack band indicated that the high temperature (120 °C) reduced the proportion of shear microcracks by 6–12%.

Apart from water-based fracturing, nonaqueous fluids, such as CO_2 and liquid nitrogen, are also an alternative as working fluid in an EGS. Those fluids are chemically inactive with the rock minerals and keep fracture conductivity during long-term operations. A series of boundary conditions, confining pressures of 5–40 MPa and temperatures of 50–300 °C, are tested on the Harcourt granite rock to study the P_b and microstructural characteristics [76]. The P_b is mainly influenced by reservoir rock temperature and decreases with temperature. Significant thermal stress is generated by a sudden temperature drop due to the injection of cold CO_2 into a heated rock. This thermal shock can weaken rock strength to create micro-cracks or secondary cracks around the main fracture. Meanwhile, due to the extremely low viscosity and low surface tension of SC–CO_2, it tends to activate the preexisting fractures during the injection. Compared to water, a combined influence of poro- and thermo-elasticity is more significantly induced by SC–CO_2.

4.4 Conclusions and Future Work

In this chapter, experimental researches on HF are reviewed. Some common experimental analysis methods, including fracture morphology analysis, injection pressure recording, AE monitoring and strain monitoring, etc. are

Figure 4.13 Schematic diagram and photograph of HF and process zone (*Source:* Patel et al. [74], Elsevier).

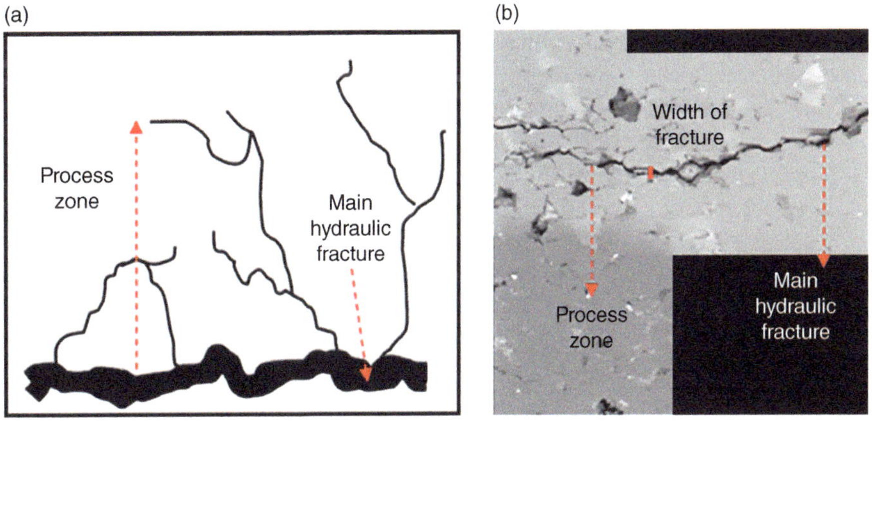

used to investigate mechanisms of HF initiation and propagation. Although numerous experimental works of hydraulic fracturing have been conducted, one key challenge is the result reproducibility due to the lack of uniform experimental standards other than the individual difference of rocks. The type of laboratory equipment used on hydraulic fracturing is of diversity in the world [9]. It is necessary to establish suggested methods, such as ISRM suggested methods for rock mechanical testing [77], for testing and measuring hydraulic fracturing features in the laboratory, providing a standard procedure for the sample preparation, parameter measurement and data analysis. Another challenge is how to properly upscale the experimental results to the on-site application. The process of HF propagation is characterized by multi-field coupling and multiscale space, complicating fracture characterization and mechanism. Most experimental researches mainly focus on a particular spatial scale (e.g. up to 30 cm) or affecting factor (e.g. testing temperature and pore pressure) to investigate the dominated mechanism. To achieve better upscaling of laboratory study, cross-scale from micro-, meso- to macro-scale methods may be the potential way to bridge laboratory and field. Meanwhile, the reservoir conditions should be considered as possible during the laboratory simulation.

With the development of unconventional reservoir stimulation, the emerging issues are challenging the current hydraulic fracturing theory. Direct observations of drilled shale cores in both Eagle Ford and Wolfcamp show HFs are of such incredible complexity that they are far beyond what current simulators or models can predict [34, 35]. The observed HF complexity and heterogeneity are characterized by numerous, widespread, closely spaced, steeply dipping, branching, and broadly parallel fractures. This phenomenon may be induced by the poroelastic effects, mixed-mode fracturing, and destressing of pressurized fractured rock mass. To address those issues, it is highly needed for advancing laboratory equipment with high precision monitoring and multiple data acquisition systems to provide experimental results as a benchmark to update analytical models and numerical simulations. Some emerging technologies, such as nuclear magnetic resonance [78] and optical fiber monitoring [20, 21], are applied to monitor fluid-driven fracture propagation and strain fields during hydraulic fracturing processes. In addition, in the laboratory study, due to the limited size of sample and injection rate, there is no effective method to simulate sand fracturing, that is adding proppant into fracture during the fracture propagation. Hence, sanding fracturing should be one of the experimental research priorities in the future.

References

1 AlTammar, M.J., Agrawal, S., and Sharma, M.M. (2019). Effect of geological layer properties on hydraulic-fracture initiation and propagation: an experimental study. *SPE Journal* 24 (2): 757–794. https://doi.org/10.2118/184871-PA.

2 Bunger, A.P. and Detournay, E. (2008). Experimental validation of the tip asymptotics for a fluid-driven crack. *Journal of the Mechanics and Physics of Solids* 56 (11): 3101–3115. https://doi.org/10.1016/j.jmps.2008.08.006.

3 Daneshy, A.A. (2003). Off-balance growth: a new concept in hydraulic fracturing. *Journal of Petroleum Technology* 55 (4): 78–85. https://doi.org/10.2118/80992-JPT.

4 Jeffrey, R.G., Bunger, A., LeCampion, B. et al. (2009). Measuring hydraulic fracture growth in naturally fractured rock. *SPE Annual Technical Conference and Exhibition*, New Orleans, Louisiana, USA.

5 Olson, J.E., Bahorich, B., and Holder, J. (2012). Examining hydraulic fracture: natural fracture interaction in hydrostone block experiments. *SPE hydraulic fracturing technology conference*, The Woodlands, Texas, USA.

6 McClure, M.W. and Horne, R.N. (2014). An investigation of stimulation mechanisms in enhanced geothermal systems. *International Journal of Rock Mechanics and Mining Sciences* 72: 242–260. https://doi.org/10.1016/j.ijrmms.2014.07.011.

7 Zang, A., Zimmermann, G., Hofmann, H. et al. (2019). How to reduce fluid-injection-induced seismicity. *Rock Mechanics and Rock Engineering* 52 (2): 475–493. https://doi.org/10.1007/s00603-018-1467-4.

8 Zhuang, L., Jung, S.G., Diaz, M. et al. (2020). Laboratory true triaxial hydraulic fracturing of granite under six fluid injection schemes and grain-scale fracture observations. *Rock Mechanics and Rock Engineering* 53 (10): 4329–4344. https://doi.org/10.1007/s00603-020-02170-8.

9 Sarmadivaleh, M. (2012). Experimental and numerical study of interaction of a pre-existing natural interface and an induced hydraulic fracture. Doctoral dissertation. Curtin University.

10 Groenenboom, J. and Romijn, R. (1997). Acoustic monitoring of hydraulic fracture growth. *First Break* 15 (9).

11 Llanos, E.M., Jeffrey, R.G., Hillis, R., and Zhang, X. (2017). Hydraulic fracture propagation through an orthogonal discontinuity: a laboratory, analytical and numerical study. *Rock Mechanics and Rock Engineering* 50 (8): 2101–2118. https://doi.org/10.1007/s00603-017-1213-3.

12 Feng, X.T., Zhang, X., Yang, C. et al. (2017). Evaluation and reduction of the end friction effect in true triaxial tests on hard rocks. *International Journal of Rock Mechanics and*

Mining Sciences 97: 144–148. https://doi.org/10.1016/j.ijrmms.2017.04.002.

13 Safari, M.R., Gandikota, R., Mutlu, U. et al. (2013). Pulsed fracturing in shale reservoirs: Geomechanical aspects, ductile-brittle transition and field implications. *Unconventional resources technology conference*, Denver, Colorado, USA.

14 Zhou, Z.L., Zhang, G.Q., Dong, H.R. et al. (2017). Creating a network of hydraulic fractures by cyclic pumping. *International Journal of Rock Mechanics and Mining Sciences* 97: 52–63.

15 Fu, C. and Liu, N. (2019). Waterless fluids in hydraulic fracturing–a review. *Journal of Natural Gas Science and Engineering* 67: 214–224. https://doi.org/10.1016/j.jngse.2019.05.001.

16 Huang, Z., Zhang, S., Yang, R. et al. (2020). A review of liquid nitrogen fracturing technology. *Fuel* 266: 117040. https://doi.org/10.1016/j.fuel.2020.117040.

17 Middleton, R.S., Carey, J.W., Currier, R.P. et al. (2015). Shale gas and non-aqueous fracturing fluids: opportunities and challenges for supercritical CO_2. *Applied Energy* 147: 500–509. https://doi.org/10.1016/j.apenergy.2015.03.023.

18 Ishida, T. (2001). Acoustic emission monitoring of hydraulic fracturing in laboratory and field. *Construction and Building Materials* 15 (5–6): 283–295. https://doi.org/10.1016/S0950-0618(00)00077-5.

19 Bunger, A.P., Kear, J., Jeffrey, R.G. et al. (2015). Laboratory investigation of hydraulic fracture growth through weak discontinuities with active ultrasound monitoring. *13th ISRM International Congress of Rock Mechanics*, Montreal, Canada.

20 Chen, R., Zaghloul, M.A., Yan, A. et al. (2016). High resolution monitoring of strain fields in concrete during hydraulic fracturing processes. *Optics Express* 24 (4): 3894–3902.

21 Yang X, Zhang G, Du X, et al. (2017). Measurement and implications of the dynamic fracture width in hydraulic fracturing using FBG strain sensors. *51st US Rock Mechanics/Geomechanics Symposium*, San Francisco, California, USA.

22 Yew, C.H. and Weng, X. (2014). *Mechanics of Hydraulic Fracturing*. Gulf Professional Publishing.

23 Lecampion, B., Desroches, J., Jeffrey, R.G. et al. (2015). Initiation versus breakdown pressure of transverse radial hydraulic fracture: theory and experiments. *13th ISRM International Congress of Rock Mechanics*, Montreal, Canada.

24 Gunarathna, G. and da Silva, B.G. (2019). Influence of the effective vertical stresses on hydraulic fracture initiation pressures in shale and engineered geothermal systems explorations. *Rock Mechanics and Rock Engineering* 52 (11): 4835–4853. https://doi.org/10.1007/s00603-019-01841-5.

25 Lu, G., Uwaifo, E. C., Ames, B. C. et al. (2015). Experimental demonstration of delayed initiation of hydraulic fractures below breakdown pressure in granite. *49th US Rock Mechanics/Geomechanics Symposium*, San Francisco, California, USA.

26 Frash, L.P., Hampton, J., Gutierrez, M. et al. (2019). Patterns in complex hydraulic fractures observed by true-triaxial experiments and implications for proppant placement and stimulated reservoir volumes. *Journal of Petroleum Exploration and Production Technology* 9 (4): 2781–2792. https://doi.org/10.1007/s13202-019-0681-2.

27 Chen, Y., Nagaya, Y., and Ishida, T. (2015). Observations of fractures induced by hydraulic fracturing in anisotropic granite. *Rock Mechanics and Rock Engineering* 48 (4): 1455–1461. https://doi.org/10.1007/s00603-015-0727-9.

28 Ratzlaff, C.W., Sondergeld, C.H., Rai, C.S. et al. (2019). SEM investigation of the fracture network (stimulated reservoir volume) induced by hydraulic fracturing in Tennessee sandstone. *53rd US Rock Mechanics/Geomechanics Symposium*, New York City, New York, USA.

29 Zhou, D., Zhang, G., Wang, Y., and Xing, Y. (2018). Experimental investigation on fracture propagation modes in supercritical carbon dioxide fracturing using acoustic emission monitoring. *International Journal of Rock Mechanics and Mining Sciences* 110: 111–119. https://doi.org/10.1016/j.ijrmms.2018.07.010.

30 Bunger, A. and Lecampion, B. (2017). *Four Critical Issues for Successful Hydraulic Fracturing Applications*. CRC Press.

31 De Pater, C.J., Cleary, M.P., Quinn, T.S. et al. (1994). Experimental verification of dimensional analysis for hydraulic fracturing. *SPE Production Facilities* 9 (4): 230–238.

32 Bunger, A.P., Jeffrey, R.G., andDetournay, E. (2005). Application of scaling laws to laboratory-scale hydraulic fractures. *40th US Symposium on Rock Mechanics (USRMS)*, Anchorage, Alaska, USA.

33 Fisher, M.K., Heinze, J.R., Harris, C.D. et al. (2004). Optimizing horizontal completion techniques in the Barnett shale using microseismic fracture mapping. *SPE Annual Technical Conference and Exhibition*, Houston, Texas, USA.

34 Ciezobka, J., Courtier, J., and Wicker, J. (2018). Hydraulic fracturing test site (HFTS)–project overview and summary of results. *Unconventional Resources Technology Conference*, Houston, TX.

35 Gale, J.F., Elliott, S.J., and Laubach, S.E. (2018). Hydraulic fractures in core from stimulated reservoirs: core fracture description of HFTS slant core, Midland Basin, West Texas. *Unconventional Resources Technology Conference*, Houston, TX.

36 Abass, H.H., Hedayati, S., and Meadows, D.L. (1996). Nonplanar fracture propagation from a horizontal wellbore: experimental study. *SPE Production and Facilities* 11 (3): 133–137. https://doi.org/10.2118/24823-PA.
37 El Rabaa, W. (1989). Experimental study of hydraulic fracture geometry initiated from horizontal wells. *SPE Annual Technical Conference and Exhibition*, San Antonio, Texas, USA.
38 Kear, J., White, J., Bunger, A. P. et al. (2013). Three dimensional forms of closely-spaced hydraulic fractures. *ISRM International Conference for Effective and Sustainable Hydraulic Fracturing*, Brisbane, Australia.
39 Weijers, L., De Pater, C.J., Owens, K.A., and Kogsbøll, H.H. (1994). Geometry of hydraulic fractures induced from horizontal wellbores. *SPE Production Facilities* 9 (02): 87–92. https://doi.org/10.2118/25049-PA.
40 Bunger, A.P., Jeffrey, R.G., Kear, J. et al. (2011). Experimental investigation of the interaction among closely spaced hydraulic fractures. *45th US Rock Mechanics/Geomechanics Symposium*, San Francisco, California, USA.
41 Rahman, M.M., Hossain, M.M., Crosby, D.G. et al. (2002). Analytical, numerical and experimental investigations of transverse fracture propagation from horizontal wells. *Journal of Petroleum Science and Engineering* 35 (3–4): 127–150. https://doi.org/10.1016/S0920-4105(02)00236-X.
42 Zhou, D., Zhang, G., Liu, Z., and Zhou, Z. (2016). Experimental investigation on propagation interferences of staged multi-cluster perforation fractures in tight sandstone. *SPE Asia Pacific Hydraulic Fracturing Conference*, Beijing, China.
43 Blanton, T.L. (1982). An experimental study of interaction between hydraulically induced and pre-existing fractures. *SPE Unconventional Gas Recovery Symposium*. https://doi.org/10.2118/10847-MS
44 Renshaw, C.E. and Pollard, D.D. (1995). An Experimentally Verified Criterion for Propagation across Unbounded Frictional Interfaces in Brittle, Linear Elastic Materials. In *International Journal of Rock Mechanics and Mining Sciences Geomechanics Abstracts* (Vol. 32, No. 3, pp. 237–249). Pergamon.
45 Daneshy, A.A. (1974). Hydraulic fracture propagation in the presence of planes of weakness. *SPE European Spring Meeting*, Amsterdam, Netherlands.
46 De Pater, C.J. and Beugelsdijk, L.J.L. (2005). Experiments and numerical simulation of hydraulic fracturing in naturally fractured rock. *40th US Symposium on Rock Mechanics (USRMS)*, Anchorage, Alaska, USA.
47 Kolawole, O. and Ispas, I. (2020). Interaction between hydraulic fractures and natural fractures: current status and prospective directions. *Journal of Petroleum Exploration and Production Technologies* 10 (4): 1613–1634. https://doi.org/10.1007/s13202-019-00778-3.
48 Fan, T.G. and Zhang, G.Q. (2014). Laboratory investigation of hydraulic fracture networks in formations with continuous orthogonal fractures. *Energy* 74: 164–173. https://doi.org/10.1016/j.energy.2014.05.037.
49 Warpinski, N.R. and Teufel, L.W. (1987). Influence of geologic discontinuities on hydraulic fracture propagation (includes associated papers 17011 and 17074). *Journal of Petroleum Technology* 39 (2): 209–220. https://doi.org/10.2118/13224-PA.
50 Zoback, M.D. and Kohli, A.H. (2019). *Unconventional Reservoir Geomechanics*. Cambridge University Press.
51 Daneshy, A.A. (1978). Hydraulic fracture propagation in layered formations. *SPE Journal* 18 (1): 33–41. https://doi.org/10.2118/6088-PA.
52 Athavale, A.S. and Miskimins, J.L. (2008). Laboratory hydraulic fracturing tests on small homogeneous and laminated blocks. *42nd US Rock Mechanics Symposium*, San Francisco, California, USA.
53 Papanastasiou, P. (1997). The influence of plasticity in hydraulic fracturing. *International Journal of Fracture* 84 (1): 61–79. https://doi.org/10.1023/A:1007336003057.
54 Funatsu, T., Seto, M., Shimada, H. et al. (2004). Combined effects of increasing temperature and confining pressure on the fracture toughness of clay bearing rocks. *International Journal of Rock Mechanics and Mining Sciences* 41 (6): 927–938. https://doi.org/10.1016/j.ijrmms.2004.02.008.
55 Papanastasiou, P. (1999). The effective fracture toughness in hydraulic fracturing. *International Journal of Fracture* 96 (2): 127. https://doi.org/10.1023/A:1018676212444.
56 Xing, Y., Zhang, G., Huang, B. (2020a). Recent advances in nonlinear fracturing characteristics of the hydraulic fracture in the deep reservoir. *IOP Conference Series: Earth and Environmental Science* (Vol. 570, No. 3, p. 032062). IOP Publishing.
57 Xing, Y., Zhang, G., and Li, S. (2020b). Thermoplastic constitutive modeling of shale based on temperature-dependent Drucker-Prager plasticity. *International Journal of Rock Mechanics and Mining Sciences* 130: 104305. https://doi.org/10.1016/j.ijrmms.2020.104305.
58 Burdine, N.T. (1963). Rock failure under dynamic loading conditions. *SPE Journal* 3 (1): 1–8.
59 Erarslan, N. and Williams, D.J. (2012). The damage mechanism of rock fatigue and its relationship to the fracture toughness of rocks. *International Journal of Rock Mechanics and Mining Sciences* 56: 15–26. https://doi.org/10.1016/j.ijrmms.2012.07.015.
60 Zhou, Z.L., Zhang, G.Q., Xing, Y.K. et al. (2019). A laboratory study of multiple fracture initiation from perforation clusters by cyclic pumping. *Rock Mechanics and Rock Engineering* 52 (3): 827–840. https://doi.org/10.1007/s00603-018-1636-5.
61 Schuring, J.R. (2002). Fracturing Technologies to Enhance Remediation. Technology Evaluation Report TE-02-02

Prepared for GWRTAC Ground-Water Remediation Technologies Analysis Center, Pittsburg PA.

62 Christiansen, C.M., Bjerg, P.L. (2010). Methods for enhanced delivery of in situ remediation amendments in contaminated clay till. PhD thesis. DTU Environment, Technical University of Denmark, Kgs Lyngby, Denmark.

63 Schmidt, B. (1966). Earth pressures at rest related to stress history. *Canadian Geotechnical Journal* 3 (4): 239–242. https://doi.org/10.1139/t66-028.

64 Murdoch, L.C. and Slack, W.W. (2002). Forms of hydraulic fractures in shallow fine-grained formations. *Journal of Geotechnical Geoenvironmental Engineering* 128 (6): 479–487. https://doi.org/10.1061/(ASCE)1090-0241(2002)128:6(479).

65 Murdoch, L.C. (1995). Forms of hydraulic fractures created during a field test in overconsolidated glacial drift. *Quarterly Journal of Engineering Geology Hydrogeology* 28 (1): 23–35. https://doi.org/10.1144/GSL.QJEGH.1995.028.P1.03.

66 Wong, R.C. and Alfaro, M.C. (2001). Fracturing in low-permeability soils for remediation of contaminated ground. *Canadian Geotechnical Journal* 38 (2): 316–327. https://doi.org/10.1139/t00-097.

67 Murdoch, L.C. (1993). Hydraulic fracturing of soil during laboratory experiments part 1. Methods and observations. *Geotechnique* 43 (2): 255–265. https://doi.org/10.1680/geot.1993.43.2.255.

68 Alfaro, M.C. and Wong, R.C. (2001). Laboratory studies on fracturing of low-permeability soils. *Canadian Geotechnical J.* 38 (2): 303–315. https://doi.org/10.1139/t00-096.

69 Pine, R.J. and Batchelor, A.S. (1984). Downward migration of shearing in jointed rock during hydraulic injections. *International Journal of Rock Mechanics and Mining Sciences Geomechanics Abstracts* 21 (5): 249–263. Pergamon. https://doi.org/10.1016/0148-9062(84)92681-0.

70 Pearson, C. (1981). The relationship between microseismicity and high pore pressures during hydraulic stimulation experiments in low permeability granitic rocks. *Journal of Geophysical Research: Solid Earth* 86 (B9): 7855–7864. https://doi.org/10.1029/JB086iB09p07855.

71 Zang, A., Yoon, J.S., Stephansson, O., and Heidbach, O. (2013). Fatigue hydraulic fracturing by cyclic reservoir treatment enhances permeability and reduces induced seismicity. *Geophysical Journal International* 195 (2): 1282–1287. https://doi.org/10.1093/gji/ggt301.

72 Bunger, A.P. and Lu, G. (2015). Time-dependent initiation of multiple hydraulic fractures in a formation with varying stresses and strength. *SPE Journal* 20 (6): 1–317. https://doi.org/10.2118/171030-PA.

73 Zang, A., Stephansson, O., Zimmermann, G. et al. (2020). Concept of fatigue hydraulic fracturing and applications in hard rock predicted from laboratory and in situ experiments. *54th US Rock Mechanics/Geomechanics Symposium*. Physical event cancelled.

74 Patel, S.M., Sondergeld, C.H., and Rai, C.S. (2017). Laboratory studies of hydraulic fracturing by cyclic injection. *International Journal of Rock Mechanics and Mining Sciences* 95: 8–15. https://doi.org/10.1016/j.ijrmms.2017.03.008.

75 Xing, Y., Zhang, G., Luo, T. et al. (2019). Hydraulic fracturing in high-temperature granite characterized by acoustic emission. *Journal of Petroleum Science and Engineering* 178: 475–484. https://doi.org/10.1016/j.petrol.2019.03.050.

76 Isaka, B.A., Ranjith, P.G., Rathnaweera, T.D. et al. (2019). Testing the frackability of granite using supercritical carbon dioxide: insights into geothermal energy systems. *Journal of CO_2 Utilization* 34: 180–197. https://doi.org/10.1016/j.jcou.2019.06.009.

77 Ulusay, R. (ed.) (2014). *The ISRM Suggested Methods for Rock Characterization, Testing and Monitoring*, 2007–2014. Springer.

78 Zhang, X., Wei, B., You, J. et al. (2021). Characterizing pore-level oil mobilization processes in unconventional reservoirs assisted by state-of-the-art nuclear magnetic resonance technique. *Energy* 236: 121549.

5

First Field Trail and Experimental Studies on $scCO_2$ Fracturing

Haiyan Zhu[1], Lei Tao[1], Shouceng Tian[2], and Haizhu Wang[2]

[1] *State Key Laboratory of Oil and Gas Reservoir Geology and Exploitation, Chengdu University of Technology, Chengdu, China*
[2] *State Key Laboratory of Petroleum Resources and Prospecting, University of Petroleum (Beijing), Beijing, China*

5.1 Introduction

China has the world's most abundant shale gas reserves. Most of shale gas reservoirs are deep under the ground and have complex geological structures with the properties of high strength and strong anisotropicity. Exploitation of these shale gas reservoirs requires large-scale hydraulic fracturing treatment to generate high permeability fracture network [1–4].

However, the conventional hydraulic fracturing treatment requires thousands of tons of water, and this could pose a big problem in the fields where water supply is limited. Moreover, dealing with the pollutions caused by flow-back fracturing fluids is very costly. Therefore, to solve these issues, liquid CO_2 fracturing and $scCO_2$ fracturing are proposed as an alternative.

When the temperature exceeds 31.26 °C and pressure is over 7.38 MPa, CO_2 will be in a supercritical state. $scCO_2$ has a near zero surface tension and its density is close to liquids. In addition, the viscosity of $scCO_2$ is low and the diffusivity is as high as gas. $scCO_2$ is nontoxic and will not interact chemically with clay. Therefore, $scCO_2$ has been widely used as working fluids in many aspects, such as cuttings removal and production enhancement [5–7].

Early in 1980s, a field test on $scCO_2$ in North America has shown that $scCO_2$ has a higher diffusivity in formations than conventional fracturing fluids [8]. Field tests on $scCO_2$ fracturing were also conducted in Changqing, Yanchang, and Jilin Oil Fields [9–11]. The CO_2 under supercritical conditions has low intermolecular forces and good mobility, which enables $scCO_2$ to flow more smoothly than the conventional fracturing fluid [12]. Because of this, $scCO_2$ can penetrate into any spaces larger than its molecular volume, thus readily connecting the preexisting fracture networks and improving the gas production rate at most.

$scCO_2$ is environmental-friendly, because (i) there is almost no pollution to water and soil, (ii) CO_2 is more accessible than water in the oil/gas fields where water supply is limited. For example, CO_2 can be collected from the exhaust of fossil fuel power plants or some factories such as biorefineries, ethylene production plants, and there is no need to capture it from air, (iii) part of CO_2 can be sequestrated underground after being injected into wellbores [13], (iv) after fracturing, CO_2 can accelerate the fluid flow-back, and (v) the dissolved CO_2 into the crude oil can reduce the viscosity, which helps improving oil production rates.

Since $scCO_2$ fracturing cannot only improve the recovery of hydrocarbons [14, 15], but also sequestrate CO_2 underground [16], there is great interest in this new environmental-friendly fracturing technology in the world. For the exploration of shale gas with $scCO_2$ fracturing, the total cost of the shale gas will depend on the source of CO_2. Sometimes the cost of the CO_2 will be higher than water, however, if the CO_2 can be collected from a power plant or factory, the cost can be reduced drastically [13]. Currently, some field tests have been conducted successfully in China [10]. However, many aspects of $scCO_2$ fracturing of shale oil and gas reservoirs still need further research, such as the interaction of shale and $scCO_2$, complex fractures initiation and propagation, the proppants-carrying capability of $scCO_2$, specialized $scCO_2$ fracturing equipment, fracturing monitoring techniques, etc. This chapter presents a comprehensive review on state of the art advances in the $scCO_2$ fracturing, and also gives introduction of a field $scCO_2$ fracturing trail of a Yanchang shale gas reservoir.

Mechanics of Hydraulic Fracturing: Experiment, Model, and Monitoring, First Edition.
Edited by Xi Zhang, Bisheng Wu, Diansen Yang, and Andrew Bunger.

5.2 Review on $scCO_2$ Fracturing

5.2.1 Shale and $scCO_2$ Interaction

5.2.1.1 Microscale Physical Changes

Okamoto et al. [17] and Vialle and Vanorio [18] found that the distribution of the pores and throats in sandstone and carbonate rocks was changed by $scCO_2$. Chiquet et al. [19] indicated that the wettability of quartz and mica in shale was altered significantly by the introduction of CO_2. Many researchers, found that $scCO_2$ eroded the surfaces of minerals (such as quartz, feldspar, some of the carbonate rocks, and illite) [20–22], which damaged the original pore structure and left irregular etched marks on these minerals' crystal surfaces [23, 24]. Wollenweber et al. [25] found that repeated CO_2 treatments can increase the permeability of shale caprock, and Jiang et al. [26] found that the specific surface area and porosity of the shale increased after being treated with $scCO_2$. The change of reservoir physical properties will affect the efficiency of fracturing treatments. Ao et al. [27] investigated $scCO_2$ effects on the physical properties (porosity, deformation, and mechanical properties) and chemical properties (mineral components) of the shale.

The experimental results of the triaxial compression, SEM, and XRD tests show that calcite and dolomite concentrations seem to decrease due to $scCO_2$ treatment. The melted minerals change the microscopic structure of the shale: more fractures and eroded minerals are generated. As shown in Figure 5.1, the SEM images for the microstructures $scCO_2$-treated samples before and after testing were quite different [28]. The primary macro-pores and micro-cracks in the shale narrowed due to the CO_2 adsorption-induced swelling (Figure 5.1, marked A). Some materials on the surface of the shale disappeared after $scCO_2$ treatment (Figure 5.1, marked B and C). This shows that $scCO_2$ can change rock microstructures and make some minerals (e.g. calcite) corrode more easily.

5.2.1.2 Microscale Chemical Changes

Lahann et al. [29] found that the concentrations of K^+, Mg^{2+}, and Ca^{2+} in the filtrate for the reaction of shale and CO_2-brine were significantly higher than that in the other experimental cases with water-treated shale and low pressure CO_2-treated shale. Many researchers (such as [5, 30, 31]) found that shale organic matter debris and kerogen decomposed when exposed to high pressure $scCO_2$. Yin et al. [28]

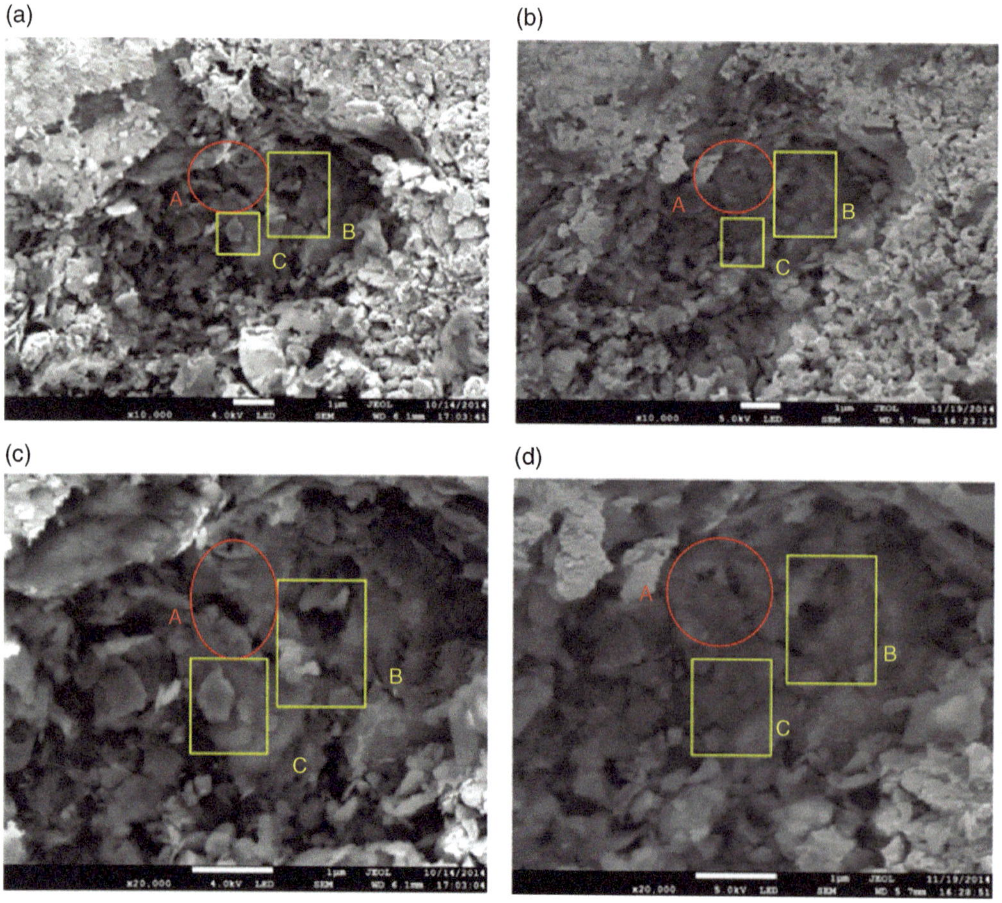

Figure 5.1 SEM images of untreated and $scCO_2$-treated sample (*Source:* Yin et al. [28], Elsevier).

found that the organic matter and some mineral components (such as montmorillonite, kaolinite, and calcite) in the shale decreased in weight because these substances dissolved in the $scCO_2$.

The $scCO_2$ lead to mineralogical changes: corrosion of mineral. Little differences (less than 3.0%) were found in the mineral components before and after the N_2-treated shale; but when $scCO_2$-treated, the amounts of calcite, dolomite and pyrite decreased, maybe leading to the formation of corrosion voids and grooves. The concentration of calcite decreased to an average of about 4.1% with a maximum of 4.8% and a minimum of 2.9% (Table 5.1). The concentration of the dolomite decreased to an average of 5.7% with a maximum of 8.9% and a minimum of 3.7%. This phenomenon can also be found in many literatures, such as Yin et al. [28], Ao et al. [27], etc.

The reactions between $scCO_2$ and water in micro-pores in shale can produce H^+ under high temperature and pressure. Shale minerals and organic matter will react with the H^+ [6, 7, 33, 34]. Calcite and dolomite minerals dissolve due to the following chemical reactions [35]:

$$CaCO_3(\text{calcite}) + H^+ \rightarrow Ca^{2+} + HCO_3^- \quad (5.1)$$

$$CaMg(CO_3)_2(\text{dolomite}) + 2H^+ \rightarrow Ca^{2+} + Mg^{2+} + 2HCO_3^- \quad (5.2)$$

According to the research done by Tian et al. [35], the dissolved ions will bind to Ca^{2+}, Mg^{2+}, and Fe^{2+} to product magnesite and ankerite.

$$Mg^{2+} + HCO_3^- \rightarrow MgCO_3(\text{magnesite}) + H^+ \quad (5.3)$$

$$Ca^{2+} + 0.3Mg^{2+} + 0.7Fe^{2+} + 2HCO_3^- \rightarrow CaMg_{0.3}Fe_{0.7}(CO_3)_2(\text{ankerite}) + 2H^+ \quad (5.4)$$

5.2.1.3 Macroscale Mechanical Changes

Kollé [36] found that $scCO_2$ jets could reduce the fracturing pressure and improve the shale gas penetration rate. Du et al. (2012) found that $scCO_2$ can reduce the threshold pressure of rock breaking. Wang et al. [37] found that $scCO_2$ jets created grid-like fractures on the end faces of shale cores in $scCO_2$ jet experiments. Zheng et al. [38] found a decrease of 7–15% in the triaxial compressive strength of sandstone in CO_2-brine-rock systems.

Lyu et al. [39] conducted triaxial tests on shale under supercritical and subcritical CO_2 conditions. Based on the results of acoustic emission monitoring, three-dimensional fracture surface scan and SEM results, it shows that the uniaxial strength and elastic modulus of shale sample are significantly reduced with the increase of CO_2 saturation. In addition, from SEM and acoustic emission monitoring results, it is found that the $scCO_2$ produced more acoustic emission signals and created more fractures.

Yin et al. [40] tested the mechanical properties of shale after 10 days of adsorption under different pressures of CO_2. The results showed that the saturation of CO_2 with a high supercritical pressure caused the decrease of shale strength.

Cai et al. [41, 42] and Huang et al. [43] conducted the experimental investigation on the rock breaking characteristics of $scCO_2$ jets, and they found that $scCO_2$ creates more micro- and macro-complex fracture morphology, larger fracture volume, and CO_2 absorption rate those that in the hydraulic fracturing.

Because of the special characteristics of Chinese shale and the complexities of the interactions between shale and $scCO_2$, the changes of shale mechanical properties are still unclear, and these are the keys to guide $scCO_2$ fracturing. Liu et al. [32] chose an outcrop of Wufeng shale and prepared cylindrical samples with seven different coring

Table 5.1 The mineral components before and after the CO_2-treated and N_2-treated shale[32].

Sample no.	State of specimen	Mineral contents (%)					
		Quartz	Feldspar	Calcite	Dolomite	Pyrite	Total clay
1-1	Before $scCO_2$	21.8	0.8	19.4	35.9	5.4	16.7
	After $scCO_2$	30.9	0.6	14.6	27.0	2.6	24.3
1-2	Before $scCO_2$	22.8	0.7	23.2	29.0	3.8	20.5
	After $scCO_2$	28.2	0.6	19.2	23.6	2.3	26.1
2-1	Before N_2	25.6	1.2	19.3	26.9	3.6	23.4
	After N_2	23.9	1.4	17.9	27.4	4.1	25.3
2-2	Before N_2	26.3	1.1	21.6	25.7	3.8	21.5
	After N_2	24.8	1.7	24.2	23.9	3.8	21.6

(*Source:* Data from Liu et al. [32]).

angles (0°, 15°, 30°, 45°, 60°, 75°, and 90°), then they carried out triaxial compressive strength tests on the samples exposed to $scCO_2$. N_2-treated samples are set as the control experiments because the injected gas will change the effective stress. The experimental group that will be tested with CO_2 injection, and the control group that will be tested with N_2 injection. There is a variation in compressive strength with changes in the coring angle: the strength increases firstly from 0° to 15° of coring angle, and decreases from 30° to 60°, then increases again from 60° to 90°, as shown in Figure 5.2.

$scCO_2$ reduces the triaxial compressive strength, and amplitude of variation in the compressive strength, with changes in the angle and fracture toughness of Wufeng shale. On both the macroscale and microscale levels, after $scCO_2$ treatment, complex chemical and physical changes in shale minerals may lead to favorable conditions for the seepage and gathering of shale gas.

$scCO_2$ injection leads more fractures on the specimen. Rock failure modes are determined by many factors, including mineralogical character, internal microscopic structure, test air conditions, load direction, and bedding plane direction. The failure modes vary from splitting failure to shear failure [44]. They studied the relationship among loading direction, bedding plane direction, and fracture morphology. The rougher surfaces are related to the failure mode, the stress field, the propagation regime, and many other factors. Therefore, in the experiments, the other factors are set almost the same for comparison. The combination of $scCO_2$ treated and the bedding plane direction caused a diversity of failure modes. In addition, the rougher surfaces due to $scCO_2$ injection have been observed in other rocks, for example in sandstone and carbonate rocks. Emberley et al. [22], Lin et al. [23], Regnault et al. [45], Busch et al. [21], Fischer et al. [46], Liu et al. [24] and Miller et al. [47] had studied the CO_2-sandstone-salt water contact under high temperature and high pressure, it was found that $scCO_2$ dissolved a variety of minerals (including quartz, feldspar, some carbonate rocks, illite, etc.), changed the original pore structure, and left irregular etch marks on the crystal surface of these minerals.

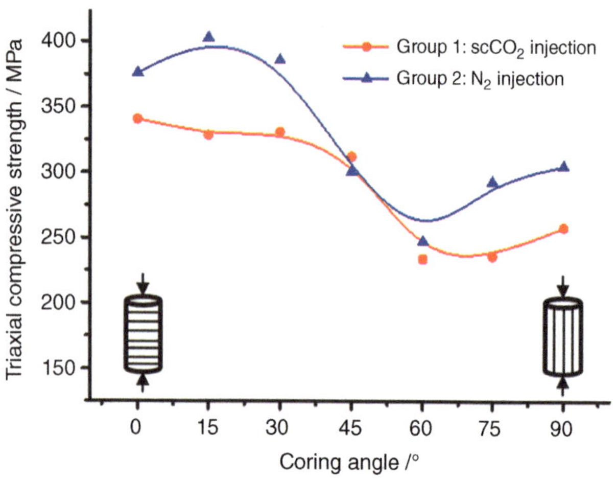

Figure 5.2 Triaxial compressive strength vs. coring angles (*Source:* Liu et al. [32]/Society of Petroleum Engineers).

The CO_2-treated specimens have more fractures after the Brazilian disk splitting tests than the control group, and Figure 5.3 shows that the failure load of the CO_2-treated specimens (5.39 kN) is lower than that of the control samples (8.27 kN). The load direction is parallel to the bedding plane of the shale samples. The fracture toughness of CO_2-treated specimens is much lower than that of the control specimens, which suggests $scCO_2$ could make the fractures in shale initiate and extend more easily.

5.2.1.4 Conclusions on the Experiments on Shale and $scCO_2$ Interaction

The shale microscopic structures and bedding planes, which are essential for the migration of gases, are damaged by the $scCO_2$, leaving many microscopic fractures, small mineral debris, and uneven mineral surfaces. On both the macro and micro scales, after $scCO_2$ treatment, complex chemical and physical changes in shale minerals may lead to favorable conditions for the seepage and gathering of shale gas.

The findings will provide basic experimental parameters for analyzing $scCO_2$ fracturing, $scCO_2$ jet breaking of rock, and carbon sequestration. The findings will also provide experimental verification for numerical simulations of $scCO_2$-treated rock. $scCO_2$ fracturing treatment creates more uniform fractures than hydraulic fracturing treatment, reducing the pressure required for fracturing shale gas reservoirs. Maybe $scCO_2$ can be used as a prepared fluid for producing more fractures before injecting the conventional fracturing fluid with proppant, and the use of $scCO_2$ would also promote the backflow rate.

5.2.2 Experiments and Numerical Simulations on $scCO_2$ Fracturing

5.2.2.1 Experiments on $scCO_2$ Fracturing

5.2.2.1.1 Cylindrical Core Samples Under Quasi – Triaxial Loading

Jia et al. [48] conducted $scCO_2$ fracturing and compared it with conventional hydraulic fracturing. The scans of fracture surfaces show that shale fracture surfaces created by $scCO_2$ are rougher and have more tortuosity than those created by conventional hydraulic fracturing. The results are also explained from the microscopic viewpoint of the move and transportation of mineral particles, as shown in

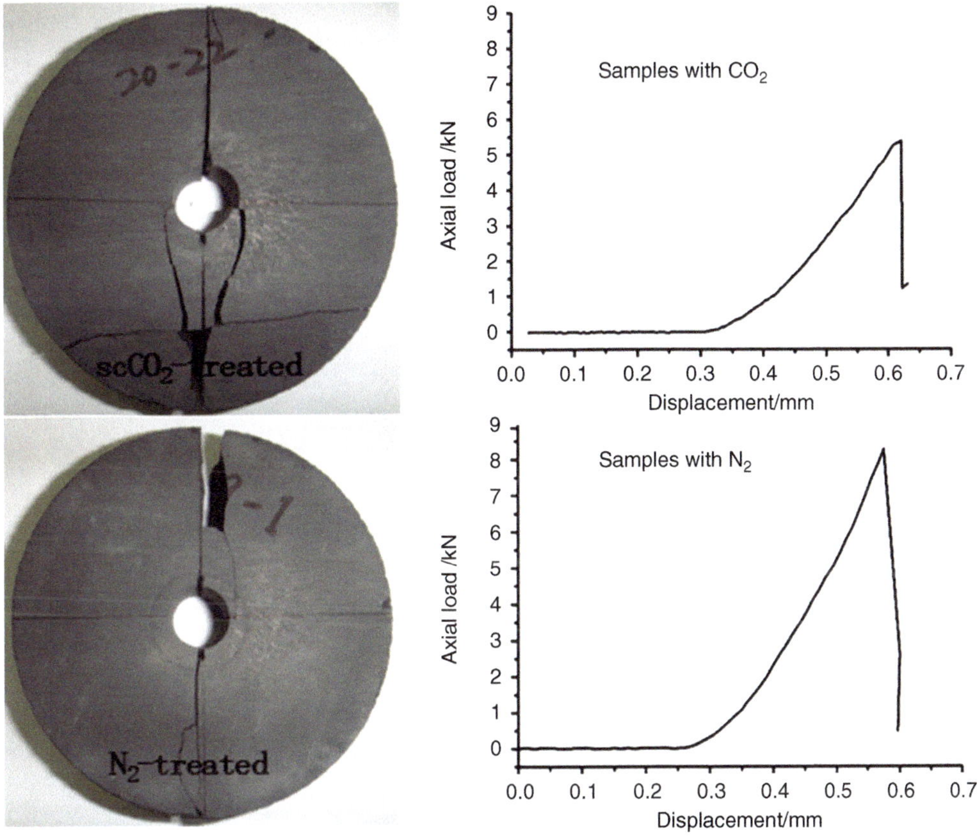

Figure 5.3 Fracture toughness test for $scCO_2$-treated and N_2-treated (*Source:* Liu et al. [32], Elsevier).

Figure 5.4. This suggest that $scCO_2$ fracturing has a better fracturing performance than water in enhancing shale permeability during shale reservoir treatment. Jiang et al. [49] carried out $scCO_2$ fracturing experiment on Longmaxi shale samples with acoustic emission monitoring and fracture CT scan, and studied the interaction between $scCO_2$ fractures and natural fractures. Zhang et al. [50] conducted $scCO_2$ fracturing experiments on cylindrical shale core samples with different bedding angles and showed that the fracture mode, fracture pressure, and fracture width can be affected by shale bedding angles. Yang et al. [51] proposed a novel full-sample fracture analysis approach to explore the morphology of the fractures and quantified their attributes during the $scCO_2$ fracturing in tight sandstone.

5.2.2.1.2 Cubic Sample Under True Triaxial Loading

Ishida et al. [52, 53] used supercritical and liquid CO_2 in hydraulic fracturing tests on 17 cm cubic granite samples. The tests show that the AE distribution of the $scCO_2$ test group was wider than that of liquid CO_2; $scCO_2$ tends to induce three-dimensional fracture propagation rather than planar fracture extension by liquid CO_2; the fracturing pressure for $scCO_2$ and liquid CO_2 injection is much lower than that of water injection.

Akihisa et al. [54] used $scCO_2$ and water to conduct fracturing simulation experiments on granite and tuff rock with a volume of 15 cm^3. The test results show that the breakdown pressure observed by $scCO_2$ fracturing is lower than that of hydraulic fracturing. In addition, $scCO_2$ fracturing readily generates complex fractures, while hydraulic fracturing produces a relatively single fracture.

Inui et al. [55] conducted $scCO_2$, water, and heavy oil fracturing tests to study the effect of different fracturing fluids on the patterns of granite fractures, and analyzed the effect of viscosity on the formation of different fractures. The results show that $scCO_2$ with the smallest viscosity tends to create shear fractures, and heavy oil with larger viscosity tends to form tensile fractures.

Li et al. [56] found that CO_2 fracturing creates nominally the most complex fracturing patterns as well as the roughest fracture surface and the greatest apparent local damage, followed by H_2O and then N_2 fracturing. Zhang et al. [57] conducted comparative tests among $scCO_2$ fracturing, liquid CO_2 fracturing, and hydraulic fracturing on shale outcrops and sandstone samples from Longmaxi Formation.

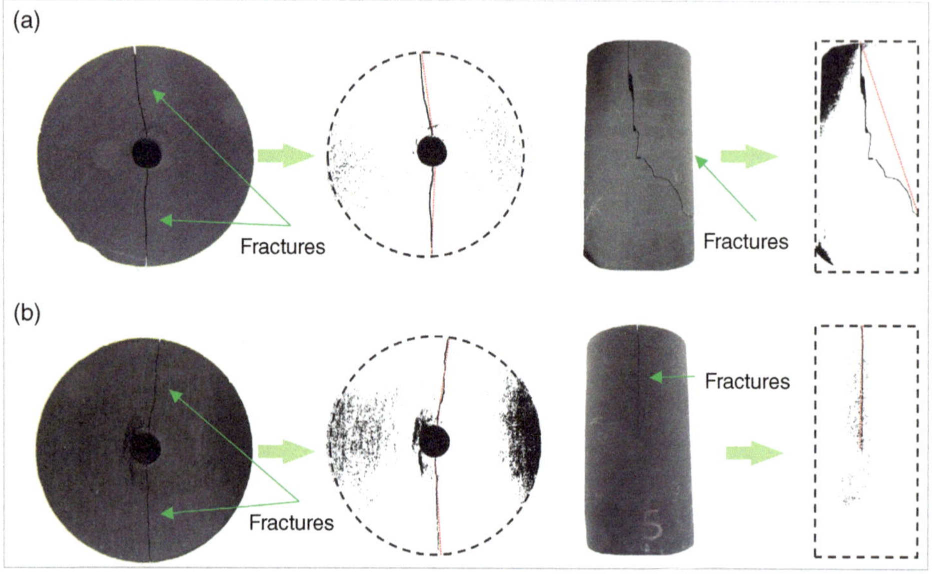

Figure 5.4 Schematic of fracture tortuosity: (a) vertical and side views of sample fractured by $scCO_2$; (b) vertical and side views of sample fractured by H_2O (*Source:* Jia et al. [48], Elsevier).

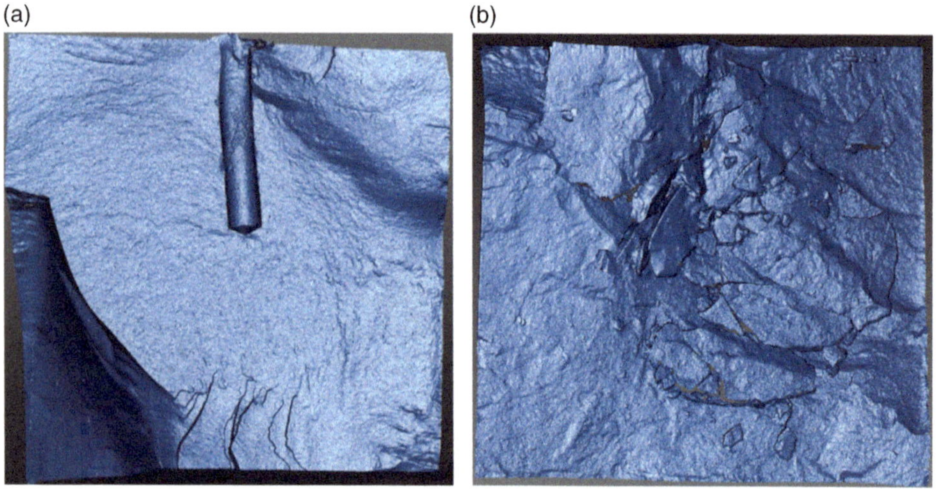

Figure 5.5 Digital radiography scans showing fractures caused by (a) hydraulic and (b) $scCO_2$ fracturing. The $scCO_2$-induced fracturing is obviously much more extensive (*Source:* Zang et al. [57], Elsevier).

The results show that both the filtration of $scCO_2$ and the increase of pore pressure can reduce the fracturing pressure of $scCO_2$ fracturing. At the same time, $scCO_2$ fracturing is easier to form microscopic pore structure and produce secondary fractures, which has better fracturing results, as shown in Figure 5.5.

Hu [58] conducted lab tests on $scCO_2$ fracturing. It is found that $scCO_2$ fracturing can produce multiple randomly distributed main fractures. The turning and bifurcation of fractures happen frequently during fracture propagation, contributing to the formation of more fractures. The geometry of fracture surface is more complex. The three special phenomena during $scCO_2$, i.e. explosion noise, exhalation, and surface temperature difference, are explained in view of expansion and phase change. It is found that the three special phenomena contribute to good effects of $scCO_2$ fracturing.

Chen et al. [59] investigated the influence of the perforation orientation on fracture initiation and propagation induced by $scCO_2$ fracturing through a series of laboratory fracturing experiments under triaxial stress states, where water was used as a fracturing fluid to compare with $scCO_2$. It was found that breakdown pressure increases with the increase of perforation angle according to the pressure curves in $scCO_2$ fracturing and hydraulic fracturing. AE energy characteristic has shown that the AE energy release rate of $scCO_2$ fracturing is larger than that of hydraulic fracturing, indicating that fractures induced by $scCO_2$ fracturing are more complex.

In summary, based on the experiments on $scCO_2$ fracturing with cylindrical and cubic core samples, and compared to conventional hydraulic fracturing and nitrogen fracturing, $scCO_2$ fracturing has a lower fracturing pressure and can generate more fractures.

The fracture initiation pressure is related to the stress field, rock properties, and the corresponding propagation regime of the hydraulic fracture [60]. According to the above studies, the $scCO_2$ can change rock microstructures, make some minerals fracture more easily, and reduce the strength of the shale ([36]; Zheng et al. 2015; [32, 39, 40]). Therefore, the fracture initiation pressure is reduced. From the results of three-dimensional acoustic emission inversion, the propagation of $scCO_2$ fracturing group can easily overcome the confinement of maximum horizontal stress on fracture propagation.

5.2.2.2 Numerical Simulations on $scCO_2$ Fracturing

Yu et al. [61] simulated the CO_2 injection into Bakken shale, which has typical reservoir and fracture properties. The simulation results show that CO_2 diffusion plays an important role in improving the recovery of tight oil reservoirs and cannot be ignored in the simulation.

Yin et al. [62] proposed a coupled THMC model based on finite element method and analyzed the changes of transient stress, pressure, and temperature and chemical concentration around an injection well. Finite element method is used to solve transient convection–diffusion equations considering heat transfer and solute transportation. At the same time, the numerical simulation on the injection of carbon dioxide saturated water in carbonate aquifer is conducted. The results show that the coupled model can analyze the stress and pressure changes in rock around the wellbore taking into account thermal and chemical effects.

Goodarzi et al. [63] built a multi-physics coupling model, which is based on heat flow and geology, in a carbon storage project. The model also considers the dynamic CO_2 injection to induce fractures. The simulation results show that injecting cold CO_2 can significantly reduce the in-situ stress, resulting in multiple fracture propagation.

Liu et al. [64] used the New Albany shale gas reservoir model to evaluate the CO_2 injection and storage capacity of shale and the influence of CO_2 injection on methane recovery. The results show the feasibility of using shale to store CO_2. About 40 000 tons CO_2 can be injected through a horizontal well (four hydraulic fracturing stages) in 5 years, while the breakthrough of CO_2 has a low possibility (less than 1% of CO_2 injection). With gas adsorption as the main storage mechanism, more than 95% of the injected CO_2 can be effectively isolated instantly.

Gou et al. [65] studied CO_2 enhanced shale gas recovery by numerical simulations. The simulation was carried out by using software TOUGH2MP-FLAC3D. The simulation results show that CO_2 moves to production well S14 along a NW-SE fault in their setup.

Currently, the numerical simulation on the interaction between $scCO_2$ and rock is mainly based on the needs for CO_2 storage. It considers the influence of long-term CO_2 injection on the mechanical properties of shale and sandstone, rather than the influence of $scCO_2$ fracturing process on the formation of complex fractures within a short time period. Currently there is no mature numerical model for fracture propagation involving the thermal – mechanical – hydraulic – chemical effects on $scCO_2$ – rock interaction.

5.3 A Field Trail on $scCO_2$ Fracturing of Continental Shale in Yanchang Oil Field

5.3.1 $scCO_2$ Fracturing Technology

Above all, $scCO_2$ fracturing currently is still in a research stage. Only a small number of field tests were conducted in the world, most of which were in China. $scCO_2$ fracturing process can be summarized as follows: use 100% CO_2 as the fracturing medium, then increase the pressure and lower the temperature, mix the proppant with the liquid CO_2, and finally use a high-pressure pump to inject the mixture into the wellbore for fracturing.

As shown in Figure 5.6, the main steps of the fracturing process are: (i) Parallel CO_2 storage tanks are connected to CO_2 booster pump truck, closed proppants mixer, fracturing pump truck, and wellhead equipment. (ii) Install temperature and pressure sensors at the bottom of the well to obtain pressure and temperature variations during CO_2 fracturing and gas testing. (iii) Conduct pump test and pressure test for high pressure pipe line, wellhead, and low pressure pipe line. If the test results meet the requirements, then move to the next step. (iv) After heating equipment or formation, CO_2 is injected to fracture the target formation. (v) Open the proppants mixing equipment to inject proppant based on the injection design, and stop pumping after completing the injection. (vi) Use microseismic to monitoring real-time fracture propagation during $scCO_2$ fracturing and subsequent hydraulic fracturing. (vii) After fracturing process is completed, shut in the well for a certain period, then open the well and pump out the

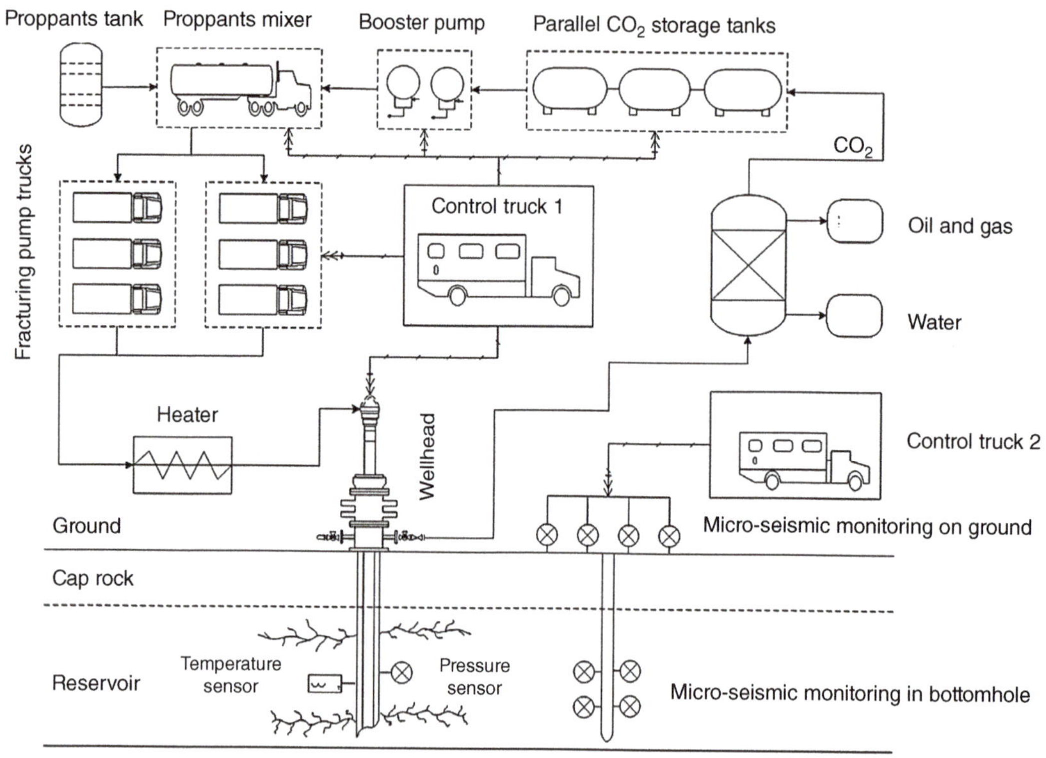

Figure 5.6 Flow chart of $scCO_2$ fracturing process.

down-hole CO_2 to the surface, at the same time, monitoring the change of CO_2 concentration at the well head.

5.3.2 $scCO_2$ Fracturing Field Test

5.3.2.1 Reservoir Properties of Test Wells

The first field test on $scCO_2$ fracturing in shale gas formation have been carried out in Yan'an X block in 2017. Shaanxi Yanchang Petroleum (Group) Co., Ltd. conducted $scCO_2$ fracturing field tests in two vertical wells (Figure 5.7). The main objective is to verify the feasibility of $scCO_2$ fracturing, such as the mechanism of increasing permeability by $scCO_2$ fracturing, the phase change during CO_2 injection, the adaptability of $scCO_2$ fracturing to shale gas reservoirs and their CO_2 storage capacity.

Shale gas in Yan'an Block X is continental sedimentation. The shale gas formation of Yan-A well is between 2456.1 and 2523.7 m (TVD), and the Yan-B well is between 2541.6 and 2609.4 m (TVD). Reservoir physical properties are porosity 2%; hydrocarbon content 0.6–7.6% (average = 4.0%); Poisson's ratio 0.19; elastic modulus 40.7 GPa. Shale gas reservoir is located in Benxi formation, the thickness of which is about 67 m, while the shale thickness is 55 m, and black shale lithology is interbedded with thin gray sandstone. Comprehensive interpretations of geological logging and electrical logging show that the two wells have good values. The perforation section is 2586–2589 m (TVD) in Benxi Formation, and the base, peak, and average of gas measurement are 0.6%, 7.6%, and 4.0%, respectively.

Figure 5.7 First field test on $scCO_2$ fracturing.

5.3.2.2 Fracturing Process and Operation Parameters

The fracturing section in the Yan-A well is at the depth 2940 m (TVD). For the Yan-A well, the conventional hydraulic jet method is used to open the window and then the $scCO_2$ fluid is used for fracturing. 8.89 cm oil tubing was employed. The cumulative liquid CO_2 injection is 363 m^3, and the volume of liquid CO_2 and water in formation is 496 m^3.The proppants volume is 5.7 m^3. The formation breakdown pressure is 58.6 MPa, the highest operation pressure is 65 MPa, and the pump-stop pressure is 8.2 MPa. For the Yan-B well, the operation parameters are: injection of liquid CO_2 is 339 m^3, the fluid volume in formation is 672 m^3 (328 m^3 for slick water, 344 m^3 for glue fluid), the proppants volume is 51.6 m^3, formation fracturing pressure is 39.5 MPa, the highest operation pressure is 56.1 MPa, and the pump-stop pressure 30.7 MPa.

$scCO_2$ friction loss along tubing and wellbore temperature profile in Yan-A well were tested, as well as proppants-carrying performance. Low-density ceramic granule (1.45 g/cm^3) was used in proppants-carrying performance test and the volume fraction of ceramic granule in proppants-carrying fluid is about 3%.

Due to the low density and viscosity of $scCO_2$, fracture tip is frequently blocked by proppants, leading to wellhead overpressure, and then the conventional fracturing fluid is used for proppants-carrying operation. The field test shows that the friction of $scCO_2$ is higher than that of traditional fracturing fluid, and the high friction is mainly in the nozzle. When the injection rate of 8-hole 6.5 mm nozzle is 2.7 m^3/min, the friction pressure drop of ordinary water jet nozzle is 19.3 MPa, and the friction pressure drop of $scCO_2$ jet nozzle is 27 MPa, which leads to frequent overpressure of ground system during operation.

CO_2 phase control at bottom hole is essential for $scCO_2$ fracturing. As shown in Figure 5.8, under the field conditions, the downhole CO_2 temperature is often above the critical temperature, which can meet the temperature requirements for $scCO_2$. After 80 minutes' injection of CO_2, wellbore temperature dropped to 25 °C. Then, 10 minutes later after stopping injection, bottom-hole temperature quickly rise to 31.1 °C, and the pressure is 52 MPa. Hence, the liquid CO_2 can become supercritical.

5.3.3 Field Test Results and Analysis

5.3.3.1 Microseismic Monitoring and Inversion of Fracture Geometry

As shown in Figures 5.9 and 5.10, the microseismic monitoring results of generated fractures in field tests show that the fracture initiation signal of $scCO_2$ fracturing is radially

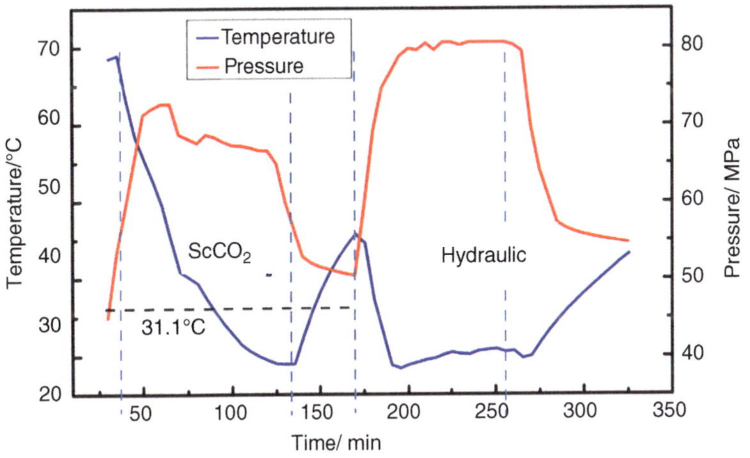

Figure 5.8 Bottom temperature and pressure monitoring results of Yan-A well during $scCO_2$ fracturing (*Source:* Hu [58]).

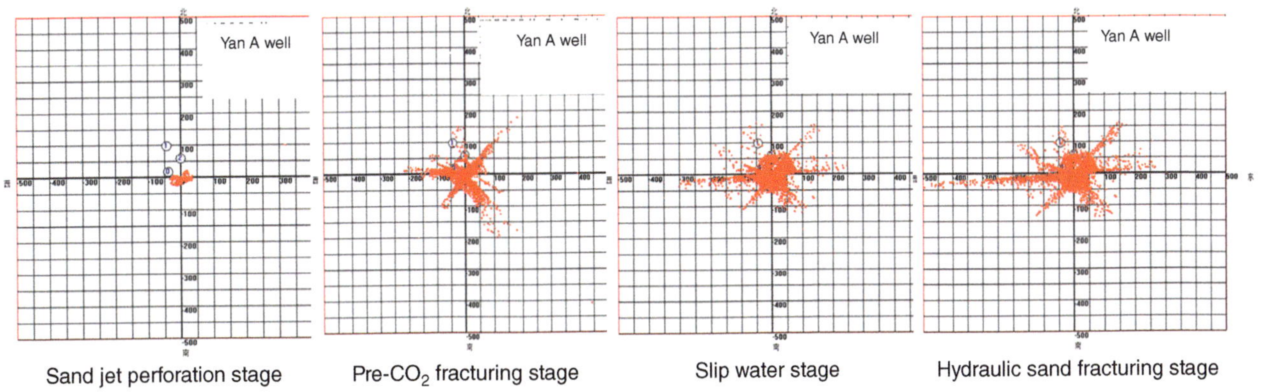

Figure 5.9 Yan-A well fractures growing process during fracturing (*Source:* Adapted from Wang et al. [66]).

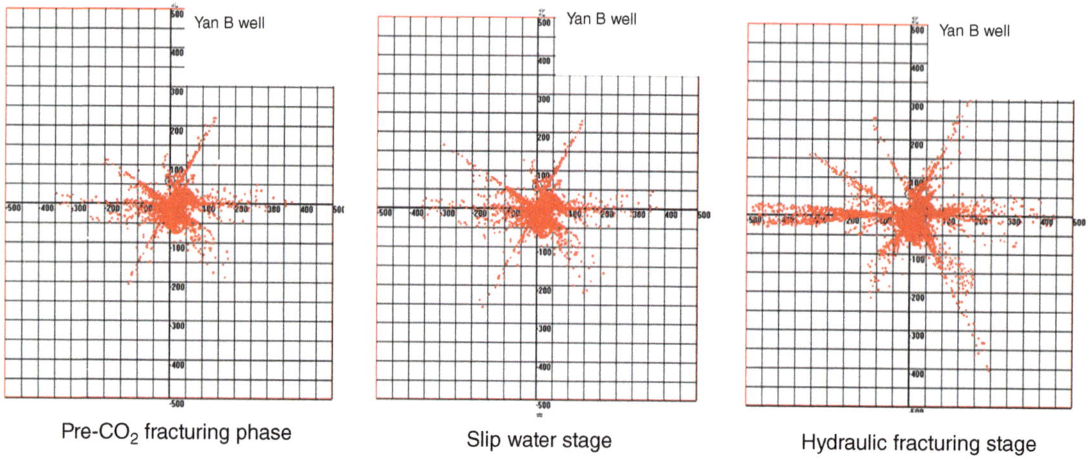

Figure 5.10 Yan-B well fractures growing process during fracturing (*Source:* Hu [58]).

dense and evenly distributed in all directions around the wellbore, while the fracture initiation signal of conventional hydraulic fracturing is only distributed along the direction of maximum horizontal stress, which further proves that $scCO_2$ fracturing can effectively break through the limit of stress barriers on fracture growth and yield complex fracture networks [66].

5.3.3.2 Production Data

The CO_2 in the pre-fracturing process is supercritical, and post injection in the reservoir, it reaches the supercritical state in a short time. High discharge rate of this fracturing fluid can help in increasing input energy and aiding gas drainage. Also most of injected CO_2 is buried in the ground after fracturing. The synergistic effect of $scCO_2$ enrichment in fracture networks and subsequent hydraulic fracturing in fracture extension effectively improve the treatment range. The average production of the well Yan-B is $4.3 \times 10^4\,m^3$ and increases by 2–2.5 times compared to the hydraulic fracturing treatment. The CO_2 concentration in produced gas is less than 2%, which is equivalent to that in the primary shale gas.

5.4 Challenges in $scCO_2$ Fracturing

5.4.1 $scCO_2$ Fracturing Mechanism Is Still Not Clear

The $scCO_2$ is highly compressible and its physical properties are extremely sensitive to pressure and temperature. These properties make $scCO_2$ – rock interaction difficult to predict. Therefore, when studying the interaction between $scCO_2$ and rocks, the coupling of multi-physics must be considered, such as stresses, pressure, temperature, and chemical effects. Furthermore, the filtration of $scCO_2$ into formations and the phase changes of $scCO_2$ under different conditions should be considered.

Due to the complexity of $scCO_2$ fracturing, the simulation software for conventional hydraulic fracturing can hardly simulate $scCO_2$ fracturing reasonably. Meanwhile, due to the limited size of experiment equipment, the experiment results can impossibly represent a real field application on $scCO_2$ fracturing. Hence, the design of $scCO_2$ fracturing highly relies on experience. Further studies on experiments and numerical methods are still needed.

5.4.2 Challenges in Proppants Carrying

During $scCO_2$ fracturing, the viscosity of liquid and $scCO_2$ is much lower than that of water-based fracturing fluid, which could lead to reduction in proppants-carrying capability. In this case, proppants tend to accumulate at the bottom of the wellbore and at the toes of fractures, eventually leading to the failure of $scCO_2$ fracturing. This problem also happened in Yan-A Well.

Potential ways to solve this problem are (i) improving pump displacement and (ii) using more viscosifier. The turbulence of fracturing fluid caused by high pump displacement will improve proppants-carrying capacity; however, it will also increase equipment cost. On the other hand, developing a new viscosifier for $scCO_2$ fracturing fluid is very hard. The reason is that liquid CO_2 is a non-polar solvent and can only dissolve in nonpolar solutes. So thickeners must also be nonpolar to be compatible with liquid CO_2, which significantly increases the difficulties in developing suitable viscosifier [67]. Therefore, increasing the proppants-carrying ability of $scCO_2$ fracturing fluid currently is an unsolved issue.

5.4.3 Challenge on the Predicting and Monitoring CO_2 Phase

The phase change of CO_2 during $scCO_2$ fracturing is very complicated. The density, viscosity, and solubility of CO_2 could change significantly with varying temperature and pressure, so the phase change of CO_2 could happen everywhere, e.g. in the tank, surface pipeline, and wellbore. Therefore, during $scCO_2$ fracturing, the phase change of CO_2 must be extensively monitored. However, there is limited research on the phase change of CO_2 in the wellbore and in the fracture.

5.4.4 Lack of Specialized Equipment for $scCO_2$ Fracturing

$scCO_2$ fracturing requires a high-power proppants mixing pump to ensure that fracturing fluid can have a high flow speed in the wellbore and thus have enough proppants-carrying capacity. The rotation speed of stirring shaft of a traditional blender truck is too low to deliver a good mixing performance. Meanwhile, the proppants-carrying performance of CO_2 is limited, and so the proppants tend to sink. Basically, the nature of $scCO_2$ fluid is quite different from water. Conventional fracturing equipment cannot meet many requirements, such as proppants mixing equipment, downhole tools, surface circulation devices, etc. It is necessary to develop specialized equipment for $scCO_2$ fracturing.

5.5 Conclusions

Experiments and numerical simulations on $scCO_2$-shale interaction show that $scCO_2$ can create solution-based pores through the reaction with minerals in shale and thus vary the macroscopic mechanical properties of shale to facilitate fracture growth. Due to the high diffusivity of $scCO_2$, the fracture initiation during $scCO_2$ fracturing is not fully controlled by the maximum horizontal stress, which can help create more complicated fracture networks and thus improve production rates. Moreover, a field test shows that $scCO_2$ can produce more radial fractures around the wellbore and increase shale gas production rate by 2–2.5 times. Overall, compared to conventional hydraulic fracturing, $scCO_2$ fracturing has more advantages, and it is highly recommended to be applied to other kinds of unconventional resources besides shale gas.

However, one should also note that the research on $scCO_2$ fracturing is particularly insufficient in regards to multiphase flow, phase change of CO_2, interaction between $scCO_2$ and proppant, and the coupling of multiple physics. The following areas are recommended for future works:

1) Numerical methods on simulating fracture propagations considering the coupling between temperature, pressure, stress, and chemistry.
2) Accurate prediction of phase change of CO_2 during $scCO_2$ fracturing.
3) Development of viscosifier for $scCO_2$ fracturing fluid.
4) Development of specialized equipment for $scCO_2$ fracturing.

Acknowledgments

This work was funded by the National Key Basic Research Program of China (973 Program, No. 2014CB239200), the National Natural Science Foundation of China (Nos. 51874253, U19A2097, and U20A20265). Thanks for Yanchang Petroleum (Group) Co., Ltd. for conducting the first field test on scCO2 fracturing in shale gas formation.

References

1 Zhu, H., Tang, X., Liu, Q. et al. (2018). 4D multi-physical stress modelling during shale gas production: a case study of Sichuan Basin shale gas reservoir, China. *Journal of Petroleum Science and Engineering* 167: 929–943.

2 Zhu, H., Zhao, Y., Feng, Y. et al. (2019). Modeling of fracture width and conductivity in channel fracturing with nonlinear Proppant-pillar deformation. *SPE Journal* 24 (3): 1288–1308.

3 Zhu, H., Shen, J., and Zhang, F. (2019). A fracture conductivity model for channel fracturing and its implementation with discrete element method. *Journal of Petroleum Science and Engineering* 172: 149–161.

4 Zhu, H., Tang, X., Song, Y. et al. (2021). An infill well fracturing model and its microseismic events barrier effect: a case in Fuling shale gas reservoir. *SPE Journal* 26 (1): 113–134.

5 Allawzi, M., Al-Otoom, A., Allaboun, H. et al. (2011). CO_2 supercritical fluid extraction of Jordanian oil shale utilizing different co-solvents. *Fuel Processing Technology* 92: 2016–2023.

6 Barati, R. and Liang, J. (2014). A review of fracturing fluid systems used for hydraulic fracturing of oil and gas wells. *Journal of Applied Polymer Science* 131: 40735.

7 Brunner, G. (2015). Supercritical process technology related to energy and future directions – an introduction. *The Journal of Supercritical Fluids* 96: 11–20.

8 Lee, J., Min, K., and Rutqvist, J. (2013). Probabilistic analysis of fracture reactivation associated with deep underground CO_2 injection. *Rock Mechanics and Rock Engineering* 46: 801–820.

9 Liu, H., Wang, F., Zhang, J. et al. (2014). Fracturing with carbon dioxide: application status and development trend. *Petroleum Exploration and Development* 41: 466–472. (In Chinese).

10 Meng, S., Liu, H., Xu, J. et al. (2016). The evolution and control of fluid phase during liquid CO_2 fracturing. *SPE Asia Pacific Hydraulic Fracturing Conference* (24–26 August 2016), Society of Petroleum Engineers, Beijing, China.

11 Wang, X., Wu, J., and Zhang, J. (2014). Application of CO_2 fracturing technology for terrestrial shale gas reservoirs. *Natural Gas Industry* 34: 64–67. (In Chinese).

12 Zhu, H., Tao, L., Liu, Q., and Lei, Z. (2016). Multiscale change characteristics of Chinese Wufeng Shale under supercritical carbon dioxide. *SPE Annual Technical Conference and Exhibition* (26–28 September 2016). Dubai, UAE: Society of Petroleum Engineers.

13 Middleton, R.S., Carey, J.W., Currier, R.P. et al. (2015). Shale gas and non-aqueous fracturing fluids: opportunities and challenges for supercritical CO_2. *Applied Energy* 147: 500–509.

14 Palmer, C. and Sito, Z. (2013). Nitrogen and carbon dioxide fracturing fluids for the stimulation of unconventional shale plays. *AGH Drilling, Oil, Gas* 30: 191.

15 Sun, J., Zou, A., and Schechter, D. (2016). Experimental and numerical studies of CO_2 EOR in unconventional liquid reservoirs with complex fracture networks. *SPE Improved Oil Recovery Conference* (11–13 April 2016). Tulsa, OK: Society of Petroleum Engineers. SPE-179634-MS. 10.2118/179634-MS

16 Busch, A., Alles, S., Gensterblum, Y. et al. (2008). Carbon dioxide storage potential of shales. *International Journal of Greenhouse Gas Control* 2: 297–308.

17 Okamoto, I., Li, X., and Ohsumi, T. (2005). Effect of supercritical CO_2 as the organic solvent on cap rock sealing performance for underground storage. *Energy* 30: 2344–2351.

18 Vialle, S. and Vanorio, T. (2011). Laboratory measurements of elastic properties of carbonate rocks during injection of reactive CO_2-saturated water. *Geophysical Research Letters* 38: L01302.

19 Chiquet, P., Broseta, D., and Thibeau, S. (2007). Wettability alteration of caprock minerals by carbon dioxide. *Geofluids* 7: 112–122.

20 Alemu, B.L., Aagaard, P., Munz, I.A., and Skurtveit, E. (2011). Caprock interaction with CO_2: a laboratory study of reactivity of shale with supercritical CO_2 and brine. *Applied Geochemistry* 26: 1975–1989.

21 Busch, A., Alles, S., Krooss, B.M. et al. (2009). Effects of physical sorption and chemical reactions of CO_2 in shaly caprocks. *Energy Procedia* 1: 3229–3235.

22 Emberley, S., Hutcheon, I., Shevalier, M. et al. (2005). Monitoring of fluid–rock interaction and CO_2 storage through produced fluid sampling at the Weyburn CO_2-injection enhanced oil recovery site, Saskatchewan, Canada. *Applied Geochemistry* 20: 1131–1157.

23 Lin, H., Fujii, T., Takisawa, R. et al. (2008). Experimental evaluation of interactions in supercritical CO_2/water/rock minerals system under geologic CO_2 sequestration conditions. *Journal of Materials Science* 43: 2307–2315.

24 Liu, F., Lu, P., Griffith, C. et al. (2012). CO_2–brine–caprock interaction: reactivity experiments on Eau Claire shale and a review of relevant literature. *International Journal of Greenhouse Gas Control* 7: 153–167.

25 Wollenweber, J., Alles, S., Busch, A. et al. (2010). Experimental investigation of the CO_2 sealing efficiency of caprocks. *International Journal of Greenhouse Gas Control* 4: 231–241.

26 Jiang, Y., Luo, Y., Lu, Y. et al. (2016). Effects of supercritical CO_2 treatment time, pressure, and temperature on microstructure of shale. *Energy* 97: 173–181.

27 Ao, X., Lu, Y., Tang, J. et al. (2017). Investigation on the physics structure and chemical properties of the shale treated by supercritical CO_2. *Journal of CO_2 Utilization* 20: 274–281.

28 Yin, H., Zhou, J., Jiang, Y. et al. (2016). Physical and structural changes in shale associated with supercritical CO_2 exposure. *Fuel* 184: 289–303.

29 Lahann, R., Mastalerz, M., Rupp, J.A., and Drobniak, A. (2013). Influence of CO_2 on New Albany Shale composition and pore structure. *International Journal of Coal Geology* 108: 2–9.

30 Xu, T., Apps, J.A., and Pruess, K. (2005). Mineral sequestration of carbon dioxide in a sandstone–shale system. *Chemical Geology* 217: 295–318. https://doi.org/10.1016/j.chemgeo.2004.12.015.

31 Angeli, M., Soldal, M., Skurtveit, E., and Aker, E. (2009). Experimental percolation of supercritical CO_2 through a caprock. *Energy Procedia* 1: 3351–3358. https://doi.org/10.1016/j.egypro.2009.02.123.

32 Liu, Q., Tao, L., Zhu, H. et al. (2018). Macroscale mechanical and microscale structural changes in Chinese Wufeng Shale with supercritical carbon dioxide fracturing. *SPE Journal* 23: 0691–0703.

33 Du, Y., Wang, R., Ni, H. et al. (2013). Dynamical analysis of high-pressure supercritical carbon dioxide jet in well drilling. *Journal of Hydrodynamics* 25: 528–534.

34 Gupta, A.P., Gupta, A., and Langlinais, J. (2005). Feasibility of supercritical carbon dioxide as a drilling fluid for deep underbalanced drilling operation. Paper presented at the

SPE Annual Technical Conference and Exhibition, Dallas, TX, October 2005. SPE-96992-MS. https://doi.org/10.2118/96992-MS.

35 Tian, H., Pan, F., Xu, T. et al. (2014). Impacts of hydrological heterogeneities on caprock mineral alteration and containment of CO_2 in geological storage sites. *International Journal of Greenhouse Gas Control* 24: 30–42.

36 Kollé, J.J. (2000). Coiled tubing drilling with supercritical carbon dioxide. *SPE/CIM International Conference on Horizontal Well Technology*, Kent, WA (6–8 November 2000). Calgary, Alberta: Tempress Technologies, Inc.

37 Wang, H., Li, G., Shen, Z. et al. (2015). Experiment on rock breaking with supercritical carbon dioxide jet. *Journal of Petroleum Science and Engineering* 127: 305–310. https://doi.org/10.1016/j.petrol.2015.01.006.

38 Zheng, H., Feng, X., and Pan, P. (2015). Experimental investigation of sandstone properties under CO2–NaCl solution-rock interactions. *International Journal of Greenhouse Gas Control* 37: 451–470. https://doi.org/10.1016/j.ijggc.2015.04.005.

39 Lyu, Q., Long, X., Ranjith, P.G. et al. (2018). Experimental investigation on the mechanical properties of a low-clay shale with different adsorption times in sub-/super-critical CO_2. *Energy* 147: 1288–1298.

40 Yin, H., Zhou, J., Xian, X. et al. (2017). Experimental study of the effects of sub- and super-critical CO_2 saturation on the mechanical characteristics of organic-rich shales. *Energy* 132: 84–95.

41 Cai, C., Kang, Y., Wang, X. et al. (2019). Experimental study on shale fracturing enhancement by using multi-times pulse supercritical carbon dioxide (SC-CO_2) jet. *Journal of Petroleum Science and Engineering* 178: 948–963.

42 Cai, C., Kang, Y., Yang, Y. et al. (2020). The effect of shale bedding on supercritical CO_2 jet fracturing: an experimental study. *Journal of Petroleum Science and Engineering* 195: 107798.

43 Huang, M., Hu, Y., Wang, X. et al. (2020). Experimental investigation on the flow and rock breaking characteristics of supercritical carbon dioxide jets. *Journal of Petroleum Science and Engineering* 187: 106735.

44 Zhu, H., Tao, L., Liu, Q. et al. (2017). Fracture characteristics and change of permeability under the influence of natural fractures: experimental study of Wufeng-Longmaxi shale. *SPE Reservoir Evaluation & Engineering* 21: 225–237.

45 Regnault, O., Lagneau, V., and Schneider, H. (2009). Experimental measurement of portlandite carbonation kinetics with supercritical CO_2. *Chemical Geology* 265 (1–2): 113–121.

46 Fischer, S., Liebscher, A., and Wandrey, M. (2010). CO_2–brine–rock interaction — First results of long-term exposure experiments at in situ P–T conditions of the Ketzin CO_2 reservoir. *Chemie der Erde - Geochemistry* 70: 155–164.

47 Miller, Q.R., Thompson, C.J., Loring, J.S. et al. (2013). Insights into silicate carbonation processes in water-bearing supercritical CO_2 fluids. *International Journal of Greenhouse Gas Control* 15: 104–118.

48 Jia, Y., Lu, Y., Elsworth, D. et al. (2018). Surface characteristics and permeability enhancement of shale fractures due to water and supercritical carbon dioxide fracturing. *Journal of Petroleum Science and Engineering* 165: 284–297.

49 Jiang, Y., Qin, C., Kang, Z. et al. (2018). Experimental study of supercritical CO_2 fracturing on initiation pressure and fracture propagation in shale under different triaxial stress conditions. *Journal of Natural Gas Science and Engineering* 55: 382–394.

50 Zhang, Y., He, J., Li, X., and Lin, C. (2019). Experimental study on the supercritical CO_2 fracturing of shale considering anisotropic effects. *Journal of Petroleum Science and Engineering* 173: 932–940.

51 Yang, B., Wang, H., Shen, Z. et al. (2021). Full-sample X-ray microcomputed tomography analysis of supercritical CO_2 fracturing in tight sandstone: effect of stress on fracture dynamics. *Energy & Fuels* 35 (2): 1308–1321.

52 Ishida, T., Aoyagi, K., Niwa, T. et al. (2012). Acoustic emission monitoring of hydraulic fracturing laboratory experiment with supercritical and liquid CO_2. *Geophysical Research Letters* 39: L16309. https://doi.org/10.1029/2012GL052788.

53 Ishida, T., Chen, Y., Bennour, Z. et al. (2016). Features of CO_2 fracturing deduced from acoustic emission and microscopy in laboratory experiments. *Journal of Geophysical Research-Solid Earth* 121: 8080–8098.

54 Akihisa, K., Hiroharu, T., and Kiyotoshi, S. (2013). Hydraulic fracturing in Inada granite and Ogino tuff using super critical carbon dioxide and water as fracturing fluids. *Journal of MMIJ* 129: 461.

55 Inui, S., Ishida, T., Nagaya, Y. et al. (2014). AE monitoring of hydraulic fracturing experiments in granite blocks using Supercritical CO_2, water and viscous oil. In: *48th US Rock Mechanics/Geomechanics Symposium*. American Rock Mechanics Association, Minneapolis, MN, June 2014, ARMA-2014-7163.

56 Li, X., Feng, Z., Han, G. et al. (2016). Breakdown pressure and fracture surface morphology of hydraulic fracturing in shale with H_2O, CO_2 and N_2. *Geomechanics and Geophysics for Geoenergy and Georesources* 2: 63.

57 Zhang, X., Lu, Y., Tang, J. et al. (2017). Experimental study on fracture initiation and propagation in shale using supercritical carbon dioxide fracturing. *Fuel* 190: 370–378.

58 Hu, Y. (2017). *Mechanism and Experimental Research of Supercritical Carbon Dioxide Fracturing Shale*. Wuhan University (In Chinese).

59 Chen, H., Hu, Y., Kang, Y. et al. (2019). Fracture initiation and propagation under different perforation orientation angles in supercritical CO_2 fracturing. *Journal of Petroleum Science and Engineering* 183: 106403.

60 Zhu, H., Deng, J., Jin, X. et al. (2015). Hydraulic fracture initiation and propagation from wellbore with oriented perforation. *Rock Mechanics and Rock Engineering* 48 (2): 585–601.

61 Yu, W., Lashgari, H.R., Wu, K., and Sepehrnoori, K. (2015). CO_2 injection for enhanced oil recovery in Bakken tight oil reservoirs. *Fuel* 159: 354–363.

62 Yin, S., Dusseault, M.B., and Rothenburg, L. (2011). Coupled THMC modeling of CO_2 injection by finite element methods. *Journal of Petroleum Science and Engineering* 80: 53–60.

63 Goodarzi, S., Settari, A., and Keith, D. (2012). Geomechanical modeling for CO_2 storage in Nisku aquifer in Wabamun Lake area in Canada. *International Journal of Greenhouse Gas Control* 10: 113–122.

64 Liu, F., Ellett, K., Xiao, Y., and Rupp, J.A. (2013). Assessing the feasibility of CO_2 storage in the New Albany Shale (Devonian-Mississippian) with potential enhanced gas recovery using reservoir simulation. *International Journal of Greenhouse Gas Control* 17: 111–126.

65 Gou, Y., Hou, Z., Liu, H. et al. (2014). Numerical simulation of carbon dioxide injection for enhanced gas recovery (CO_2-EGR) in Altmark natural gas field. *Acta Geotechnica* 9: 49–58.

66 Wang, H., Li, G., Zheng, Y. et al. (2020). Research status and prospects of supercritical CO, fracturing technology. *Acta Petrolei Sinica* 41: 116–126. (In Chinese).

67 Sun, B. and Sun, W. (2015). Research progress and prospectives of supercritical CO_2 thickening technology. *Journal of China University of Petroleum (Edition of Natural Science)* 39: 76–83. (In Chinese).

6

An Unstructured Moving Element Mesh for Hydraulic Fracture Modeling

John Napier[1] *and Emmanuel Detournay*[2]

[1] *Department of Mining Engineering, University of Pretoria, Hatfield, Pretoria, South Africa*
[2] *Department of Civil, Environmental, and Geo-Engineering, University of Minnesota, Minneapolis, MN, USA*

6.1 Introduction

A number of severe challenges have to be overcome when constructing models of hydraulic fractures [1, 2]. They originate from the moving boundary nature of the problem and the degeneracy of the nonlinear equations near the crack tip; but also crucially from the competition between different dissipative processes in the tip region: namely, the flow of viscous fracturing fluid, the creation of fracture surfaces in the solid, and the leak-off of fluid from the fracture.

While state-of-the-art planar fracture growth simulators exist now, capable of accurately assessing how the vertical growth of a hydraulic fracture outside the hydrocarbon bearing layer in conventional reservoir applications is affected by stress or material contrast between horizontal layers, there are numerous other applications that require the capability to model nonplanar propagation of hydraulic fractures. Existing numerical algorithms capable of conducting such simulations generally rely on a volume discretization of the domain, either based on a lattice representation [3], or on finite elements with continuous remeshing to adapt to the fracture growth [4], or on extended finite elements that use enrichment functions to capture the displacement jump and also the tip asymptotic behavior in the elements containing the front [5, 6]. The domain-based computational algorithms, especially the lattice model, have the ability to mimic complex geometrical cases, but at the expense of large computational cost.

This paper describes a novel computational scheme to simulate the propagation of fluid-driven fractures in an impermeable elastic medium. Although the algorithm is currently restricted to plane cracks, it can be extended to simulate nonplanar fracture propagation. The scheme, which involves only variables defined on the fracture, relies on representing the crack surface by an unstructured tessellation of triangular elements, with the outer edge of the propagating grid approximating the crack front at a given time. The algorithm combines the displacement discontinuity method [7] and a finite volume scheme [8] to solve the elasticity and the lubrication equations on the unstructured mesh.

A significant part of the logic behind the propagation algorithm is inspired by the implicit level set algorithm [1, 2, 9], which also relies on a combination of displacement discontinuity and finite volume methods. However, this algorithm assumes that the fracture plane is known a priori and uses a fixed computational grid with the propagating front tracked as a level set of the aperture field [1, 2, 9–12]. Although computationally very efficient, the fixed grid scheme cannot be extended to nonplanar cracks, as the direction of propagation of the front is evolving during fracture growth. The use of an expanding computational grid, as described here, requires constructing a new logic for crack advance that properly accounts for the multiscale tip asymptote within the framework of the unstructured mesh. Correctly accounting for the dissipation mechanisms at the tip is critical, as they control the speed of propagation of the moving boundary and ultimately affect the regime of propagation [13, 14].

6.2 Discrete Model of a Planar Hydraulic Fracture

We consider the propagation of a planar fracture driven by the injection of a fluid in a rock medium at volumetric rate $Q(t)$ starting at time $t = 0$. The evolving geometry of the hydraulic fracture is described by crack front $\mathcal{C}(t)$, which encloses the crack footprint $\mathcal{A}(t)$, see Figure 6.1 Enough time is presumed to have elapsed since injection began,

Mechanics of Hydraulic Fracturing: Experiment, Model, and Monitoring, First Edition.
Edited by Xi Zhang, Bisheng Wu, Diansen Yang, and Andrew Bunger.

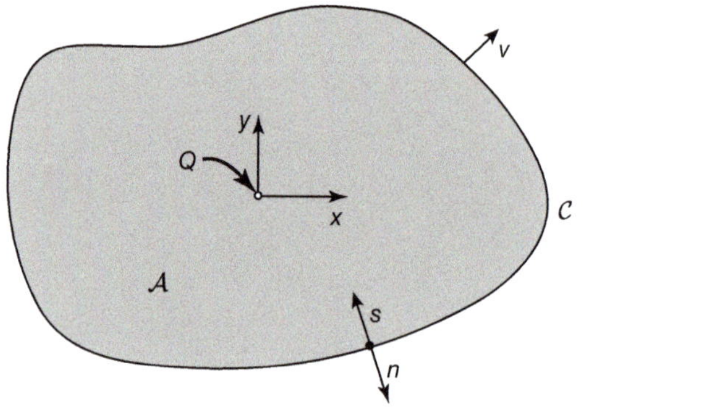

Figure 6.1 Planar hydraulic fracture with crack edge $\mathcal{C}(t)$ enclosing footprint $\mathcal{A}(t)$.

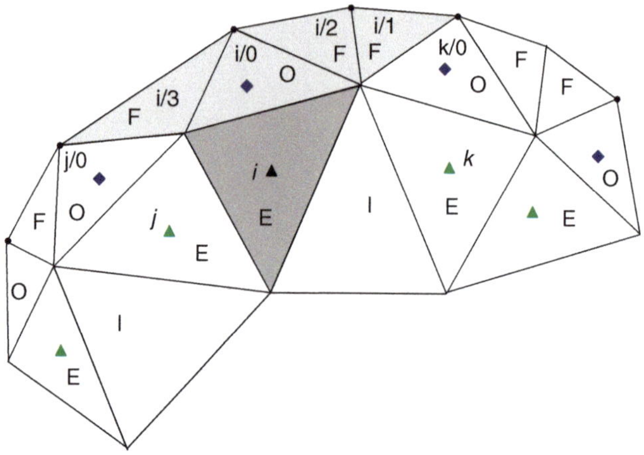

Figure 6.2 Channel and tip elements. Channel elements are either interior (I) or edge (E) elements, while tip elements are either offset edge (O) or filler elements (F). An I-element shares edges only with other channel elements (of subtypes I or E), while an E-e element has a common edge with an O-element. An F-element only shares edges with other tip elements.

so the dimensions of the footprint are large enough that fluid injection can be viewed as occurring at a point. The injection point conveniently serves as the origin for position vector x.

To simplify the problem, it is assumed that (i) the rock is homogeneous, linearly elastic, brittle, and impermeable and of infinite extent; (ii) the fracturing fluid is Newtonian with viscosity μ and is incompressible; the orientation of the minimum in situ compressive principal stress, $\sigma_0(x)$, is uniform; the propagating crack is always in limit equilibrium; the crack is fully filled by the fracturing fluid. The hydraulic fracturing problem thus can be formulated as follows. Given the fluid and solid properties, far-field stress and injection history, determine how front $\mathcal{C}(t)$, crack aperture field $w(x, t)$, and fluid pressure $p_f(x, t)$ evolve from initial conditions described by $\mathcal{C}(0)$, $w(x, 0)$, and $p_f(x, 0)$.

Here we directly introduce the model in a discrete formulation of the hypersingular integral equation of elasticity and the nonlinear partial differential Reynolds equation of lubrication theory, see [15] for details. A key feature of the discrete model is the distinction between fixed channel elements representing the bulk of the fracture and adjustable tip elements at the periphery of the crack. Different sets of equations apply to these elements [15].

6.2.1 Unstructured Mesh

The hydraulic fracture is discretized with a mesh of triangular elements, consisting of a grid of "channel" elements encircled by a ring of "tip" elements whose outer edge define the fracture front. The channel elements are fixed while the geometry of the tip elements evolves during the iterative search of the solution at a given time t. Furthermore, the tip elements are located in the tip asymptotic umbrella and their apertures are computed according to the tip asymptote, a function of the local front velocity.

At time t, the hydraulic fracture is discretized by a mesh of $n(t)$triangular elements, consisting of $n_c(t)$ channel and $n_f(t)$ tip elements. Each triangular cell T_i is characterized by a uniform aperture w_i and a net pressure p_i defined at the element centroid.

By reference to Figure 6.2, there are two subtypes of channel and tip elements. Channel elements are either interior (I) or edge (E) elements, while tip elements are either offset edge (O) or filler elements (F). An I-element shares edges only with other channel elements (of subtypes I or E), while an E-element has a common edge with an O-element. An F-element only shares edges with other tip elements. The filler elements bridge the space between two offset edge elements with at most two F-elements used to fill the gap. Filler elements have an edge on the crack front, while offset elements have only one vertex located on the front. Thus, a stencil of up to 5 tip elements is associated with each edge channel element: 1 O-element and 0–4 F-elements (up to 2 fillers on each side of an O-element). Figure 6.2 illustrates an example with four tip elements (with light gray filling) attached to edge channel element i (with dark gray filling).

6.2.2 Discrete Elasticity Equation

The assumption of homogeneity, linear elasticity, and infinite extent of the rock makes it possible to represent the crack – a surface of discontinuity of the displacement field, by a continuous distribution of dislocation dipoles. This representation leads to the formulation of a singular integral equation relating the net fluid pressure $p(x, t) = p_f(x, t) - \sigma_0(x)$ to the crack aperture $w(x, t)$, with $x \in \mathcal{A}(t)$.

The discrete form of the elastic integral equations is obtained by assembling the solution for all the triangular elements used to tessellate the crack surface. The elastic solution for a triangular Volterra dislocation, which is characterized by a uniform displacement discontinuity, was first constructed by Yoffe [16] through superposition of angular dislocations. Since this original contribution, several articles have been written on the triangular dislocation (or on the more general case of a polygonal dislocation) in infinite or semi-infinite elastic media [17–21].

Balancing the normal stress at the centroid (x_i, y_i) of element T_i leads to a linear dependence of the net pressure p_i at the centroid of element T_i to the aperture of all the elements of the mesh

$$p_i = -E' \sum_{j=1}^{n} C_{ij} w_j \tag{6.1}$$

An explicit expression for the elastic influence coefficient C_{ij} can be found elsewhere [15].

6.2.3 Discretized Lubrication Equations for Channel Elements

The discrete form of the lubrication equation is formulated by first integrating the continuity equation on a triangular element T_i, and then invoking Poiseuille law followed by the application of Gauss theorem. These considerations show that Reynolds equation integrated over the element can be written as

$$\mu' \frac{\mathrm{d}w_{\mathrm{i}}}{dt} = A_{jj}(w)p_i + \sum_{j \in \mathcal{J}_i} A_{ij}(w)p_j \tag{6.2}$$

where $\mathcal{J}_i$ is the index set containing the element numbers of all the neighbors of element i, which are used to calculate the normal pressure gradients $\psi_{i,\,k}$, $k = 1, 2, 3$ across the three edges of cell i. The matrix $\boldsymbol{A}$ is sparsely populated as the off-diagonal elements A_{ij} are zero if cells i and j are not hydraulically connected, i.e. $j \notin \mathcal{J}_i$. The non-zero elements A_{ii} and $A_{ij}, j \in \mathcal{J}_i$ depend only on the aperture of cell i and of its neighbors with a common edge. Conservation of fluid across edge common to element i and j is guaranteed by the "symmetry" $\|T_i\| A_{ij} = \|T_j\| A_{ji}$.

Because of the skewness of the triangular cells, calculation of the pressure gradient in a direction normal to an edge requires a correction to the gradient calculated by the divided difference of the pressure at the centroids of the two cells sharing that edge. This issue, which arises when computing flow on an unstructured mesh, has been the subject of numerous efforts. Here we have followed the approach originally proposed by Mathur and Murthy [8].

6.2.4 Tip Elements

The elements at the periphery of the crack are treated differently from channel elements for a number of reasons. First, the geometry of these elements evolves due to movement of the front, second the aperture field in the tip elements takes a particular asymptotic form that is a function of the local crack front velocity. The use of an asymptotic aperture implies that there is not any direct elastic dependence between the pressure in the tip elements and the aperture in the elements of the grid; in other words the discrete elasticity eq. (6.1) is not used for assessing the pressure at the centroid of the tip elements. The general approach is inspired by the original algorithm described in [2].

The universal tip asymptote for a hydraulic fracture propagating in an impermeable solid can be written as

$$\hat{w}(s) = w_* \hat{\Omega}\left(\frac{s}{\ell_*}\right), \tag{6.3}$$

where ℓ_* and w_* are scales, which depend explicitly on the local front velocity v ($\ell_* \sim v^{-2}$, $w_* \sim v^{-1}$). The universal function $\hat{\Omega}(\xi)$ reduces in the near-field to the toughness asymptote $\xi^{1/2}$ and in the far-field to the viscosity asymptote $\xi^{2/3}$ [22]. For computational purposes, the function $\hat{\Omega}(\xi)$ can be approximated in the intermediate range by a best fit function [15]. Alternatively, $\hat{\Omega}(\xi)$ can be computed using the fast method described by [23]. The tip opening asymptote $\hat{w}(s)$ is used not only to determine the local position of the front as a function of the aperture in an edge channel element, but also to calculate the volume of fluid stored in the tip elements. This volume is required for the continuity equation of these elements and for the calculation of their elastic influence on the other elements of the mesh.

From a numerical point of view, the transition is legislated by the tip length scale ℓ_* – a function of the local front velocity, in relation to the size d of the tip element. If ℓ_*/d is small, the tip element is inside the viscosity asymptotic umbrella, but if ℓ_*/d is large, it is dominated by the toughness asymptote.

Although the discrete elasticity eq. (6.1) is not used to evaluate the pressure in the tip elements, these elements do affect the pressure in the channel elements. A constant opening, equal to the element volume divided by element area is thus assigned to the tip elements.

6.3 Time-Marching Algorithm

This section presents the algorithm that is used to advance the solution over one time step. It describes the sequential updation of the aperture of the channel elements.

A distinction is made between interior and edge channel elements, as the aperture update of the edge channel involves a local readjustment of the crack front.

6.3.1 Iteration Loops

Each time step advance involves computing a new footprint of the hydraulic crack and a new opening at each cell of the mesh. The solution is obtained via an iterative procedure, which loops repetitively on all the channel elements and updates their aperture using both elasticity and lubrication equations. Because of the sequential nature of the algorithm, the aperture of each cell is updated one by one, in terms of the aperture of all the other elements, by solving a nonlinear equation. Different equations are used to update the opening of an interior and an edge channel element. For an interior element, the update relies entirely on the discrete elasticity equations and on the discrete lubrication equation that links an interior element to its direct neighbors. For an edge channel element, the volume of fluid exchanged with its offset fringe neighbor involves a local adjustment of the crack front position.

Advancing the solution of a hydraulic fracturing problem over one time step necessitates therefore two nested loops in our algorithm: the outer loop sequentially updates the element aperture and footprint until convergence is achieved; the inner loop solves iteratively the nonlinear equation governing the aperture of any cell. It is worth emphasizing that the pressure in the tip elements are essentially devices to draw fluid from the edge channel elements into the tip elements so as to ensure that the volume of fluid in these elements corresponds to the volume calculated according to the tip asymptotic solution.

6.3.2 Local Front Update

The variation of the volume of the tip elements over a time step reflects the movement of the crack front as well as the changing nature of the tip asymptote driven by the local front velocity. The iterative procedure devised to update the vertex R and the adjacent fill vertex D is outlined in Figure 6.3, by considering again the group of four tip elements associated with edge channel element i. The tip element volumes V_l, $l = 0, \ldots, 3$ are calculated on the assumption that the tip elements and the fixed parent element are inside the tip asymptotic umbrella.

With the position of vertex R updated at a new iteration, the footprint of offset element $i0$ is altered and thus the volume of this element can be calculated, in accordance with the tip asymptote appropriate to the updated front velocity. The front velocity at the fill vertex D is calculated as the average of the front velocities at the neighboring offset vertices R and Q. Finally, the volumes of the tip elements are

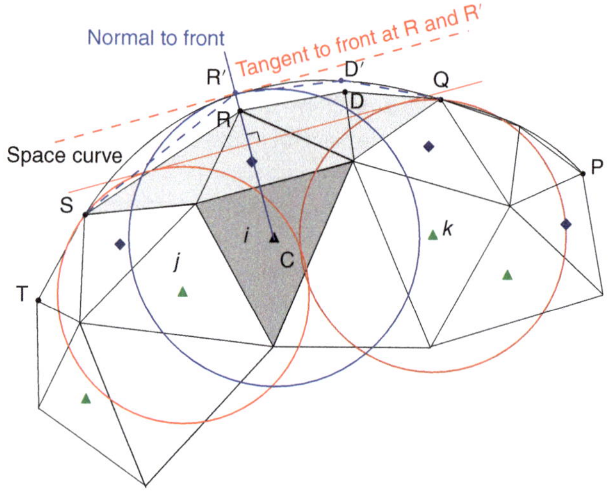

Figure 6.3 Front update. The line PQDRST corresponds to the current crack front. The vertex R is adjusted to a new position R′ during the current time step. This position is determined by the revised opening width at the fixed element center C and is positioned along the direction of a local estimate of the crack front normal vector. The crack edge velocity is estimated using the incremental advance distance RR′ and the time step size. The revised vertex position R′ is used to construct an estimate of the adjacent fill vertex movement from D to D′ using a cubic space curve interpolation passing through vertices P, Q, R′, and S. The centers of the fixed triangles adjacent to the current edge are depicted using triangular-shaped symbols and the centers of the moving edge elements are depicted using diamond-shaped symbols.

converted into constant aperture for these elements, which is used in the elasticity eq. (6.1).

6.3.3 Generation of a New Ring of Tip Elements

To avoid untoward distortion of the fringe region caused by the continued advance of the crack front, a decision must eventually be made to promote the tip elements to the status of channel elements and to create a new ring of tip elements. In the implementation of the moving edge scheme, the renewal procedure is triggered at the end of a completed time step iteration if the advance distance of any offset vertex on the crack front exceeds a specified maximum value, A_{max}. The advance distance of point R in Figure 6.3, for example, is measured as the difference $D_2 - D_1$, where D_2 is the distance between the center C of the adjacent permanent element and point R and D_1 is the perpendicular distance between the center C and the adjacent permanent element edge. The advance distance of each offset vertex is checked at the end of each time step and is compared to the specified maximum advance distance parameter, A_{max}.

The edge renewal process is triggered if any offset vertex distance is found to exceed the specified maximum advance distance. The edge renewal procedure comprises the following steps: (i) generate permanent I and E-type element entries corresponding to each offset and fill element in the temporary tip element region, (ii) construct a new representation of the offset edge data structure and minimum edge offset distances to reflect the new population of E-type elements, and (iii) set initial aperture and pressure values in the new I and E elements. The standard time step iteration procedure is resumed once the new set of permanent elements has been generated, using the new set of E-type elements to control the replaced moving crack edge.

6.3.4 Crack Surface Remeshing

One notable feature of the renewal process is that a range of sizes of elements can be generated around the crack perimeter depending on the local variations in the edge advance velocity. In the case of axisymmetric problems these advances are uniform but may vary when considering problems such as buoyancy-driven dyke formation, heavy fluid sequestration or the propagation of hydraulic fractures which are bounded vertically by stress or toughness barriers. An undesirable consequence of the basic renewal algorithm in these cases is the generation of large numbers of small elements in the regions where the edge advance velocity is relatively low. This can lead to unnecessarily arduous computational costs and may contribute to poor solution iterative stability.

One strategy to address this problem is to allow for remeshing of the crack surface as an option during the edge renewal process. This possibility has been implemented in the unstructured mesh solution algorithm. The remeshing procedure is initiated if any of the offset vertex advance distances are found to fall below a specified minimum advance distance parameter, $A_{\min}$.

The remeshing proceeds by identifying a replacement region adjacent to the new crack edge perimeter and offset from the perimeter by a lower bound distance set by $A_{\max}$. All permanent elements that fall within or overlap the designated border edge region offset distance, together with the temporary element configuration define the region to be re-meshed. All permanent elements falling outside the identified border edge region are retained. New vertex positions are generated within the border edge region between the outer offset vertices and fill element crack edge segments and the inner fixed element region. The vertex insertion is carried out by scanning the crack edge segments and determining the offset distance from the crack edge to the inner fixed element boundary in the local inward normal direction to the crack edge tangent. Care is taken to space the vertices approximately according to the specified $A_{\max}$ distance measure. Once the new vertices have been assigned, a new element mesh is generated from the inner fixed element boundary toward the outer crack edge border using a Delaunay triangulation procedure. The Delaunay algorithm is a standard device in automated mesh generation routines ensuring that the generated triangles are as "equiangular" as possible. The algorithm as implemented has the special feature of allowing the mesh to be generated, if necessary, on a nonplanar surface to facilitate future implementation of curvilinear crack surface propagation.

Once the new element mesh is constructed it is necessary to transfer all key flow variables from the old mesh to the new mesh. These variables include the flow width values at the centers of the new elements and the crack edge velocities on the outer crack border. The flow width values are determined in two stages. In the first stage all new elements having one or two vertices on the crack edge are identified. The element opening volume and average width are calculated using the crack edge velocity estimated from the most recent offset vertex configuration at the start of the edge element renewal procedure together with the assigned asymptotic crack tip shape. Volumes and average widths are calculated using the procedure described in Section 6.2.4. In the second stage, flow volumes and opening width values in the remaining newly assigned border region elements are inferred using a least squares estimation of the flow width distribution in the vicinity of each new element center. The least squares estimate is based on the values of all existing (old) flow element width values and the new element widths that are assigned in the stage one process. The values are selected within a specified radius from the center of the new element. The metric for the critical selection distance is set as a multiple of the square root of the mean area of all the newly assigned elements. (This ratio is chosen to have a value of 2.5 which empirically appears to yield robust local estimates of the width values.) For example, in a regular equilateral triangle mesh where the edge length is g the mean triangle size would be $\bar{g} = 3^{1/4}(g/2)$ and the selection radius is $\sim 1.645g$. The width estimation function is assumed to be a bi-quadratic polynomial function. The bi-quadratic function requires at least six surrounding points to be available. A final adjustment is made to the estimated width values in the new border region elements to ensure that the total fluid volume in the re-meshed region is equal to the existing fluid volume in this region. The remeshing procedure is illustrated in the section on numerical simulations.

6.3.5 General Solution Algorithm Logic

The general logic for the solution algorithm steps can be summarized as follows:

1) Advance the flow simulation time step counter. Exit if the simulation is complete.

2) Check the current temporary edge element region crack tip advance distances. Replace the temporary edge elements with fixed crack elements if the advance distance exceeds the specified maximum value. (Alternatively, generate a new border region element mesh if re-meshing is required.)
3) Perform the next iteration loop to solve the current element configuration flow state.
4) Process each fixed flow element and calculate the elastic stress influence at the element center induced from all external fixed elements and temporary edge elements.
5) If the current element is internally located (i.e. is not adjacent to a temporary moving edge element), calculate the equilibrium flow balance and update the element opening width. Calculate the adjusted solution error. If the current element has an adjacent moving edge offset element, search for the equilibrium asymptotic crack tip position of the moving edge element. Calculate the local fringe region fill element configuration and calculate the fixed border element solution error.
6) Return to step 4 if the fixed element scan is not completed.
7) When the element scan is completed, review the maximum recorded element solution error. Return to step 3 if the specified error tolerance is not satisfied.
8) Return to step 1 when the specified error tolerance is satisfied in all elements.

6.4 Numerical Simulations: Stress Barriers

6.4.1 Description of Experiment

To illustrate this study, we consider the propagation of a hydraulic fracture across two symmetric stress barriers, see Figure 6.4a. In the stimulation treatments of deep hydrocarbons reservoirs, the dominant mechanisms for containing the growth of the fracture to the pay zone is the stress contrast between the hydrocarbon-bearing permeable layer (e.g. sandstone) and the adjacent quasi-impermeable layers of rock (typically shales), the higher stress in the cap rocks resulting in part from the time-dependent deformation that tend to decrease the horizontal deviatoric stress. However, the numerical simulations reported here are done in the context of a laboratory experiment [24], in which fluid was injected into the interface between two transparent blocks of polymethylmethacrylate (PMMA) in forced contact. The surface of the mating blocks has been machined in such a way to create a stress contrast between the middle zone and the adjacent ones, when the PMMA blocks are pressed against each other.

The parameters characterizing this laboratory experiment are [24]: viscosity $\mu = 30.2$ Pa·s ($\mu' = 362$ Pa·s), Young modulus $E = 3300$ MPa, Poisson ratio $\nu = 0.4$ ($E' = 3930$ MPa), far-field normal stress in the pay-zone (middle layer) $\sigma_o = 2.2$ MPa, far-field normal stress in the barrier (adjacent layers) $\sigma_o + \Delta\sigma_o = 6.5$ MPa, and height of pay zone $H = 50$ mm. The fluid was injected for about 600 s at a nominal rate $Q_o = 1.7$ mm^3/s. However, because of compressibility effect, the actual injection rate was not constant and previous efforts to simulate the experiments have adopted the pumping schedule $Q_o = 2$ mm^3/s for $0 < t < 243$ s, and $Q_o = 1.3$ mm^3/s thereafter [9, 24]. Since the experiment involves the opening of a preexisting interface, toughness $K_{Ic} = 0$ and the fracture propagates strictly in the viscosity-dominated regime.

6.4.2 Numerical Simulations (no Remeshing)

The numerical simulation was carried out using 250 time steps with a step size $\Delta t = 1$ s; for simplicity, the injection rate was maintained at the constant value $Q_o = 2$ mm^3/s during the 250 s of simulation. To eliminate numerical

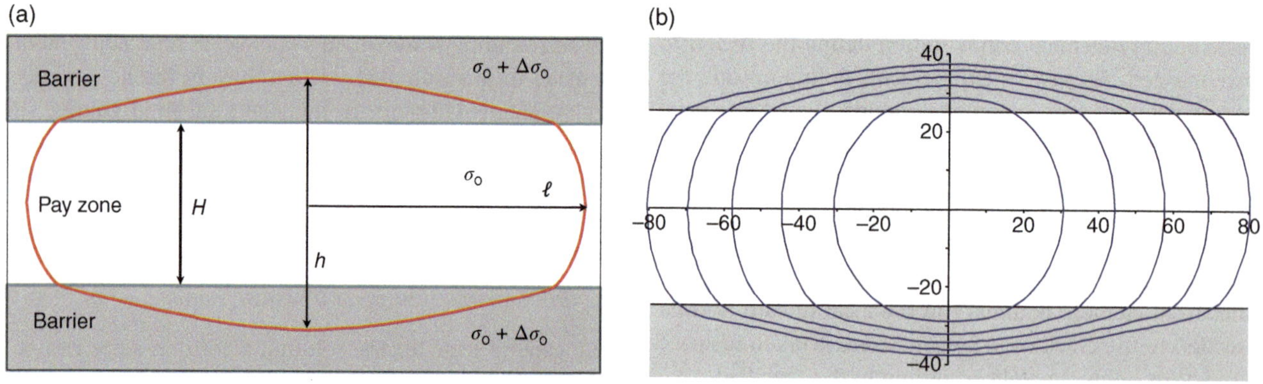

Figure 6.4 (a) Problem Description (*Source:* [9]/With permission of Elsevier). (b) Footprint at $t = 50, \ldots, 250$ s at intervals of 50 s computed for $K' = 0$ (length units in mm).

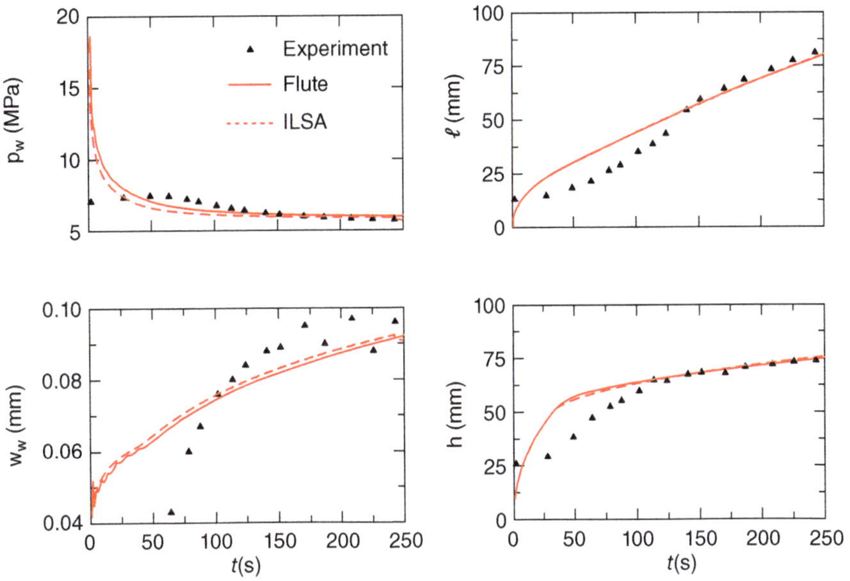

Figure 6.5 Comparison between experimental data (black triangles) and simulation results obtained with ILSA (dashed curves) and with Flute (continuous curves). (*Source:* Adapted from [9, 24]).

instabilities caused by the discontinuity in the far-field stress, the stress was assumed to vary linearly in 4 mm thick transition zone straddling the stress discontinuity interface. (This transition zone is represented in dark gray in Figure 6.4.) The starter crack located half-way between the two barriers is based on a 13 elements configuration. Here the starter crack has a nominal radius of 3 mm, with a 1 mm side length for the central square element. The injection is spread across the central square element of the starter crack. Plots of the position of the crack front, calculated at 50 s intervals until $t = 250$ s, can be found in Figure 6.4b.

6.4.3 Comparison with Experimental Results and Other Simulations

To verify this set of simulations on the propagation of a hydraulic fracture in the presence of stress barriers, we compare the simulation conducted with $K' = 0$ with experimental data and published numerical results obtained using an implicit level set algorithm (ILSA) [9]. This algorithm operates on a fixed regular grid and is also based on a combination of displacement discontinuity and finite volume methods [2, 9].

All these results are compiled in Figure 6.5, which illustrates the evolution of injection pressure p_w, inlet-fracture aperture w_w, half-length ℓ and height h (both defined in Figure 6.1a) during the first experiment [24], the continuous curves the predictions of Flute, and the dashed curves the results of ILSA [9]. There is good agreement between the zero-toughness numerical predictions and the experimental data when $t \gtrsim 125$; the poor match between experiment and simulation in the early phase of the experiment has been explained by fluid compressibility effect, which is responsible for an initial variable injection rate [24]. Furthermore, the numerical predictions of Flute and ILSA are in excellent agreement; in particular the graphs of $\ell(t)$ obtained by both algorithms cannot be distinguished at the scale of the plot shown in Figure 6.5. The most visible difference between the two sets of simulations is in the evolution of the borehole pressure. The injection pressure, which is calculated at the center of the element acting as a fluid source is sensitive, however, to the size of this element in view of the logarithmic variation of the pressure in the vicinity of the injection region.

6.4.4 Illustration and Assessment of the Element Re-Meshing Strategy

To illustrate the performance of the element remeshing strategy as part of the periodic edge renewal process, consider the zero toughness case grid configuration illustrated in Figure 6.6 at time 150 s. The base case (no remeshing), shown in Figure 6.3a corresponds to the maximum and minimum advance parameters given by $A_{\max} = 3.2$ and $A_{\min} = 0$ mm. This simulation was repeated with ($A_{\max}$, $A_{\min}$) equal to (i) (3.2 mm, 1.6 mm), (ii) (4.8 mm, 2.4 mm), and (iii) (6.4 mm, 3.2 mm). The choice $A_{\min} = A_{\max}/2$ appears to provide a reasonable threshold to trigger the remeshing algorithm. Figure 6.6b and c illustrate the resulting element grid configurations for case (i) and case (iii), respectively at time 150 s and evidently replicate the outer border of Figure 6.6a fairly closely. Figure 6.6d shows the comparison between all three remeshed cases and the border region when no remeshing occurs at time 300 s showing good agreement. The most important

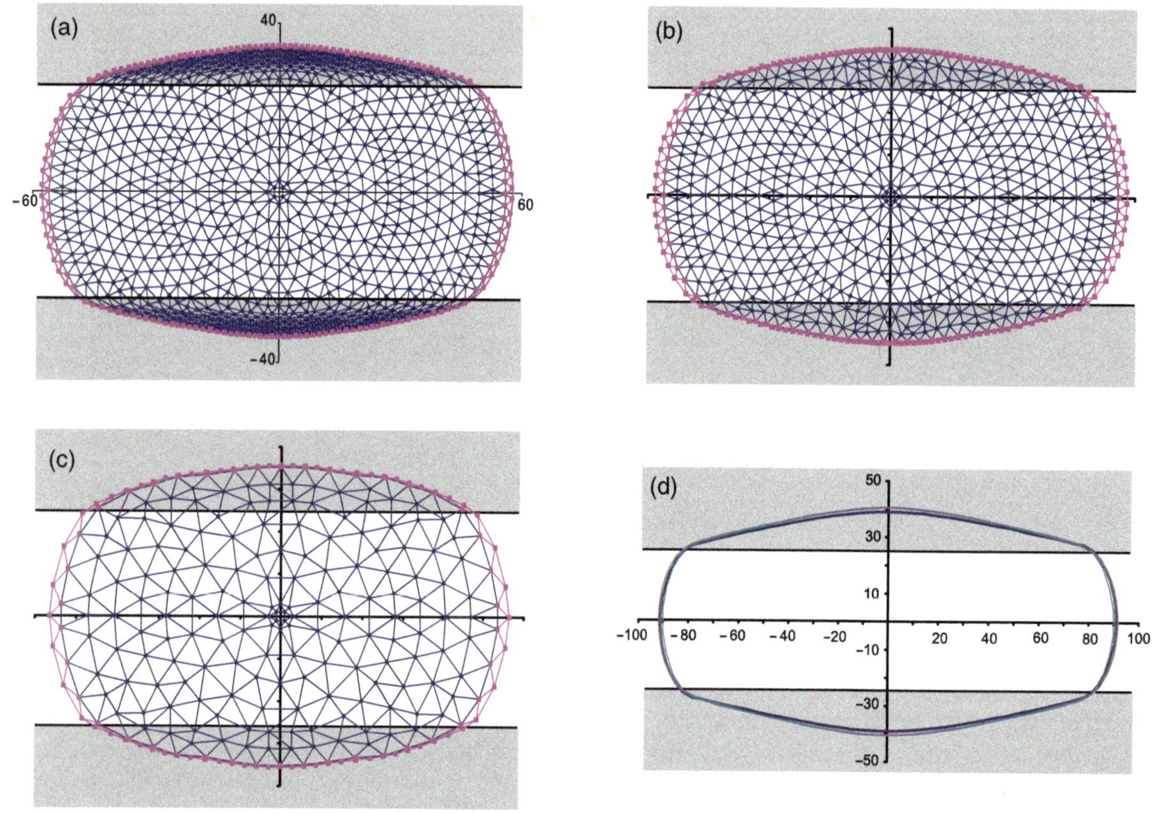

Figure 6.6 Comparison between different computational meshes obtained at time t = 150 s for K' = 0 (a–c). Without remeshing and A_{max} = 3.2 mm (a). Periodic grid remeshing with A_{max} = 3.2 mm and A_{min} = 1.6 mm (b); and with A_{max} = 6.4 mm and A_{min} = 3.2 mm (c). Crack footprint comparison at time 300 s with remeshing and different maximum advance distances (d). All length units in mm.

consequence is that the simulation computation time is reduced from a normalized value of 1000 for the base case to 155.8, 30.6, and 9.8 for cases (i), (ii), and (iii), respectively. The number of fixed elements in the base case is 3933. This number is reduced to 2181, 1039, and 607 for cases (i), (ii) and (iii) respectively.

Figure 6.7 compares the time evolution of the maximum horizontal and maximum vertical crack edge half-lengths indicating good agreement for each mesh density although it is apparent that in the coarsest case (c) the horizontal distance is slightly under-estimated and that in all cases the vertical edge advance is slightly overestimated when compared to the base case.

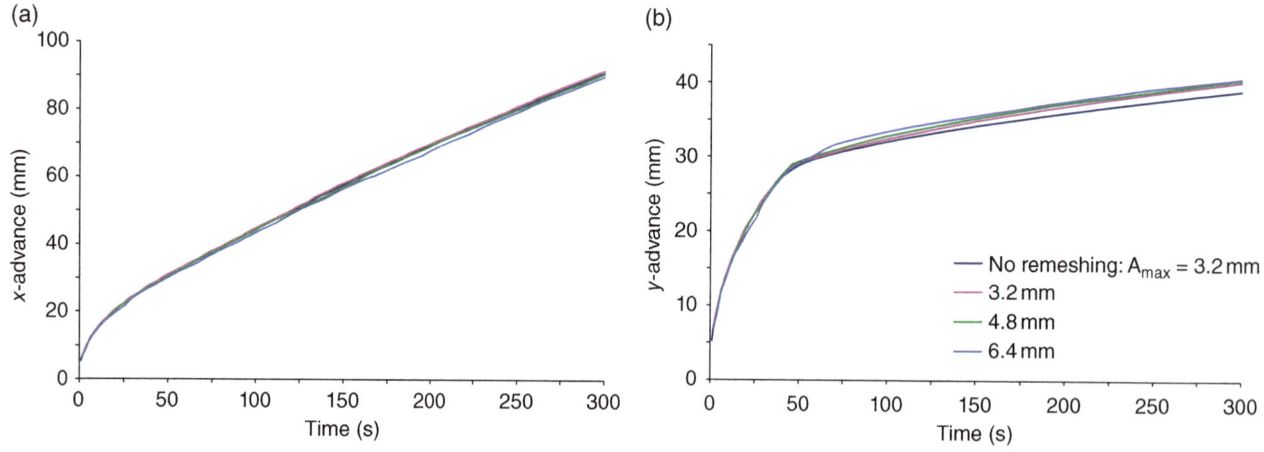

Figure 6.7 Comparison of maximum horizontal (a) and vertical (b) advance of the crack edge from the central injection point with remeshing and different values of the maximum edge renewal distance parameter A_{max}.

6.5 Conclusions

A novel computational scheme has been developed to solve moving boundary problems associated with the propagation of fluid-driven fractures in an impermeable elastic medium. The crack surface is represented by an unstructured triangular element tessellation. The elastic interactions between the crack opening displacement and the surrounding medium are calculated using the displacement discontinuity boundary element method with constant opening width values in each element.

The unstructured triangular element mesh provides a flexible representation of general crack footprint configurations and provides a basic mesh growth strategy to enable the future simulation of nonplanar fracture surface geometries.

A special feature of the moving crack edge scheme is to include a region of adjustable tip elements adjacent to the crack edge. The moving vertices of the tip elements are fully coupled to the iterative solution algorithm to accommodate both the edge location and the appropriate fluid volume balance during each solution time step cycle. Special care is taken to include appropriate asymptotic expressions for the characteristic crack opening shapes in the vicinity of the crack edge to determine the correct crack edge velocity for viscous dominated and for intermediate viscous-toughness asymptotic regimes.

The moving edge scheme includes the periodic replacement of the temporary fringe element region with a fixed element mesh. An element remeshing procedure can be triggered as well if there is a significant variation in the crack edge advance distance along the crack front. The element remeshing operation is carried out within a specified border region adjacent to the crack edge and yields a more uniformly sized element mesh with attendant benefits in overall computational efficiency.

The approach is shown to concur with experimental results of planar fracture propagation in a strip-like "pay zone" region that is bounded by two parallel stress barriers. The performance of the scheme is both accurate and robust to a range of choices of time-step size and mesh resolution.

It is planned to extend the proposed moving edge technique to nonplanar fracture propagation and, if possible, to investigate fracture intersection problems. A major future challenge is the simulation of crack edge disintegration in mixed mode propagation and the treatment of closely parallel fracture interactions.

Acknowledgments

JN would like to acknowledge with much gratitude the ongoing support received from the MTS visiting Professorship in Geomechanics at the University of Minnesota. The MTS Professorship has enabled the current research to be completed during several fruitful technical visits to the University of Minnesota, which have facilitated the development of the numerical scheme that is described in the paper. Partial support of ED was provided by the T.W. Bennett Chair in Mining Engineering and Rock Mechanics. This support is also gratefully acknowledged.

References

1 Peirce, A.P. (2016). Implicit level set algorithms for modelling hydraulic fracture propagation. *Philosophical Transactions of the Royal Society A: Mathematical, Physical and Engineering Sciences* 374 (2078): https://doi.org/10.1098/rsta.2015.0423.

2 Peirce, A.P. and Detournay, E. (2008). An implicit level set method for modeling hydraulically driven fractures. *Computer Methods in Applied Mechanics and Engineering* 197: 2858–2885. https://doi.org/10.1016/j.cma.2008.01.013.

3 Damjanac, B., Detournay, C., and Cundall, P.A. (2016). Application of particle and lattice codes to simulation of hydraulic fracturing. *Computational Particle Mechanics* 3 (2): 249–261. https://doi.org/10.1007/s40571-015-0085-0.

4 Salimzadeh, S., Usui, T., Paluszny, A., and Zimmerman, R.W. (2017). Finite element simulations of interactions between multiple hydraulic fractures in a poroelastic rock. *International Journal of Rock Mechanics and Mining Sciences* 99: 9–20. https://doi.org/10.1016/j.ijrmms.2017.09.001.

5 Gupta, P. and Duarte, C.A. (2016). Coupled formulation and algorithms for the simulation of non-planar three-dimensional hydraulic fractures using the generalized finite element method. *International Journal for Numerical and Analytical Methods in Geomechanics* 40 (10): 1402–1437. https://doi.org/10.1002/nag.2485.

6 Gupta, P. and Duarte, C.A. (2018). Coupled hydromechanical-fracture simulations of nonplanar three-dimensional hydraulic fracture propagation. *International Journal for Numerical and Analytical Methods in Geomechanics* 42 (1): 143–180. https://doi.org/10.1002/nag.2719.

7 Crouch, S.L. and Starfield, A.M. (1983). *Boundary Element Methods in Solid Mechanics*. London: George Allen and Unwin.

8 Mathur, S.R. and Murthy, J.Y. (1997). A pressure-based method for unstructured meshes. *Numerical Heat Transfer, Part B: Fundamentals* 31 (2): 195–215. https://doi.org/10.1080/10407799708915105.

9 Peirce, A.P. (2015). Modeling multi-scale processes in hydraulic fracture propagation using the implicit level set algorithm. *Computer Methods in Applied Mechanics and Engineering* 283: 881–908. https://doi.org/10.1016/j.cma.2014.08.024.

10 Dontsov, E.V. and Peirce, A.P. (2017). A multiscale implicit level set algorithm (ilsa) to model hydraulic fracture propagation incorporating combined viscous, toughness, and leak-off asymptotics. *Computer Methods in Applied Mechanics and Engineering* 313: 53–84. https://doi.org/10.1016/j.cma.2016.09.017.

11 Zia, H. and Lecampion, B. (2019). Explicit versus implicit front advancing schemes for the simulation of hydraulic fracture growth. *International Journal for Numerical and Analytical Methods in Geomechanics* 43 (6): 1300–1315. https://doi.org/10.1002/nag.2898.

12 Zia, H., Lecampion, B., and Zhang, W. (2018). Impact of the anisotropy of fracture toughness on the propagation of planar 3d hydraulic fracture. *International Journal of Fracture* 211 (1): 103–123. https://doi.org/10.1007/s10704-018-0278-7.

13 Detournay, E. (2016). Mechanics of hydraulic fractures. *Annual Review of Fluid Mechanics* 48: 311–339. https://doi.org/10.1146/annurev-fluid-010814-014736.

14 Garagash, D. (2009). Scaling of physical processes in fluid-driven fracture: perspective from the tip. In: *IUTAM Symposium on Scaling in Solid Mechanics, IUTAM Bookseries*, vol. 10, 91–100. Dordrecht: Springer https://doi.org/10.1007/978-1-4020-9033-29.

15 Napier, J. and Detournay, E. (2020). An unstructured mesh algorithm for simulation of hydraulic fracture. *Journal of Computational Physics* 419 (109): 691. https://doi.org/10.1016/j.jcp.2020.109691.

16 Yoffe, E.H. (1960). The angular dislocation. *Philosophical Magazine: A Journal of Theoretical Experimental and Applied Physics* 5 (50): 161–175. https://doi.org/10.1080/14786436008243299.

17 Jeyakumaran, M., Rudnicki, J.W., and Keer, L.M. (1992). Modeling of slip zones with triangular dislocation elements. *Bulletin of the Seismological Society of America* 82 (5): 2153–2169.

18 Kuriyama, K. and Mizuta, Y. (1993). Three-dimensional elastic analysis by the displacement discontinuity method with boundary division into triangular leaf elements. *International Journal of Rock Mechanics and Mining Sciences & Geomechanics Abstracts* 30 (2): 111–123. https://doi.org/10.1016/0148-9062(93)90704-H.

19 Meade, B.J. (2007). Algorithms for the calculation of exact displacements, strains, and stresses for triangular dislocation elements in a uniform elastic half space. *Computers & Geosciences* 33 (8): 1064–1075. https://doi.org/10.1016/j.cageo.2006.12.003.

20 Nikkhoo, M. and Walter, T.R. (2015). Triangular dislocation: an analytical, artefact-free solution. *Geophysical Journal International* 201 (2): 1119–1141.

21 Nikolskiy, D.V., Mogilevskaya, S.G., and Labuz, J.F. (2015). Boundary element analysis of non-planar three-dimensional cracks using complex variables. *International Journal of Rock Mechanics and Mining Sciences* 76: 44–54. https://doi.org/10.1016/j.ijrmms.2015.02.007.

22 Garagash, D.I., Detournay, E., and Adachi, J.I. (2011). Multiscale tip asymptotics in hydraulic fracture with leak-off. *Journal of Fluid Mechanics* 669: 260–297. https://doi.org/10.1017/S002211201000501X.

23 Dontsov, E.V. and Peirce, A.P. (2015). A non-singular integral equation formulation to analyse multiscale behaviour in semi-infinite hydraulic fractures. *Journal of Fluid Mechanics* 781: https://doi.org/10.1017/jfm.2015.451.

24 Jeffrey, R.G. and Bunger, A.P. (2007). A detailed comparison of experimental and numerical data on hydraulic fracture height growth through stress contrast. *Society of Petroleum Engineers Journal* 14 (3): 413–422.

7

Study of Hydraulic Fracture Interference with a Lattice Model

C. Detournay[1], B. Damjanac[1], M. Torres[1], and Y. Han[2]

[1] Itasca Consulting Group, Minneapolis, MN, USA
[2] Aramco Services Company: Aramco Research Center, Houston, TX, USA

7.1 Introduction

The manifestation of fracture interference is ubiquitous in nature. A walk in a rocky countryside anywhere in the world offers fascinating examples of fractures on a straight or curving path, crossing or being arrested by each other. When fractures are engineered for specific purposes, such as in hydraulic fracturing for oil and gas stimulation or preconditioning of orebodies in cave mining operations, it is of interest to study the types of interactions and clarify the underlying mechanisms to enhance the safety and efficiency of the operations. Researchers have performed both laboratory and large-scale experiments to study the topic of fracture interference, and many examples of their findings are included in the literature and in this book chapter [1]. Numerical modeling, based on proper physics, is an alternative investigation tool that offers the advantages of unrestricted experimentation and repeatability. The qualitative aspect of the findings may uncover surprising results that demand experimental confirmation and may lead to increased understanding.

This chapter gives a short inventory of fracture interference cases that were observed using the numerical simulation code *XSite*. A brief outline of the code logic and capabilities is provided in Section 7.2. Five cases of fracture interference, in increasing order of complexity, are described in Section 7.3. A summary of observational findings concludes the chapter in Section 7.4.

7.2 *XSite* Code Overview

The code *XSite* [2] is a three-dimensional numerical simulator developed by Itasca Consulting Group [3–5] to simulate the propagation of hydraulic fractures in conventional and naturally fractured reservoirs. The code uses a lattice of springs and masses to represent a brittle material that may develop macroscopic fractures that consist of chains of spring breakages, each of which is regarded as a microfracture. Each broken spring has a fluid pressure node associated with it, and neighboring nodes are linked by "pipes" that conduct fluid as the simulation evolves. Mechanical and flow models are fully coupled: fracture permeability depends on aperture via deformation of the mechanical model, fluid pressure affects deformation and strength of the solid model, and deformation of the solid model affects fluid pressures. By injecting fluid into a "cluster" of nodes, a hydraulic fracture may propagate in the lattice in a direction that is determined by the local stress tensor (which is related to the forces carried by the lattice springs). The code allows fluid injection/production from multiple boreholes and multiple clusters, as well as preexisting joints derived from the user-specified discrete fracture network (DFN). In this case, hydraulic fracture propagation is modeled as intact-rock failure in tension and slip and opening on joints.

Additional code features include proppant transport, nonsteady matrix flow, Newtonian or power-law fluid, arbitrary initial stress by layer, generation of microseismicity data associated with the formation of microcracks and slipping on preexisting joints, and heat propagation and associated rock deformation and damage.

7.3 Numerical Studies of Fracture Interference

As a preamble, we note that hydraulic fractures follow the principle of minimal energy: they tend to propagate in the direction normal to the local minimum compressive stress

Mechanics of Hydraulic Fracturing: Experiment, Model, and Monitoring, First Edition.
Edited by Xi Zhang, Bisheng Wu, Diansen Yang, and Andrew Bunger.

in virgin rock and to take easy open paths in initially fractured media. Manifestations of hydraulic fracture interference with other fractures can be understood by appealing to the stress shadow effect. The stress shadow is the augmented compressive stress in the rock surrounding a fracture caused by the pressure exerted by the fluid in the fracture or by the residual fracture aperture when it closes on the proppant. The stress shadow modifies the virgin stress state and thereby impacts the direction and amount of new or simultaneous fracture growth.

Five numerical simulations of fracture interference are described in increasing order of complexity in the following sections: (i) the interaction of a single hydraulic fracture with a preexisting joint, (ii) the attraction/deflection of two sequential hydraulic fractures, (iii) the interaction of hydraulic fractures arising from borehole injection into multiple clusters, (iv) the interference caused by a preexisting fracture network on the propagation of hydraulic fractures from multiple clusters along a horizontal well, and (v) hydraulic fracture interference in a two-well zipper-stage treatment.

7.3.1 Interaction of a Hydraulic Fracture with a Natural Fracture

The importance of three-dimensional analysis of hydraulic fracturing is illustrated in the example simulation of the interaction between a hydraulic fracture and a preexisting natural fracture [6, 7]. In this example, the propagation of a horizontal hydraulic fracture is simulated (the vertical stress is the minimum principal stress) by injection of fluid in the vicinity of a vertical penny-shaped preexisting fracture in an otherwise homogeneous rock. Three states of the fracture geometry in plan-view are shown in Figure 7.1. The hydraulic fracture starts propagating as a penny-shaped fracture. When the hydraulic fracture reaches the preexisting vertical fracture (Figure 7.1a), its propagation in that direction is apparently arrested. Subsequently, the hydraulic fracture continues to propagate on both sides around the preexisting fracture (Figure 7.1b) until the two sides merge, completely enclosing the preexisting fracture (Figure 7.1c). The side-view of the fractures is shown in Figure 7.2. In this view, which corresponds to the most reasonable plane of a 2D model for this problem, it appears that the hydraulic fracture has propagated through the preexisting fracture. As the 3D model shows, the actual mechanism of interaction is very different.

7.3.2 Interaction of Two Hydraulic Fractures

Preconditioning of orebodies by hydraulic fracturing is a technique used to increase the efficiency of cave mining operations. Typically, arrays of hydraulic fractures are generated from boreholes drilled sub-parallel to the minimum stress direction. This borehole orientation promotes fracture growth transverse to the borehole axis. An issue of practical importance is to determine how closely hydraulic fractures can be spaced to allow the most effective preconditioning of a volume of rock mass treated per borehole. However, the interaction between closely spaced hydraulic fractures can lead to deflection of their path. Bunger et al. [8] have derived a flow chart for determining if a hydraulic fracture in plane strain satisfies sufficient conditions for neglecting curving. However, the chart does not cover the 3D case considered here. Numerical simulations with *XSite* show evidence of either attractive or repulsive 3D interaction between a new hydraulic fracture and a previously placed one, depending on their spacing [9]. The reason for the behavior is explored in a simple framework using the results of *FLAC* simulations [10].

7.3.2.1 Numerical Study

The simulations consist of propagating sequentially two horizontal hydraulic fractures located at a distance, *d*, apart along a vertical borehole. The generation of the first (bottom) fracture is followed with shut-in and fluid pressure dissipation, leaving behind locked-in stresses in the rock mass due to incomplete fracture closure. Different values

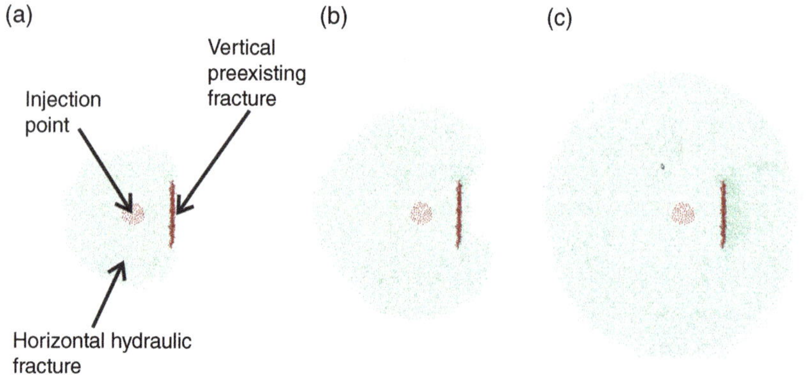

Figure 7.1 Three occasions of interaction between a horizontal hydraulic fracture with a vertical preexisting natural fracture shown in plan-view. (a) 200 s, (b) 530 s, and (c) 1198 s.

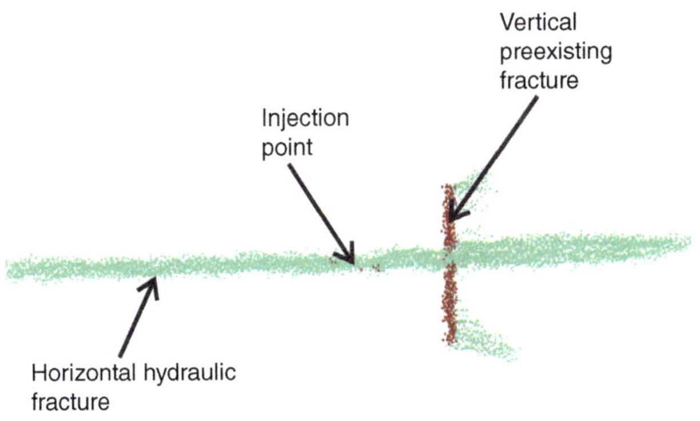

Figure 7.2 Side-view of the interaction between a horizontal hydraulic fracture with a vertical preexisting natural fracture after 1500 s of injection.

of the distance between injection points are considered, and the path of the second fracture is analyzed to study the impact of residual stresses on fracture propagation.

The *XSite* model in Figure 7.3 shows the borehole, the injection points (pictured as spheres), and the resolution domains used in the simulation.

The rock mass is assumed to be elastic and of negligible permeability, and the fracturing fluid is water. The first (bottom) fracture, induced after 100 s of injection, is penny-shaped. Vertical side views of the fractures generated at the end of the second episode of injection are presented for three different spacings in Figures 7.4–7.6.

Spacing of 2.5 m – In Figure 7.4, the second hydraulic fracture (top location) is seen to curve toward the first in the direction of the intermediate principal in situ stress (green axis on the plot). It has a semicylindrical shape, elongated in the direction of the maximum principal stress (red axis in the plot).

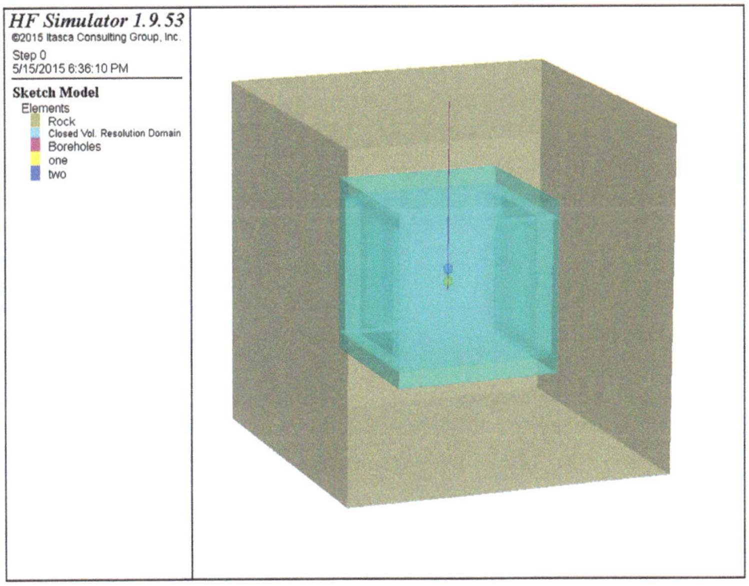

Figure 7.3 *XSite* model showing borehole, injection points, and fine-resolution domains.

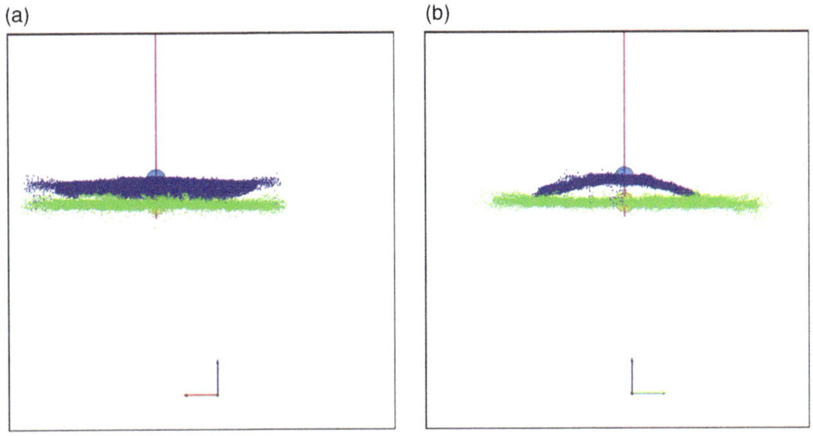

Figure 7.4 Side views of the simulated fractures at the end of the simulation, parallel (a) and perpendicular (b) to the maximum in situ stress – 2.5 m spacing.

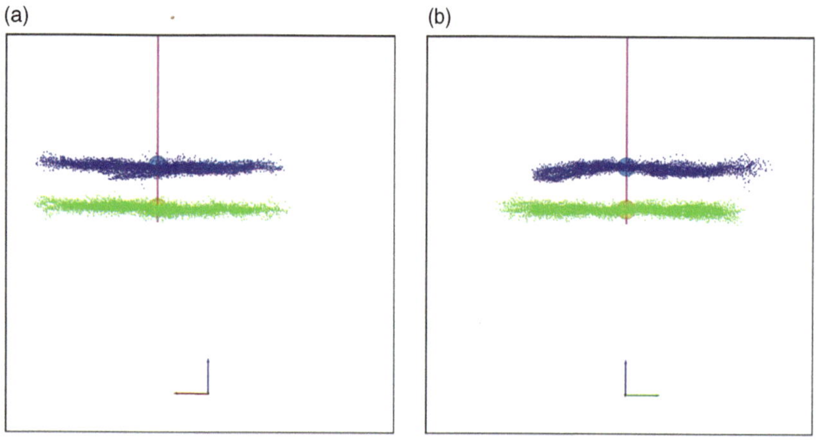

Figure 7.5 Side views of the simulated fractures at the end of the simulation, parallel (a) and perpendicular (b) to the maximum in situ stress – 4.5 m spacing.

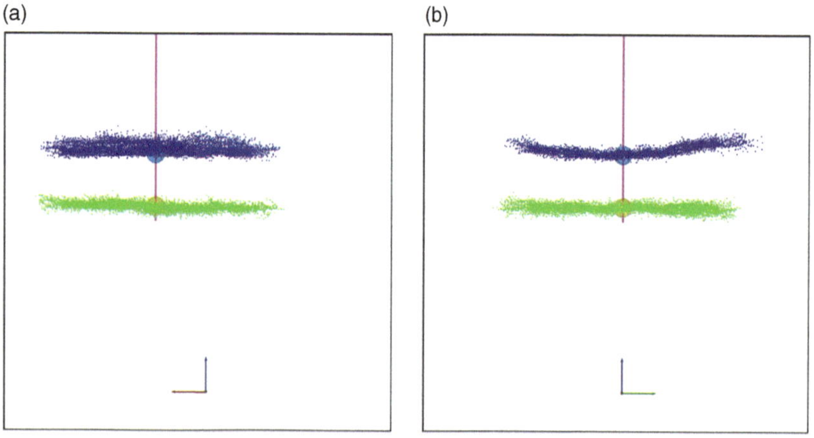

Figure 7.6 Side views of the simulated fractures at the end of the simulation, parallel (a) and perpendicular (b) to the maximum in situ stress – 5.5 m spacing.

Spacing of 4.5 m – In Figure 7.5, the fracture interaction appears to be weaker than in the 2.5 m spacing case. The second fracture is approximatively penny-shaped with a slight dip toward the first fracture.

Spacing of 5.5 m – In Figure 7.6, the fracture interaction is still evident. However, there is a shift in behavior in this case. Indeed, the second fracture is propagating away from the first.

7.3.2.2 Interpretation of Results

The simulations show the attraction or repulsion of the second fracture from the first one. The change of path direction of a second fracture as a function of distance from the first is somewhat counterintuitive. To analyze the situation in simple terms, we look at the equilibrium state of a penny-shaped crack of radius 50 m loaded by a 1 MPa pressure, and a shorter penny-shaped crack above it, loaded to 2 MPa, see Figure 7.7. We use the code *FLAC* in axisymmetric mode for the simulations.

We plot the angle of the maximum principal stress (PS) to the horizontal in the zone at the tip of the second crack as a

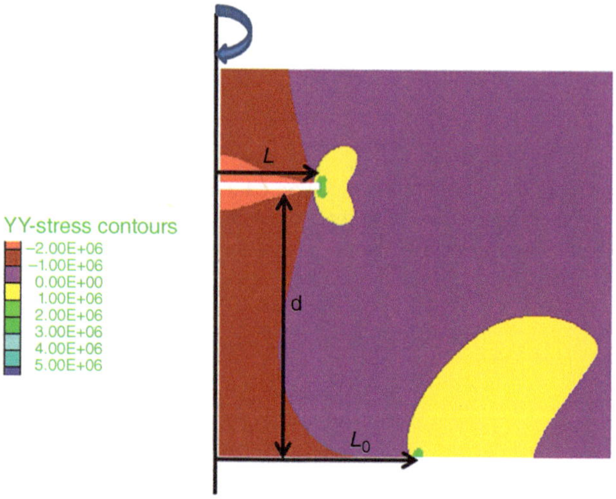

Figure 7.7 2D *FLAC* axisymmetric elastic model: L_0 and L denote two pressurized slots at a distance of d apart. The contours are of the vertical stress component (compression negative).

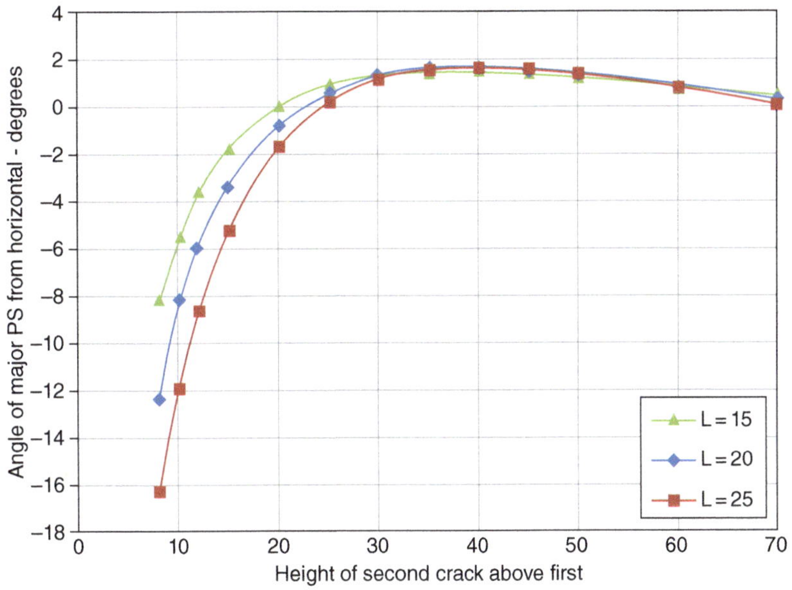

Figure 7.8 Angle of major principal stress at second crack tip versus height of second crack above the first for three lengths of the second crack [m] – $\sigma_h = 3$ MPa.

function of the distance of the second crack from the first – this should be the angle that a new crack will tend to follow (perpendicular to minor principal stress). The results for the two pressurized crack problems with second crack lengths, *L*, of 15, 20, and 25 m and a horizontal stress of 3 MPa are shown in Figure 7.8.

All graphs in Figure 7.8 show a transition from tending downwards to upwards for the conditions of the *FLAC* simulations. This trend is similar to that observed in the HF simulation results documented above. Note that the *FLAC* simulations are greatly simplified and do not consider fracture evolution. As mentioned above, the effect of fracture distance on different modes of their interaction (i.e. curving toward or away from the previous fractures) has been investigated by Bunger et al. [8, 11]. Using a plane strain (2D) model, they identified the same trends observed in these simulations.

On the other hand, the fracture interaction resulting in curving of the fracture in the direction of the intermediate principal stress (i.e. the smaller horizontal principal stress) could be a possible explanation for the elongated shape of fractures at the Cadia East Trial site, approximately two times longer in the direction of the larger horizontal principal stress than in the perpendicular direction [12].

7.3.3 Interaction of Hydraulic Fractures in Injection of Multiple Clusters

This example studies the interaction between hydraulic fractures generated from multiple clusters in a horizontal well. The *Xsite* model in Figure 7.9 contains two layers: the bottom layer is about 35 m below the borehole, and its toughness is much larger than that of the top layer. Fluid injection in the borehole is modeled (with no addition of proppant) at a rate of 0.0025 m^3/s for a total duration of 1000 s.

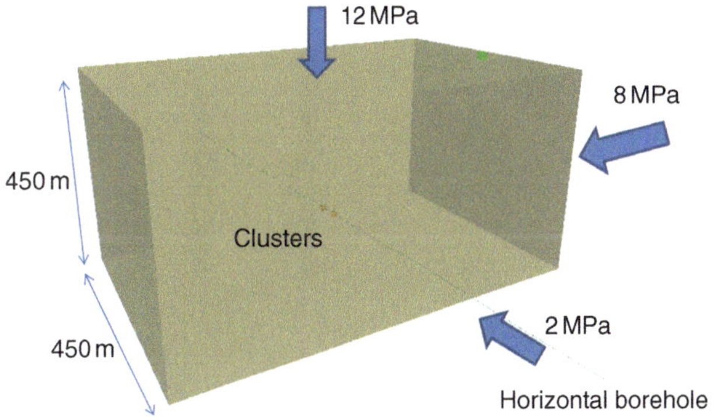

Figure 7.9 *Xsite* model showing the horizontal borehole and two clusters.

The hydraulic fracture propagated using a single cluster is shown at four simulation times in Figure 7.10.

The fracture evolves as a penny shape in early times. Later, the high toughness layer interferes with the downwards fracture growth, and the fracture shape features a flat base.

The example of two clusters 30 m apart is shown in Figure 7.11.

Two quasi-parallel fractures of similar shape and dimensions are generated; there is little interference between the growth of the two fractures for the 30 m cluster spacing case.

Hydraulic fractures from two to six clusters 15 m apart are shown in Figures 7.12–7.14.

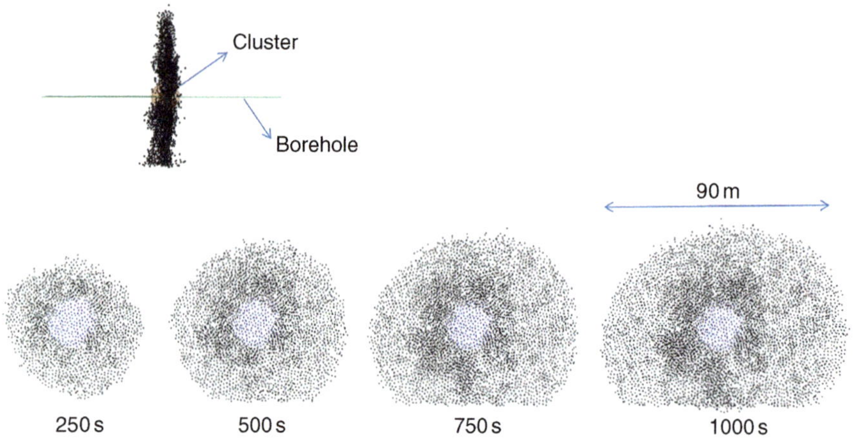

Figure 7.10 Hydraulic fracture generated from a single cluster.

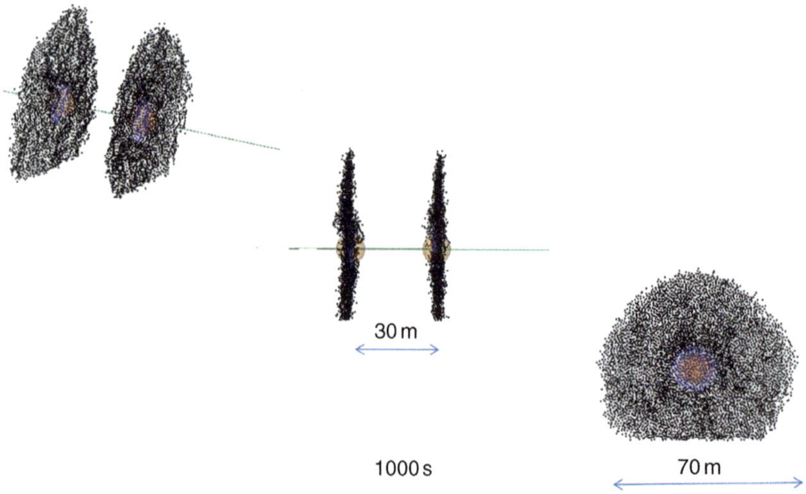

Figure 7.11 Hydraulic fractures from two clusters 30 m apart.

Figure 7.12 Hydraulic fractures from two and three clusters 15 m apart, and from two clusters 30 m apart.

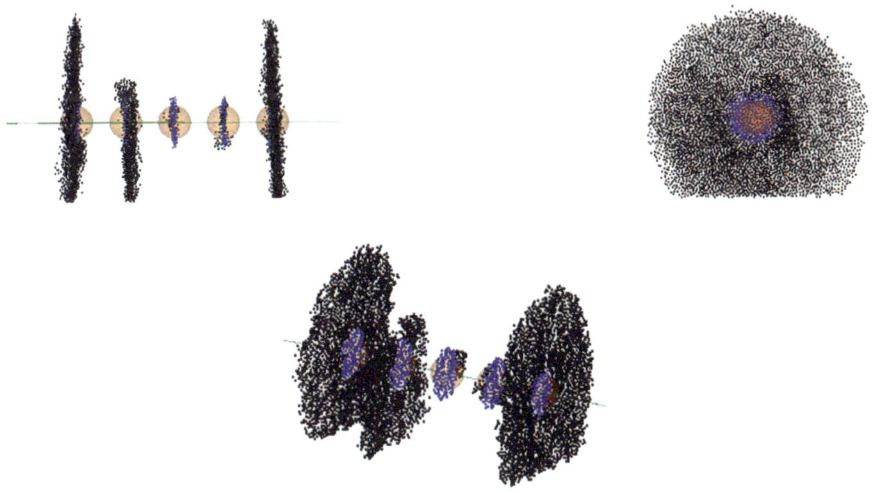

Figure 7.13 Hydraulic fractures from five clusters 15 m apart.

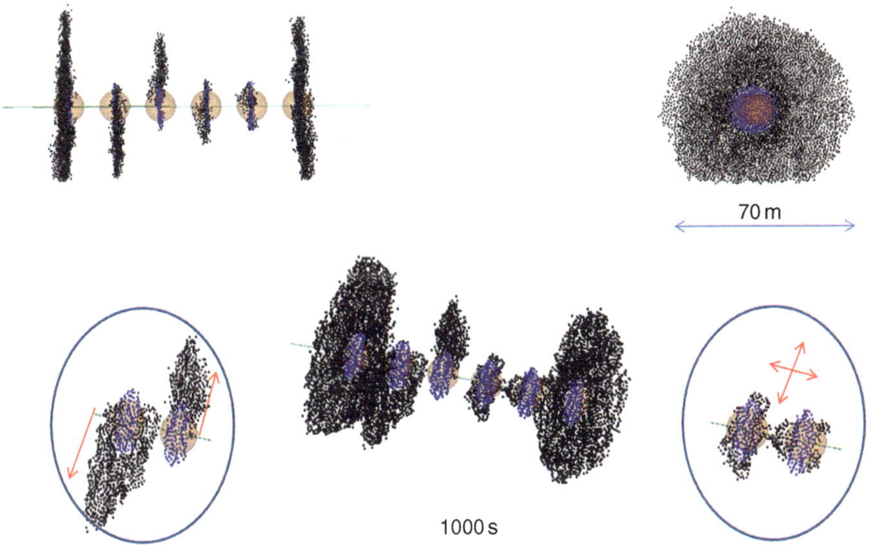

Figure 7.14 Hydraulic fractures from six clusters 15 m apart.

The two fractures from two clusters 15 m apart are seen to grow apart, although slightly. When three clusters separated by 15 m are used, the growth of the middle fracture is impeded by the edge fractures. This growth repression is another manifestation of the stress-shadow effect. The fluid in the fracture exerts a pressure in the surrounding rock. Although the rock near the edge of the fracture is in tension, there is also a bulb region surrounding the fracture surfaces that experiences compressive stresses from the newly created crack field (see, e.g. yellow and brown area in Figure 7.7). It is the combination of the augmented compressive stress and the presence of slight inhomogeneity that inhibits the formation of the middle fracture in this case.

A growth repression of the middle fractures is also observed in the simulations with five and six clusters, seen in Figures 7.13 and 7.14, respectively.

In all cases, the growth of the edge-fractures is promoted. This is because they have only a one-sided stress-shadow impediment, compared to two for the middle fractures. A further interesting trend for the middle fracture is that they tend to grow in opposite direction from adjacent fractures (the fracture grows preferentially in the direction of the minimal stress shadow).

7.3.4 Interaction of Hydraulic Fractures in Fractured Medium

In this example [13, 14], injection is simulated from a horizontal borehole with four clusters. The four-layer numerical model is sketched in Figure 7.15. The target layer, Point Pleasant, contains a network of preexisting fractures, and the full reservoir extends to the top of the Utica layer.

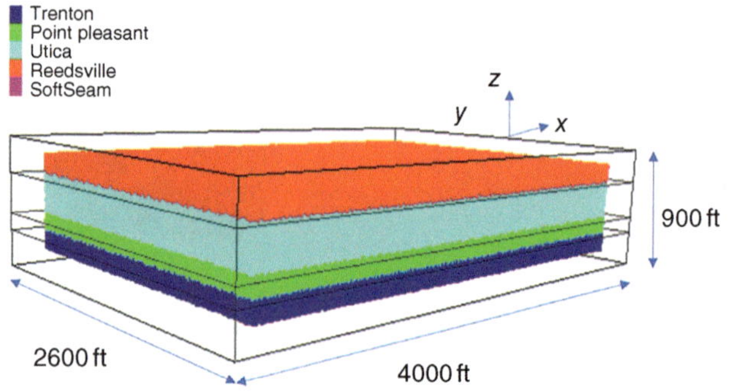

Figure 7.15 *XSite* four-layer model.

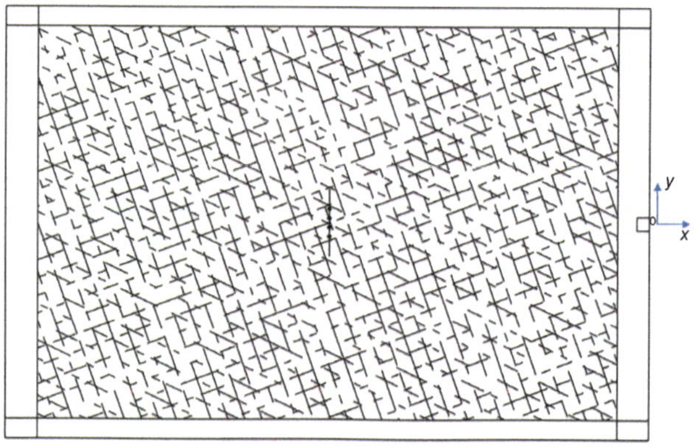

Figure 7.16 Plan view of DFN in target layer.

The horizontal stress normal to the borehole axis in the top (Reedsville) and bottom (Trenton) surrounding layers is much larger than in the producing layers; they act as stress barriers in the fracturing process.

The DFN present in the target layer is plotted in plan view in Figure 7.16, together with a borehole segment and the four clusters (marked as dots).

Borehole fluid injection is modeled for a total of 90 minutes. The fluid migrates from the clusters into the DFN, and new fractures are also created in the intact rock in the target layer. Overall, a system of about three major fractures, oriented in the direction perpendicular to the minimum principal stress, can be identified in the left plot in Figure 7.17 as well as in the top right plot in Figure 7.18.

New quasi-planar fractures are generated simultaneously in the homogeneous reservoir layer above the target layer; see right plot of Figure 7.17 for a side view and bottom right plot of Figure 7.18 for a map view. These fractures propagate vertically to the top of the reservoir, first as a swarm from which three major fractures emerge. Fracture interference manifests itself in various forms in this example. It is interesting to note that only three fractures are propagated out of the four clusters below, that these fractures grow in opposite lateral directions (a manifestation of the stress shadow effect), and that their location is off-centered compared to the location of the clusters located below in the initially fractured layer.

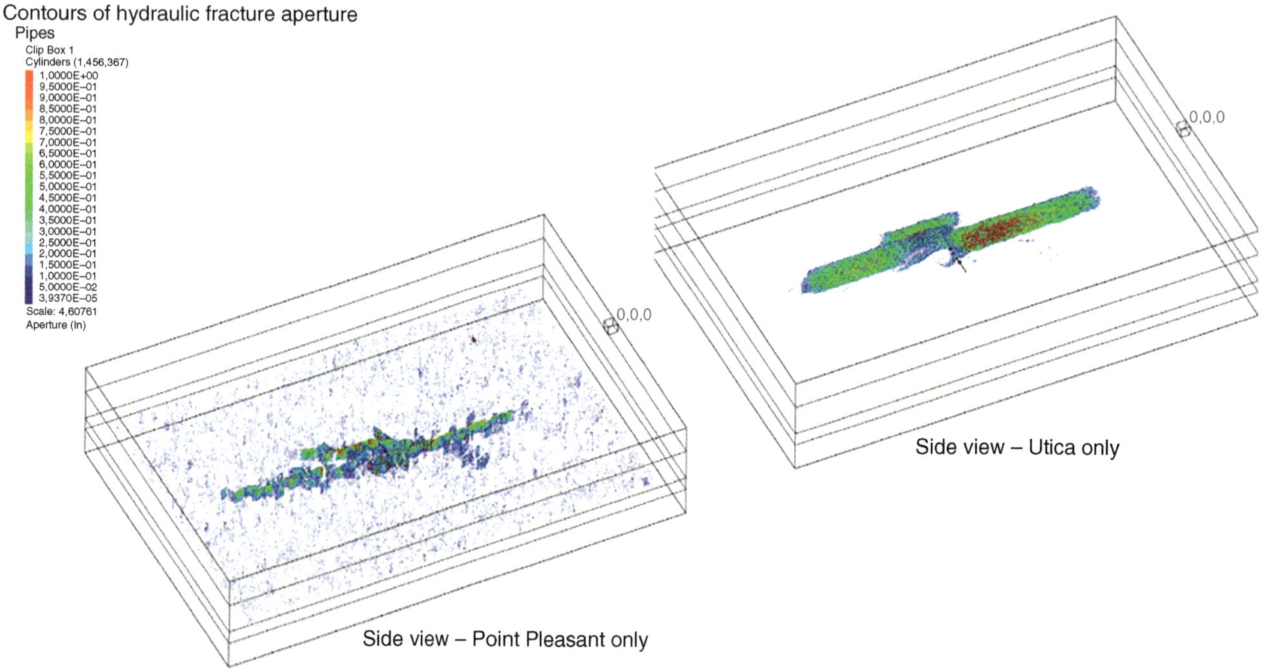

Figure 7.17 Hydraulic fractures from four clusters – side view – with DFN.

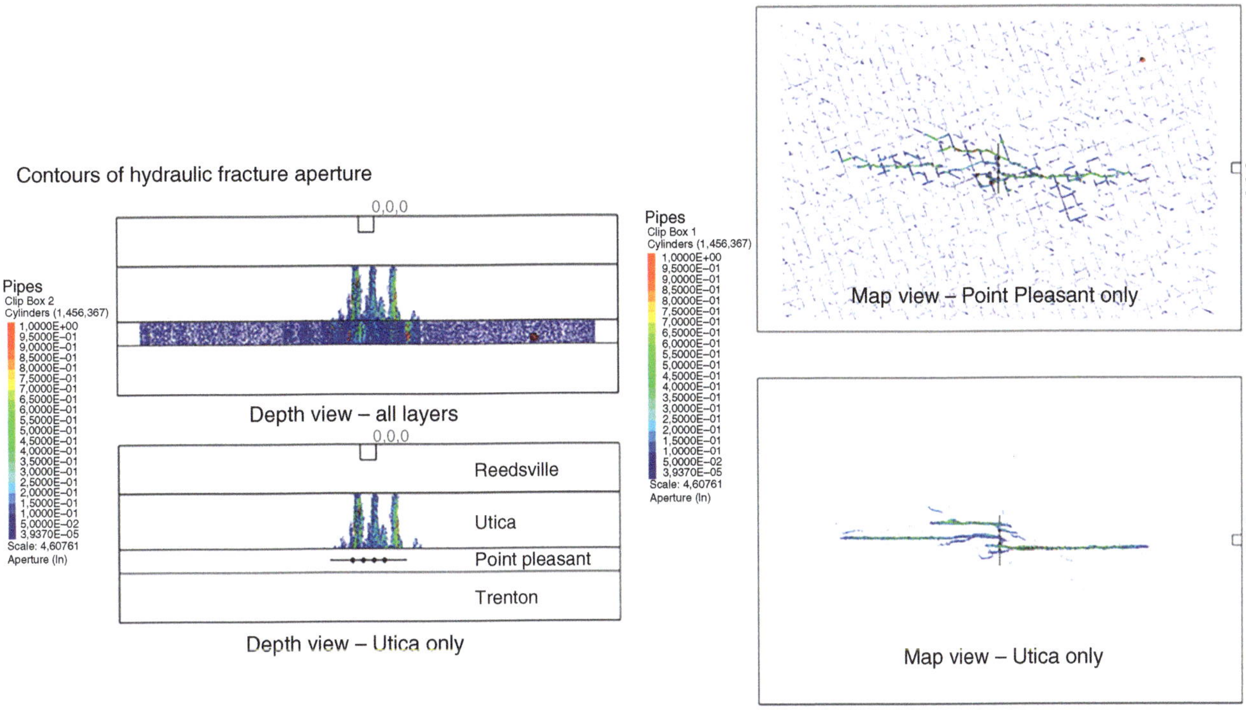

Figure 7.18 Hydraulic fractures from four clusters - depth and map views - with DFN.

This simulation is interesting because it shows the hydraulic fracture pattern that could develop in the homogeneous reservoir when the fluid injection source at the base of the layer is diffused by the DFN below (and not directly constrained by the location of the injection clusters).

7.3.5 Interaction of Hydraulic Fractures in Zipper-Stage Injection

A complex system of two horizontal wells with six alternating stages of four clusters is considered in the four-layer example sketched in Figure 7.19.

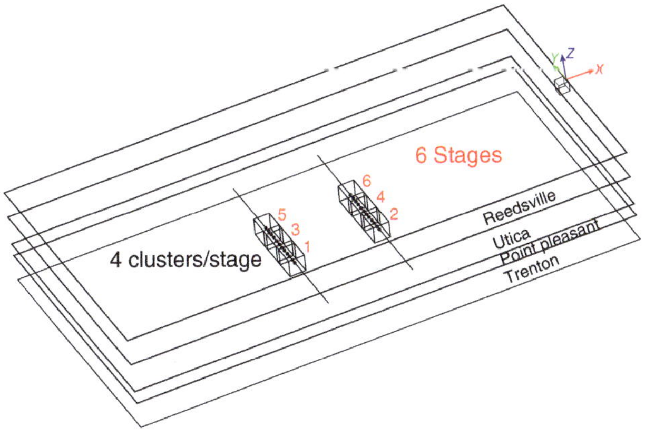

Figure 7.19 Two wells with six zipper stages (enclosed in fine resolution boxes) and four clusters (marked as dots) per stage in the four-layer *XSite* model.

The layers are homogeneous in this case, with the top and bottom layers acting as stress barriers. The two layers sandwiched in the middle form the reservoir, and the lower of the two is the target layer (where the boreholes are located). Borehole injection of a proppant loaded fluid is simulated sequentially in alternating 45-minute stages between the two wells. A map view of the fractures generated at the end of each stage (of four clusters) is shown in Figure 7.20.

As the simulation results show, there is a strong interference between fractures generated in each stage, but also between the two wells.

7.4 Afterword

The results of five numerical simulations are used to explore possible modes of interference among hydraulic and preexisting fractures. The first example shows how a vertical preexisting fracture interfered with a horizontal hydraulic fracture in the 3D simulation. While a fracture arrest or crossing would have been expected in a 2D simulation, the 3D simulation shows the hydraulic fracture wrapping around the penny-shaped crack after a temporary arrest.

A common intuitive assumption concerning the propagation of two hydraulic fractures induced *sequentially* is that they would remain planar. The second example shows that

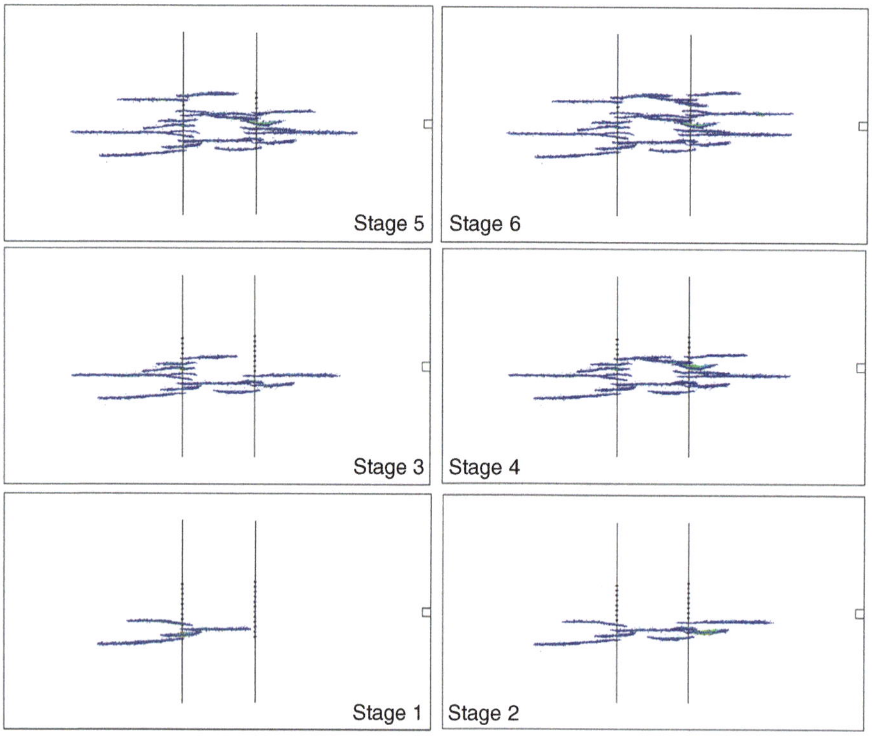

Figure 7.20 Map view of hydraulic fractures in both Utica and Point Pleasant at the end of the six stages for a total of 270 minutes of injection (conventional stimulation).

attraction and or repulsion between the two fractures could also happen, depending on the distance between initiation points and residual aperture, among other factors, a mechanism analyzed thoroughly by Bunger et al. [8].

In the third example, borehole injection is used to stimulate fracture growth in multiple clusters. With a borehole oriented in the direction parallel to the minimum principal stress, it would not be unreasonable to expect similar and *simultaneous* fracture growth from the clusters. However, fracture development is impacted by adjacent fractures (or by the lack thereof if they are positioned at the end of the stage), and their growth can be suppressed or forced to evolve away from that of their neighbors, depending on cluster distance and small inhomogeneities in the system that cause a differential delay in fracture initiation.

The simulations in the fourth example show the diversion of hydraulic fracture paths caused by the presence of preexisting fractures in a layered reservoir. The diffuse domain invaded by the fracturing fluid in the fractured layer is a compositional assembly of preexisting fracture segments and newly formed segments in the rock matrix, oriented on average in the direction perpendicular to the minimum principal stress. The fracturing fluid emerging from this fractured system into a homogeneous formation above it can initiate fractures there in a free manner, independent of the imposed location of the clusters.

In this example, a swarm of parallel fractures is seen to develop in the homogeneous formation, from which only a few grow to a respectable size. Evidently the physics at play is complex and many aspects of the simulation are invitations for more inquisition!

The final example shows the complex fracture pattern that could develop in a two-well zipper stage injection system, including many of the interference features observed earlier, plus the interplay of fractures generated from different wells.

In most instances of fracture interference documented here, the stress shadow effect plays a dominant role. This effect is controlled in turn by multiple factors, including the in situ stress state, the rock mass heterogeneity, the viscosity/toughness fracture propagation regime that encompasses rock properties, injection rate, and fracturing fluid characteristics [15], the proppant type, the connectivity and properties of preexisting fractures, and more. Each of these factors make for a rich and boundless domain of research and investigation using experimental field and laboratory techniques, analytical analysis, and numerical simulations.

References

1 Fu, W., Savitski, A., Damjanac, B., and Bunger, A. (2019). Three-dimensional lattice simulation of hydraulic fracture interaction with natural fractures. *Computer and Geotechnics* 107: 214–234.

2 Itasca Consulting Group, Inc. (2019). *XSite (Version 3.0).* Minneapolis: Itasca.

3 Cundall, P.A. (2011). Lattice method for Brittle, jointed rock. Continuum and distinct element numerical modeling in geomechanics. *Proceedings, 2nd International FLAC/DEM Symposium*, Melbourne, pp. 11–19. Minneapolis: Itasca International, Inc, February 14–16, 2011.

4 Damjanac, B., Detournay, C., Cundall, P.A., and Varun (2013). Chapter 41: three-dimensional numerical model of hydraulic fracturing in fractured rock masses. In: *Effective and Sustainable Hydraulic Fracturing - 2013.* https://doi.org/10.5772/45724 (ed. A.P. Bunger, J. McLennan and R. Jeffrey). InTech.

5 Damjanac, B. and Cundall, P. (2016). Application of distinct element methods to simulation of hydraulic fracturing in naturally fractured reservoirs. *Computers and Geotechnics* 71: 283–294.

6 Damjanac, B., Detournay, C., and Cundall, P. (2016). Application of particle and lattice codes to simulation of hydraulic fracturing. *Computational Particle Mechanics* 3 (2): 249–261. https://doi.org/10.1007/s40571-015-0085-0.

7 Damjanac, B., Detournay, C., and Cundall, P. (2020). Numerical simulation of hydraulically driven fractures. In: *Modelling Rock Fracturing Processes* (ed. B. Shen, O. Stephansson and M. Rinne). Cham: Springer https://doi.org/10.1007/978-3-030-35525-8_20.

8 Bunger, A.P., Zhang, X., and Jeffrey, R.G. (2011). Parameters effecting the interaction among closely spaced hydraulic fractures. *SPE Journal* 13: 140426.

9 Detournay, C., Cundall, P., and Damjanac, B. (2016). A study of fracture interference in 3D with *XSite*, in applied numerical modeling in geomechanics — 2016. *Proceedings, 4th Itasca Symposium on Applied Numerical Modeling*, Lima (March 2016), pp. 607–616. Minneapolis: Itasca, March 7–9, 2016.

10 Itasca Consulting Group, Inc. (2020). *FLAC — Fast Lagrangian Analysis of Continua (Version 8.1).* Minneapolis: Itasca.

11 Bunger, A.P., Jeffrey, R.G., Kear, J. et al. (2011). Experimental investigation of the interaction among closely spaced hydraulic fractures. *Proceedings of ARMA 2011.* ARMA 11-318. 45th U.S. Rock Mechanics / Geomechanics Symposium, San Francisco, California, June 26–29, 2011.

12 NewCrest Mining Limited (2010). Cadia East Hydraulic Fracture Trial Report. NewCrest Mining Limited, Internal Report. Melbourne, Victoria.

13 Cipolla, C., Gilbert, C., Sharma, A., and LeBas, J. (2018). Case History of Completion Optimization in the Utica. *SPE-189838-MS.*

14 Detournay, C., Torres, M., Han, Y., & Abousleiman, Y. N. (2019) Numerical analysis of the impact of a discrete fracture network on a single well hydraulic fracturing treatment. *American Geophysical Union, Fall Meeting 2019*, abstract #H34A-07.

15 Detournay, E. (2004). Propagation regimes of fluid-driven fractures in impermeable rocks. *International Journal of Geomechanics*, 10.1061/(ASCE) 1532-3641 4 (1): 35.

8

The Tipping Point: How Tip Asymptotics Can Enhance Numerical Modeling of Hydraulic Fracture Evolution

A. Peirce

Department of Mathematics, University of British Columbia, Vancouver, British Columbia, Canada

8.1 Introduction

Hydraulic fractures (HFs) are a class of brittle fractures that propagate in pre-stressed solid media due to the injection of a viscous fluid. On a geological scale, HF occurs as a kilometers-long vertical dike that brings magma from deep underground chambers to the earth's surface [39]. Industrial applications of HF include: accelerating the waste remediation process, waste disposal, or preconditioning in rock mining [41]. One of the most common industrial applications is reservoir stimulation to enhance the recovery of hydrocarbons to create permeable pathways [14].

Inspired by the seminal 1987 paper by Spense and Sharp [38] and the 1994 discovery of the so-called "SCR viscous asymptote" [7], the beginning of this millennium heralded intense research into the propagation of HF in the special cases of plane strain and radial HF. See [8] for an excellent review of this corpus of research, which established the pivotal role of the asymptotic behavior of the solution in the vicinity of the fracture tip and the complex multiscale structure of the solution, in which two or more physical processes compete at multiple length and time scales.

However, progress on incorporating this multiscale structure into numerical models has lagged behind these developments. Despite significant progress made since the first algorithms were developed in the 1970's [5], many 3D HF algorithms still use propagation criteria based on the LEFM of dry cracks [2]. This lag in development is due to the intrinsic complexity of the coupled HF equations, which involve a degenerate hypersingular integro-delay-PDE along with a singular free boundary problem. It was not clear how to develop numerical models of HF in 3D elastic media with arbitrarily shaped boundaries able to capture the multiscale structure without prohibitively expensive and complex re-meshing. In this chapter we describe how the tip asymptotic behavior is an essential tool in the construction of numerical schemes that can locate the unknown fracture boundary and capture the multiscale behavior on a coarse mesh [13, 31, 33, 34].

In Section 8.2, we state the governing equations for the elastic medium, fluid flow, boundary conditions, and tip asymptotics; in Section 8.3, we discuss discretization of the DD equations, solving the coupled equations, and locating the free boundary using the implicit level set method; in Section 8.4 we provide numerical examples; in Section 8.5 we provide concluding remarks.

8.2 Mathematical Model

8.2.1 Assumptions

The equations governing the propagation of a HF in a reservoir have to account for the dominant physical processes taking place during the treatment, namely: deformation of the rock; creation of new fracture surfaces, flow of fracturing fluid in the crack, and leak-off of fracturing fluid into the reservoir. We consider the governing equations for hydraulic fracture propagation that involve continuum descriptions of fracture in the solid medium and fluid flow within the fracture. For the solid medium we discuss the displacement discontinuity (DD) boundary integral formulation and assume: (i) that the rock is homogeneous, linear elastic, brittle, permeable, and of infinite extent whose stiffness is characterized by the Young's modulus E and Poisson's ratio ν, while the energy required to break the rock is represented by the fracture toughness K_{Ic}, (ii) the flow within the fracture is laminar and that fracturing fluid is incompressible and Newtonian with a dynamic viscosity μ_f, (iii) Carter's leak-off [4] is used to model loss of fluid

---

This paper is dedicated to Dr. Rob Jeffrey on the occasion of his 70th birthday for his tireless dedication to conducting and promoting excellent fundamental scientific research into hydraulic fracture.

Mechanics of Hydraulic Fracturing: Experiment, Model, and Monitoring, First Edition.
Edited by Xi Zhang, Bisheng Wu, Diansen Yang, and Andrew Bunger.

to the reservoir and characterized by the coefficient C_L, (iv) the fracture evolves in a plane orthogonal to the minimum in situ confining stress field $\sigma_0(x, y)$, and (v) high confinement conditions prevail so the fluid front rapidly coalesces with the crack front [16].

The solution of the hydraulic fracture problem involves determining the fracture aperture $w(x, y, t)$, the fluid pressure $p_f(x, y, t)$, the fluid flux $\mathbf{q}(x, y, t)$, and the position of the front $\mathcal{C}(t)$, where t denotes the time and x, y are coordinates referenced to the injection point (see Figure 8.1). The solution depends on the injection rate $Q(t)$, the far-field compressive stress perpendicular to the fracture plane $\sigma_0(x, y)$ (a known function of position), and the four material parameters μ', C', E', and K' defined as

$$\mu' = 12\mu_f, \qquad C' = 2C_L, \qquad E' = \frac{E}{1-\upsilon^2}, \qquad K' = 4\left(\frac{2}{\pi}\right)^{\frac{1}{2}} K_{Ic} \tag{8.1}$$

Here E' is the plane strain modulus, and the alternate viscosity μ', leak-off C', and toughness K' are introduced to keep the equations uncluttered by numerical factors.

8.2.2 Governing Equation

8.2.2.1 Elasticity

In view of the rock homogeneity and linear elasticity assumptions, the elasticity equation that relates the fracture aperture w to the compressive stress field along the crack (which is related to the fluid pressure p_f) can be condensed into a single hypersingular integral equation [6]

$$p_f(x, y, t) = \sigma^h(x, y) - \frac{E'}{8\pi}\int_{\mathcal{A}(t)} \frac{w(x', y', t)dx'dy'}{\left[(x'-x)^2 + (y'-y)^2\right]^{3/2}}, \tag{8.2}$$

where $\mathcal{A}(t)$ denotes the fracture footprint, i.e. the area enclosed by the crack front $\mathcal{C}(t)$ (see Figure 8.1) and $\sigma^h(x, y)$ is the prescribed in situ geological stress field.

8.2.2.2 Fluid Transport

By combining Poiseuille's law $\mathbf{q} = -\frac{w^3}{\mu'}\nabla p_f$ and the continuity equation $\frac{\partial w}{\partial t} + \nabla\cdot\mathbf{q} + \frac{C'H(t-t_0(x, y))}{\sqrt{t-t_0(x, y)}} = Q(t)\delta(x, y)$, we obtain the Reynolds lubrication equation

$$\frac{\partial w}{\partial t} = \frac{1}{\mu'}\nabla\cdot\left(w^3\nabla p_f\right) - \frac{C'H(t-t_0(x, y))}{\sqrt{t-t_0(x, y)}} + Q(t)\delta(x, y) \tag{8.3}$$

where the trigger-time $t_0(x, y)$ represents the time at which the unknown fracture front passes the point $\boldsymbol{x} = (x, y)$. This constitutes a history term $t_0(x, y)$, which is not known a priori, so that (8.3) is a delay partial differential equation.

8.2.2.3 Boundary and Propagation Conditions

For this nonlinear problem there exist multiple equilibrating and volume-conserving width and pressure fields, so imposing the appropriate boundary and propagation conditions is the key to selecting the desired global solution [9].

8.2.2.3.1 *Finite Lag Between Fluid and Fracture Fronts*

When the fluid front $\mathcal{C}_f(t)$ is distinct from and entirely contained within the fracture front $\mathcal{C}(t)$ there is a finite lag

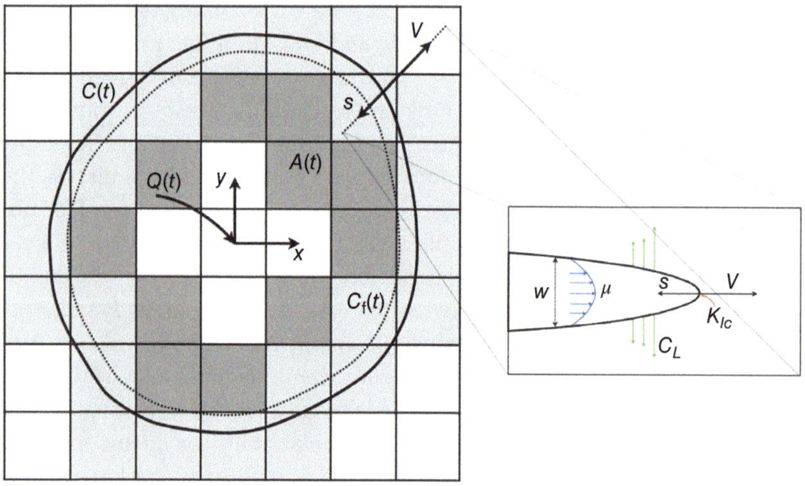

Figure 8.1 The planar fracture footprint $\mathcal{A}(t)$ inscribed within curve $\mathcal{C}(t)$ moving with normal velocity V and driven by the fluid contained within the dashed curve $\mathcal{C}_f(t)$. The xy coordinate system is centered on the point source.

between the two fronts and the boundary and propagation conditions are:

$$p_f(\boldsymbol{x}, t) = 0 \text{ and } \mathbf{V_f}(\boldsymbol{x}, t) = \mathbf{q}(\boldsymbol{x}, t)/w(\boldsymbol{x}, t) \text{ for } \boldsymbol{x} \in \mathcal{C}_f(t), \tag{8.4}$$

$$w(\boldsymbol{x}, t) = 0 \text{ and } K_I(\boldsymbol{x}, t) = K_{Ic} \text{ for } \boldsymbol{x} \in \mathcal{C}(t), \tag{8.5}$$

In each of (8.4)–(8.5) the first conditions prescribe boundary conditions on the fluid pressure and width fields, while the second conditions in (8.4)–(8.5), respectively provide the Stefan condition for the moving fluid front $\mathcal{C}_f(t)$ and the propagation condition that the stress intensity factor match the fracture toughness at all points of $\mathcal{C}(t)$.

8.2.2.3.2 Coalescent Fluid and Fracture Fronts

For most HF treatments the confining stress σ_0 is large compared to the characteristic driving stress $\mu' VE'^2/K'^2$ causing the lag to vanish rapidly [16]. Rather than resolve this vanishingly small lag with a fine mesh it is more efficient to establish the appropriate boundary conditions for the limit $\mathcal{C}_f(t) \to \mathcal{C}(t)$. For an impermeable medium, as $\boldsymbol{x} \in \mathcal{C}_f \to \mathcal{C}$ the corresponding width $w(\boldsymbol{x}, t) \to 0$, thus from the second equation in (8.4), in order that $|\mathbf{V_f}|$ remain finite, it follows that $\mathbf{q}(\boldsymbol{x}, t) \to 0$. The limit for a permeable medium is more subtle [9] but yields the same result, so that (8.4)–(8.5) reduce to

$$w(\boldsymbol{x}, t) = 0 \text{ and } K_I(\boldsymbol{x}, t) = K_{Ic} \text{ and } \boldsymbol{q}(\boldsymbol{x}, t) = 0 \text{ for } \boldsymbol{x} \in \mathcal{C}(t), \tag{8.6}$$

From the propagation condition $K_I = K_{Ic}$ and LEFM it follows [35] that

$$w \sim \frac{K'}{E'} s^{1/2} \tag{8.7}$$

where s (see Figure 8.1a) denotes the distance from the crack front $\mathcal{C}(t)$. Since the front velocity $\boldsymbol{V}$ for an impermeable medium is equal to the average fluid velocity as $s \to 0$ it follows that

$$\boldsymbol{V} = \frac{1}{\mu'} \lim_{s\to 0} w^2 \nabla p_f, \tag{8.8}$$

which shows that $\boldsymbol{V}$ is the limit of an indeterminate form, which is difficult to evaluate numerically. For a permeable medium, in the region of dominant leak-off the average fluid velocity tends to infinity as $s \to 0$, and the crack front velocity needs to be determined by detailed asymptotic analysis [9]. Thus, front location algorithms based on knowing the front velocity $\boldsymbol{V}$ explicitly [37], cannot be applied directly to this singular free boundary problem.

8.2.2.4 Tip Asymptotics, Vertex Solutions, and Generalized Asymptotes

A propagating HF with zero lag is governed by two competing dissipative processes: one associated with viscous energy losses and the other with the energy expended in breaking the rock, which is characterized by the fracture toughness; and two competing fluid-balance components: fluid storage in the fracture and storage in the surrounding rock – associated with leak-off. Limiting regimes can be identified in which just one of the dissipative processes and one of the storage mechanisms is dominant. Associated with each of these limiting regimes are the so-called vertex solutions with reference to vertices of a multi-process phase space within which the dissipative and storage mechanisms compete. For example, for the storage-viscosity regime known as the m-vertex, leak-off is negligible compared to the fluid stored in the fracture and the energy expended breaking the rock is negligible compared to viscous dissipation, so that $K' \approx 0 \approx C'$. A HF having radial symmetry and for which $K' \neq 0 \neq C'$ will change its regime of propagation over time during which multiple length and time scales are important (see [8, 35] for further details). The key tool to characterize this multiprocess multiscale environment is an analysis of the tip asymptotics and the so-called semi-infinite fracture problem.

8.2.2.4.1 Tip Asymptotics and Semi-Infinite Crack Problem

Provided the fracture front $\mathcal{C}$ is sufficiently smooth that a finite radius of curvature can be defined at each point along its perimeter and assuming a homogeneous elastic medium it can be shown [31] that the governing Eqs. (8.2), (8.3), and (8.7) for the aperture w and net pressure p in the vicinity of the fracture front reduce to those for a semi-infinite fluid-driven fracture steadily propagating in a state of plane-strain at a constant velocity V and characterized by zero lag (see [16]). Introducing a moving coordinate s whose origin is located on the moving fracture tip and pointing toward the interior of the fracture and changing variables $w(\boldsymbol{x}, t) \to \hat{w}(s, t)$ and $p_f(\boldsymbol{x}, t) - \sigma^h(x, y) \to \hat{p}(s, t)$, (see [31]), yields:

$$\frac{\hat{w}^3}{\mu'} \frac{d\hat{p}}{ds} = V\hat{w} + 2C'V^{1/2}s^{1/2}, \ \hat{p} = \frac{E'}{4\pi} \int_0^\infty \frac{d\hat{w}}{dz} \frac{dz}{s-z},$$
$$\lim_{s\to 0} \frac{\hat{w}}{s^{1/2}} = \frac{K'}{E'} \tag{8.9}$$

The consequence of this reduction is profound in that the tip asymptotic solution for a *finite fracture* at any time is given by the solution of the stationary semi-infinite crack in a state of plane strain whose constant tip velocity corresponds to the *current* propagation speed of the finite fracture. The tip solution is thus autonomous. Note that the spatial variation of the far-field stress can be ignored when viewed at the tip scale, unless the stress field is discontinuous (in which case, the tip solution outlined here is not relevant).

8.2.2.4.2 Vertex Solutions

Assuming the fracture width is given by a power law $\hat{w} = As^{\alpha}$ and recognizing that in this case the integral operator (8.9) reduces to a Mellin Transform, it follows [25] that

$$\hat{p} = \frac{E'A\alpha}{4\pi}\int_0^{\infty}\frac{z^{\alpha-1}}{s-z}dz = \frac{1}{4}E'A\alpha\cot(\pi\alpha)s^{\alpha-1},\ 0<\alpha<1 \tag{8.10}$$

Thus, for this range of values of α, the power law is an eigenfunction of the integral operator in (8.10).

This power law solution pair can be used to identify the three limiting regimes of propagation also known as vertex solutions, namely: the viscosity dominated solution, the leak-off dominated solution, and the toughness dominated solution. Choosing $\alpha \neq 1/2$ and substituting the power law behavior $\hat{w} = As^{\alpha}$ and the corresponding power law behavior for $\hat{p}$ given by (8.10) into the lubrication equation in (8.9) we can obtain two different vertex solutions by identifying two distinct dominant balances. Choosing the dominant balance between the term on left side and the first term on the right, we find that $\alpha = 2/3$ and matching coefficients yield the so-called m-vertex solution [7]

$$w_m = \beta_m\left(\frac{\mu' V}{E'}\right)^{1/3} s^{2/3},\ p_m = O\left(s^{-1/3}\right),\ \beta_m = 2^{1/3}3^{5/6} \tag{8.11}$$

Choosing a dominant balance between the term on the left and the second term on the right of (8.9a), we obtain that $\alpha = 5/8$ and matching coefficients we obtain the so-called $\tilde{m}$-vertex solution [23]

$$w_{\tilde{m}} = \beta_{\tilde{m}}\left(\frac{4\mu'^2 C'^2 V}{E'^2}\right)^{1/8} s^{5/8},\quad p_{\tilde{m}} = O\left(s^{-3/8}\right),$$
$$\beta_{\tilde{m}} = 4/\left(15\left(\sqrt{2}-1\right)\right)^{1/4} \tag{8.12}$$

Finally, if $\alpha = 1/2$ then the toughness asymptote (8.7) is the so-called k-vertex solution w_k and matching the term on the left of (8.9a) with the first term on the right we find that $p_k = O(\log(s))$.

8.2.2.4.3 Generalized Asymptotes

From the vertex solutions described above, we see that for $s \ll 1$ there is a hierarchy of powers

$$s^{1/2} \gg s^{5/8} \gg s^{2/3} \tag{8.13}$$

that establishes the region moving away from the fracture tip that will be occupied by the different solutions. For example, closest to the tip the toughness solution (provided $K' > 0$) will always be dominant. Then provided $C'>0$, as one moves further from the tip, the solution will transition to the leak-off asymptote characterized by the 5/8 power law. The length scale at which this transition occurs is determined by the relative magnitudes of the coefficients of the two solutions. Finally, the leak-off solution will transition to the viscous solution characterized by the 2/3 power which prevails farthest from the tip. Thus, we can see that there are multiple physical processes possible that can be active at vastly different length scales. If only two of these processes are present, then the same ranking in (8.13) determines the regions occupied by the two processes relative to the tip.

Garagash [15] was the first to consider a two process solution involving the transition from the toughness to the viscous dominated regimes ($C' = 0$) by solving the semi-infinite crack problem (8.9) with the boundary condition at infinity $\hat{w} \xrightarrow{s\to\infty} w_m(s)$. The numerical solution to this connection problem (shown in Figure 8.4b) involves the generalized asymptote $\hat{\Omega}$ in terms of which the asymptotic solution can be expressed in the form:

$$w_{mk} = \frac{K'^4}{E'^3\mu' V}\hat{\Omega}\left(\frac{s}{l_{mk}}\right),\quad \ell_{mk} = \frac{K'^6}{E'^4\mu'^2 V^2} \tag{8.14}$$

where ℓ_{mk} is the transition length scale. The m and k vertex solutions as well as the m–k transition solution have all been observed experimentally for penny-shaped fractures [3, 8].

The corresponding viscous to leak-off transition ($K' = 0$) involving the transition length scale $\ell_{m\tilde{m}} = \frac{2^6 C'^6 E'^2}{\mu'^2 V^5}$ has been considered by [1]. Having established the two-process "edge solutions" Garagash et al. [17] have provided the numerical solution for the generalized tip asymptotics for an HF in which all three processes are active $K' > 0$, $\mu' > 0$, $C' > 0$. This generalized asymptote involves multiple length scales, which are characterized by the dimensionless parameter $\chi = \left(\frac{\ell_{m\tilde{m}}}{\ell_{mk}}\right)^{1/6} = \frac{2C'E'}{K'V^{1/2}}$. Recently [11] a simpler non-singular formulation of the hypersingular integral Eq. (8.9) has made it possible to solve the multiscale problem using standard numerical techniques and yields an approximate formulation in terms of a separable first order ordinary differential equation (ODE). The closed form solution of this ODE yields not only all the vertex solutions but also the following approximate solution [11, 13] for $\tilde{w} := \frac{E'w}{K's^{1/2}}$ as a function of $\tilde{s} = \left(\frac{s}{\ell_{mk}}\right)^{1/2}$, which is defined implicitly in terms of a given function g_δ as follows:

$$\frac{\tilde{s}}{\tilde{w}^3} = g_\delta\left(\frac{1}{\tilde{w}}, \frac{\chi}{\tilde{w}}\right), \tag{8.15}$$

which is able to capture the complete multiscale solution within a fraction of a percent. A contour map of $\tilde{w}$ as a function of $\tilde{s}$ and χ is shown in Figure 8.3, in which the transitions between the m, k, and $\tilde{m}$ can be clearly observed. This generalized asymptote has been used to model 3-process multiscale planar [13] and multi-planar [12] hydraulic fractures.

8.3 Discretization, Coupled Equations, and the Multiscale ILSA Scheme to Locate the Free Boundary

In this section, we briefly describe the discretization of the DD equations on a structured rectangular mesh. Since the resulting coupled evolution equations are extremely stiff (for which explicit schemes require time step restrictions $\Delta t \leq O(\Delta x^3)$ [2, 33]) the L-stable backward Euler implicit time stepping is often used to evolve the solution. Having established the importance of the tip asymptotic behavior in governing the propagation of HF, we discuss a specialized numerical scheme that has been developed to locate the free boundary and to accommodate this multiple length scale behavior that can span 5–20 decades without having to resort to mesh refinement. In a recent comparative study [22] this class of schemes was shown to provide accurate solutions on a relatively coarse fixed structured mesh.

8.3.1 Discretization

8.3.1.1 Displacement Discontinuity Formulation for Planar Fractures

We assume [31] that the fracture will grow within a rectangular region that has been tessellated into a fixed uniform rectangular mesh $\cup\Delta\mathcal{A}_{m,n}$ with dimensions Δx and Δy in the two coordinate directions (similar to the fracture shown in Figure 8.1). The fracture footprint $\mathcal{A}(t)$ is then covered by rectangular elements $\Delta\mathcal{A}_{m,n}$ such that $\mathcal{A} \subseteq \cup\Delta\mathcal{A}_{m,n}$. Constant DD elements are used for the elasticity computations [6] along with collocation at element centers to yield a fully populated linear system relating the element net pressures p to the element widths w

$$\boldsymbol{p} = \boldsymbol{p}_f - \boldsymbol{\sigma}_0 = \boldsymbol{C}\mathbf{w}, \tag{8.16}$$

in which, for a homogeneous elastic medium, $\boldsymbol{C}$ assumes the convolution form

$$C_{k-m,l-n} = -\frac{E'}{8\pi}\left[\frac{\sqrt{(x_k-x)^2+(y_l-y)^2}}{(x_k-x)(x_l-x)}\right]_{x=x_m-\frac{\Delta x}{2},\, y=y_n-\frac{\Delta y}{2}}^{x=x_m+\frac{\Delta x}{2},\, y=y_n+\frac{\Delta y}{2}} \tag{8.17}$$

Consistent with the constant DD discretization with collocation at element centers, the lubrication Eq. (8.3) is discretized via the finite volume method to yield a five node finite difference stencil [35, 40] and the Backward Euler scheme to obtain

$$\boldsymbol{w} - \boldsymbol{w}_{t-\Delta t} = \Delta t\mathcal{A}(\boldsymbol{w})\boldsymbol{p}_f + \boldsymbol{S}, \tag{8.18}$$

where $\boldsymbol{S}$ is a vector of source/sink terms and $\mathcal{A}(w)$ is the difference operator defined by

$$[\mathcal{A}(\boldsymbol{w})\boldsymbol{p}]_{k,l} = \frac{q_{k+\frac{1}{2},l} - q_{k-\frac{1}{2},l}}{\Delta x} + \frac{q_{k,l+\frac{1}{2}} - q_{k,l-\frac{1}{2}}}{\Delta y}, \tag{8.19}$$

and $q_{k\pm\frac{1}{2},l}$ are the fluxes along the vertical edges of the element defined as $q_{k\pm\frac{1}{2},l} = \pm w^3_{k\pm\frac{1}{2},l}\left(\frac{p_{k\pm1,l} - p_{k,l}}{\Delta x}\right)$, where $w_{k\pm\frac{1}{2},l}$ are the corresponding edge widths defined by $w_{k\pm\frac{1}{2},l} = \left(\frac{w_{k\pm1,l} + w_{k,l}}{2}\right)$. The fluxes and widths along the horizontal edges of the element are defined analogously. Zero flux boundary conditions are implemented in tip elements by removing those terms associated with the element faces having zero boundary fluxes from the difference operator.

Solving the coupled Eqs. (8.16) and (8.18) for $\boldsymbol{p}_f$ and $\boldsymbol{w}$ as primary variables requires the inversion of the operator $\mathcal{A}(\boldsymbol{w})$, which is rank deficient owing to the Neumann boundary condition on $\boldsymbol{p}_f$ due to the zero flux boundary condition (8.6) that holds in the zero lag case. In order to ensure that a solution exists in this case it is necessary to impose the solvability condition

$$\int_{\mathcal{A}} (w - w_{t-\Delta t})dxdy + \Delta\mathcal{L} = Q_0\Delta t, \tag{8.20}$$

where $\Delta\mathcal{L}$ is the volume leaked from the fracture over the current time step and (8.20) is obtained by integrating (8.3) in space over the region $\mathcal{A}$ and in time over $[t, t+\Delta t]$, using the divergence theorem, and imposing the zero flux boundary condition (8.6). To avoid the complexity of having to impose the solvability condition (8.20) we use (8.16) to eliminate $\boldsymbol{p}_f$ from (8.18) to obtain the following evolution equation for the width $\boldsymbol{w}$ at time t

$$\boldsymbol{w} = \boldsymbol{w}_{t-\Delta t} + \Delta t\mathcal{A}(\boldsymbol{w})(\boldsymbol{C}w + \boldsymbol{\sigma}_0) + \boldsymbol{S} \tag{8.21}$$

Once the width has been determined, the corresponding net and fluid pressures can be obtained directly from (8.16).

This coupled system of nonlinear equations is typically solved by Picard, Fixed-Point, or Newton iterative methods or combinations thereof [39]. The large system of dense linear equations that need to be solved during each of these nonlinear iterations are also typically solved iteratively, which facilitate acceleration methods such as: the FFT [27] and Fast Multipole Method [28] to perform $\boldsymbol{C}$ matrix–vector products; and preconditioners that significantly reduce the number of linear iterations based on multigrid [29] or Incomplete LU factorizations of localized Jacobian matrices [30].

8.3.2 Locating the Free Boundary Using the Implicit Level Set Algorithm (ILSA)

Since there are an infinite number of volume preserving-equilibrating pairs (w, p_f) that are solutions to the nonlinear Eqs. (8.16) and (8.18), each associated with a slightly different footprint $\mathcal{A}$, it is important to select the correct solution by determining the appropriate fracture boundary $\mathcal{C}$. Due to the singular free boundary problem characteristic of HF propagating with zero lag, the local fluid velocity in the vicinity of the tip, which requires the numerical evaluation of a distinguished limit, cannot be used to locate the free boundary $\mathcal{C}$.

Indeed, the algorithm to locate the free boundary has to incorporate the singular tip asymptotics discussed in Section 8.2.2.4. We now describe the Implicit Level Set Algorithm (ILSA) [13, 31, 33] that was developed to use the tip asymptotics to locate the fracture free boundary. It is notionally convenient to separate the DD elements representing the fracture into two sets, namely: the tip elements $\mathcal{A}^t$ and the channel elements $\mathcal{A}^c$. The tip elements are the partially filled elements colored by light shading in Figure 8.1, while the channel elements comprise those elements surrounded by the tip elements that are completely filled with fluid. We also identify the ribbon of channel elements on the boundary of the channel region $\mathcal{A}^c$ that share at least one side with a tip element (see the dark shaded elements in Figure 8.1), which we refer to as survey elements.

Steps in the ILSA procedure:

The steps in the ILSA algorithm are illustrated in Figure 8.2 in which width cross sections $w(x, 0, t)$ are plotted. The blue curves represent the exact width (dashed at time step t-Δt and solid at time step t). The black rectangles represent the piecewise constant DD approximations to the crack width. The first row of sub-figures represents the procedure used by the ILSA scheme in the first front iteration, which starts with the injection of a volume of fluid $Q_0\Delta t$ over the interval $[t$-$\Delta t, t]$ and assumes that the front starts

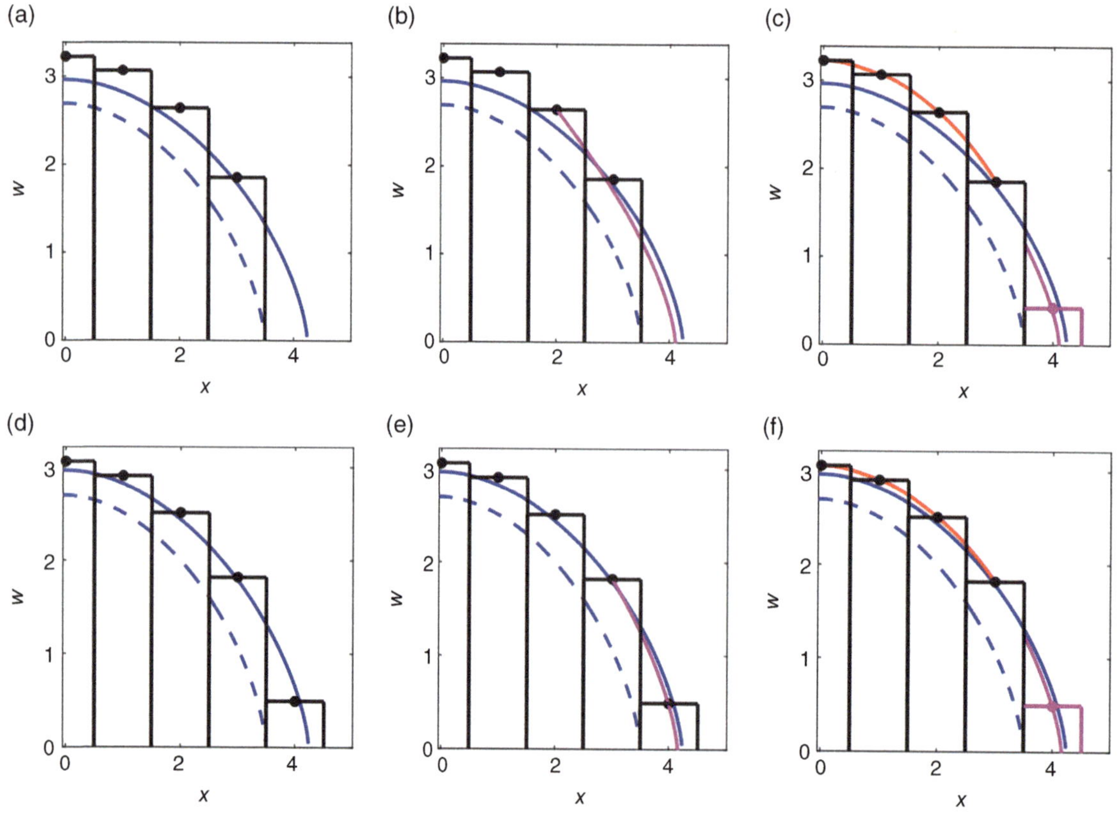

Figure 8.2 Steps used by the ILSA scheme to locate the fracture free boundary using the tip asymptote.

at the position reached at the end of the previous step t-Δt. The second row of sub-figures represents the procedure used by the ILSA scheme assuming that the front has been moved to the position determined by the prevailing asymptote at the end of the first front iteration.

(i) Determine Equilibrium Pressures and Widths for the Current Footprint $\mathcal{A}$

Given a trial footprint (initially taken to be that at the previous time step $\mathcal{A}(t-\Delta t)$ see Figure 8.2a) the coupled Eqs. (8.21) are used to determine the width w (and hence p_f) that corresponds to the fluid flux into the well-bore over the current time step. In the first front iteration (see Figure 8.2a) the crack widths are treated as the primary variable over the whole fracture $\mathcal{A}$. In the second and subsequent front iterations, since estimates of the asymptotic tip widths are known (see step (iv) below), the fluid pressures are treated as the primary unknown in the tip region $\mathcal{A}^t$ and the widths are treated as the primary unknown in the channel region $\mathcal{A}^C$ (see Figure 8.2d);

(ii) Use tip Asymptote to Find Shortest Distances from Elements in $\partial\mathcal{A}^C$ to the New Trial Crack Front $\mathcal{C}$

Given the new width estimate w at each of the survey elements in $\partial\mathcal{A}^C$ the current trial fracture widths are used to invert the appropriate asymptotic relation from Section 8.2.2.4 to determine the shortest distance s from the centers of these elements to the free boundary $\mathcal{C}$. In particular, the third equation in (8.9) can be inverted to obtain s given w_k, (8.11) can be inverted to obtain s given w_m, (8.12) can be inverted to obtain s given $w_{\tilde{m}}$, the generalized asymptote (8.14) can be inverted to find s given w_{mk}, and given $w_{mk\tilde{m}}$ the generalized asymptote (8.15) can be used to find s. This process is illustrated in Figure 8.2b and e in which the asymptote is represented by the magenta curve, which terminates at the center of the survey element (just adjacent to the tip element) on the one end and, in the case of this cross section, the other end terminates at the new estimate of the fracture front position.

We observe that all the tip asymptotes (except for the toughness asymptote (8.9)) involve the unknown front velocity V, which is so crucial to locating the free boundary. To resolve this issue, if s is the shortest distance from the current collocation point to the trial free boundary, then in each asymptote involving the front velocity explicitly we replace V by its difference quotient approximate

$$V = \left(\frac{s - s_{t-\Delta t}}{\Delta t}\right) \tag{8.22}$$

The generalized mk-asymptote (8.14) has been implemented for plane strain fractures [19] and for planar fractures [33]. For each survey point in $\partial\mathcal{A}^C$, the trial width w_{mk} is used to find s from the nonlinear equation

$$w_{mk} = \frac{K'^4\Delta t}{E'^3\mu'(s - s_{t-\Delta t})}\hat{\Omega}\left(\frac{E'^4\mu'^2(s - s_{t-\Delta t})^2 s}{K'^6\Delta t^2}\right).$$

The generalized $mk\tilde{m}$-asymptote (8.15) has been implemented in planar [13] and multi-planar HF models [12]. For each survey point in $\partial\mathcal{A}^C$ the trial width w_{mkc} is used to determine s from

$$\frac{s^2(s - s_{t-\Delta t})\mu'}{E' w_{mkc}^3\Delta t} = g_\delta\left(\frac{K' s^{1/2}}{E' w_{mkc}}, \frac{2C' s^{1/2}\Delta t^{1/2}}{w_{mkc}(s - s_{t-\Delta t})^{1/2}}\right).$$

(iii) Solve the Eikonal Equation to Locate the New Trial Crack Front $\mathcal{C}$

In Figure 8.2, the fracture font is parallel to the sides of the rectangular elements. However, for an arbitrarily shaped fracture boundary this is not the case for all elements of the structured mesh. For this situation, the solution of the eikonal equation is used to reconstruct the front position from the set of shortest distances from the survey elements in $\partial\mathcal{A}^c$ to the new trial front $\mathcal{C}(t)$, which establishes the following initial conditions

$$\mathcal{T}^0(x, y) = -s \text{ for all } (x, y) \in \partial\mathcal{A}^C \tag{8.23}$$

to construct the signed distance function $\mathcal{T}(x, y)$ by solving the eikonal equation

$$|\nabla\mathcal{T}| = 1. \tag{8.24}$$

The negative sign in the initial condition (8.23) enforces the sign convention that $\mathcal{T}(x, y) < 0$ for all points (x, y) that lie within the fracture boundary curve $\mathcal{C}(t)$, while points for which $\mathcal{T}(x, y) > 0$ lie outside $\mathcal{C}(t)$. Moreover, because the initial condition (8.23) defines the distance to the free boundary, the fracture boundary curve $\mathcal{C}(t)$ is defined by the level set $\mathcal{T}(x, y) = 0$;

(iv) Set the Tip Volumes from the Asymptote to Impose the Tip Asymptotics in a Weak Form

Having determined a new estimate for the location of the free boundary $\mathcal{C}(t)$, the volume of fluid in each tip element can be determined by integrating the applicable tip asymptote backward from the front to opposite vertex of a triangular region shown in Figure 8.3

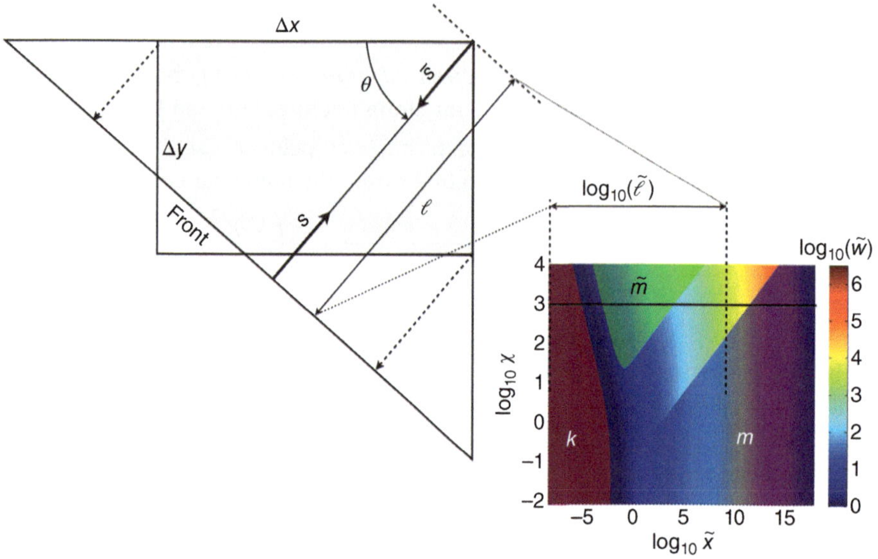

Figure 8.3 Scheme to integrate the asymptotic solution, backwards from the front, over the partially filled rectangular tip element using the integral over a triangle.

$$\mathcal{V}_{\blacktriangle}(\ell) = \frac{2}{\sin 2\theta}\int_0^{\ell} w_a(\ell - \bar{s})\bar{s}d\bar{s}, \tag{8.25}$$

where w_a is one of the asymptotes w_m, w_k, $w_{\tilde{m}}$, w_{mk}, or $w_{mk\tilde{m}}$, ℓ is the distance from the front to the farthest interior corner of a tip element, and θ is the angle that the local outward normal to the front makes with the tip element edge. From the mapping of the span of $s \in (0, \ell)$ within the element to the interval of length $\tilde{\ell}$ in the $\tilde{w}$ map shown in Figure 8.3, we observe that the model is able to capture, in a weak sense, in that all the multiscale behavior from the computational element scale right down to the finest length scale is included as an integral over the applicable asymptote. The volume over a triangular region $\mathcal{V}_{\blacktriangle}(\ell)$ can be used to construct the fracture volume within a partially filled rectangular element as follows:

$$\mathcal{V}_{\blacksquare} = \mathcal{V}_{\blacktriangle}(\ell) - \mathcal{V}_{\blacktriangle}(\ell - \Delta x \cos\theta) - \mathcal{V}_{\blacktriangle}(\ell - \Delta y \cos\theta).$$

In order to evaluate these integrals, only the first two moments of the tip asymptote w_a are required. For the special cases $\theta = 0, \pi/2$, the process is much simpler in that integral $\mathcal{V}_{\blacksquare}$ can be obtained directly by evaluating the integral of the tip asymptote w_a.

The average width w^t in the tip element is set to

$$w^t = \frac{\mathcal{V}_{\blacksquare}}{\Delta x \Delta y}.$$

In Figure 8.2c and f the asymptotic widths are indicated by the magenta curves while the average tip widths in the tip element are represented by the magenta rectangles.

Since the width in the tip element is determined from the asymptotics once the front position is set, the fluid pressure in the tip element now becomes the primary unknown. The equation for the fluid pressure comes from integrating the Reynolds Eq. (8.3) over the tip element and applying the zero flux boundary condition. This is precisely the procedure used in the finite volume method and results in an expression for the change in fluid volume over the time step involving $\mathcal{V}_{\blacksquare}$, which should match the flux of fluid across the tip element sides that are shared with neigboring elements in the channel region and exclude any exchange of fluid with neighboring elements outside the fracture because of the zero flux boundary condition. Thus, in spite of the zero flux boundary condition, because the tip width is known from the multiscale asymptotic solution this procedure provides a way to determine the fluid pressure at the center of the tip element, which is consistent with the solvability condition (8.20);

(v) Convergence Check

If the iterations on the fracture front $\mathcal{C}$ have converged then proceed to the next time step, if not return to (i) with the new estimate of $\mathcal{A}$

We observe that the concept of a partially filled element with tip asymptotics imposed in a weak sense is remarkably effective at representing a fracture with an arbitrary front on a structured mesh. The ability to capture the tip asymptotics in a weak form makes it possible to account for multiscale effects that act at length scales that are many orders of magnitude smaller than the computational mesh size. Modeling such multiscale behavior directly would require prohibitively fine meshes. Given that the material parameters are assumed to be fixed, the tip velocity V is the primary determinant of the region in phase space in which a particular asymptote finds itself. Since the ILSA scheme is able to autonomously determine the front velocity V from the asymptote, it is able to automatically adjust the point in phase space to determine the propagation regime of the fracture front in the neighborhood of the survey points.

In a comparative study [22], this class of schemes was shown to provide accurate solutions on a relatively coarse fixed structured mesh. Recent enhancements to the ILSA strategy [42] have achieved reduced front iterations by using the velocity at the previous step to determine an initial estimate of the front position in the current time step. The ILSA strategy has also been implemented in XFEM HF models [18–20]. Moreover, since ILSA scheme treats the width as the primary variable in the determination of the fracture front position, it is an ideal candidate to combine with the Extended Kalman Filter to develop a scheme to monitor the evolution of the fracture footprint from remote tiltmeter measurements [26, 32, 36].

8.4 Numerical Results

In this section, we present numerical solutions that illustrate the performance of the ILSA strategy for three different examples. In the first two examples we consider a fracture that propagates in an elastic medium in which there are positive jumps in the confining stress field located symmetrically with respect to the fluid source. For the first example, we compare the ILSA solution using the m-vertex solution (8.11) to the results from a laboratory experiment. In the second example, we consider the same experimental symmetric stress jump situation, but demonstrate the influence of toughness using the ILSA scheme along with the mk-asymptote (8.14). In the third example, we consider a hydraulic fracture propagating in a homogeneous elastic medium in which there are jumps in the confining stress field – positive in the layer above and negative in the layer below the fluid source.

8.4.1 Symmetric Stress Barrier: *m*-Vertex Solution vs Experiment and the Effect of Toughness

In this section, we compare the ILSA m-vertex solution with the results of a laboratory experiment [21] in which an HF was initiated at the center of a layer of height $H = 50\,\text{mm}$ bounded symmetric stress jumps $\Delta\sigma = 4.3$ MPa in σ_0, which is given by

$$\sigma_0 = \begin{cases} 2.2\,MPa \text{ for } |y| < 25\,mm \\ 6.5\,MPa \text{ for } |y| > 25\,mm \end{cases}.$$

The ILSA model uses the following parameter values reported for the experiment:

$$E = 3.3\,\text{MPa},\ \nu = 0.4,\ \mu = 30.2\,\text{Pa.s},\ K' = 0,\ C' = 0,$$
$$Q_0 = 0.0017\,\text{ml/s}$$

Since the fracture takes place between two un-bonded impermeable blocks of PMMA, the fracture in the experiment propagates in the m-vertex viscous regime. In Figure 8.4a, the experimental (blue) and ILSA (black) blade-like fracture footprints and fracture width cross-sections show remarkable agreement. In Figure 8.4b, we compare the ILSA results using the m-vertex and mk-asymptote $K' = 0.75\,\text{MPa.m}^{1/2}$ at $t = 600\,s$, while all the other parameters are kept the same. In this plot, the DD mesh is denoted by the thin grid lines and the thick horizontal lines represent the interface across which the confining stress field jumps. The colored survey elements indicate the different length scale active at each point of the fracture boundary, which corresponds to color bar located below the $\hat{\xi}$ axis in the plot of the generalized asymptote $\hat{\Omega}$ vs $\hat{\xi}$.

8.4.2 A Stress Drop: Distinct Propagation Regimes Along the Periphery

In this section, we illustrate the ILSA solution using the generalized mkc-asymptote (8.15) for a hydraulic fracture propagating in a confining stress field σ_0 that drops discontinuously across two interfaces a distance $H = 20\,m$ apart. The material parameters used in this example are

$$E = 9.5\,\text{GPa},\ \nu = 0.2,\ \mu = 0.1\,\text{Pa.s},$$
$$K' = 3.192\,\text{MPa.m}^{1/2},$$
$$C' = 0.521\times10^{-5}\,\text{m/s}^{1/2},\ Q_0 = 0.01\,\text{m}^3/\text{s},$$

while the confining stress field is defined by

$$\sigma_0 = \begin{cases} 7.25\,\text{MPa for } y > 10\text{ m} \\ 7\,\text{MPa for } |y| < 10\text{ m} \\ 6.5\,\text{MPa for } y < -10\text{ m} \end{cases} \tag{8.26}$$

The dimensionless leak-off parameter [24] associated with this parameter set is given by $\phi = \dfrac{\mu'^3 E'^{11} C'^4 Q_0}{K'^{14}} = 10^{-4}$. The footprints provided by the ILSA scheme using the mkc-asymptote are provided at selected time steps in in Figure 8.5. The survey elements are colored according to the location of the corresponding asymptotic solution in the parametric mkc-triangle. Black circular markers inside the parametric triangle show the positions of all survey elements for the footprint that corresponds to the last time instant $t = 3600$ s.

8.5 Conclusions

In this chapter, we have described how to build efficient numerical schemes to model HF propagation that exploit the recent advancements in the analysis of the multiscale tip asymptotic behavior that occurs on multiple time scales as the different physical processes of viscous dissipation, energy required to break the rock associated with

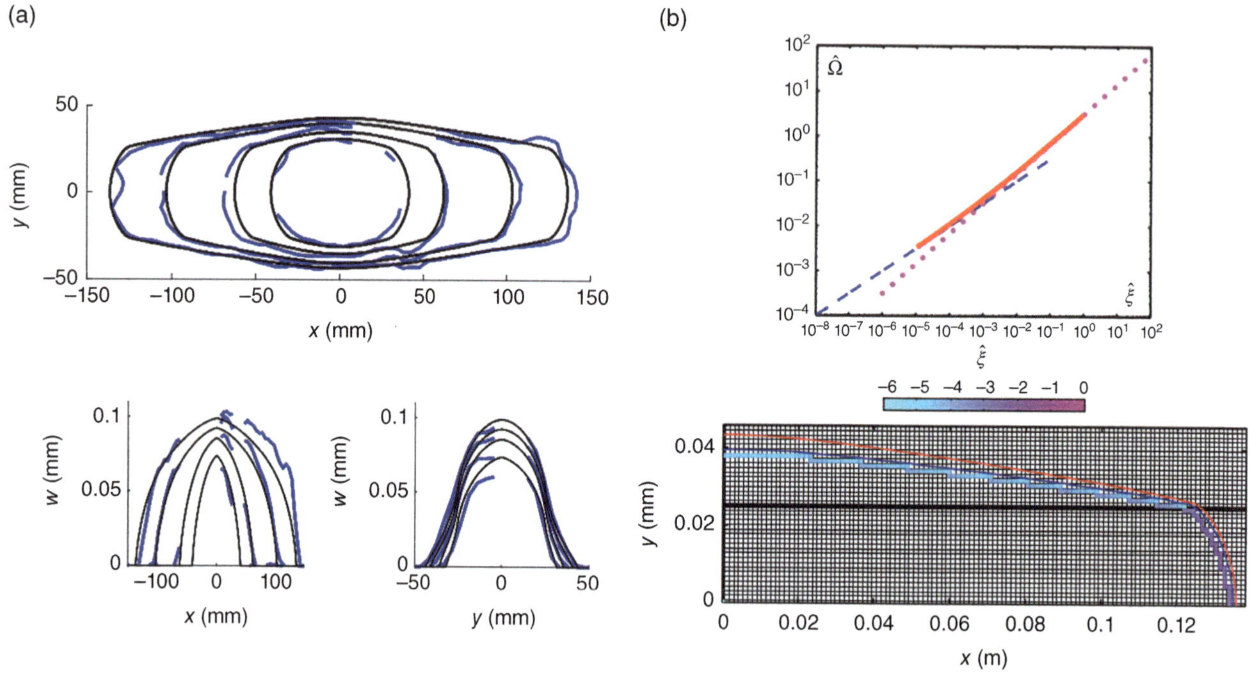

Figure 8.4 Experimental and ILSA footprints of the blade-shaped fracture formed by a symmetric stress jump in σ_0 at times t = 88, 171, 375, 603 s; Left-Bottom: Experimental and ILSA fracture widths (left cross section y = 0) and (right cross section x = 0) at the same time steps; Right: Fracture footprint $\mathcal{C}(t)$ at time t = 600 s for K′ = 0 and 0.75 MPa.m$^{1/2}$. The M-vertex footprint encompasses the MK footprint, which is sampled at the corresponding time.

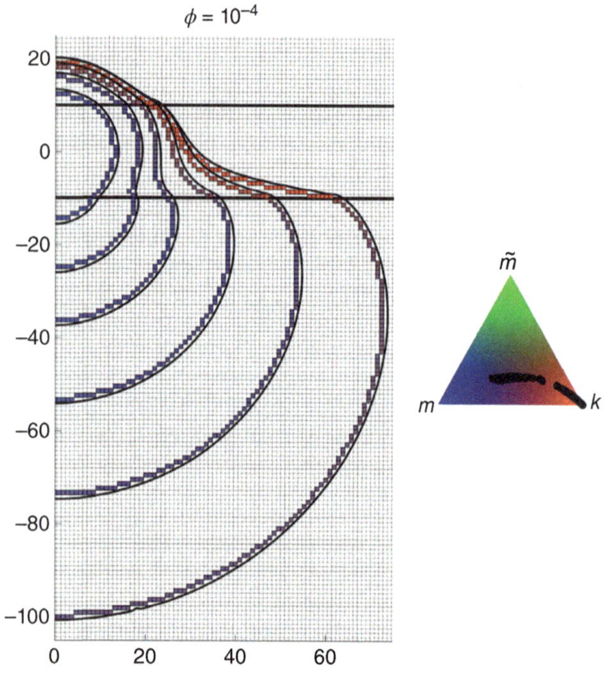

Figure 8.5 Hydraulic fracture footprints calculated for the stress drop geometry defined in (8.26) for $\phi = 10^{-4}$ at the time instants t = {112.5, 225, 450, 900, 1800, 3600} s.

toughness, and diffusive leak-off of fluid into the porous rock. As a starting point we briefly summarize the governing equations, boundary and propagation conditions, and the limiting tip asymptotic behavior needed in the construction of these algorithms. We then illustrate in detail the process of iterative location of the free boundary and the capacity of the ILSA scheme to capture, in a weak sense, the multiscale behavior relevant at the computational scale by the integration of the tip asymptote down to the finest scale active in the problem. We conclude by providing some numerical results illustrating the performance of the algorithm for the following problems: viscosity dominated propagation compared with results from a symmetric stress jump experiment; two-process (viscous-toughness) propagation for the same symmetric stress jump as imposed in the experiment; three-process (viscous-toughness-leak-off) propagation subject to a confining stress field that drops discontinuously across two interfaces. The ILSA-mk results provided in this numerical example were used to calibrate and develop the Extended Pseudo-3D model described in [10].

8.6 Acknowledgment

The author gratefully acknowledges grant support from NSERC (Canada) and the BC Oil and Gas Commission.

References

1 Adachi, J.I. and Detournay, E. (2008). Plane-strain propagation of a fluid-driven fracture in a permeable medium. *Engineering Fracture Mechanics* 75: 4666–4694.

2 Adachi, J., Siebrits, E., Peirce, A., and Desroches, J. (2007). Computer simulation of hydraulic fractures. *International Journal of Rock Mechanics & Mining Systems* 44 (5): 739–757.

3 Bunger, A.P. and Detournay, E. (2008). Experimental validation of the tip asymptotics for a fluid-driven crack. *Journal of the Mechanics and Physics of Solids* 56: 310115.

4 Carter, E.D. (1957). Optimum fluid characteristics for fracture extension. In: *Drilling and Production Practices* (ed. G.C. Howard and C.R. Fast), 261270. Tulsa, OK: American Petroleum Institute.

5 Clifton, R.J. and Abou-Sayed, A.S. (1979). On the computation of the three-dimensional geometry of hydraulic fractures. *Paper Presented at the Symposium on Low Permeability Gas Reservoirs*, (20–22 May 1979). Richardson Texas USA: Society of Petroleum Engineers. SPE 7943

6 Crouch, S.L. and Starfield, A.M. (1983). *Boundary Element Methods in Solid Mechanics*. London: George Allen and Unwin.

7 Desroches, J., Detournay, E., Lenoach, B. et al. (1994). The crack tip region in hydraulic fracturing. *Proceedings. Mathematical and Physical Sciences* 447: 3948.

8 Detournay, E. (2016). Mechanics of hydraulic fractures. *Annual Review of Fluid Mechanics* 48: 31139.

9 Detournay, E. and Peirce, A. (2014). On the moving boundary conditions for a hydraulic fracture. *International Journal of Engineering Science* 84: 147–155.

10 Dontsov, E. and Peirce, A. (2015). An enhanced pseudo-3D model for hydraulic fracturing accounting for viscous height growth, non-local elasticity, and lateral toughness. *Engineering Fracture Mechanics* 42: 116–139.

11 Dontsov, E. and Peirce, A. (2015). A non-singular integral equation formulation to analyse multiscale behaviour in semi-infnite hydraulic fractures. *Journal of Fluid Mechanics* 781: R1.

12 Dontsov, E. and Peirce, A. (2016). Implementing a universal tip asymptotic solution into an Implicit Level Set Algorithm (ILSA) for multiple parallel hydraulic fractures. Submitted to Proceedings of Amer. Rock Mech. Assoc. (ARMA2016) Houston, ARMA 16-268.

13 Dontsov, E. and Peirce, A. (2017). A multiscale implicit level set algorithm (ILSA) to model hydraulic fracture propagation incorporating combined viscous, toughness, and leak-off asymptotics. *Comp. Meth. in Appl. Mech. and Eng.* 313: 53–84.

14 Economides, M.J. and Nolte, K.G. (ed.) (2000). *Reservoir Stimulation*. New York: Wiley.

15 Garagash D. (1998). Near-Tip Processes of Fluid-Driven Fractures. Ph.D. thesis. University of Minnesota, Minneapolis.

16 Garagash, D.I. and Detournay, E. (2000). The tip region of a fluid-friven fracture in an elastic medium. *Journal of Applied Mechanics* 67 (1): 183–192.

17 Garagash, D., Detournay, E., and Adachi, J. (2011). Multi-scale tip asymptotics in hydraulic fracture with leak-off. *Journal of Fluid Mechanics* 669: 260297.

18 Gordeliy, E. and Peirce, A.P. (2013). Coupling schemes for modeling hydraulic fracture propagation using the XFEM. *Computer Methods in Applied Mechanics and Engineering* 253: 305–322.

19 Gordeliy, E. and Peirce, A.P. (2013). Implicit level set schemes for modeling hydraulic fractures using the XFEM. *Computer Methods in Applied Mechanics and Engineering* 266: 125–143.

20 Gordeliy, E. and Peirce, A.P. (2015). Enrichment strategies and convergence properties of the XFEM for hydraulic fracture problems. *Computer Methods in Applied Mechanics and Engineering* 283: 474–502.

21 Jeffrey, R.G. and Bunger, A.P. (2009). A detailed comparison of experimental and numerical data on hydraulic fracture height growth through stress contrasts. *SPE Journal* 14 (3): 413–422.

22 Lecampion, B., Peirce, A.P., Detournay, E. et al. (2013). The impact of the near-tip logic on the accuracy and convergence rate of hydraulic fracture simulators compared to reference solutions. In: *Effective and Sustainable Hydraulic Fracturing*, vol. 43 (ed. A.P. Bunger, J. McLennan and R. Jeffrey) ISBN 978-953-51-1137-5, 855–873. Intech.

23 Lenoach, B. (1995). The crack tip solution for hydraulic fracturing in a permeable solid. *Journal of the Mechanics and Physics of Solids* 43: 102543.

24 Madyarova, M. (2003). Fluid-driven penny-shaped fracture in permeable rock. Ph.D. thesis. University of Minnesota.

25 Oberhettinger, F. (1970). *Tables of Mellin Transforms*. NY: Springer-Verlag.

26 Pandurangan, V., Peirce, A., Chen, Z.R., and Jeffrey, R.G. (2017). Tiltmeter mapping of measured non-symmetric hydraulic fracture growth in a conglomerate-sandstone formation using the implicit level set algorithm and the extended Kalman filter. *SPE Journal* https://doi.org/10.2118/186100-PA.

27 Peirce, A., Spottiswoode, S., and Napier, J.A.L. (1992). The spectral boundary element method: a new window on boundary elements in rock mechanics. *International*

Journal of Rock Mechanics and Mining Sciences & Geomechanics Abstracts 29 (4): 379–400.

28 Peirce, A. and Napier, J.A.L. A spectral multipole method for efficient solution of large-scale boundary element. *International Journal for Numerical Methods in Engineering* 38: 4009.

29 Peirce, A.P. and Siebrits, E. (2005). A dual mesh multigrid preconditioner for the efficient solution of hydraulically driven fracture problems. *International Journal of Numerical Methods in Engineering* 63 (13): 1797–1823.

30 Peirce, A. (2006). Localized Jacobain ILU preconditioners for hydraulic fractures. *International Journal of Numerical Methods in Engineering* 65 (12): 1935–1946.

31 Peirce, A. and Detournay, E. (2008). An implicit level set method for Modeling hydraulically driven fractures. *Computer Methods in Applied Mechanics and Engineering* 197: 2858–2885.

32 Peirce, A. and Rochinha, F. (2012). An integrated extended Kalman filter implicit level set algorithm for monitoring planar hydraulic fracture. *Inverse Problems* 28: 015009.

33 Peirce, A. (2015). Modeling multi-scale processes in hydraulic fracture propagation using the implicit level set algorithm. *Computer Methods in Applied Mechanics and Engineering* 283: 881–908.

34 Peirce, A.P. (2016). Implicit level set algorithms for modelling hydraulic fracture propagation. *Philosophical Transactions of the Royal Society A* 374: 20150423.

35 Rice, J.R. (1968). Chap. 3. In: *Mathematical Analysis in the Mechanics of Fracture, Fracture, an Advanced Treatise* (ed. H. Liebowitz), 191–311. New York: Academic Press.

36 Rochinha, F. and Peirce, A. (2010). Monitoring hydraulic fractures: state estimation using an extended Kalman filter. *Inverse Problems* 26 (2): 025009, 17p.

37 Sethian, J.A. (1999). *Level Set Methods and Fast Marching Methods: Evolving Interfaces in Computational Geometry Fluid Mechanics Computer Vision and Materials Science.* Cambridge: Cambridge University Press.

38 Spence, D.A. and Sharp, P. (1985). Self-similar solutions for Elastohydro dynamic cavity flow. *Proceedings of the Royal Society of London. A* 400: 289313.

39 Spence, D.A., Sharp, P.W., and Turcotte, D.L. (1987). Buoyancy-driven crack propagation: a mechanism for magma migration. *Journal of Fluid Mechanics* 174: 135153.

40 Siebrits, E. and Peirce, A.P. (2002). An efficient multi-layer planar 3D fracture growth algorithm using a fixed mesh approach. *International Journal for Numerical Methods in Engineering* 53: 691717.

41 van As, A. and Jeffrey, R.G. (2000). Caving induced by hydraulic fracturing at North Parkes mines. In: *Pacific Rocks 2000*, 353–360. Rotterdam: Balkema.

42 Zia, H. and Lecampion, B. (2019). Explicit versus implicit front advancing schemes for the simulation of hydraulic fracture growth. *International Journal for Numerical and Analytical Methods in Geomechanics* 43 (6): 13001315.

9

Plasticity: A Mechanism for Hydraulic Fracture Height Containment

Panos Papanastasiou

Department of Civil and Environmental Engineering, University of Cyprus, Nicosia, Cyprus

9.1 Introduction

One of the most important questions of hydraulic fracture design is whether the fracture will remain vertically contained, achieving a long penetration in the target zone. A contained fracture will minimize the risk of mixing hydrocarbons with water of the surrounding layers and maximize the flow of hydrocarbon in the wellbore through the long conductive path created by the propped fracture. The most obvious explanation for the observed fracture containment in reservoirs is the horizontal stress difference between the reservoir and the barrier cap rock. It is expected that mechanical properties of fractured and surrounding layers have an influence on the fracture height. In order to predict fracture containment, the hydraulic fracturing simulators assume that the barrier shale layers either are under higher horizontal stress, often called closure stress, or have higher fracture toughness than the reservoir rock. In many cases, even though such conditions do not exist in the barrier layer, it was observed that the hydraulic fractures still remain contained and often they do not even reach the shale-reservoir interface. In practice the cased wellbore is perforated in the lower part of the reservoir to minimize the risk of hydraulic fracture to reach the upper barrier layer.

Over the past half-century of the development of hydraulic fracturing technique, many laboratory experiments [1–4], analytical and numerical modeling [5–9], and field tests [10] were carried out on fracture containment. The mechanisms for fracture containment were divided into three groups: (i) stress contrast with higher values in the adjacent layers, (ii) rock property contrast with higher resistance to fracturing in the barrier layers, and (iii) weak bonding interfaces [11–16]. By the end of the last century, the effect of closure-stress contrast dominated as the main reason for fracture containment [17, 18]. The closure-stress contrast means the differences in the horizontal in situ stress between adjacent layers which is expected to be higher in the stiffer shale layers. This case is predominately referred to hydraulic fracture stimulation in conventional sandstone reservoirs which have lower horizontal minimum in situ stress than the adjacent shale layers. In unconventional shale reservoirs, the stress-contrast mechanism is not in place to control fracture height which is still a critical parameter in hydraulic fracture design. Instead, it is assumed that fracture height is limited by weak interfaces. In multilayer formation, fracture propagation in the vertical direction arrests when the fracture tip terminates at layer interfaces that are weaker than the caprock. This fracture height containment mechanism is independent of the closure-stress contrast mechanism discussed above.

A different mechanism is presented here which can explain the fracture height containment in both unconventional shale reservoirs and in conventional reservoir sandstones without the presence of a stress-contrast. This finding is a result of hydraulic fracture simulation studies in weak formations which were carried out originally to investigate the high-net pressures that are often observed in field operations and the differences between simulators and field measurements. This issue was a highly debatable in the nineties [19–22]. These studies investigated the influence of plastic deformation in hydraulic fracturing of weak rocks using a coupled elastoplastic hydraulic fracturing model based on non-linear finite element analysis and robust fracture propagation criterion. The findings were verified by experimental work of van Dam et al. [23]. It has been found that the pumping energy dissipates in plastic deformation around the propagating fracture tip providing a shielding mechanism to the cracking line resulting in a significant increase of the rock's effective fracture toughness [20]. As a result, a fracture that propagates in a weak, plastically deformed formation (plastic fracture) is shorter and wider than a fracture that propagates in a strong elastic

Mechanics of Hydraulic Fracturing: Experiment, Model, and Monitoring, First Edition.
Edited by Xi Zhang, Bisheng Wu, Diansen Yang, and Andrew Bunger.

formation (elastic fracture) of the same fracture volume [19]. Hence, a higher pumping pressure is needed to propagate a plastic fracture than an elastic fracture [19, 20, 22]. Other relevant applications of this work can be found in assessing the risk of induced hydraulic fracturing in CO_2 geological storage [24]. In related recent studies, Garagash and Osiptsov [25] examined the effects of nonuniform initial stress state on apparent fracture toughness and Wrobel et al. [26] investigated the redirection of a crack driven by viscous fluid, taking into account plastic effects in the process zone.

This chapter is structured as follows: In the next section, we present finite element method (FEM) results that show that the effective fracture toughness depends on the fracture propagation direction. In Section 9.3, we compare the balance effect of effective fracture toughness vs closure stress for vertical and horizontal fracture propagation. In Section 9.4, we present a new brittleness index which predicts brittle to ductile fracture propagation and how it varies with propagation direction. The main conclusions are drawn in Section 9.5.

9.2 The Dependence of the Effective Fracture Toughness on Propagation Direction

The results of the studies which examined the influence of plastic deformation in hydraulic fracturing netpressure and fracture dimensions were obtained using a coupled elastoplastic hydraulic fracturing model based on finite element analysis [19–22]. Figure 9.1 illustrates the involved processes and mechanisms that were modeled by the coupled elastoplastic hydraulic fracturing model. The injected viscous fluid pressurizes the fracture surfaces to create an opening and when critical stress condition is reached at the crack-tip, the rock splits and the fracture propagates in the direction perpendicular to the minimum in situ stress (closure stress), $\sigma_{3,\,\mathrm{min}}$. The fluid-front during propagation may not reach the crack-tip forming a dry zone, called fluid-lag. The existence of the fluid-lag is important as it

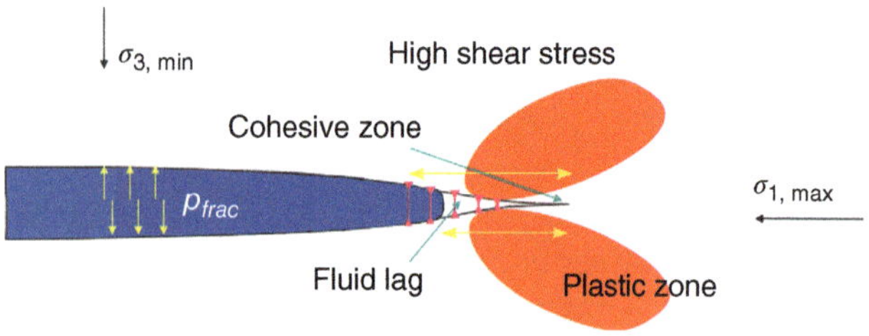

Figure 9.1 Illustration of the involved processes in hydraulic fracture propagation in weak rocks (*Source:* Adapted from Papanastasiou [27]).

affects the stress concentration near the tip and hence the deformation and propagation conditions. If the rock is weak, relative to the stress field, it may yield around the tip due to high shear stresses. High shear stresses develop with the fracture pressurization because the stress perpendicular to the fracture plane decreases, whereas the stress parallel to the propagation direction remains nearly unchanged. The formed plastic zones dissipate the energy that is provided to fracture the rock, shielding the rock from splitting. Straight ahead of the fracture-tip along the propagation direction the rock enters in a tensile mode following a strain-softening behavior until the displacement perpendicular to the fracture plane reaches a critical value at which a complete separation takes place and the fracture propagates. Branching of the fracture front is possible at the microscale causing surface roughness but in the macroscale the fracture will align itself perpendicular to the minimum in situ stress [23].

The fracture propagation was modeled utilizing the following assumptions for building the plane-strain numerical model:

a) Plasticity theory linking the aperture with fracture pressurization in the presence of in situ stresses.
b) Lubrication theory for the fluid flow inside the crack channel.
c) Fluid leak off was ignored assuming an impermeable medium.
d) Propagation criterion based on cohesive interface elements of zero thickness along the propagation direction.

Gravity force, which potentially contributes to the upward fracture growth, was ignored in the FEM model but comparison with the influence of gravity on closure stress is presented in the next section.

The unknown values of the effective fracture toughness were determined during propagation by calculating the path independent J-integral along different contours that do not cross the plastic zones [28]. It was found that the path-independent J-integral and hence the effective fracture toughness reach a limit value after a propagation length, required for the active plastic zones to be fully developed. Once the crack-tip advances, the plastic zones unload elastically near the body of the fracture. Interaction at the pore-scale between plasticity and fracturing which can explain the mechanism of fracture toughness was studied in Tjioe and Borja [29]. A significant increase of the rock fracture toughness with confining pressure was measured in experiments on Kimachi sandstone by Funatsu et al. [30]. The increase was approximately 470% at 9 MPa confinement over its value at atmospheric pressure.

It has been found that the plastic yielding around the crack-tip of a propagating fracture provides an effective shielding on the rock splitting process, resulting in a significant increase of the rock apparent or effective fracture toughness [20]. We will present here how the effective fracture toughness depends on the propagation direction. Table 9.1 presents value of the effective fracture toughness determined from the J-integral during propagation for varying (i) rock strength in terms of the uniaxial compressive strength, σ_C, and the uniaxial tensile strength, σ_T, and (ii) in situ stresses in terms of the maximum, σ_1, and minimum, σ_3, field stresses. The case of first row corresponds to propagation in a hydrostatic field or to the fracture front propagating horizontally and the cases of second and third rows correspond to propagation in a non-hydrostatic field or the fracture front propagates vertically.

The main conclusion from Table 9.1, relevant to this chapter, is that for the same rock strength parameters the calculated effective fracture toughness is a strong function of the relevant field stress anisotropy. In a normal stress environment, when the fracture front propagates horizontally the stress-field near the front is in a plane-strain condition with the active stress field (σ_1, σ_3) to be nearly isotropic (line 1). When the fracture front advances in the vertical direction, the active stress field (σ_1, σ_3) is anisotropic (lines 2, 3). Plastic yielding does not develop for a hydrostatic field, even for the weakest formation characterized by $\sigma_c/\sigma_T = 20/2$ (column 1), and the effective fracture toughness remains equal to the constitutive definition of the interface elements, $K_{IC} = 2\,\text{MPa}\sqrt{\text{m}}$. For a strong formation $\sigma_c/\sigma_T = 60/6$ (column 3), plastic yielding does not develop even in a highly anisotropic stress-field and the effective fracture toughness maintains its initial elastic value. It is noted here that the tensile strength of the material, σ_T, enters in the propagation criterion but its influence on the elevated effective fracture toughness appears to be small compared to the influence of the rock shear strength. Of course, the two strength parameters of rocks are interrelated.

Table 9.1 Values of the effective fracture toughness [in MPa $m^{1/2}$] for different values of insitu stresses and rock strength.

		Rock strength (MPa)		
		$\sigma_C = 60$ $\sigma_T = 6$	$\sigma_C = 20$ $\sigma_T = 6$	$\sigma_C = 20$ $\sigma_T = 2$
In situ stresses (MPa)	$\sigma_1 = 30$, $\sigma_3 = 30$ $\frac{\sigma_1}{\sigma_3} = 1$	2.0	2.0	2.0
	$\sigma_1 = 45$, $\sigma_3 = 30$ $\frac{\sigma_1}{\sigma_3} = 1.5$	2.0	4.60	7.31
	$\sigma_1 = 60$, $\sigma_3 = 30$ $\frac{\sigma_1}{\sigma_3} = 2$	2.0	7.03	15.48

Source: Adapted from Papanastasiou [20].

In other words, the effective fracture toughness in hydraulic fracturing depends both on rock strength and stress field and cannot be considered in isolation as a function of one parameter. The relevant stress field for fracturing depends clearly on the propagation direction. Furthermore, it is mentioned here for completeness, the effective fracture toughness increases also with the pumping parameters (flow rate × fluid viscosity) [20]. The main conclusion is that the effective fracture toughness depends strongly on the stress-field relevant to fracture plane which depends on the direction the fracture front propagates. Apparently, when the fracture front advances vertically, the rock will resist more to splitting due to high effective fracture toughness, whereas when the front advances horizontally, the resistance to splitting is less as there is no elevation of the effective fracture toughness. This finding suggests that a planar fracture is likely to propagate horizontally remaining contained in the target zone.

9.3 Effective Fracture Toughness vs. Closure Stress

Contrary to the conclusion above, one may claim that the decrease of the closure stress with fracture growth in the vertical direction could be a reason for a fracture to propagate vertically than horizontally where the closure stress remains unchanged. Hence, the issue here is which of the resistance mechanisms will dominate, the effective fracture toughness or the closure stress. The answer to this question can be found by comparing the results for relevant parameters.

By definition, the net pressure in hydraulic fracturing is defined by

$$P_{net} = P_f - \sigma_{min} \tag{9.1}$$

where P_f is the fluid pressure in the fracture and σ_{min} is the closure stress.

For a uniform loaded fracture of length α, the stress intensity factor K_I according to linear elastic fracture mechanics is given by

$$K_I = P_{net}\sqrt{\pi\alpha} \tag{9.2}$$

Combining (9.1) and (9.2) gives

$$K_I = \left(P_f - \sigma_{min}\right)\sqrt{\pi\alpha} \tag{9.3}$$

A fracture will propagate when the stress intensity factor is equal with material fracture toughness or the effective fracture toughness.

$$K_I = K_{EFT} \tag{9.4}$$

Combining (9.3) and (9.4) and solving for the fluid pressure gives

$$P_f = \sigma_{min} + \mathrm{K}_{\mathrm{EFT}}/\sqrt{\pi\alpha} \tag{9.5}$$

The vertical decrease of the closure stress is given by the local stress gradient of the horizontal stress which is approximately 15.8 kPa/m (0.7 psi/ft). As a fracture propagates vertically upward the resistance decreases by 15.8 KPa/m but soon after propagation the resistance had increased proportionally to K_{EFT}. It was found that after a fracture propagation of the order of a meter, the effective fracture toughness has been fully developed [20].

As fracture propagates horizontally the closure stress does not change and the resistance due to fracture toughness is small as $K_{\mathrm{EFT}} = K_{\mathrm{IC}}$. As an example, we plotted in Figure 9.2 the fluidpressure that is needed for propagating a fracture vertically and horizontally using the data of last column of Table 9.1 which correspond to a weak formation at 1900 m depth. It is clear that much lower pressure is needed to propagate the fracture horizontally after it is initiated at wellbore. Although the pressure needed for propagating the fracture vertically appears to be lower after the fracture length (or fracture height in this case) exceeds 60 m, this height might never be reached as the fracture is likely to propagate horizontally under lower pressure.

These results and the conclusions are strictly valid for hydraulic fracturing in a normal stress environment where the in situ vertical stress is the maximum and one of the horizontal stresses is the minimum. In the case of a strike-slip stress regime, the opposite results are expected with highly effective fracture toughness to be mobilized when the fracture propagates horizontally rather than vertically. In such a case, the fracture will not remain contained in the target zone unless another fracture containment mechanism is in place at the barrier layer or interface. Finally, in the case of hydraulic fracturing in a reverse faulting stress regime the vertical stress is the minimum and hence the fracture plane will be horizontal.

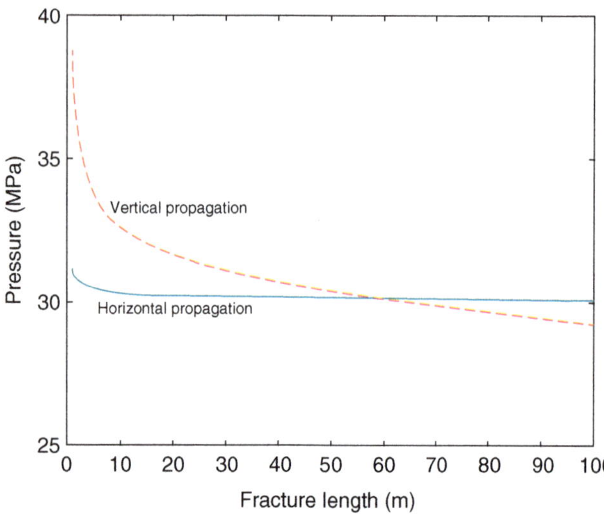

Figure 9.2 Fluidpressure for propagating a planar fracture horizontally and vertically.

In a normal stress environment and in the absence of stress barrier, the assumed penny-shape fracture geometry may not be valid if the formation is weak. The fracture is expected to be of an elliptical shape with a long horizontal axis. Accurate prediction of the fracture geometry requires a 3D planar HF model based on plasticity in order to account properly for propagation direction dependent effective fracture toughness.

9.4 A New Brittleness Index Defines Fracture Containment

As it was presented in Section 9.2, the generated effective fracture toughness depends both on rock strength and in situ stresses. It was also mentioned that the effective fracture toughness increases with the pumping parameters (flow rate × fluidviscosity) as they affect the stress concentration around the tip and hence the generated plastic zones [20]. For the sake of discussion, we will ignore the influence of the pumping parameters assuming a uniform pressurized fracture. In such a case, the influence of the rock strength and in situ stress field can be combined in a single parameter given by

$$\mathrm{BI} = \frac{(\sigma_1 - \sigma_3)}{2\mathrm{c}\cos\varphi + (\sigma_1 + \sigma_3)\sin\varphi} \tag{9.6}$$

where c and φ are the material cohesion and friction, respectively. The definition of BI in (9.6) is valid for the general case of a slow propagation in a frictional material where the deformation takes place under drained conditions governed by the effective stresses.

In the case of a fast propagation in impermeable formation the deformation is undrained and is governed by the total stresses, $\varphi = 0$, and the BI is simplified to

$$\mathrm{BI} = \frac{(\sigma_1 - \sigma_3)}{2\mathrm{c}} \tag{9.7}$$

The dominant single parameter (9.6) or (9.7) was derived from a model based on dislocation theory for

Mohr–Coulomb material that accounts for small- and large-scale yielding that develops around the crack-tip [31, 32]. Small-scale yielding is valid when the irreversible deformation is contained in a small region near the tip which is much smaller than the fracture length. In the large-scale yielding, the size of the plastic zone becomes important relative to the fracture length and the full solution of the problem is needed. In the dislocation model, it is assumed that the plastic yielding describes the irreversible deformation in the form of distributed normal and shear dislocations. The effect of plasticity through the distributed dislocations was further replaced by super-dislocations that were placed in the effective centers of the complete distributed slip processes. The analytical model enables the identification of the dominant parameter (9.6) or (9.7) which serves as a new brittleness index that defines the regimes of brittle to ductile fracture propagation and the limit beyond which no fracture can propagate in mode-1 due to the dissipation of energy by ductile deformation [33]. A summary on different definitions of the brittleness index can be found in Holt et al. [34] and Feng et al. [35]. This new definition is the only brittleness index definition that combines both rock strength and in situ stresses. As it is well known, the transition of a rock behavior from brittle to ductile depends strongly on the confining stress.

The parameter BI in (9.6) or (9.7), which dictates the scale of plastic yielding for a Mohr–Coulomb and von Mises material, respectively, can vary between 0 and 1. For value 0, the material is elastic or brittle during fracturing. For increasing value of BI the scale of plastic yielding increases. The asymptotic limit value of BI parameter is 1 which corresponds to the case that the material yields everywhere requiring infinite energy for mode-1 fracture propagation.

As an example, we present in Figure 9.3 the results of the dislocation model for a hydraulic fracture at the state of propagation for different values of the brittleness index. The dimensionless quantities shown in the y-axis are the center of dislocation $l^{1/2}/[K_{el}/c]$, the strength of the dislocation, $b/[(1-\nu^2)K_I^2/(E\ c)]$, the crack opening displacement, $\delta/[(1-\nu^2)\mathrm{Kel}^2/(E\ c)]$, and the dislocation force, $F/[(1-\nu^2)K_I^2/E]$, where K_I is the stress intensity factor, c is the material cohesion, and E, ν are the Elastic modulus and Poisson ratio, respectively. The dislocation length and strength are proportional to the size of the plastic zones. For short fractures, the size of the plastic zones could be comparable to the fracture length. The crack opening displacement and the dislocation force are related to the strain energy release rate and hence are good indicators of the effective fracture toughness.

The propagation regimes are clearly dependent on the orientation of the propagating fracture front with respect to the in situ stress field. We assume that the stress field is isotropic (horizontal stresses are equal) and the vertical stress is the maximum. For a vertical plane fracture that its front propagates in the horizontal direction the parameter BI will be close to 0 and the fracture will propagate in a nearly brittle mode for which the conditions for linear elastic fracture mechanics (LEFM) hold. If the fracture front propagates vertically, as the vertical in situ stress is greater than the horizontal stresses, the value of BI will be much greater than 0 with a limit value approaching 1 for a relatively weak formation. As the value of BI increases toward 1 the size of the plastic zones becomes important relative to the fracture length. For values of BI approaching 1, the energy dissipated in the plastic zones grows abruptly suggesting that a large portion of the supplied pumping energy for fracturing is lost in the plastic deformation, elevating the effective fracture toughness for mode-1 propagation. In this case, the pressurized fracture will inflate like a balloon shape of a maximum width.

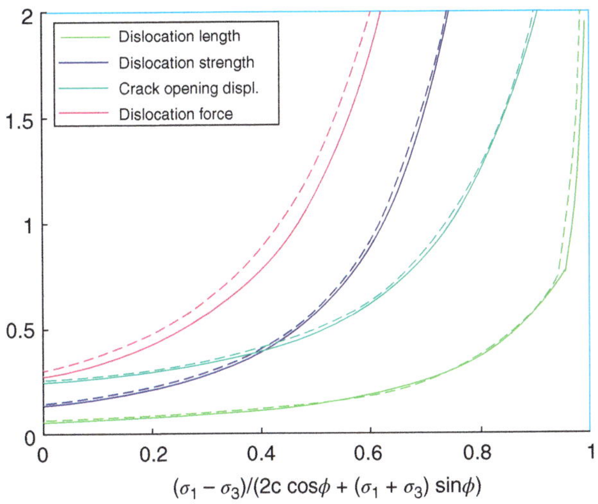

Figure 9.3 Fracture parameters vs brittleness index for drained frictional analysis for small- (solid lines) and large-scale yielding (dashed lines).

In Figure 9.3, the results to the left of the curves correspond to a fracture front that propagates horizontally. As moving to the right, the results correspond to a fracture front that propagates vertically. It is obvious that much higher energy is dissipated in plastic deformation around the tip of the fracture that propagates vertically elevating accordingly the effective fracture toughness.

9.5 Conclusions

In this chapter, we presented a potential mechanism for hydraulic fracture containment in the target zone. The study was based on earlier work on the modeling of hydraulic fracturing in weak rocks and comparison here in particular hydraulic fracturing growth in the vertical direction vs hydraulic fracturing in the horizontal direction. The findings are based on both analytical and FEM models of hydraulic fracture propagation in weak formations described by the Mohr–Coulomb plasticity material.

A mechanism based on plasticity explains why in a normal stress environment a hydraulic fracture propagates horizontally remaining contained in the target zone even in the absence of a barrier layer of higher closure stress or a weak interface. The effective fracture toughness due to plasticity shielding around the fracture tip is higher if the fracture front propagates vertically, than horizontally. When the fracture front propagates vertically the relevant stress field for fracturing is highly deviatoric inducing large plastic yielding and hence high effective fracture toughness. Propagation of the fracture front in the horizontal direction takes place in a nearly isotropic stress field for which the near tip deformation is characterized as elastic or by small-scale yielding contained at the tip and hence there is no significant increase of the fracture toughness. The regimes of brittle to ductile propagation in mode-1 is governed by an introduced new definition of a brittleness index. This brittleness index is a combination of rock strength and in situ stresses and hence its value depends on the propagation direction.

Acknowledgments

This work was funded by Schlumberger and the European Regional Development Fund and the Republic of Cyprus through the Research Promotion Foundation (RPF). Currently it is funded by the Cyprus RPF, RESTART 2016-2020 PROGRAMMES, Excellence Hubs, Project EXCELLENCE/1216/0481.

References

1 Beugelsdijk, L.J.L., de Pater, C.J., and Sato, K. (2000). Experimental hydraulic fracture propagation in a multi-fractured medium. *Proceedings of the SPE Asia Pacific Conference on Integrated Modelling for Asset Management*, Yokohama.

2 Warpinski, N.R., Schmidt, R.A., and Northrop, D.A. (1982). In-situ stresses: the predominant influence on hydraulic fracture containment. *Journal of Petroleum Technology* 34: 653–664.

3 Warpinski, N.R., Clark, J.A., Schmidt, R.A., and Huddle, C.W. (1982). Laboratory investigation on the effect of in situ stresses on hydraulic fracture containment. *Society of Petroleum Engineers Journal* 22: 333–340.

4 Xing, P., Bunger, A., Yoshioka, K. et al. (2016). Experimental study of hydraulic fracture containment in layered reservoirs. *Proceedings of the 50th US Rock Mechanics/Geomechanics Symposium*, Houston, TX.

5 Daneshy, A. (1978). Hydraulic fracture propagation in layered formations. *Society of Petroleum Engineers Journal* 18: 33–41.

6 Fung, R., Vilayakumar, S., and Cormack, D.E. (1987). Calculation of vertical fracture containment in layered formations. *SPE Formation Evaluation* 2: 518–522.

7 Guo, J., Luo, B., Lu, C. et al. (2017). Numerical investigation of hydraulic fracture propagation in a layered reservoir using the cohesive zone method. *Engineering Fracture Mechanics* 186: 195–207.

8 Simonson, E., Abou-Sayed, A., and Clifton, R. (1978). Containment of massive hydraulic fractures. *Society of Petroleum Engineers Journal* 18: 27–32.

9 Van Eekelen, H. (1982). Hydraulic fracture geometry: fracture containment in layered formations. *Society of Petroleum Engineers Journal* 22: 341–349.

10 Teufel, L.W. and Clark, J.A. (1984). Hydraulic-fracture propagation in layered rock: experimental studies of fracture containment. *Society of Petroleum Engineers Journal* 24 (1): SPE-9878-PA. https://doi.org/10.2118/9878-PA.

11 Chuprakov, D.A. and Prioul, R. (2015). Hydraulic fracture height containment by weak horizontal interfaces. *Proceedings of the SPE Hydraulic Fracturing Technology Conference*, 1–17. Woodland, TX, USA, 3–5 February. SPE-173337-MS. 10.2118/173337-MS.

12 Tang, J. and Wu, K.A. (2018). 3-D model for simulation of weak interface slippage for fracture height containment in shale reservoirs. *International Journal of Solids and Structures* 144: 248–264.

13 Wang, H., Liu, H., Wu, H.A., and Wang, X.X. (2015). A 3D numerical model for studying the effect of interface shear failure on hydraulic fracture height containment. *Journal of Petroleum Science and Engineering* 133: 280–284.

14 Weng, X., Chuprakov, D., Kresse, O. et al. (2018). Hydraulic fracture-height containment by permeable weak bedding interfaces. *Geophysics* 83: MR137–MR152.

15 Xu, W., Prioul, R., Berard, T. et al. (2019). Barriers to hydraulic fracture height growth: a new model for sliding interfaces. *Proceedings of the SPE Hydraulic Fracturing Technology Conference and Exhibition*, Woodlands, TX.

16 Zheng, Y., Liu, J., and Zhang, B. (2019). An investigation into the effects of weak interfaces on fracture height

containment in hydraulic fracturing. *Energies* 12 (7): 3245. https://doi.org/10.3390/en12173245.

17 Huang, J., Ma, X., Shahri, M. et al. (2016). Hydraulic fracture growth and containment design in unconventional reservoirs. *Proceedings of the 50th US Rock Mechanics/ Geomechanics Symposium*, Houston, TX.

18 Wasantha, P., Konietzky, H., and Xu, C. (2019). Effect of in-situ stress contrast on fracture containment during single- and multi-stage hydraulic fracturing. *Engineering Fracture Mechanics* 205: 175–189.

19 Papanastasiou, P. (1997). The influence of plasticity in hydraulic fracturing. *International Journal of Fracture* 84: 61–79.

20 Papanastasiou, P. (1999). The effective fracture toughness in hydraulic fracturing. *International Journal of Fracture* 96: 127–147.

21 Papanastasiou, P. and Thiercelin, M. (1993). Influence of inelastic rock behaviour in hydraulic fracturing. *International Journal of Rock Mechanics and Mining Sciences* 30 (7): 1241–1247.

22 Sarris, E. and Papanastasiou, P. (2013). Numerical modelling of fluid-driven fractures in cohesive poro-elastoplastic continuum. *International Journal for Numerical and Analytical Methods in Geomechanics* 37 (12): 1822–1846.

23 Van Dam, D.B., Papanastasiou, P., and de Pater, C.J. (2000). Impact of rock plasticity on hydraulic fracture propagation and closure. *Journal of SPE Production and Facilities* 17 (3): 149–159.

24 Papanastasiou, P., Papamichos, E., and Atkinson, C. (2016). On the risk of hydraulic fracturing in CO_2 geological storage. *International Journal for Numerical and Analytical Methods in Geomechanics* 40: 1472–1484.

25 Garagash, I.A. and Osiptsov, A.A. (2020). Effects of nonuniform initial stress state on apparent fracture toughness. *Engineering Fracture Mechanics* 226: 106837.

26 Wrobel, M., Piccolroaz, A., Papanastasiou, P., and Mishuris, G. (2019). Redirection of a crack driven by viscous fluid taking into account plastic effects in the process zone. *Geomechanics for Energy and the Environment* 26: 100147.

27 Papanastasiou, P. (1999). An efficient algorithm for propagating fluid-driven fractures. *Computational Mechanics* 24: 258–267.

28 Rice, J.R. (1968). A path-independent integral and the approximate analysis of strain concentration by notches and cracks. *Journal of Applied Mechanics Transaction ASME* 35: 379–386.

29 Tjioe, M. and Borja, R.I. (2015). On the pore-scale mechanisms leading to brittle and ductile deformation behavior of crystalline rocks. *International Journal for Numerical and Analytical Methods in Geomechanics* https://doi.org/10.1002/nag.2357.

30 Funatsu, T., Seto, M., Shimada, H. et al. (2004). Combined effects of increasing temperature and confining pressure on the fracture toughness of clay bearing rocks. *International Journal of Rock Mechanics and Mining Sciences* 41 (6): 927–938. https://doi.org/10.1016/j.ijrmms.2004.02.008.

31 Papanastasiou, P. and Atkinson, C. (2000). Representation of crack-tip plasticity in pressure sensitive geomaterials. *International Journal of Fracture* 102: 271–286.

32 Papanastasiou, P. and Atkinson, C. (2006). Representation of crack-tip plasticity in pressure-sensitive geomaterials: large scale yielding. *International Journal of Fracture* 139: 137–144.

33 Papanastasiou, P. and Atkinson, C. (2015). The brittleness index in hydraulic fracturing. ARMA 15-489. In 49th US Rock Mechanics/Geomechanics Symposium 2015 (Vol. 4, pp. 3149–3155), 29 June–1 July 2015, San Francisco, California, American Rock Mechanics Association

34 Holt, R.M., Fjær, E., Stenebraten, J.F., and Nes, O.M. (2015). Brittleness of shales: relevance to borehole collapse and hydraulic fracturing. *Journal of Petroleum Sciences and Engineering* 04 (2015): 131. https://doi.org/10.1016/j.petrol.2015.04.006.

35 Feng, R., Zhang, Y., Rezagholilou, A. et al. (2020). Brittleness index: from conventional to hydraulic fracturing energy model. *Rock Mechanics and Rock Engineering* 53: 739–753.

10

Turbulent Flow Effects on Propagation of Radial Hydraulic Fracture in Permeable Rock

E.A. Kanin[1], D.I. Garagash[2], and A.A. Osiptsov[1]

[1] *Project Center for Energy Transition and ESG, Skolkovo Institute of Science and Technology (Skoltech), Moscow, Russian Federation*
[2] *Department of Civil and Resource Engineering, Dalhousie University, Halifax, Nova Scotia, Canada*

10.1 Introduction

Hydraulic fractures (HFs) are tensile cracks that form and propagate in the solid media with the preexisting confining stress due to the high-pressure fluid injection. HFs manifest in nature as magma-driven dykes [1–4] and water-driven cracks in glacier beds [5, 6]. They can also be human-made, when hydraulic fracturing treatment of oil and gas wells is utilized to enhance production of hydrocarbons [7, 8]. HFs ensure a larger hydraulically connected area between the rock and the well. In recent times, the horizontal drilling with the multistage hydraulic fracturing is frequently applied, especially in the development of formations with low permeability and porosity.

When the water-based fracturing fluid is used in reservoir stimulation, the injection rate can be set relatively high. It is required to compensate for the undesirable effect of the relatively low viscosity of the HF fluid on the proppant settling and in order to create a crack aperture sufficiently large for the proppant placement [9]. High injection rate of low-viscous fluid leads to the onset of the turbulent flow regime in the part of the fracture adjacent to the wellbore. The remaining part of the fracture, i.e. the region from the transition boundary to the fracture front, is occupied by laminar flow (Figure 10.1). Since Reynolds number for the plane channel flow depends on the aperture, the laminar flow regime always exists near the fracture front where the aperture value tends to be zero.

Turbulent flow effects were first discussed in the context of hydraulic fracture propagation by Perkins and Kern [10] for the Perkins–Kern–Nordgren (PKN) fracture geometry. Further, Nilson [11] analyzed a model for a plane strain gas-driven fracture, and this work discusses the flow regime transformation inside the crack channel from turbulent to laminar in the direction of the fracture front. The limiting propagation regimes associated with turbulent flow inside magma dykes were studied by Emerman et al. [12], Lister and Kerr [13]. Tsai and Rice [5] considered a plane strain HF with fully turbulent flow in application to natural water-driven fracture in glaciers. Different aspects of the turbulent flow regime impact on the HF growth with PKN geometry have been investigated in the following works [14–18]. Ames and Bunger [14] highlighted that the calibration of the fully laminar crack model on the field data can lead to the incorrect estimation of Carter's leak-off coefficient if the crack is driven by turbulent flow. It happened because the fully laminar model overestimates crack size and underestimates pressure at the wellbore. The analyses of either fully turbulent or turbulent–laminar flow influence on the crack propagation have also been carried out for fractures with different geometries including a plane strain [19, 20], radial [21–23], planar 3D [24] fracture. In the enumerated works, researchers studied not only the turbulent flow in a crack channel with smooth walls but also with rough walls. For example, Zia and Lecampion [17] derived turbulent smooth and turbulent rough limiting propagation regimes for a PKN fracture, while similar regimes for a radial hydraulic fracture were obtained by Lecampion and Zia [21]. In addition, Dontsov and Peirce [24] and Zolfaghari and Bunger [22] demonstrated the evolution of the radial, PKN, and plane strain crack parameters for different values of the crack wall relative roughness. By constructing the Planar3D fracture model with turbulent–laminar flow, Dontsov and Peirce [24] were able to investigate how crack evolves in the layered formation, e.g. symmetric stress barrier and stress drop cases. The aforementioned works consider finite crack models; however, the laminar-to-turbulent flow regime transformation has also been studied in the asymptotic near-tip region model. For example, Dontsov [25] developed a model for a HF tip for the pure water case, while the slickwater case was examined by Lecampion and Zia [21]; Kanin et al. [26].

Mechanics of Hydraulic Fracturing: Experiment, Model, and Monitoring, First Edition.
Edited by Xi Zhang, Bisheng Wu, Diansen Yang, and Andrew Bunger.

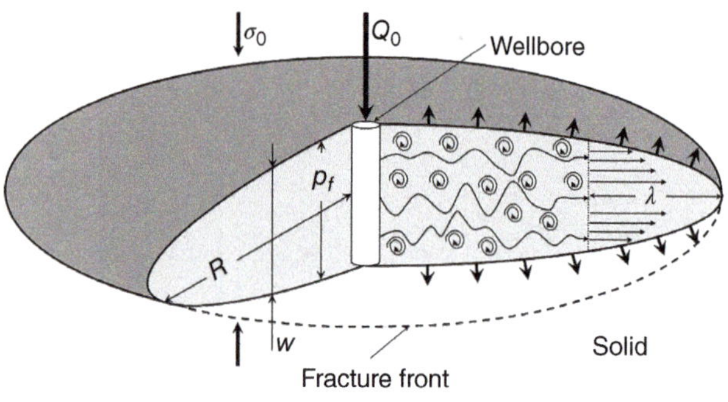

Figure 10.1 A radial hydraulic fracture model with the laminar–turbulent channel flow and leak-off.

The fluid injection into the reservoir at a relatively high volumetric flow rate requires significant energy consumption for fluid pumping [27], especially during the simultaneous growth of multiple cracks, i.e. multistage HF treatment. The limitation of the operational costs can be achieved by adding the specific polymers to pure water resulting in a mixture called slickwater. The used additives slightly increase the fluid viscosity and considerably decrease the friction, up to 70% compared to that of the pure water [28]. The experiments of Virk [29, 30] demonstrate how the flow friction factor declines depending on the used polymer additive type and its concentration. It was also determined in this work that the friction reduction has the so-called maximum drag reduction (MDR) or Virk's asymptote, which can be achieved at a relatively small concentration of the polymers.

This chapter considers a radial hydraulic fracture propagation in a permeable reservoir. The model accounts for the flow regime transformation from laminar to turbulent when Reynolds number for the plane channel flow becomes greater than the critical value. The HF fluid is slickwater, and the MDR asymptote is used to approximate its frictional behavior in the turbulent flow regime. The fluid exchange between the crack and permeable formation is assumed in the form of the leak-off process governed by the pressure-independent, one-dimensional Carter's law [31]. The main aim of the present examination is to analyze the combined effects of the laminar–turbulent flow of the HF fluid with drag reduction agents and the fluid leak-off into the ambient rock. For the problem formulation, we utilize the original model framework of Lecampion and Zia [21], where the authors discuss a radial HF driven by the laminar–turbulent flow of slickwater in impermeable rock.

Hydraulic fractures in the field may have complex geometries, and one can simulate their evolution via more sophisticated numerical approaches only, e.g. planar or nonplanar three-dimensional models. However, the proposed model has several useful features as well as other studies on a crack with simple geometry (KGD, PKN, and radial). First of all, it provides a reliable physical and mechanical framework that allows understanding the impact of various physical phenomena (in our case, laminar–turbulent slickwater flow inside the crack channel) on hydraulic fracture growth. Secondly, it can be used to guide and validate the robustness of more complex/realistic numerical fracture models.

We organize the chapter in the following way. Section 10.2 outlines the problem formulation and governing equations, while Section 10.3 introduces the numerical scheme. The solution examples corresponding to the typical field cases are considered in Section 10.4. Further, we investigate the limiting propagation regimes in Section 10.5, namely, revisit the existing limiting solutions and derive the new ones. Finally, Section 10.7 describes the results of the problem parameter space investigation, which is accomplished by using the problem formulation in dimensionless variables introduced in Section 10.6.

10.2 Model Formulation

10.2.1 Problem Definition

We examine a radial (penny-shaped) hydraulic fracture propagation from a fluid point source along the plane perpendicular to the far-field confining stress direction. The model is axisymmetric, i.e. all fracture properties are the functions of the distance to the source r and time t. The sketch of the model is presented in Figure 10.1.

We consider a constant volumetric rate Q_0 of fluid injection such that the injected volume is $V_{inj}(t) = Q_0 t$. The rock is assumed to be a homogeneous, linear-elastic solid characterized by Young's modulus E and Poisson's ratio ν. We assume that the size of the region near the fracture tip, where the rock failure occurs, is small compared to the other lengthscales realized in the model (e.g. linked with the viscous fluid flow inside the fracture, leak-off into the formation, etc.), which is generally true when the in situ confining stress σ_o is much smaller than the rock tensile strength [32]. Therefore, we can apply the linear elastic fracture mechanics (LEFM) theory to model the quasi-static crack propagation in the solid media with toughness K_{Ic} [33]. The fracture surface is loaded by the fluid pressure $p_f(r, t)$ from the internal side. The considered radial crack model is fully described by evolution in time of the fracture radius $R(t)$, opening $w(r, t)$ and net pressure $p(r, t) = p_f(r, t) - \sigma_o$ profiles, and efficiency parameter $\eta(t) = V_{crack}(t)/V_{inj}(t)$. The latter allows evaluating the partition of the injected fluid volume $V_{inj}(t)$ between that of the crack $V_{crack}(t)$

and the fluid volume leaked into the rock $V_{leak-off}(t)$.

The HF fluid is considered to be slickwater with viscosity μ and density ρ which behaves like a Newtonian fluid in the laminar flow regime. When turbulent flow occurs, slickwater rheological response is modeled by the MDR asymptote which we will discuss further. The lubrication theory [34] controls the fluid flow inside the fracture. We neglect the fluid lag [35–37], assuming that its maximum value $\sim \mu V E^2 \sigma_o^{-3}$ expressed in terms of the crack tip velocity V, is small compared to the fracture radius. In other words, the crack and the HF fluid fronts coincide in the model. Reynolds number for the plane channel flow, $Re = \rho v w/\mu$, is increasing with distance from the crack front where $Re = 0$. The laminar-to-turbulent flow regime transformation happens at distance λ from the tip (this value is a part of the solution), where Reynolds number equals to the critical value, $Re = Re_c$. Consequently, the laminar flow domain of extent λ is adjacent to the front, i.e. observed within the interval $r \in (R-\lambda, R)$ along which $Re < Re_c$, while the non-laminar flow, i.e. transient and fully turbulent, happens inside the domain $r \in (0, R-\lambda)$ characterized by $Re > Re_c$. We implement the flow regime transformation into the model via the usage of the fluid friction factor $f(Re)$ as detailed further in this section.

The ambient reservoir is characterized by porosity ϕ and permeability k. For simplicity, we utilize Carter's leak-off law [31] to govern the fluid exchange process between the crack channel and formation. According to this law, the leak-off rate $g(r, t)$ is proportional to the inverted square root of the "exposure" time, i.e. the period between the current time moment and when the crack tip passes the considered point on the fracture plane. We consider the case in which the pore fluid has the same properties as that of the slickwater (viscosity μ, the compressibility c_t). Here, the proportionality (or Carter's) coefficient is the following [38]: $C_L = k(\sigma_o - p_o)/(\mu\sqrt{\pi c})$, where p_o is the far-field pore pressure, and $c = k/(\phi c_t \mu)$ is the diffusivity coefficient. When it is required to take into account the filter-cake building or (and) the different properties of the HF and pore fluids in the pressure-independent approximation of the fluid exchange process, the reader can find the appropriate expressions for Carter's coefficient in Economides et al. [8].

10.2.2 Governing Equations

In this section, the governing equations are formulated for the evolution of unknown crack radius $R(t)$, opening $w(r, t)$, and net pressure $p(r, t)$ profiles. In doing so, we make use of the following set of material parameters in common with the laminar HF models:

$$E' = \frac{E}{1-\nu^2}, K' = 4\sqrt{\frac{2}{\pi}}K_{Ic}, \mu' = 12\mu, C' = 2C_L, \tag{10.1}$$

and the injection rate Q_0. In Eq. (10.1), E' is the plane strain elastic modulus, K' and μ' are toughness and viscosity parameters, and C' is the leak-off parameter. Additional problem parameters related to turbulent flow will be introduced further in Section 10.2.2.2.

10.2.2.1 Crack Elasticity

The elasticity equation links the net fluid pressure $p(r, t)$ and aperture $w(r, t)$ [39–41]:

$$p(r, t) = -\frac{E'}{2\pi R(t)}\int_0^{R(t)} M\left(\frac{r}{R(t)}, \frac{s}{R(t)}\right)\frac{\partial w(s, t)}{\partial s}ds, \tag{10.2}$$

where integral kernel $M(\rho, s)$ is defined as:

$$M(\rho, s) = \begin{cases} \frac{1}{\rho}\mathrm{K}\left(\frac{s^2}{\rho^2}\right) + \frac{\rho}{s^2-\rho^2}\mathrm{E}\left(\frac{s^2}{\rho^2}\right), \rho > s, \\ \frac{s}{s^2-\rho^2}\mathrm{E}\left(\frac{\rho^2}{s^2}\right), \rho < s, \end{cases} \tag{10.3}$$

and $K(x)$ and $E(x)$ are the complete elliptic integrals of the first and the second kind, respectively.

10.2.2.2 Fluid Flow

Fluid flow inside the crack channel is described by width-averaged continuity equation for an incompressible fluid [34]:

$$\frac{\partial w}{\partial t} + \frac{1}{r}\frac{\partial(rq)}{\partial r} + g(r, t) = 0, g(r, t) = \frac{C'}{\sqrt{t-t_0(r)}}, \tag{10.4}$$

where $q(r, t) = w(r, t)v(r, t)$ is the flow rate, $v(r, t)$ is the width-averaged fluid flow velocity, and $g(r, t)$ is the fluid leak-off rate given by Carter's law. The latter is defined in terms of the time $t_0(r)$ when the fracture tip reaches distance r from the source and can be obtained as the inverse of the fracture radius function, i.e. $R(t_0(r)) = r$.

We determine $v(r, t)$ from the width-averaged momentum conservation equation in which the inertial terms are neglected (see Lecampion and Zia [21] for details):

$$v|v| = -\frac{w}{\rho f}\frac{\partial p}{\partial r}, \tag{10.5}$$

where f is the Fanning friction factor. By comparing Eq. (10.5) with Poiseuille's law, $v = -(w^2/\mu')(\partial p/\partial r)$, we obtain the friction factor for laminar flow ($Re < Re_c$) in plane channel: $f^{lam} = 12/Re$. Using the definition of the normalized friction factor $\tilde{f} = f/f^{lam}$, we rewrite Eq. (10.5):

$$v = -\frac{w^2}{\mu' \tilde{f}} \frac{\partial p}{\partial r}. \tag{10.6}$$

The combination of the continuity (10.4) and momentum conservation (10.6) yields Reynolds equation:

$$\frac{\partial w}{\partial t} = \frac{1}{\mu'} \frac{1}{r} \frac{\partial}{\partial r} \left(\frac{r w^3}{\tilde{f}} \frac{\partial p}{\partial r} \right) - \frac{C'}{\sqrt{t - t_0(r)}}. \tag{10.7}$$

In the turbulent flow regime ($Re > Re_c$), the slickwater frictional behavior is governed by the MDR asymptote that is a phenomenological relation proposed by Virk [29, 30] on the basis of the experiments:

$$\frac{1}{\sqrt{f}} = 19 \log_{10} \left(Re_d \sqrt{f} \right) - 32.4, \tag{10.8}$$

where $Re_d = \rho v d/\mu$ is Reynolds number for pipe flow (d is a pipe diameter). In order to "translate" (10.8) to the "crack channel" flow geometry, the relation between apparent pipe Re_d and channel Re is postulated by equating corresponding expressions for the laminar friction factors for a channel, $f^{lam} = 12/Re$, and a pipe, $f_d^{lam} = 16/Re_d$, yielding $Re_d = \frac{4}{3} Re$ [42].

Finally, the implicit Eq. (10.8) is approximated by explicit power-law relation suggested by Lecampion and Zia [21]:

$$f = f_0 Re_d^{-n} = f_0' Re^{-n}, \tag{10.9}$$

where $f_0' = f_0 (4/3)^{-n}$, and the numerical values for f_0 and n will be provided later.

Behavior of the friction f and normalized friction $\tilde{f} = f/f^{lam}$ in the entire range of Reynolds number values is approximated by piecewise functions:

$$f = \begin{cases} 12/Re, & Re \leq Re_c, \\ f_0' Re^{-n}, & Re > Re_c, \end{cases} \qquad \tilde{f} = \begin{cases} 1, & Re \leq Re_c, \\ f_0'' Re^{1-n}, & Re > Re_c, \end{cases} \tag{10.10}$$

with prefactor $f_0'' = f_0'/12$. We estimate the critical value of Reynolds number Re_c for the slickwater case from the intersection of the laminar and turbulent branches, to ensure the continuity of $f(Re)$: $Re_c = \left(12/f_0'\right)^{1/(1-n)}$.

To summarize the discussion on the slickwater frictional behavior, we plot Figure 10.2. Here, we show the friction factor f dependence on the Reynolds number Re for the channel flow during the laminar and turbulent flow regimes in the traditional (a) and Prandtl–Karman (b) notations. Function $f(Re)$ defined by Eq. (10.10) is shown by the combination of blue (laminar part) and red (turbulent part) solid lines. In addition to Virk's asymptote (green line), we show Blasius correlation (yellow line) for pure water flow in a channel with smooth walls. Using the green and yellow curves, one can observe how the slickwater drag reduction agents drastically reduce the friction compared to that of water. The power-law form of the MDR asymptote (10.9) (red line) closely approximates the exact function (10.8) within the range $Re \in (10^3, 1.5 \cdot 10^4)$ with the relative error of less than 5%.

Numerical values of the parameters in rheological relations (10.9), (10.10) for the slickwater (power-law approximation of MDR) and pure water (Blasius correlation) cases are taken after [21], correspondingly:

$$\begin{aligned} &\text{slickwater}: f_0 = 1.78,\ n = 0.7,\ f_0' = 1.46,\ f_0'' = 0.122,\ Re_c = 1132.6; \\ &\text{water}: f_0 = 0.079,\ n = 0.25,\ f_0' = 0.074,\ f_0'' = 0.006,\ Re_c = 1650. \end{aligned} \tag{10.11}$$

The critical Reynolds number for the channel flow of Newtonian fluid (pure water) is evaluated from the corresponding parameter for the pipe flow $(Re_d)_c = 2200$: $Re_c = 2200 \cdot 3/4 = 1650$. One can notice that the laminar-to-turbulent transition for slickwater occurs at a smaller Reynolds number value than that for pure water.

10.2.2.3 Fracture Propagation

According to the LEFM theory, the quasi-static fracture propagation is realized when the stress intensity factor matches the solid toughness: $K_I = K_{Ic}$. Alternatively, this condition can be expressed in terms of asymptotic behavior of the crack opening $w(r, t)$ near the tip [43]:

$$w = \frac{K'}{E'} \sqrt{R - r}, \quad r \to R, \tag{10.12}$$

where elasticity E' and toughness K' parameters are defined in (10.1).

10.2.2.4 Boundary Conditions

At the fracture inlet ($r = 0$), the volumetric flow rate is prescribed:

$$\lim_{r \to 0} 2\pi r q(r, t) = \lim_{r \to 0} 2\pi r w(r, t) v(r, t) = Q_0, \tag{10.13}$$

while at the fracture tip ($r = R(t)$), the zero crack opening and the no-flow conditions are applied:

$$w(R, t) = 0, \quad q(R, t) = 0. \tag{10.14}$$

10.2.2.5 Global Fluid Volume Balance

Global fluid balance follows from the integration of the continuity Eq. (10.4) with respect to time and distance from the source and application of the flux boundary conditions (10.13) and (10.14),

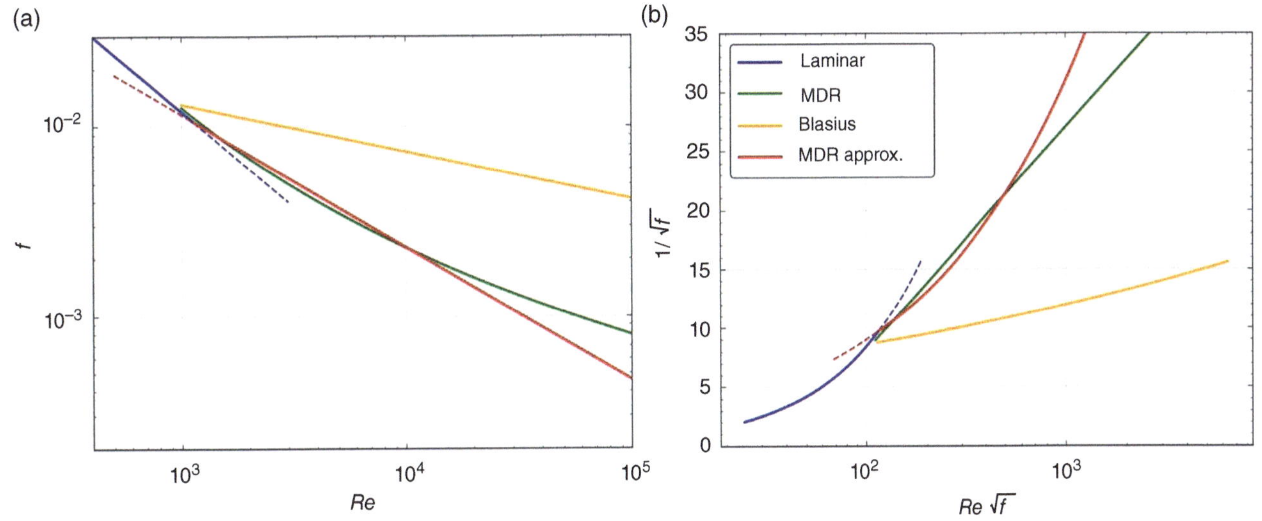

Figure 10.2 The friction factor dependence on the Reynolds number in ordinary (a) and Prandtl–Karman (b) coordinates for channel flow. The laminar branch is shown by blue color, while the turbulent branches for the slickwater flow described by the original Virk's MDR asymptote (10.8) [30] and its power-law approximation (10.9) [21] are depicted by solid green and red lines, respectively. Blasius correlation for turbulent flow of Newtonian fluid (pure water in our case) in a channel with smooth walls is plotted by solid yellow line. (For interpretation of the references to colour in this figure legend, the reader is referred to the web version of this article.) (*Source:* Adapted from Virk [30] and Lecampion and Zia [21])

$$\underbrace{Q_0 t}_{V_{inj}} = \underbrace{2\pi \int_0^R r w(r, t) dr}_{V_{crack}} + \underbrace{4\pi C' \int_0^R r\sqrt{t - t_0(r)} dr}_{V_{leak-off}}, \tag{10.15}$$

and corresponds to the partition of the injected fluid volume (V_{inj}) between the crack (V_{crack}) and porous rock ($V_{leak-off}$).

10.3 Solution Approach

The current section describes the numerical algorithm utilized in the solution of the radial hydraulic fracture propagation problem. The method is based on the Gauss–Chebyshev quadrature and Barycentric Lagrange interpolation techniques. Viesca and Garagash [44] review these techniques and apply them to the solution of various fracture propagation problems with coupled physics. Liu et al. [45] adapt the methodology to the radial fracture propagation. It is important to mention that the numerical scheme does not require an explicit implementation of the near tip region asymptote (if different from the classical LEFM one) compared to other approaches which use the "tip logic," e.g. [46–51]. In present work, we follow the method developed by Liu et al. [45] and extend it to account for different HF fluid rheology and for the fluid leak-off.

Firstly, we introduce the dimensionless distance from the source as $\xi = r/R(t)$, $\xi \in [0, 1]$, and rewrite the system of governing equations through ξ variable using the transformation of the time and coordinate derivatives: $\partial/\partial t|_r = \partial/\partial t|_\xi - [\xi V(t)/R(t)] \cdot \partial/\partial\xi|_t$, $\partial/\partial r|_t = R(t)^{-1}\partial/\partial\xi|_t$, $V = dR/dt = \dot{R}$. The application of Gauss–Chebyshev quadrature implies the problem consideration on the spatial coordinate segment $\xi \in [-1, 1]$. In this regard, the spatiotemporal fracture characteristics and governing equations are extended symmetrically to negative ξ. Such modified form of the elasticity Eq. (10.2) is given by:

$$p(\xi, t) = -\frac{E'}{4\pi R(t)} \int_{-1}^{1} G(\xi, s) \frac{\partial w(s, t)}{\partial s} ds, \; \xi \in [-1, 1], \tag{10.16}$$

where the corresponding form of the integral kernel is provided by Liu et al. [45]:

$$G(\xi, s) = \begin{cases} \text{sign}(\xi s)\left[\frac{1}{\xi}K\left(\frac{s^2}{\xi^2}\right) + \frac{1}{s-\xi}E\left(\frac{s^2}{\xi^2}\right)\right], & |\xi| > |s|, \\ \frac{1}{s-\xi}\text{E}\left(\frac{\xi^2}{s^2}\right), & |\xi| < |s|. \end{cases} \tag{10.17}$$

Spatial derivative of the opening profile is written as:

$$\frac{\partial w}{\partial \xi} = \mathcal{W}(\xi) F(\xi, t), \mathcal{W}(\xi) = \frac{1}{\sqrt{1-\xi^2}}, \; \xi \in [-1, 1],$$

where the singular weight function $\mathcal{W}(\xi)$ is chosen in accordance with the fracture opening behavior near the tip, i.e. $\partial w/\partial\xi \sim 1/\sqrt{1-|\xi|}, \xi \to \pm 1$, and $F(\xi, t)$ is an unknown regular function. Since $w(\xi, t)$ and $\mathcal{W}(\xi)$ are even functions, function $F(\xi, t)$ should be odd.

Next, the coordinate domain $\xi \in [-1, 1]$ is discretized using two sets of nodes connected with the selected weight function (the details are provided by Viesca and Garagash [44]): primary $\boldsymbol{s} = \{s_j\}_{j=1}^{N} = \{\cos[\pi(j-1/2)/N]\}_{j=1}^{N}$ and complementary $\boldsymbol{z} = \{z_i\}_{i=1}^{M} = \{\cos(\pi i/N)\}_{i=1}^{M}, M = N-1$, sets. Hereafter we utilize bold symbols for vectors. Primary and complementary nodes are the roots of the Chebyshev's polynomials of the first and the second kind. We further make use of the values of $F(\xi, t)$ at the primary nodes: $\boldsymbol{F} = \{F(s_j, t)\}_{j=1}^{N}$, while the crack aperture $w(\xi, t)$, net pressure $p(\xi, t)$ and fluid flow velocity $v(\xi, t)$ are evaluated at the complementary nodes: $\boldsymbol{w} = \{w(z_i, t)\}_{i=1}^{M}$, $\boldsymbol{p} = \{p(z_i, t)\}_{i=1}^{M}$, $\boldsymbol{v} = \{v(z_i, t)\}_{i=1}^{M}$.

Let us turn to the discretization of the governing equations. We focus on the Reynolds and global fluid balance equations since the remaining equations (elasticity, propagation criterion, and boundary conditions) are identical to those in Liu et al. [45] and given by their Eqs. (3), (10), (8). Integrating Reynolds Eq. (10.7) in space from a given position along the crack ξ to the tip $\xi = 1$, applying discretization and substituting discretized elasticity (Eq. (41) of Liu et al. [45]), we arrive to

$$-\frac{\boldsymbol{z}^2}{2}\frac{\partial \boldsymbol{w}}{\partial t} + \frac{1}{2}\frac{\partial}{\partial t}\left[\mathbb{S} \times (\boldsymbol{s}^2\boldsymbol{F})\right] + \frac{V}{R}(\mathbb{S} \times (\boldsymbol{s}^2\boldsymbol{F}))$$
$$-\frac{\boldsymbol{zwv}}{R} + C'\mathbb{R} \times \frac{\boldsymbol{z}}{\sqrt{t - t_0(\boldsymbol{z}R)}} = 0, \boldsymbol{w} = \mathbb{S} \times \boldsymbol{F},$$
$$\boldsymbol{v} = -\frac{\boldsymbol{w}^2}{\mu'\tilde{f}R}\boldsymbol{p}_\xi, \boldsymbol{p}_\xi = \frac{E'}{4R}\mathbb{D} \times (\mathbb{G}_2 \times \boldsymbol{F}),$$
$$\tilde{f} = \begin{cases} 1, \ \boldsymbol{Re} \le Re_c, \\ f_0''\boldsymbol{Re}^{1-n}, \ \boldsymbol{Re} > Re_c, \end{cases} \qquad \boldsymbol{Re} = \frac{12\rho\boldsymbol{wv}}{\mu'}$$

Here, we utilize the sign "×"for the matrix multiplication. Components of the matrices $\mathbb{S}, \mathbb{D}, \mathbb{G}_2$ are provided by Liu et al. [45] in their Eqs. (34), (39), (43), while matrix $\mathbb{R}$ is given by:

$$\mathbb{R} = \{R_{ii'}\} = \left\{\sum\nolimits_{k=0}^{M-1}[\Psi_k(1) - \Psi_k(z_i)]B_{ki'}\right\},$$
$$\Psi_k(z) = \cos[(k+1)\theta]/(k+1), \theta = \arccos z,$$
$$B_{ki'} = 2/N \cdot \sin(\pi i'/N)\sin(\pi i'(k+1)/N).$$

The rate form of the global fluid balance Eq. (10.15) can be discretized as follows:

$$\frac{Q_0}{\pi}\left(\frac{1}{R^2} - \frac{2tV}{R^3}\right) = -\mathbb{S}_H \times (\boldsymbol{s}^2\dot{\boldsymbol{F}}) + 2C'\mathbb{R}_H$$
$$\times \frac{\boldsymbol{z}}{\sqrt{t - t_0(\boldsymbol{z}R)}} - \frac{2C'V}{R^3}\int_0^t \frac{R^2(s)ds}{\sqrt{t-s}},$$

where the matrix $\mathbb{S}_H$ is defined by Liu et al. [45] in their Eq. (38), while $\mathbb{R}_H$ is given by:

$$\mathbb{R}_H = \left\{\sum\nolimits_{k=0}^{m-1}[\Psi_k(1) - \Psi_k(0)]B_{ki'}\right\}.$$

In the right-hand side of the global fluid balance equation we have the integral with respect to time, and it can be evaluated via application of the Simpson's rule.

The discretized Reynolds equations (with the elasticity equation and boundary conditions already taken into account in it) together with the global fluid balance and propagation condition can be combined into a system of ordinary differential equations (ODEs):

$$\mathbb{A}(\boldsymbol{X}, t)\frac{d\boldsymbol{X}}{dt} = \boldsymbol{B}(\boldsymbol{X}, t), \ \boldsymbol{X} = \{R, \boldsymbol{F}\},$$

where $\boldsymbol{X}$ is the solution vector. Once the above system of ordinary differential equations is integrated, the crack radius evolution is known, $R = R(t)$, while the solution for opening, net pressure profiles, and efficiency is evaluated from $F(\xi, t)$ using the following representation:

$$\boldsymbol{w} = \mathbb{S} \times \boldsymbol{F}, \boldsymbol{p} = \frac{E'}{4R}\mathbb{G}_2 \times \boldsymbol{F}, \eta(t) = -\frac{\pi R^2}{Q_0 t}\{\mathbb{S}_A \times (\boldsymbol{s}^2\boldsymbol{F})\},$$

where matrix $\mathbb{S}_A$ is defined in Liu et al. [45] by their Eq. (37).

Due to the form of the elasticity equation [45], the number of primary nodes N should be odd. Since function $F(\xi, t)$ is odd, the target vector $\boldsymbol{F}$ has the form: $\boldsymbol{F} = \{F_1, ...F_{(N-1)/2}, 0, -F_{(N-1)/2}, ..., -F_1\}$. The total number of the independent unknown parameters is $(N-1)/2 + 1$ which constrains the system of governing ODEs to be composed of the Reynolds equations at the complementary nodes $z_1, ..., z_{(N-1)/2-1}$, the global fluid balance equation and the propagation condition. We choose $N = 101$, and the initial condition is represented by the storage-viscosity-turbulent limiting propagation regime originally developed by Lecampion and Zia [21], and further discussed in Section 10.5. The time computational domain is discretized on a logarithmic scale, and we apply ODE solver for each time step. The numerical algorithm is implemented in Python programming language, the system of ODEs is solved via *solve_ivp* function of SciPy library.

10.4 Solution Examples for Typical Field Applications

In this section, we demonstrate the results of the simulations for a radial hydraulic fracture growth in which the model parameters are close to the typical field cases. We would like to achieve the following objectives. First of all, we investigate how the laminar-to-turbulent flow regime transformation inside the fracture channel changes the problem solution compared to the fully laminar flow case for both impermeable and permeable reservoirs. Moreover, we perform calculations not only for slickwater but also for pure water driven cracks to examine the influence of the frictional behavior on the fracture parameters.

The injection volumetric flow rate is assumed to be a fairly high $Q_0 = 0.1$ m^3/s (e.g. see representative ranges in the reviews [9, 52]). For simplicity, we consider a situation when the polymer molecules do not change the solvent (pure water) viscosity, i.e. they modify the frictional behavior during turbulent flow only (the MDR asymptote instead of the Blasius approximation). The following rock and fluid properties are chosen:

$$E' = 20\,\text{GPa}, K_{Ic} = 1\,\text{MPa}\cdot\sqrt{\text{m}}, \sigma_o = 10\,\text{MPa}, p_o = 6\,\text{MPa},$$
$$\mu = 1\,\text{cP}, c_t = 10^{-3}\,\text{MPa}^{-1}, \rho = 10^3\,\text{kg/m}^3,$$
$$k = 10\,\text{mD (permeable rock)}, \phi = 20\%. \tag{10.18}$$

Using the set of parameters (10.18), one can estimate the Carter's leak-off number corresponding to the permeable rock case, $C' = 2.04\cdot 10^{-4}\text{m}/\sqrt{\text{s}}$, while $C' = 0$ for impermeable rock.

Figure 10.3a–c present the turbulent-laminar solutions for the evolution of the fracture radius $R(t)$, opening at the wellbore $w(0, t)$ and net pressure $p(0.1, t)$ at the distance $r = 0.1$ m normalized by the corresponding laminar solutions (Figure 10.4). As a result, the laminar solutions in these charts are simply unity along the entire time domain. The choice of the location for the pressure monitoring is conditioned by the pressure singularity at $r = 0$ in the point fluid source model and by the value of a typical borehole radius. We also note that, with exception of very early propagation time (not shown), $R(t) \gg 0.1$ m, and the point source modeling is deemed appropriate.

Using these figures, one can notice that the laminar-to-turbulent flow regime transformation inside the crack

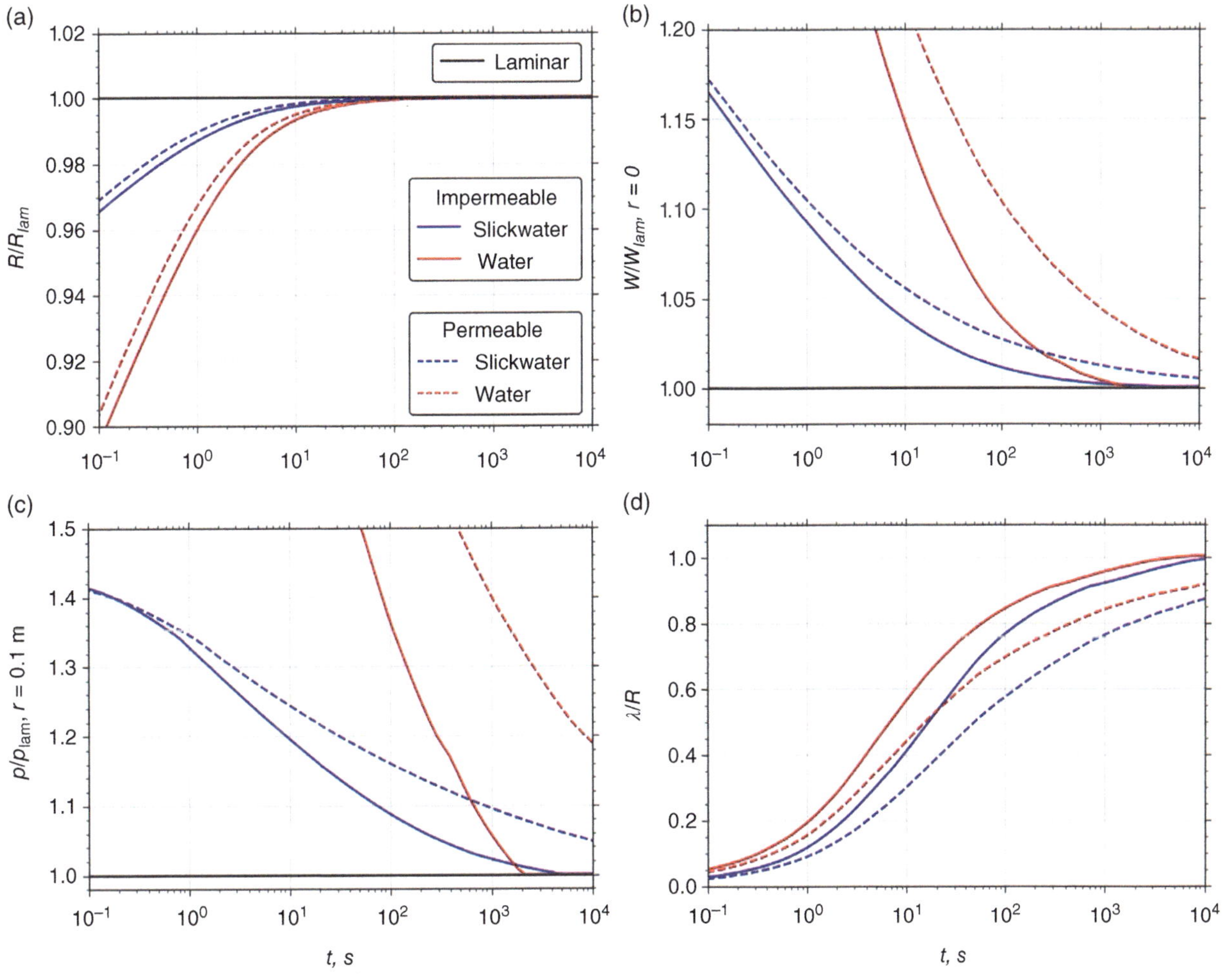

Figure 10.3 Evolution of fracture radius $R(t)$, (a), opening at the wellbore $w(0, t)$, (b), and net pressure $p(0.1, t)$ at the distance $r = 0.1$ m, (c), in the turbulent-laminar solutions, normalized by the corresponding solutions of the fully laminar model (Figure 10.4). Evolution of length $\lambda(t)$ of the laminar flow domain, (d), normalized by fracture radius $R(t)$ (the laminar flow domain extends distance λ from the crack tip, see Figure 10.1). The slickwater (water) solutions are presented by blue (red) color, while solid (dashed) lines indicate impermeable (permeable) rock case. (For interpretation of the references to colour in this figure legend, the reader is referred to the web version of this article.)

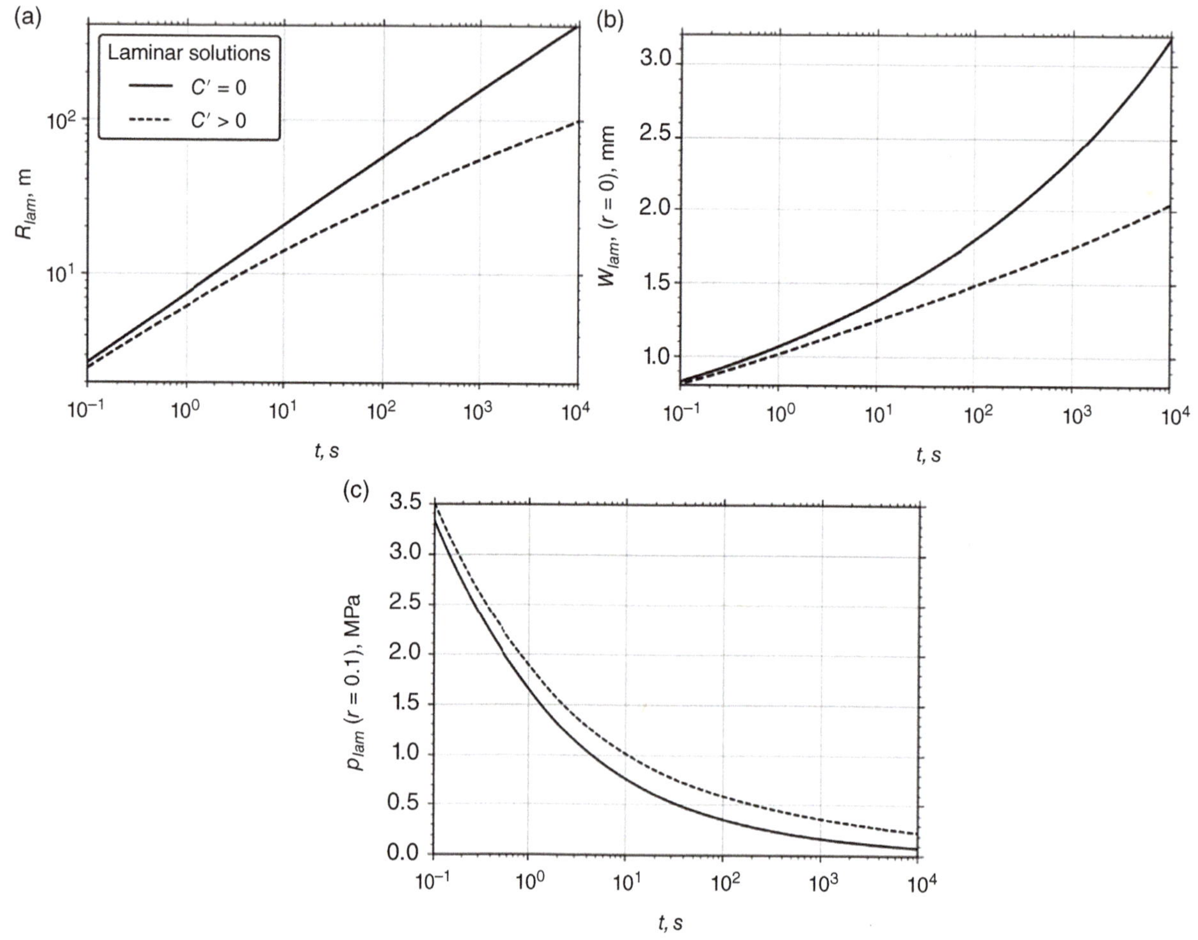

Figure 10.4 Evolution of the fracture radius $R(t)$, (a), maximum opening $w(0, t)$, (b), and net pressure $p(0.1, t)$ at the distance $r = 0.1$ m in the laminar solutions. The solid (dashed) lines indicate the impermeable (permeable) rock case.

channel affects the fracture parameters during an initial period of the propagation, leading to a shorter radius, larger opening at the wellbore, and pressure values compared to the fully laminar model. Turbulent effects eventually become negligible, and the turbulent–laminar and fully laminar solutions become nearly identical. It is important to highlight that the threshold time is different for radius, opening, and pressure for a particular choice of the governing parameters of the model.

Let us define a time moment t_* when a time-dependent fracture characteristic $\mathcal{A}(t)$ (radius, opening, pressure) corresponding to the turbulent–laminar solution can be closely approximated by $\mathcal{A}_{lam}(t)$ from the fully laminar solution as follows: $|\mathcal{A}(t_*) - \mathcal{A}_{lam}(t_*)| / |\mathcal{A}_{lam}(t_*)| = 5\%$. We summarize the values t_* for the crack parameters depicted in Figure 10.3 for all considered cases, namely, slickwater (blue lines) and pure water (red lines) fracturing in the impermeable (solid lines) and permeable (dashed lines) reservoirs, in Table 10.1.

It is evident from Table 10.1 that the noticeable difference in the fracture radius values $R(t)$ and $R_{lam}(t)$ appears on the timescale less than one second, suggesting that in practice, the deviation of the fracture size in the turbulent–laminar solution from that assuming fully laminar flow can be neglected. The flow regime transformation impacts opening at the wellbore during approximately 5 and 15 s for slickwater fracturing in an impermeable and permeable rock, correspondingly. These quantities for pure water fracturing are around 1 and 12 minutes, respectively. The interpretation of the results for $w(0, t)$ leads to the following conclusions: (i) the leak-off process prolongs the duration of the turbulent flow regime influence on the crack aperture near the wellbore, and (ii) the turbulence effects are more pronounced for the pure water case since they continue during much longer period of time. However, even when the crack geometries (radius and aperture) are approximately the same in the turbulent–laminar and fully laminar models, the pressure values near the wellbore ($r = 0.1$ m) can still differ significantly. This observation means that larger amount of energy is required to create a hydraulic fracture driven by turbulent–laminar flow compared to the value predicted by the fully laminar model.

Table 10.1 The table contains the time moments (in seconds) starting from which the relative differences between the fracture characteristics $R(t)$, $w(0, t)$, $p(0.1, t)$ estimated by the turbulent–laminar and fully laminar solutions are less than 5%.

		t_*, s			
HF fluid		**Slickwater**		**Water**	
Rock type		**Impermeable**	**Permeable**	**Impermeable**	**Permeable**
A	$R(t)$	0.03	0.02	0.7	0.5
	$w(0,t)$	5.5	14.2	70.5	715.3
	$p(0.1, t)$	308	8933	1079	67 383

From Figure 10.3 and Table 10.1, we identify for the slickwater case that $p \approx p_{\mathrm{lam}}$ (less 5% difference) after 5 minutes from the fracture initiation in an impermeable formation, while in a permeable rock, this period reaches 2.5 hours. In turn, when the HF fluid is pure water, the alignment of the pressure values occurs after 0.5 and 18 hours (beyond typical HF treatment duration in the field), correspondingly.

Figure 10.3d shows the extent $\lambda(t)$ of the region $r \in (R-\lambda, R)$ with laminar flow as a fraction of the crack radius $R(t)$. It is an increasing function on time evolving from $\lambda/R \approx 0$ at early time and approaching $\lambda/R \approx 1$ at large time. In other words, the turbulent flow regime is realized along the entire fracture at the beginning of the propagation, while the laminar regime is spatially dominant at large time. For example, we observe in the permeable rock case that near the end of a typical fluid injection ($t \sim 10^4$s) only a small part of the crack $\sim$0.1R near the wellbore is occupied by turbulent flow, whereas the remainder of the crack $\sim$0.9R supports laminar flow. Turbulent flow spatial extent is yet smaller for the case of impermeable rock. By comparing slickwater and water injection cases in Figure 10.3d, we notice that the spatial domain with turbulent flow is larger for slickwater HF, i.e. λ/R at all times is smaller, yet the water HF corresponds to larger deviations of $R(t)$, $w(0, t)$, $p(0.1, t)$ from the laminar solution.

We now discuss how the spatial distributions of the fracture aperture and net pressure along the fracture evolve with time. These distributions are shown in Figure 10.5 for three time moments $t = \{1, 10, 100\}$ s. As it has been already mentioned, a radial crack driven by turbulent–laminar flow has a shorter radius compared to the fully laminar model, at early time, the opening and pressure profiles for $t = 1$ s in Figure 10.5 confirm this fact. In the subsequent time moments, i.e. $t = 10$ and 100 s, the difference between $R(t)$ and $R_{\mathrm{lam}}(t)$ values is imperceptible. The turbulent–laminar solutions become very close to their laminar 'analogs' starting from a certain distance from the fluid source. In turn, in the remainder of the crack, i.e. along the zone adjacent to the wellbore, a considerable difference, decreasing over time, is observed where the turbulent–laminar crack has a wider opening and larger net pressure. In general, the HF in a permeable rock (Figure 10.5b and d) has a smaller volume due to the leak-off compared to the crack propagation in the impermeable reservoir, which is manifested in a shorter radius and aperture in the leak-off cases. On the other hand, the net pressure profile has greater values in the cases with leak-off.

Using Figure 10.5c and d, one can observe the presence of the pressure singularities at the wellbore and the fracture front. For example, the pressure behavior near the wellbore ($r \ll R$) is governed by $p \sim |\log r|$ for the fully laminar crack, while $p \sim r^{-3/10}$ and $p \sim r^{-3/4}$ are applicable for slickwater and pure water fracturing, respectively, in the turbulent-laminar model. The near-tip region of a hydraulic fracture ($R - r \ll R$) has multiscale nature (see, e.g. [53]) such that the dominant tip singularity depends on both fracture length and propagation speed. It can take form of either toughness asymptote $p \sim -|\log r|$ or storage-viscosity asymptote [53]. The latter is given by $p \sim -(R-r)^{-1/3}$ [54] in the laminar flow regime and by $p \sim -(R-r)^{-7/27}$ for slickwater or $p \sim -(R-r)^{-1/9}$ for pure water in the turbulent flow regime [21, 26].

10.5 Limiting Propagation Regimes

Two different mechanisms control the propagation regime of a finite hydraulic fracture. The first one regulates the total dissipated energy distribution between the creation of new fracture surfaces at the tip and viscous fluid flow inside the fracture channel. The second mechanism is related to the partitioning of the injected fluid volume between the fracture and host permeable rock (due to leak-off). During the crack growth, the allocation of the dissipated energy and injected volume can change over time resulting in the realization of the limiting propagation regimes characterized by one dissipation (out of two) and one fluid balance (out of two) mechanisms. The

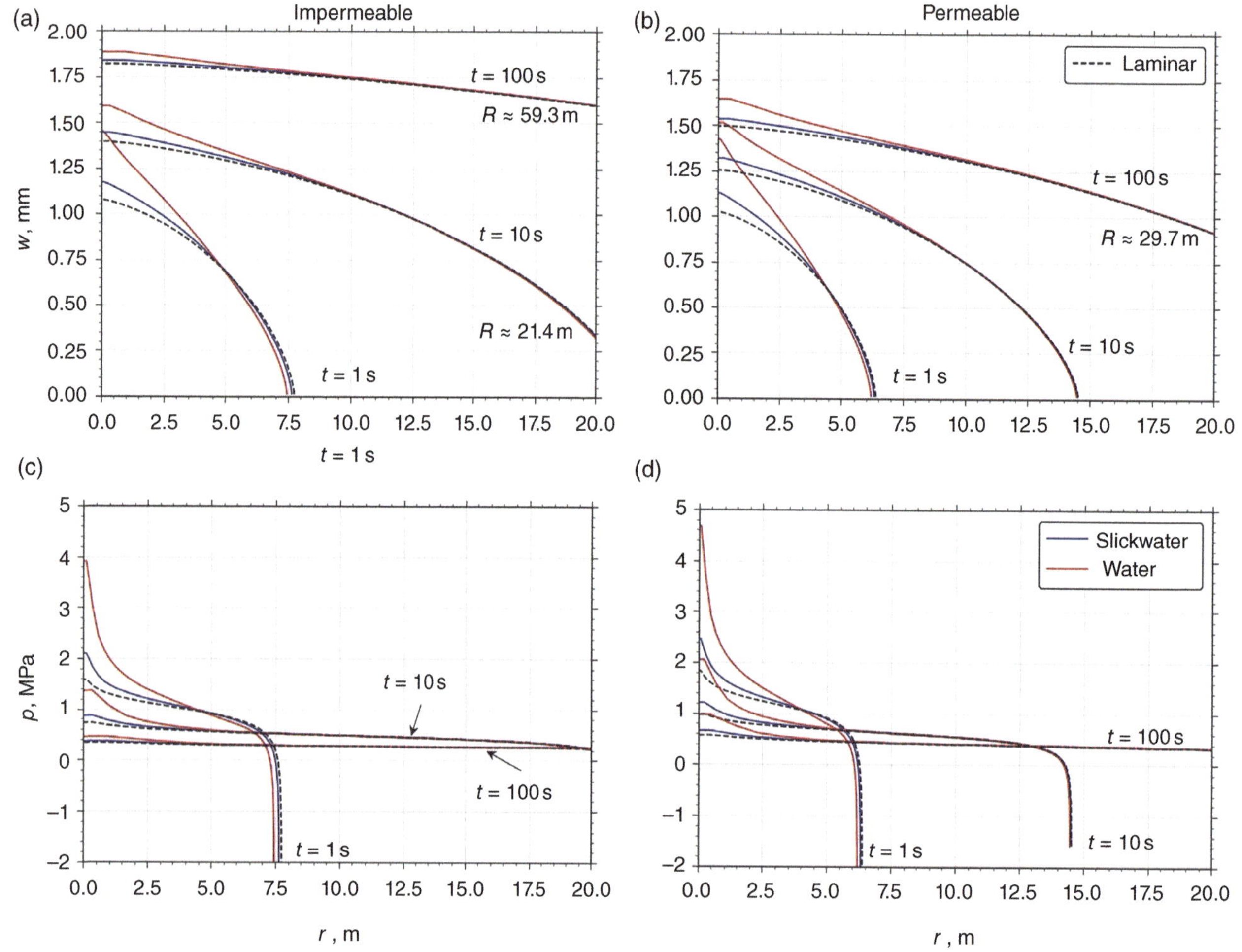

Figure 10.5 The spatial variations of the fracture opening (top raw) and net fluid pressure (bottom raw) at the time moments $t = \{1, 10, 100\}$ s. The figures (a) and (c) are related to the impermeable rock case, while (b) and (d) show the discussed crack characteristics during the growth at the permeable formation. The solution profiles for the slickwater (water) fracturing are depicted by solid blue (red) lines. The fully laminar solutions are presented by black dashed lines. (For interpretation of the references to colour in this figure legend, the reader is referred to the web version of this article.)

partitioning of the dissipated energy is influenced by the viscosity μ' and toughness K' parameters, while the leak-off parameter C' affects the distribution of the injected volume.

In the present radial fracture model, we distinguish six limiting propagation regimes (also known as the vertex solutions or vertices). Four of them can be observed in the penny-shaped HF model with laminar flow realized along the whole fracture channel (see a review paper of [55] and references therein):

- $\boldsymbol{M}$ – storage-viscosity-laminar – $C' = K' = 0$, $Re \le Re_c$;
- $\tilde{\boldsymbol{M}}$ – leak-off-viscosity-laminar – $C' \to +\infty$, $K' = 0$, $Re \le Re_c$;
- **K**–storage-toughness – $C' = M' = 0$;
- $\tilde{\boldsymbol{K}}$ – leak-off-toughness – $C' \to +\infty$, $M' = 0$.

The toughness dominated regimes K and $\tilde{K}$ do not depend on the flow regime type. Therefore, the remaining two limiting regimes emerge in the viscosity-dominated case, when the entire fracture is occupied by turbulent flow:

- $\boldsymbol{T}$ – storage-viscosity-turbulent – $C' = K' = 0$, $Re > Re_c$;
- $\tilde{\boldsymbol{T}}$ – leak-off-viscosity-turbulent – $C' \to +\infty$, $K' = 0$, $Re > Re_c$;

Although the laminar flow always exists near the hydraulic fracture front, the above two regimes correspond to the limit when the laminar domain becomes negligibly small.

Savitski and Detournay [41] derived solutions for M and K vertices, while $\tilde{K}$ and $\tilde{M}$ solutions are given by Bunger et al. [56], and by Madyarova [57], Peirce and Detournay [50], respectively. Moreover, Dontsov [46] present approximate solutions for all laminar limiting regimes. The semi-analytical and approximate solutions for the storage-viscosity-turbulent regime, T-vertex, are found by Lecampion and Zia [21].

Further, we summarize scalings in different limiting regimes of the solution following [55] for the laminar regimes and [21] for the turbulent ones. Firstly, we introduce the dimensionless radius γ, opening Ω, net pressure Π and fluid velocity $\mathcal{V}$ as follows:

$$R(t) = L(t)\gamma(\mathcal{P}), w(r,t) = \epsilon(t)L(t)\Omega(\xi,\mathcal{P}),$$
$$p(r,t) = E'\epsilon(t)\Pi(\xi,\mathcal{P}), v(r,t) = L(t)\mathcal{V}(\xi,\mathcal{P})/t \tag{10.19}$$

where $\xi = r/R(t)$ is the normalized distance from the source, $\mathcal{P} = \{\mathcal{P}_1, \mathcal{P}_2, \mathcal{P}_3\}$ are dimensionless evolution parameters depending on time, material parameters (10.1), fluid density ρ and injection rate Q_0. $L(t)$ is a lengthscale of the same order as the crack radius, and $\epsilon(t)$ is a small dimensionless parameter.

Next, we substitute formulas (10.19) into the governing equations and obtain their normalized form:

- *Elasticity*:

$$\Pi = -\frac{1}{2\pi\gamma}\int_0^1 M(\xi,s)\frac{\partial\Omega}{\partial s}ds; \tag{10.20}$$

- *Reynolds*:

$$\mathcal{G}_v\left(\left[\frac{\dot{\epsilon}t}{\epsilon} + \frac{\dot{L}t}{L}\right]\Omega + \dot{\Omega}t - \xi\left(\frac{\dot{L}t}{L} + \frac{\dot{\gamma}t}{\gamma}\right)\frac{\partial\Omega}{\partial\xi}\right) + \frac{\mathcal{G}_c}{\sqrt{1-\tau_0(\xi L\gamma)}} = -\frac{1}{\gamma\xi}\frac{\partial(\xi\Omega\tilde{\mathcal{V}})}{\partial\xi};$$
$$\tilde{\mathcal{V}} = G_v\mathcal{V} = \begin{cases} -\left(\mathcal{G}_m^{lam}\right)^{-1}\Omega^2\gamma^{-1}\nabla\Pi, \ \mathcal{R}|\mathcal{V}|\Omega \le Re_c, \\ -\left(\mathcal{G}_m^{turb}\right)^{-1}\mathrm{sign}(\nabla\Pi)\left[\Omega^{n+1}\gamma^{-1}|\nabla\Pi|\right]^{1/(2-n)}, \ \mathcal{R}|\mathcal{V}|\Omega > Re_c; \end{cases} \tag{10.21}$$

- *Propagation condition:*

$$\Omega = \mathcal{G}_k\sqrt{\gamma}\sqrt{1-\xi}, \xi \to 1; \tag{10.22}$$

- *Boundary conditions:*

$$\lim_{\xi\to 0}\xi\Omega\mathcal{V} = \frac{\mathcal{G}_v^{-1}}{2\pi\gamma}, \Omega(1,t) = 0, \Omega(1,t)\mathcal{V}(1,t) = 0; \tag{10.23}$$

- *Global fluid balance:*

$$\frac{1}{\pi\gamma^2} = 2\mathcal{G}_v\int_0^1 \Omega\xi d\xi + 4\mathcal{G}_c\int_0^1\sqrt{1-\tau_0(\xi\gamma L)}\xi d\xi; \tag{10.24}$$

where $\tau_0(t) = t_0(r)/t$ is the dimensionless inverse radius function, $\nabla\Pi = \partial\Pi/\partial\xi$ is the pressure gradient, and $\mathcal{R} = 12\rho\epsilon L^2/(\mu' t)$ is the characteristic Reynolds number. We also introduce five dimensionless groups:

$$\mathcal{G}_v = \frac{\epsilon L^3}{Q_0 t}, \mathcal{G}_c = \frac{C'L^2}{Q_0\sqrt{t}}, \mathcal{G}_k = \frac{K'}{E'\epsilon\sqrt{L}}, \mathcal{G}_m^{lam} = \frac{\mu' Q_0}{E'L^3\epsilon^4},$$
$$\mathcal{G}_m^{turb} = \left(\frac{f_0''(12\rho)^{1-n}\mu'^n Q_0^{2-n}}{E'L^{4-n}\epsilon^4}\right)^{1/(2-n)} = \left(f_0''\mathcal{G}_m^{lam}(\mathcal{R}/\mathcal{G}_v)^{(1-n)}\right)^{1/(2-n)}. \tag{10.25}$$

Numbers $\mathcal{G}_v$ and $\mathcal{G}_c$ quantify the fluid volume stored inside the fracture and leaked into the rock, respectively, in reference to the total injected volume. Numbers $\mathcal{G}_k$, $\mathcal{G}_m^{lam}$, and $\mathcal{G}_m^{turb}$ with the meaning of the nondimensional rock toughness, fluid viscosity, and equivalent viscosity of turbulent flow, correspondingly, relate to the partition of the dissipated energy between solid and fluid.

To derive scalings linked to the limiting regimes (i.e. lengthscale $L(t)$ and factor $\epsilon(t)$ in Eq. (10.19)), we assign one out of two storage numbers ($\mathcal{G}_v$, $\mathcal{G}_c$) and one out of three dissipation numbers ($\mathcal{G}_m^{lam}$, $\mathcal{G}_m^{turb}$, $\mathcal{G}_k$) equal to one. The remaining three groups constitute the evolution parameters $\mathcal{P}$. When all three of these parameters tend to zero in a given scaling, one can observe the emergence of the corresponding vertex solution.

We consider the storage-viscosity-turbulent scaling $\mathcal{G}_v = \mathcal{G}_m^{turb} = 1$. The following formulas for the lengthscale and small parameter follow:

$$L_t = \left(\frac{E't^4Q_0^{n+2}}{(12\rho)^{1-n}f_0''\mu'^n}\right)^{1/(8+n)}, \epsilon_t = \frac{Q_0 t}{L_t^3}. \tag{10.26}$$

while three nondimensional evolution parameters $\mathcal{P}$ (i.e. toughness $\mathcal{G}_k$, leak-off $\mathcal{G}_c$, and laminar-viscosity $\mathcal{G}_m^{lam}$) are:

$$\mathcal{K}_t = K'\left(\frac{t^{4-2n}}{(12\rho)^{5-5n}\left(f_0''\right)^5E'^{11+2n}\mu'^{5n}Q_0^{6-3n}}\right)^{1/(16+2n)},$$
$$\mathcal{C}_t = C'\left(\frac{E'^4t^{8-n}}{(12\rho)^{4-4n}\left(f_0''\right)^4\mu'^{4n}Q_0^{8-2n}}\right)^{1/(16+2n)},$$
$$\mathcal{M}_t = \left(\frac{E'^{1-n}\mu'^{8-8n}t^{4-4n}}{(12\rho)^{9-9n}\left(f_0''\right)^9Q_0^{6-6n}}\right)^{1/(8+n)}. \tag{10.27}$$

The solution in the storage-viscosity-turbulent regime (T-vertex) is then obtained from the normalized Eqs. (10.20)–(10.24) in the above scaling when the evolution parameters (10.27) are set to zero. The solution is computed numerically using the method defined in Section 10.3, and is given below for the time evolution of the fracture radius, opening at the wellbore, and pressure at the half-radius:

$$R_t(t) = 0.759L_t, w_t(0,t) = 1.11\left(\frac{(f_0'')^2(12\rho)^{2-2n}\mu'^{2n}Q_0^{4-n}t^n}{E'^2}\right)^{1/(8+n)},$$
$$p_t(R_t(t)/2,t) = 0.5\left(\frac{(f_0'')^3E'^{n+5}\mu'^{3n}(12\rho)^{3-3n}Q_0^{2-2n}}{t^{4-n}}\right)^{1/(8+n)}. \quad (10.28)$$

The storage-viscosity-laminar scaling $\mathcal{G}_v = \mathcal{G}_m^{lam} = 1$ and corresponding scales L_m and ϵ_m can be alternatively obtained by setting $n = f_0'' = 1$ in expressions (10.26) for L_t, ϵ_t. Fracture characteristics in the M-vertex solution ($\mathcal{G}_k = \mathcal{G}_c = \mathcal{G}_m^{turb} = 0$) have the following form [41]:

$$R_m(t) = 0.699\left(\frac{E'Q_0^3t^4}{\mu'}\right)^{1/9}, w_m(0,t) = 1.188\left(\frac{\mu'^2Q_0^3t}{E'^2}\right)^{1/9},$$
$$p_m(R_m(t)/2,t) = 0.62\left(\frac{E'^2\mu'}{t}\right)^{1/3}. \quad (10.29)$$

where, as before, the numerical prefactors are evaluated from the full numerical solution of the problem.

Next, we look at the leak-off-viscosity-turbulent scaling $\mathcal{G}_c = \mathcal{G}_m^{turb} = 1$ and derive the formulas for the lengthscale and small parameter provided below:

$$L_{\tilde{t}} = t^{1/4}\sqrt{\frac{Q_0}{C'}}, \epsilon_{\tilde{t}} = \frac{(12\rho)^{(1-n)/4}(f_0'')^{1/4}C'^{(4-n)/8}\mu'^{n/4}}{E'^{1/4}Q_0^{n/8}t^{(4-n)/16}}. \quad (10.30)$$

Fracture characteristics in the $\tilde{T}$-vertex solution for the leak-off-viscosity-turbulent regime ($\mathcal{G}_v = \mathcal{G}_k = \mathcal{G}_m^{lam} = 0$) can be obtained in the following form:

$$R_{\tilde{t}}(t) = \frac{\sqrt{2}}{\pi}L_{\tilde{t}}, w_{\tilde{t}}(0,t) = \frac{(12\rho)^{\frac{1-n}{4}}(f_0'')^{1/4}\mu'^{n/4}Q_0^{(4-n)/8}t^{n/16}}{E'^{1/4}C'^{n/8}},$$
$$p_{\tilde{t}}(R_{\tilde{t}}(t)/2,t) = 0.75\left(\frac{E'^{12}(f_0'')^4C'^{8-2n}\mu'^{4n}(12\rho)^{4-4n}}{Q_0^{2n}t^{4-n}}\right)^{1/16}, \quad (10.31)$$

where the numerical prefactors (for opening and pressure) are computed from the full numerical solution of the governing Eqs. (10.20)–(10.24) in the $\tilde{T}$-scaling when the evolution parameters equal zero; the coefficient for the fracture radius is found analytically (exact).

The substitution $n = f_0'' = 1$ in Eqs. (10.30) for $L_{\tilde{t}}, \epsilon_{\tilde{t}}$ leads to the leak-off-viscosity-laminar scaling $\mathcal{G}_c = \mathcal{G}_m^{lam} = 1$ and corresponding scales $L_{\tilde{m}}$ and $\epsilon_{\tilde{m}}$. The fracture properties in the $\tilde{M}$-vertex solution ($\mathcal{G}_k = \mathcal{G}_v = \mathcal{G}_m^{turb} = 0$) are given below:

$$R_{\tilde{m}}(t) = R_{\tilde{t}}(t), w_{\tilde{m}}(0,t) = 1.05\left(\frac{\mu'^4Q_0^6t}{C'^2E'^4}\right)^{1/16},$$
$$p_{\tilde{m}}(R_{\tilde{m}}(t)/2,t) = 0.84\left(\frac{E'^{12}C'^6\mu'^4}{Q_0^2t^3}\right)^{1/16}, \quad (10.32)$$

where the numerical coefficients for the opening and pressure are calculated from the full numerical solution of the problem.

The toughness dominated vertex solutions (K and $\tilde{K}$) do not depend on the realized flow regime inside the fracture channel since the corresponding energy dissipation in the fluid flow is assumed negligibly small. For the completeness, we write out expressions for the radius, opening at the wellbore and pressure (uniform along the whole crack) in these regimes [41, 56]:

$$R_k = \left(\frac{3}{\sqrt{2}\pi}\right)^{2/5}\left(\frac{E'Q_0t}{K'}\right)^{2/5}, w_k(0,t) = \left(\frac{3}{8\pi}\right)^{1/5}\left(\frac{K'^4Q_0t}{E'^4}\right)^{1/5},$$
$$p_k(t) = \left(\frac{\pi^6}{2^{17}3}\cdot\frac{K'^6}{tE'Q_0}\right)^{1/5}; \quad (10.33)$$

$$R_{\tilde{k}}(t) = R_{\tilde{t}}(t), w_{\tilde{k}}(0,t) = \frac{1}{2^{1/4}\sqrt{\pi}}\left(\frac{K'^8Q_0^2t}{C'^2E'^8}\right)^{1/8},$$
$$p_{\tilde{k}}(t) = \frac{\pi^{3/2}}{2^{15/4}}\frac{C'^{1/4}K'}{Q_0^{1/4}t^{1/8}}. \quad (10.34)$$

10.6 Normalization of the Governing Equations

Normalized solution (10.19) in one of the vertex-scalings of Section 10.5 depends on three nondimensional evolution parameters which quantify the departure from the corresponding vertex solution. Either of these three parameters can be regarded as nondimensional "time" in the problem. Alternatively, the so-called "mixed" scalings have also been used [46, 56–58] to compute solution evolution in time between the limiting regimes (vertices).

One of such scalings, the mk-scaling [55], is defined in terms of the timescale t_{mk} quantifying the solution transition period between M and K regimes. Specifically, t_{mk} is the time moment when the K-vertex and M-vertex lengthscales are the same, i.e. $L_k(t) = L_m(t)$ at $t = t_{mk}$. The corresponding length $L_{mk} = L_m(t_{mk}) = L_k(t_{mk})$ and nondimensional small parameter $\epsilon_{mk} = \epsilon_m(t_{mk}) = \epsilon_k(t_{mk})$ are used to define the mk-scaling:

$$\xi = \frac{r}{R}, \tau = \frac{t}{t_{mk}}, \gamma = \frac{R}{L_{mk}}, \Omega = \frac{w}{\epsilon_{mk} L_{mk}}, \Pi = \frac{p}{E' \epsilon_{mk}},$$
$$\mathcal{V} = \frac{v}{L_{mk}/t_{mk}}, \tag{10.35}$$

where the characteristic scales evaluated as:

$$t_{mk} = \sqrt{\frac{E'^{13}\mu'^{5}Q_0^3}{K'^{18}}}, L_{mk} = \left(\frac{E' Q_0^3 t_{mk}^4}{\mu'}\right)^{1/9}, \epsilon_{mk} = \left(\frac{\mu'}{E' t_{mk}}\right)^{1/3}. \tag{10.36}$$

We note that either vertex scalings (Section 10.5) or the alternative "mixed" mk-scaling can be utilized to obtain general solution and its evolution in time. In the following, we apply the mk-scaling, since it allows for more direct interpretation of the normalized solution (since the scales L_{mk} and ϵ_{mk} are constants compared to the time-dependent vertex scales $L(t)$ and $\epsilon(t)$).

The substitution of (10.35), (10.36) into the system of the governing equations yields their normalized form:

- *Elasticity*:

$$\Pi = -\frac{1}{2\pi\gamma}\int_0^1 M(\xi, s)\frac{\partial \Omega}{\partial s} ds;$$

- *Reynolds*:

$$\frac{\partial \Omega}{\partial \tau} - \frac{\xi\dot{\gamma}}{\gamma}\frac{\partial \Omega}{\partial \xi} + \frac{1}{\xi\gamma}\frac{\partial}{\partial \xi}(\xi\Omega\mathcal{V}) + \frac{\phi^{1/4}}{\sqrt{\tau - \tau_0(\xi L_{mk}\gamma)}} = 0,$$
$$\mathcal{V} = -\frac{\Omega^2}{\tilde{f}\gamma}\frac{\partial \Pi}{\partial \xi}, \quad \tilde{f} = \begin{cases} 1, & \mathcal{R}|\mathcal{V}|\Omega \le Re_c, \\ f_0''(\mathcal{R}|\mathcal{V}|\Omega)^{1-n}, & \mathcal{R}|\mathcal{V}|\Omega > Re_c; \end{cases}$$

- *Propagation condition:*

$$\Omega = \sqrt{\gamma}\sqrt{1-\xi}, \xi \to 1;$$

- *Boundary conditions:*

$$\lim_{\xi\to 0} \xi\Omega\mathcal{V} = \frac{1}{2\pi\gamma}, \Omega(1, t) = 0, \Omega(1, t)\mathcal{V}(1, t) = 0;$$

- *Global fluid balance:*

$$\frac{\tau}{\pi\gamma^2} = 2\int_0^1 \Omega\xi d\xi + 4\phi^{1/4}\int_0^1 \sqrt{\tau - \tau_0(\xi\gamma L_{mk})}\xi d\xi.$$

The normalized solution of the above set of equations depends on nondimensional spatial coordinate ξ, time τ, and two dimensionless numbers, leak-off ϕ and characteristics Reynolds number $\mathcal{R}$:

$$\phi = \frac{C'^4 E'^{11}\mu'^3 Q_0}{K'^{14}}, \mathcal{R} = \frac{12K'^4\rho}{E'^3\mu'^2}. \tag{10.37}$$

Let us evaluate the ranges for the governing parameters ϕ and $\mathcal{R}$ corresponding to typical field applications. We choose the following intervals for the dimensional model parameters:

- geomechanics:
  - plane-strain elastic modulus: $E' = 10 \div 50$ GPa,
  - rock toughness: $K_{Ic} = 0.5 \div 2.5\,\text{MPa}\cdot\sqrt{\text{m}}$,
  - far-field confining stress: $\sigma_o = 10 \div 30$ MPa;
- reservoir:
  - permeability: $k_r = 0.1 \div 100$ mD,
  - porosity: $\phi_r = 5 \div 25\%$,
  - ratio of the pore pressure and confining stress: $p_o/\sigma_o = 0.4 \div 0.95$;
- fluid:
  - viscosity: $\mu = 1 \div 5$ cP,
  - total compressibility: $c_t = 10^{-3}\,\text{MPa}^{-1}$,
  - density: $\rho = 10^3\,\text{kg/m}^3$;
- volumetric injection rate: $Q_0 = 0.01 \div 0.1\,\text{m}^3/\text{s}$.

We vary each parameter independently and evaluate the ranges: $\phi \in [1.1 \cdot 10^{-18}, 6.7 \cdot 10^{14}]$, $\mathcal{R} \in [0.2, 3.4 \cdot 10^5]$. The plane-strain modulus and fracture toughness have the most considerable impact on the dispersion of ϕ and $\mathcal{R}$. Typical duration of the fluid injection $t_{\text{end}} = 1$ hour, which results in the following range for the nondimensional injection time: $\tau_{\text{end}} \in [2.5 \cdot 10^{-6}, 3 \cdot 10^8]$.

10.7 Problem Parameter Space Analyses

Let us investigate the parametric space of the model for a radial crack driven by turbulent-laminar flow of slickwater in both impermeable and permeable reservoirs. We perform the analyses using the problem formulation in the dimensionless form (Section 10.6) where the parameter space is three dimensional with the axes: time τ, leak-off number ϕ, and characteristics Reynolds number $\mathcal{R}$.

First, we determine the applicability domains of the limiting propagation regimes (Section 10.5) and present them as the regime maps. Such analysis is useful to frame the general solution inside the parameter space resulting in better understanding of the propagation conditions as a function of time. Similar to Dontsov [46], we utilize the following criterion for the determination of the validity zone of the considered limiting solution i:

$$\sqrt{\left(1-\frac{\Omega_i(0,\ \tau)}{\Omega(0,\ \tau)}\right)^2+\left(1-\frac{\gamma_i(\tau)}{\gamma(\tau)}\right)^2}<0.01, \tag{10.38}$$

where the subscript i can be $M, \tilde{M}, K, \tilde{K}, T, \tilde{T}$. In other words, the measure of the relative difference between the general numerical solution and given limiting case i is taken to be less than 1%, (10.38), for the limiting solution to be considered a valid approximation, and the fracture to be said to propagate in the corresponding limiting regime.

Further, we demonstrate the solution series for different time-dependent crack characteristics such as $\gamma(\tau)$, $\Omega(0, \tau)$, $\Pi(0.5, \tau)$, $\eta(\tau)$ by fixing one (out of two) control parameters (e.g. $\mathcal{R}$) while the remaining (e.g. ϕ) is varied.

10.7.1 Zero Leak-Off Case (Impermeable Rock)

Lecampion and Zia [21] have already considered a radial crack propagation in an impermeable formation; however, the authors did not include the regime map for the corresponding reduced parametric space ($\tau, \phi = 0, \mathcal{R}$), and we would like to fill this gap. Figure 10.6a presents the applicability domains of the vertex solutions observed in the current model, i.e. storage-viscosity-turbulent (T), storage-viscosity-laminar (M), and toughness (K) regimes. The problem solution evolves from the T- to K-vertex, and the M-vertex emerges during an intermediate propagation time if $\mathcal{R} < 10^2$. For larger values of the characteristic Reynolds number, the validity zone of the T-vertex expands, while the M-region shrinks. The laminar-to-turbulent flow transformation affects the boundary of the toughness-dominated regime (K) in the range $\mathcal{R} > 10^6$, leading to its shift to larger time.

Further, we would like to clarify how exactly we determine the applicability boundaries shown in Figure 10.6a. We estimate the borders numerically via criterion (10.38) at sample points at small and large $\mathcal{R}$ and then fit the evaluated points by the appropriate analytical functions derived from the transition timescales between the limiting regimes. As an example, let us consider the validity zone of the T-vertex solution shown by orange color in Figure 10.6a. It is bounded by two power-law functions of time which are straight lines in the log–log scale. The first boundary located in the region $\mathcal{R} < 10^4$ relates to the TM-transition, i.e. between T and M vertices. We expect that it corresponds to $t \sim t_{mt}$, where the time-scale t_{mt} is the solution of $L_t(t) = L_m(t)$. This boundary expressed in the normalized time is $\tau \sim t_{mt}/t_{mk} = \mathcal{R}^{9/4}$, where the proportionality coefficient is found from the fitting procedure at small $\mathcal{R}$. The functional dependence for the second curve is obtained from the TK transition analysis and has the following form: $\tau \sim t_{tk}/t_{mk} = \mathcal{R}^{(5n-5)/(2n-4)}$, where the numerical prefactor is found from a similar fitting procedure at large $\mathcal{R}$. Both curves are extended until the intersection point, bounding the domain depicted in Figure 10.6a. The full usage of criterion (10.38) in the entire parametric space of the solution yields a very similar region but with a smoother boundary (without sharp corner point evident in the domain approximation shown in Figure 10.6a).

We also display two time bounds in Figure 10.6a. The first one $\tau_0(\mathcal{R}, \phi = 0)$ (dashed black line) indicates the time moment when the spatial extent of the laminar flow domain is a small fraction $\lambda = \lambda_0 = R/25$ of the fracture radius. As a result, the crack can be approximately considered as fully turbulent for $\tau < \tau_0(\mathcal{R}, \phi = 0)$ (and $\lambda < \lambda_0$). One can observe that $\tau_0(\mathcal{R}, \phi = 0)$ is an increasing function of the characteristic Reynolds number $\mathcal{R}$, i.e. the time interval, within which the length of the laminar flow region is small, and turbulent effects dominate, grows with an increase of $\mathcal{R}$. For $\mathcal{R} < 10^3$, the dashed line is very close to the border of the T-vertex validity domain. The second bound $\tau_\infty(\mathcal{R}, \phi = 0)$ (dotted black line) corresponds to the time moment following which the crack geometry (radius and aperture) in the turbulent-laminar case is approximately the same as the fully laminar model provides. In other words, one can interpret the fracture as fully laminar for $\tau > \tau_\infty(\mathcal{R}, \phi = 0)$. This bound is computed via Eq. (10.38) where i should be understood as a fully laminar solution. The function $\tau_\infty(\mathcal{R}, \phi = 0)$ is an increasing one (similarly to $\tau_0(\mathcal{R}, \phi = 0)$), and for $\mathcal{R} < 10^2$, the dotted line coincides with the boundary of the M-vertex domain.

Next, we discuss the time bound $\tau_0(\mathcal{R}, \phi = 0)$ in the context of the tip element concept. Many numerical models for hydraulic fracture growth are based on the "tip logic" [46, 49–51, 59]. It means that specialized near-tip region model is applied within such algorithms to determine the fracture front position at each time step (propagation criterion) and to describe crack characteristics (aperture, pressure) near the tip. The fracture tip model should include all physical phenomena that can be realized during the fracture evolution and resolve their influence on the propagation. It is assumed that the near-tip region model is valid along small distance from the fracture front, and in the case of 1D models (plane strain or penny-shaped) the typical tip element length equals λ_0. When we analyze the turbulent-tolaminar flow transformation within the crack channel, two different options for the near-tip region model exist. Firstly, when $\lambda > \lambda_0$ or for $\tau > \tau_0(\mathcal{R}, \phi = 0)$ laminar flow occurs along

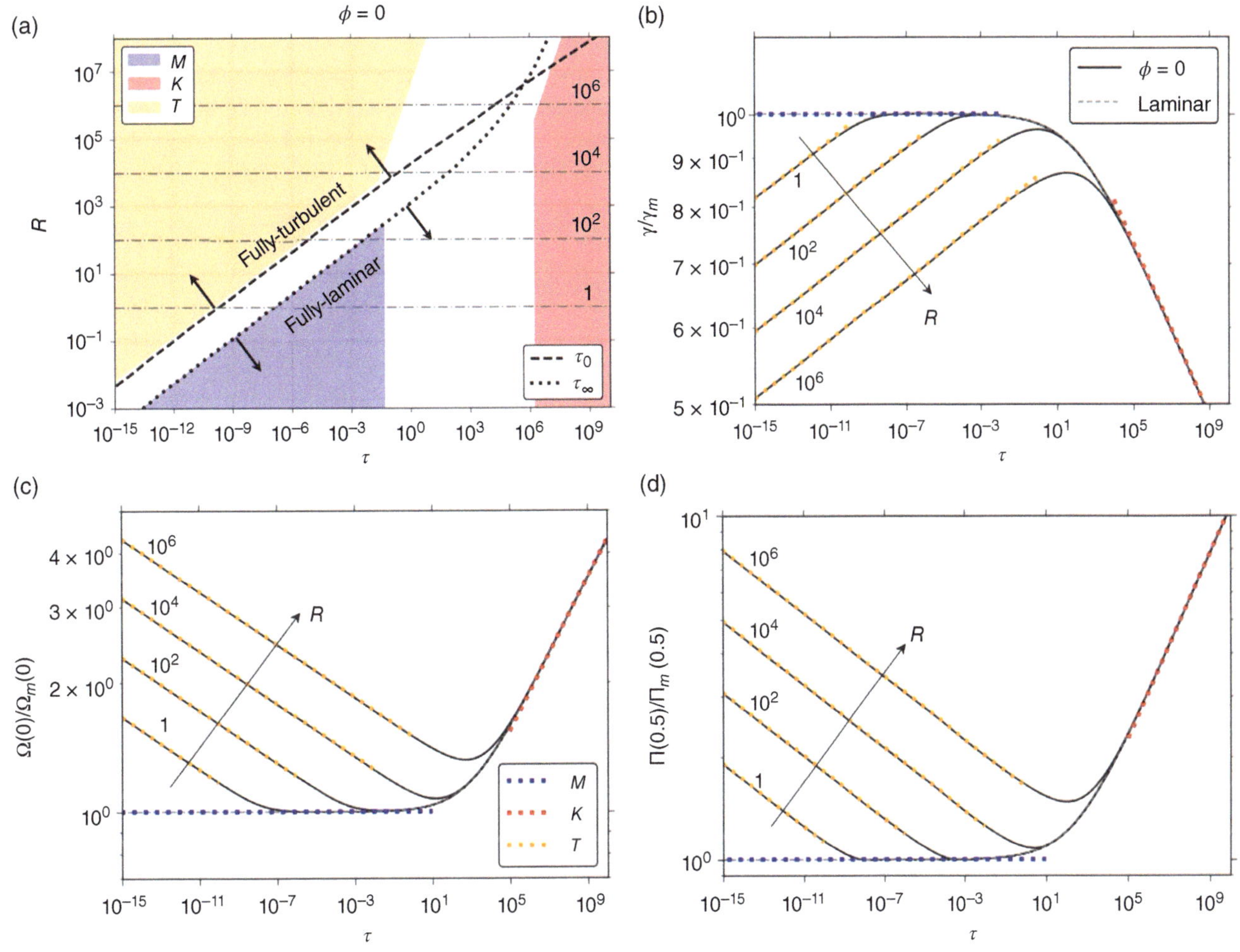

Figure 10.6 The problem solution for a radial fracture in an impermeable rock ($\phi = 0$) driven by turbulent-laminar flow of slickwater. (a) illustrates the regime map in the coordinates (τ, $\mathcal{R}$), where the colored zones denote the validity domains of the limiting propagation regimes. Two supplemental time bounds are shown: (i) time $\tau_0(\mathcal{R})$ (dashed black line) at which the transition between laminar and turbulent flow within the crack channel is located at small distance $\lambda_0 = R/25$ from the tip (such that laminar flow spatial domain is small $\lambda < \lambda_0$ for $\tau < \tau_0(\mathcal{R})$), and (ii) time $\tau_\infty(\mathcal{R})$ (dotted black line) past which the fracture radius and opening are approximately given by the fully-laminar solution. The time-dependent fracture characteristics (radius $\gamma(\tau)$, opening at the wellbore $\Omega(0, \tau)$, and pressure at the half-radius $\Pi(1/2, \tau)$) normalized by the storage-viscosity-laminar limiting solution (M-vertex) are shown in (b)–(d) for the set of values of $\mathcal{R} = \{1, 10^2, 10^4, 10^6\}$. The fully laminar solution is depicted by dashed gray line, and the colored dotted lines present the vertex solutions. In (a), the gray dash-dotted lines underline the $\mathcal{R}$-trajectories demonstrated in (b)–(d). (For interpretation of the references to colour in this figure legend, the reader is referred to the web version of this article.)

the entire tip element, one can apply the model of Garagash et al. [53] or its approximate and computationally efficient version [60]. On the other hand, when $\lambda < \lambda_0$ or for $\tau < \tau_0(\mathcal{R}, \phi = 0)$, it is recommended to utilize the laminar–turbulent tip model developed by Lecampion and Zia [21].

Figure 10.6b–d demonstrate the time-dependent fracture characteristics such as radius $\gamma(\tau)$, opening at the wellbore $\Omega(0, \tau)$, and pressure at the half-radius $\Pi(1/2, \tau)$ normalized by the storage-viscosity-laminar limiting solution (M-vertex). The turbulent–laminar solutions for $\mathcal{R} = \{1, 10^2, 10^4, 10^6\}$ are contrasted to the fully laminar solution profiles shown by gray lines. The flow regime transformation inside the fracture channel leads to a shorter radius, wider aperture, and greater pressure values at the beginning of the propagation compared to the fully laminar case. The time span when the turbulent–laminar and the fully laminar solutions differ significantly grows with an increase of the characteristic Reynolds number.

10.7.2 Nonzero Leak-Off Case (Permeable Rock)

Figure 10.7 shows the regime map for the fully laminar model ($\mathcal{R} = 0$) with nonzero leak-off ϕ in the coordinates (τ, ϕ), while Figure 10.8 demonstrates its modifications for the turbulent–laminar model for different values of

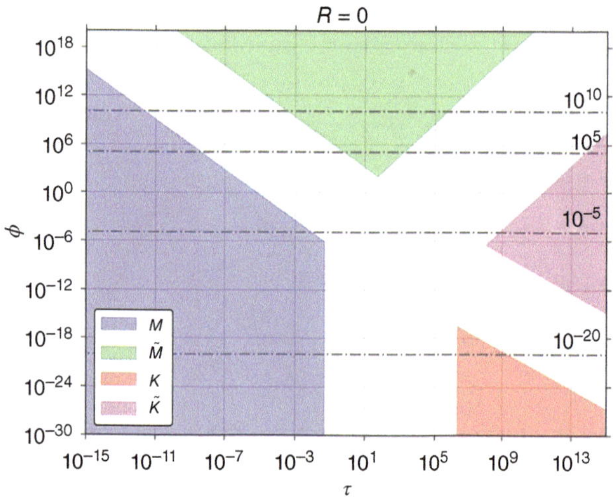

Figure 10.7 The parameter space of the model for a radial hydraulic fracture driven by laminar flow ($\mathcal{R} = 0$) in a permeable rock ($\phi > 0$). The coordinate system (τ, ϕ) is applied to build the regime map in which the applicability domains of the limiting propagation regimes are filled by different colors. The gray dash-dotted lines highlight the considered solution trajectories discussed in the current section.

the characteristic Reynolds number: $\mathcal{R} = \{1, 10^2, 10^4, 10^6\}$.

In the latter case, the problem solution evolves from the storage-viscosity-turbulent (T) solution at early time to the leak-off-toughness ($\tilde{K}$) limiting solution at large time. Moreover, the solution can approach M,K,$\tilde{M}$, or $\tilde{T}$ vertices at intermediate times depending on the values of the governing parameters ϕ,$\mathcal{R}$ (e.g. see trajectories ϕ = const in Figure 10.8b shown by dash-dotted grey lines). By looking at (a) and (b) in Figure 10.8, one can notice that the validity regions of the turbulent limiting propagation regimes, T and $\tilde{T}$, expand with an increase of $\mathcal{R}$ and gradually reduce the applicability zones of the laminar regimes, M and $\tilde{M}$, until their complete disappearance (see (c) and (d) in Figure 10.8). Similar behavior is observed for the toughness-dominated regimes, K and $\tilde{K}$, from the regime map in Figure 10.8d; however, they do exist for all $\mathcal{R}$ and large enough τ.

In section 10.7.1, we have already discussed the physical meaning of the temporal bounds τ_0 and τ_∞, and here, we analyze their behavior with the alteration of the governing parameters $\mathcal{R}$ and ϕ (see Figure 10.8). For $\mathcal{R}$ = const, both timescales are independent of ϕ when it is small. However, for large values of the leak-off number ϕ, we determine that τ_0, $\tau_\infty \sim \phi^2$. When the characteristic Reynolds number grows, the bounds shift towards larger time since the turbulent flow regime prevails inside the fracture channel during longer time period for greater values of $\mathcal{R}$. We should also mention that it is recommended to apply the turbulent–laminar fracture tip model [26] as a propagation criterion in more complex numerical models for HF growth (such as Planar3D) to simulate the crack evolution when the laminar flow domain is small, corresponding to the parametric zone bounded by the dashed black line in Figure 10.8, i.e. when $\tau < \tau_0(\mathcal{R}, \phi)$.

Figure 10.9 demonstrates the crack characteristics in the turbulent–laminar ($\mathcal{R} = 10^2$) and fully laminar (dashed grey lines) cases for different values of the leak-off number $\phi = \{10^{-20}, 10^{-5}, 10^5, 10^{10}\}$. Since the characteristic Reynolds number is constant, we focus on the impact of the leak-off on the propagation of a radial crack driven by turbulent–laminar flow. The storage-viscosity-laminar (M) limiting solution is utilized to normalize the time-dependent properties such as the radius $\gamma(\tau)$, opening at the wellbore $\Omega(0, \tau)$ and pressure at the half-radius $\Pi(1/2, \tau)$ in (a)–(c) in Figure 10.9. It is evident from Figure 10.9a that larger values of leak-off number ϕ lead to shorter time duration over which the fracture radius differs significantly from the fully laminar case (i.e. when turbulent effects on the fracture run-out are significant). The situation is opposite for the opening at the wellbore and pressure at the half-radius (see Figure 10.9b and c) for which increase of the leak-off extends the influence of the turbulent flow effects. Finally, the fracture efficiency is roughly independent of the laminar-to-turbulent flow regime transformation in all considered cases, as exposed in Figure 10.9d.

10.8 Conclusions

In this work, we have analyzed how the turbulent–laminar flow inside the fracture channel impacts the propagation of a radial hydraulic fracture in a permeable formation. We account for the flow regime transformation from the laminar to turbulent moving away from the fracture tip. Both flow regimes always coexist in the model – the turbulent at the wellbore and the laminar one near the tip. The fluid exchange process between the fracture and ambient rock is modeled in the form of Carter's leak-off law. The hydraulic fracturing fluid is slickwater, i.e. the water-based fluid with polymeric additives increasing the solvent viscosity and significantly modifying the turbulent flow frictional behavior. The latter is governed by the phenomenological MDR (Virk's) asymptote. We carry out several simulations of the radial crack growth for the values of the model parameters corresponding to typical field cases. The problem solution is found numerically via the Gauss–Chebyshev quadrature and barycentric interpolation techniques. The

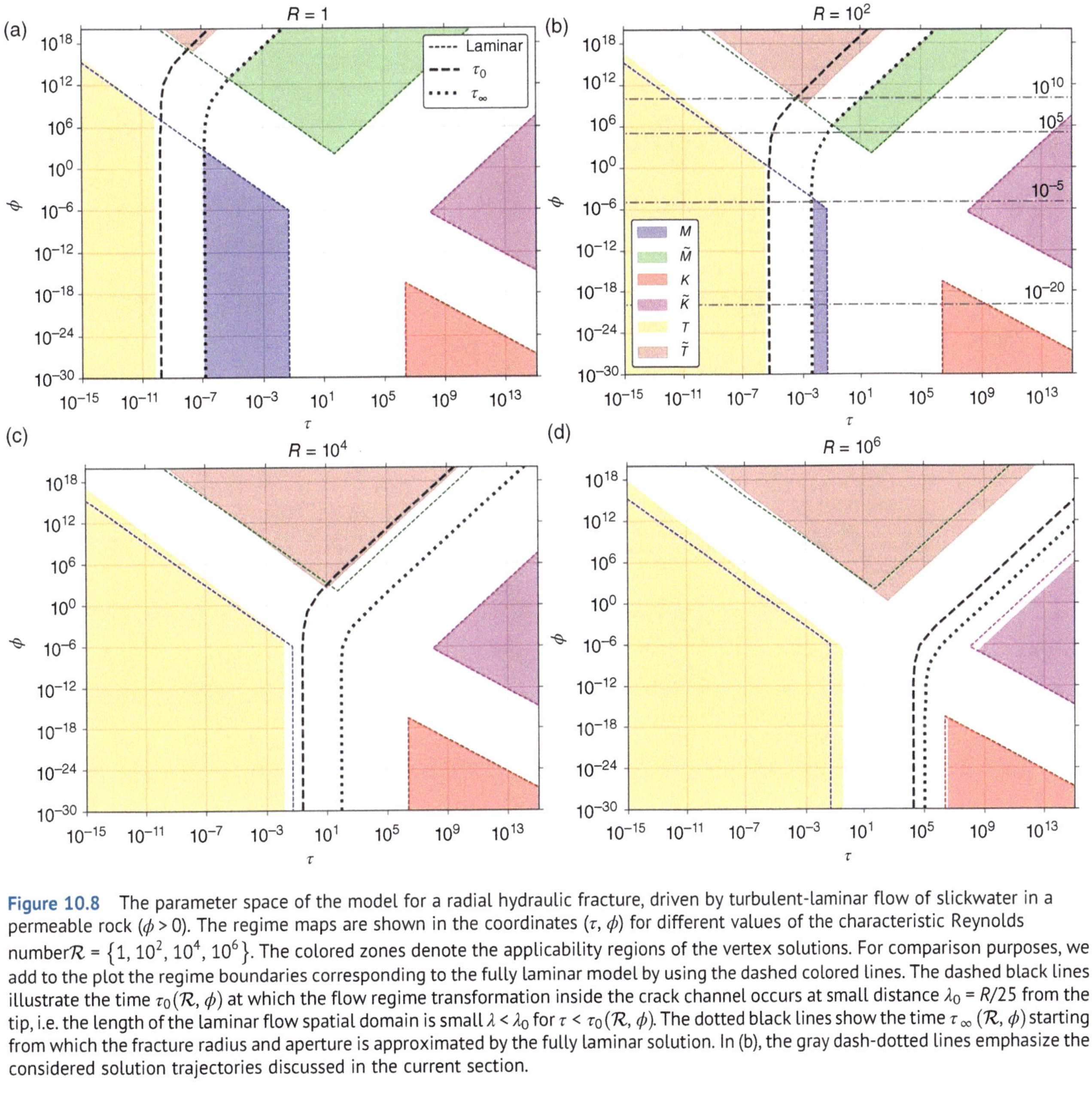

Figure 10.8 The parameter space of the model for a radial hydraulic fracture, driven by turbulent-laminar flow of slickwater in a permeable rock ($\phi > 0$). The regime maps are shown in the coordinates (τ, ϕ) for different values of the characteristic Reynolds number$\mathcal{R} = \{1, 10^2, 10^4, 10^6\}$. The colored zones denote the applicability regions of the vertex solutions. For comparison purposes, we add to the plot the regime boundaries corresponding to the fully laminar model by using the dashed colored lines. The dashed black lines illustrate the time $\tau_0(\mathcal{R}, \phi)$ at which the flow regime transformation inside the crack channel occurs at small distance $\lambda_0 = R/25$ from the tip, i.e. the length of the laminar flow spatial domain is small $\lambda < \lambda_0$ for $\tau < \tau_0(\mathcal{R}, \phi)$. The dotted black lines show the time $\tau_\infty(\mathcal{R}, \phi)$ starting from which the fracture radius and aperture is approximated by the fully laminar solution. In (b), the gray dash-dotted lines emphasize the considered solution trajectories discussed in the current section.

obtained results demonstrate that turbulence changes the crack characteristics near the wellbore zone during the initial propagation period. E.g. in the slickwater fracturing, the turbulent–laminar solution differs significantly from the fully laminar one during a couple of seconds, tens of seconds, and several minutes in terms of the radius, opening at the wellbore, and pressure, respectively. We reveal that the leak-off process prolongs the turbulence effects, and they continue much longer for pure water fracturing than for the slickwater driven crack. Although the fracture geometries (radii and aperture profiles) in the models with and without turbulent flow are practically the same after several minutes of the propagation, the pressure near the wellbore has larger values in the turbulent–laminar case during tens of minutes, meaning a greater amount of power is required for the fluid pumping than the laminar model predicts. The latter confirms that slickwater HF propagation, especially in permeable rock, remains in the turbulent–laminar regime on the timescale of typical treatment duration.

We have analyzed the limiting propagation regimes, characterized by the dominance of a single energy

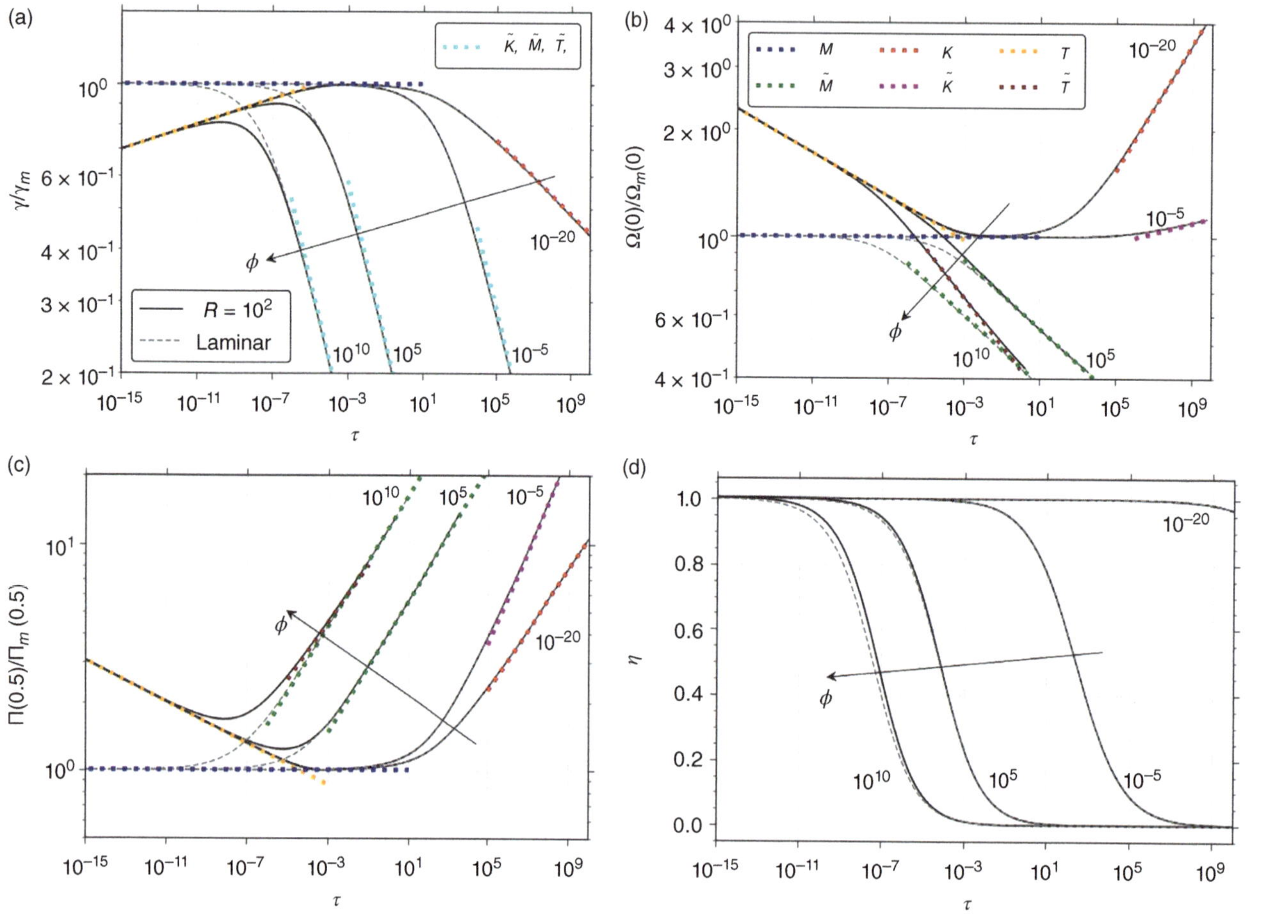

Figure 10.9 The time-dependent characteristics of a radial hydraulic fracture driven by turbulent–laminar flow of slickwater in a permeable rock ($\phi > 0$): (a) radius $\gamma(\tau)$, (b) opening at the wellbore $\Omega(0, \tau)$, (c) pressure at the half-radius $\Pi(1/2, \tau)$, and (d) efficiency $\eta(\tau)$. The properties in (a)–(c) are normalized by storage-viscosity-laminar (M) limiting solution. The solution profiles corresponding to $\mathcal{R} = 10^2$ and $\phi = \{10^{-20}, 10^{-5}, 10^5, 10^{10}\}$ are shown. The analogous profiles from the fully laminar solutions are depicted by the gray dashed lines. We plot the vertex solutions by the colored dotted lines; in (a), the leak-off dominated regimes ($\tilde{K}$,$\tilde{M}$,$\tilde{T}$) have the same color since the fracture radii in these regimes are governed by the same relation (e.g. see Eq. (10.31)).

dissipation and a single fluid storage mechanism, and derive corresponding limiting solutions semi-analytically. Then, we convert the system of governing equations into the dimensionless form in which the solution depends on two numbers, normalized leak-off $\phi = C'^4 E'^{11} \mu'^3 Q_0 / K'^{14}$ and characteristic Reynolds $\mathcal{R} = 12K'^4\rho/(E'^3\mu'^2)$, respectively, in addition to the dimensionless time and distance from the source (spatiotemporal properties). Using the normalized variables, we explore the problem parameter space by determining the applicability domains of the limiting regimes and looking at the variation of the crack characteristics with the change of the values of the governing parameters. We introduce the temporal bounds, $\tau_0(\mathcal{R}, \phi)$ and $\tau_\infty(\mathcal{R}, \phi)$, which can be used to determine the duration of the approximately fully turbulent fracture propagation $\tau < \tau_0$ and approximately fully laminar one for $\tau > \tau_\infty$.

In the current chapter, we have not accounted for the roughness of the crack surface, and such analysis is among plans for the future extension of the model. The proposed approach can be used as a benchmark solution for the numerical simulators of more realistic (complex) fracturing problems which include the laminar-to-turbulent flow regime transformation, e.g. turbulent–laminar Planar3D model.

Acknowledgments

Support by the Natural Sciences and Engineering Research Council of Canada, Discovery Program grant no. 05743, to D.I. Garagash is gratefully acknowledged.

References

1 Dontsov, E. (2016). Propagation regimes of buoyancy-driven hydraulic fractures with solidification. *Journal of Fluid Mechanics* 797: 1–28.

2 Lister, J.R. (1990). Buoyancy-driven fluid fracture: the effects of material toughness and of lowviscosity precursors. *Journal of Fluid Mechanics* 210: 263–280.

3 Rivalta, E., Taisne, B., Bunger, A., and Katz, R. (2015). A review of mechanical models of dike propagation: Schools of thought, results and future directions. *Tectonophysics* 638: 1–42.

4 Spence, D. and Turcotte, D. (1985). Magma-driven propagation of cracks. *Journal of Geophysical Research: Solid Earth* 90 (B1): 575–580.

5 Tsai, V.C. and Rice, J.R. (2010). A model for turbulent hydraulic fracture and application to crack propagation at glacier beds. *Journal of Geophysical Research: Earth Surface* 115 (F3): F03007 1–F03007 18.

6 van der Veen, C.J. (2007). Fracture propagation as means of rapidly transferring surface meltwater to the base of glaciers. *Geophysical Research Letters* 34 (1): L01501.

7 Economides, M., Oligney, R., and Valkó, P. (2002). *Unified fracture design: bridging the gap between theory and practice*. Orsa Press.

8 Economides, M.J., Nolte, K.G., and Ahmed, U. (1989). *Reservoirstimulation*, vol. 2. Cliffs, NJ: PrenticeHallEnglewood.

9 Barati, R. and Liang, J.-T. (2014). A review of fracturing fluid systems used for hydraulic fracturing of oil and gas wells. *Journal of Applied Polymer Science* 131 (16).

10 Perkins, T. and Kern, L. (1961). Widths of hydraulic fractures. *Journal of Petroleum Technology* 13 (9): 937–949.

11 Nilson, R. (1981). Gas-Driven Fracture Propagation. *J. Appl. Mech.* 48 (4): 757–762.

12 Emerman, S.H., Turcotte, D., and Spence, D. (1986). Transport of magma and hydrothermal solutions by laminar and turbulent fluid fracture. *Physics of the Earth and Planetary Interiors* 41 (4): 249–259.

13 Lister, J.R. and Kerr, R.C. (1991). Fluid-mechanical models of crack propagation and their application to magma transport in dykes. *Journal of Geophysical Research: Solid Earth* 96 (B6): 10049–10077.

14 Ames, B.C., Bunger, A. (2015). Role of turbulent flow in generating short hydraulic fractures with high net pressure in slickwater treatments. *SPE Hydraulic Fracturing Technology Conference*. Society of Petroleum Engineers, The Woodlands, TX. https://doi.org/10.2118/SPE-173373-MS.

15 Anthonyrajah, M., Mason, D., and Fareo, A. (2013). Propagation of a pre-existing turbulent fluid fracture. *International Journal of Non-Linear Mechanics* 54: 105–114.

16 Kano, M., Zolfaghari, N., Ames, B., and Bunger, A. (2015). Solution for a pkn hydraulic fracture driven by turbulent fluid with large leakoff. *Hydraulic Fracturing Journal* 2 (1): 34–38.

17 Zia, H. and Lecampion, B. (2017). Propagation of a height contained hydraulic fracture in turbulent flow regimes. *International Journal of Solids and Structures* 110: 265–278.

18 Zolfaghari, N., Meyer, C.R., and Bunger, A.P. (2017). Blade-shaped hydraulic fracture driven by a turbulent fluid in an impermeable rock. *Journal of Engineering Mechanics* 143 (11): 04017130.

19 Tsai, V.C. and Rice, J.R. (2012). Modeling turbulent hydraulic fracture near a free surface. *Journal of Applied Mechanics* 79 (3): 1–5.

20 Zolfaghari, N., Dontsov, E., and Bunger, A. (2018). Solution for a plane strain rough-walled hydraulic fracture driven by turbulent fluid through impermeable rock. *International Journal for Numerical and Analytical Methods in Geomechanics* 42 (4): 587–617.

21 Lecampion, B. and Zia, H. (2019). Slickwater hydraulic fracture propagation: near-tip and radial geometry solutions. *Journal of Fluid Mechanics* 880: 514–550.

22 Zolfaghari, N. and Bunger, A. (2018). Approximate semi-analytical solution for a penny-shaped roughwalled hydraulic fracture driven by turbulent fluid in an impermeable rock. *International Journal of Solids and Structures* 143: 144–154.

23 Zolfaghari, N. and Bunger, A. (2019). Numerical model for a penny-shaped hydraulic fracture driven by laminar/turbulent fluid in an impermeable rock. *International Journal of Solids and Structures* 158: 128–140.

24 Dontsov, E. and Peirce, A. (2017). Modeling planar hydraulic fractures driven by laminar-to-turbulent fluid flow. *International Journal of Solids and Structures* 128: 73–84.

25 Dontsov, E. (2016). Tip region of a hydraulic fracture driven by a laminar-to-turbulent fluid flow. *Journal of Fluid Mechanics* 797: 1–12.

26 Kanin, E., Garagash, D., and Osiptsov, A. (2020). Turbulent flow effects in a slickwater fracture propagation in permeable rock. In: *ECMOR XVII*, vol. 2020, 1–18. European Association of Geoscientists & Engineers. https://doi.org/10.3997/2214-4609.202035216.

27 Yang, B., Zhao, J., Mao, J. et al. (2019). Review of friction reducers used in slickwater fracturing fluids for shale gas reservoirs. *Journal of Natural Gas Science and Engineering* 62: 302–313.

28 Nieuwstadt, F.T., Westerweel, J., and Boersma, B.J. (2016). *Turbulence: Introduction to Theory and Applications of Turbulent Flows*. Springer.

29 Virk, P. (1971). Drag reduction in rough pipes. *Journal of Fluid Mechanics* 45 (2): 225–246.

30 Virk, P.S. (1975). Drag reduction fundamentals. *AIChE Journal* 21 (4): 625–656.

31 Carter, E. (1957). Optimum fluid characteristics for fracture extension. In: *Drilling and Production Practices, API* (ed. G.C. Howard and C.R. Fast), 261–270.

32 Garagash, D.I. (2019). Cohesive-zone effects in hydraulic fracture propagation. *Journal of the Mechanics and Physics of Solids* 133: 103727.

33 Rice, J.R. (1968). Mathematical analysis in the mechanics of fracture. *Fracture: An Advanced Treatise* 2: 191–311.

34 Batchelor, G.K. (1967). *An Introduction to Fluid Dynamics*. Cambridge: Cambridge University Press.

35 Detournay, E. and Garagash, D. (2003). The near-tip region of a fluid-driven fracture propagating in a permeable elastic solid. *Journal of Fluid Mechanics* 494: 1–32.

36 Garagash, D. and Detournay, E. (2000). The tip region of a fluid-driven fracture in an elastic medium. *Journal of Applied Mechanics* 67 (1): 183–192.

37 Kanin, E.A., Garagash, D.I., and Osiptsov, A.A. (2020). The near-tip region of a hydraulic fracture with pressure-dependent leak-off and leak-in. *Journal of Fluid Mechanics* 892: A31. https://doi.org/10.1017/jfm.2020.193.

38 Collins, R.E. (1976). *Flow of Fluids Through Porous Materials*. Research & Engineering Consultants.

39 Arin, K. and Erdogan, F. (1971). Penny-shaped crack in an elastic layer bonded to dissimilar half spaces. *International Journal of Engineering Science* 9 (2): 213–232.

40 Cleary, M. and Wong, S. (1985). Numerical simulation of unsteady fluid flow and propagation of a circular hydraulic fracture. *International Journal for Numerical and Analytical Methods in Geomechanics* 9 (1): 1–14.

41 Savitski, A. and Detournay, E. (2002). Propagation of a penny-shaped fluid-driven fracture in an impermeable rock: asymptotic solutions. *International Journal of Solids and Structures* 39 (26): 6311–6337.

42 Jones, O. Jr. (1976). An improvement in the calculation of turbulent friction in rectangular ducts. *Journal of Fluids Engineering* 98: 173–180.

43 Irvin, G.R. (1957). Analysis of stresses and strains near the end of a crack traversing a plate. *J. Appl. Mech.* 24: 361–364. https://doi.org/10.1115/1.4011547.

44 Viesca, R.C. and Garagash, D.I. (2018). Numerical methods for coupled fracture problems. *Journal of the Mechanics and Physics of Solids* 113: 13–34.

45 Liu, D., Lecampion, B., and Garagash, D.I. (2019). Propagation of a fluid-driven fracture with fracture length dependent apparent toughness. *Engineering Fracture Mechanics* 220: 106616.

46 Dontsov, E. (2016). An approximate solution for a penny-shaped hydraulic fracture that accounts for fracture toughness, fluid viscosity and leak-off. *Royal Society Open Science* 3 (12): 160737.

47 Dontsov, E. and Peirce, A. (2017). A multiscale implicit level set algorithm (ilsa) to model hydraulic fracture propagation incorporating combined viscous, toughness, and leak-off asymptotics. *Computer Methods in Applied Mechanics and Engineering* 313: 53–84.

48 Kanin, E.A., Dontsov, E.V., Garagash, D.I., and Osiptsov, A.A. (2020). A radial hydraulic fracture with pressure-dependent leak-off. *Journal of the Mechanics and Physics of Solids* 143: 104062.

49 Peirce, A. (2015). Modeling multi-scale processes in hydraulic fracture propagation using the implicit level set algorithm. *Computer Methods in Applied Mechanics and Engineering* 283: 881–908.

50 Peirce, A. and Detournay, E. (2008). An implicit level set method for modeling hydraulically driven fractures. *Computer Methods in Applied Mechanics and Engineering* 197 (33–40): 2858–2885.

51 Zia, H. and Lecampion, B. (2020). Pyfrac: A planar 3d hydraulic fracture simulator. *Computer Physics Communications* 255: 107368.

52 Barbati, A.C., Desroches, J., Robisson, A., and McKinley, G. H. (2016). Complex fluids and hydraulic fracturing. *Annual Review of Chemical and Biomolecular Engineering* 7: 415–453.

53 Garagash, D.I., Detournay, E., and Adachi, J.I. (2011). Multiscale tip asymptotics in hydraulic fracture with leak-off. *Journal of Fluid Mechanics* 669: 260–297. https://doi.org/10.1017/S002211201000501X.

54 Desroches, J., Detournay, E., Lenoach, B. et al. (1994). The crack tip region in hydraulic fracturing. *Proceedings of the Royal Society of London. Series A: Mathematical and Physical Sciences* 447 (1929): 39–48.

55 Detournay, E. (2016). Mechanics of hydraulic fractures. *Annual Review of Fluid Mechanics* 48: 311–339.

56 Bunger, A.P., Detournay, E., and Garagash, D.I. (2005). Toughness-dominated hydraulic fracture with leak-off. *International Journal of Fracture* 134 (2): 175–190.

57 Madyarova, M.V. (2003). Fluid-driven penny-shaped fracture in elastic medium. Master's thesis. University of Minnesota.

58 Adachi, J.I. and Detournay, E. (2008). Plane strain propagation of a hydraulic fracture in a permeable rock. *Engineering Fracture Mechanics* 75 (16): 4666–4694.

59 Dontsov, E. (2017). An approximate solution for a plane strain hydraulic fracture that accounts for fracture toughness, fluid viscosity, and leak-off. *International Journal of Fracture* 205 (2): 221–237.

60 Dontsov, E. and Peirce, A. (2015). A non-singular integral equation formulation to analyse multiscale behaviour in semi-infinite hydraulic fractures. *Journal of Fluid Mechanics* 781: 1–11.

11

Analysis of a Constant Height Hydraulic Fracture

E.V. Dontsov

ResFrac Corporation, 555 Bryant St., #185 Palo Alto, CA, USA

11.1 Introduction

Hydraulic fractures are most often created underground for the purpose of enhancing oil and gas recovery from reservoirs [1]. Nowadays, complex numerical models are readily used to model such a phenomenon, which can consider propagation of a single or multiple hydraulic fractures, interaction between fractures, proppant transport, thermal effects, chemical effects, multiphase flow effects, etc. Yet, understanding of the basic mechanisms and behavior of hydraulic fractures is better obtained by using simple models that focus on a particular aspect of the problem, while ignoring the rest of the complexity. This is the case for this study, in which a relatively simple geometry of a constant height hydraulic fracture is investigated.

There is a variety of "simple" hydraulic fracture models. The word "simple" here essentially means that the mathematical formulation for the problem is one-dimensional. As a result of such mathematical simplicity, it is possible to investigate the problem in greater details and to better understand structure of the solution. The first example is the problem of a semi-infinite hydraulic fracture, which is the model for the tip region of a finite planar fracture [2–4]. The limiting analytic solutions for the problem were first obtained in [5, 6] for the case of domination of viscosity and leak-off, respectively. The last limiting solution that corresponds to the domination of toughness stems from classical linear elastic fracture mechanics (LEFM) [7]. General solution for the problem that considers Newtonian fluid, Carter's leak-off, and toughness was obtained in [3]. The rapid approximation for this solution was later derived in [8], which then was subsequently used as a propagation condition in a planar hydraulic fracture model [9]. It is also worth mentioning multiple extensions of the model that consider power-law fluids [10], Herschel–Bulkley fluids [11], Carreau fluids [12], the effect of fluid lag [13], the effect of turbulent flow [14], the effect of cohesive zone [15], the effect of proppant [16], and the effect of pressure dependent leak-off [17]. This model already shows how multiple physical effects can be understood by considering a relatively simple scenario.

Plane strain and radial hydraulic fractures can be considered "simple" geometries, since the mathematical formulation for them is also one-dimensional. In contrast to the semi-infinite fracture, for which there are three limiting solutions, the finite fractures have four limiting cases. Note that this is only the case for the model that considers the effects of toughness, viscosity, and Carter's leak-off, i.e. ignoring the effects of fluid lag, cohesive zone, complex fluid rheology, etc. The aforementioned four cases are defined by the competition between fracture toughness and fluid viscosity, and fluid storage inside the fracture or inside the formation. Therefore, these limits are defined as: storage toughness, storage viscosity, leak-off toughness, and leak-off viscosity, see e.g. review paper [18]. Early studies focused predominantly on finding solutions for these limiting cases [19–22]. The first analysis for a complete problem is given by Hu and Garagash [23] for the case of a plane strain fracture and [24] for the case of a radial geometry. An alternative analysis for these problems was introduced in [25, 26]. The rapid approximate solution for the global problem is constructed by using the semi-infinite or tip solution [8] and global volume balance. Such rapid solution allowed to scan the whole parametric space and to construct map of the solutions that determines regions of applicability of the limiting solutions. Similar concept was later used to analyze the effect of anisotropy and power-law rheology on the parametric space in [27], to investigate the effect of cohesive zone in [15], and to construct an ultrafast hydraulic fracturing simulator in [28].

The last, but not least, "simple" geometry is a constant height hydraulic fracture or Perkins–Kern–Nordgren

Mechanics of Hydraulic Fracturing: Experiment, Model, and Monitoring, First Edition.
Edited by Xi Zhang, Bisheng Wu, Diansen Yang, and Andrew Bunger.

(PKN) fracture [29, 30]. Technically, it is possible to also include pseudo-3D model [31–33] into the list of "simple" cases. But it is noticeably more complex in the sense that its parametric space is larger, which makes the analysis more sophisticated. The classical PKN model does not consider the effect of fracture toughness, only viscosity and leak-off. The viscosity and leak-off dominated solutions can be found in [1]. Analysis of the tip region, again in the absence of toughness, is done in [34]. Laboratory observations that confirm the predicted near-tip behavior are summarized in [35]. Note that there is also analysis of the effects of turbulent flow [36, 37] for this fracture geometry. All of the above studies do not consider the effect of fracture toughness on the solution. There are two possibilities to include toughness into the model. The first was proposed by Nolte [38], in which the pressure boundary condition at the tip is taken from that for a uniformly pressurized radial fracture. This condition was later revised in [39], where authors proposed to alter it slightly based on energetic considerations. The second approach is to replace the local elasticity relation used in the classical PKN model by a more accurate non-local expression, that can be combined with the standard LEFM propagation criterion [40]. While the second approach is shown to be more accurate [40], it is more difficult to do analysis with it. Therefore, the model with local elasticity and a finite pressure boundary condition at the tip is used in this study.

As is evident from the above literature review, the parametric spaces for all "simple" models except the PKN case were thoroughly analyzed for the case when the model considers the simultaneous interplay between toughness, viscosity, and leak-off. This is probably because the procedure to include the effect of toughness into PKN model was introduced relatively recently. To fill the gap, this study aims to investigate the parametric space for the tip region of a PKN fracture, i.e. extend the work of Kovalyshen and Detournay [34] to include the effect of fracture toughness. In addition to that, the bigger goal is to thoroughly investigate the parametric space for the finite fracture and to determine all the limiting solutions and their applicability zones in the parametric space. In particular, the structure of this study is the following. Section 11.2 outlines the governing equations for the problem. Then, the problem of the tip region is analyzed in Section 11.3 and the parametric map for the problem is constructed. After that, Section 11.4 presents vertex solutions for a finite PKN fracture. The parametric map for the finite fracture is constructed in Section 11.5 using the fast approximate solution that is benchmarked against the numerical solution. Finally, Section 11.6 presents application examples and Section 11.7 summarizes the results.

11.2 Governing Equations

Consider the problem of a constant height hydraulic fracture or simply PKN fracture depicted in Figure 11.1a. Let the fracture be vertical and occupy the (x, y) plane, whereby y is the vertical coordinate, while x is the horizontal coordinate. Let H be the fracture height and l be the half-length. The height H is assumed to be constant for the whole fracture, while the length is a function of time. It is also assumed that $H \ll l$, which allows to simplify the governing equations to be one-dimensional.

Various solutions for the problem and their asymptotes are schematically shown in Figure 11.1b. The black line indicates the most accurate solution that utilizes nonlocal elasticity relation. It has the classical tip asymptotic behavior stemming from the problem of a semi-infinite plane strain hydraulic fracture (dashed red line). The thick dashed line indicates solution that employs the local

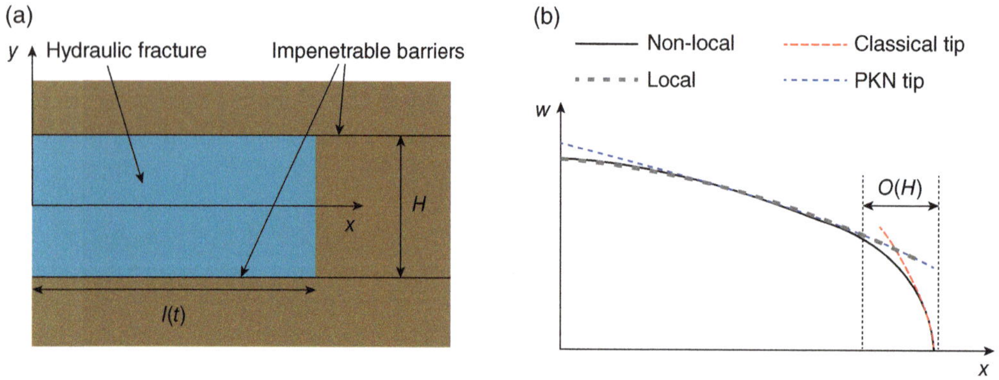

Figure 11.1 Schematics of a constant height hydraulic fracture (a). Schematics of the variation of fracture width along the x-axis for different models (b).

elasticity together with the pressure boundary condition at the tip to capture the effect of fracture toughness. This leads to the discontinuous behavior of fracture width at the tip. Even though it is not physical, this boundary condition actually signifies the behavior at some distance from the tip $O(H)$. Indeed, as shown in [40], the solutions with local and non-local elasticity approach each other further away from the fracture tip, while the differences are concentrated in the $O(H)$ region near the tip. The global fracture characteristics, such as length and width at the wellbore are captured accurately, as soon as $H \ll l$. Further, one may also consider the tip asymptotics for PKN fracture, which is schematically shown by the thin dashed line. It is valid at the distances that are commensurate with the fracture length l, which is in contrast to the classical asymptote (dashed line in the $O(H)$ region) that is accurate at the distances on the order of the fracture height H for the geometry under consideration. To simplify the analysis, the model with local elasticity is considered in this study.

By using classical assumptions of PKN model [29, 30], each vertical fracture cross section is assumed to be elliptical, and the pressure is determined based on the local plane strain elasticity assumption. This implies that

$$w(x,y) = \frac{4}{\pi}\,\overline{w}(x)\sqrt{1-\left(\frac{2y}{H}\right)^2},$$

$$p(x) = \frac{2\,E'\overline{w}(x)}{\pi H}, \quad \overline{w}(x) = \frac{1}{H}\int_{-H/2}^{H/2} w(x,y)dy, \tag{11.1}$$

where $w(x, y)$ denotes fracture opening, $\overline{w}(x)$ is the effective width, $p(x)$ is the fluid pressure, and $E' = E/(1-\nu^2)$ is the plane strain Young's modulus. Given the assumption $H \ll l$, the vertically-averaged lubrication equation for the case of Newtoniafracturing fluid becomes

$$\frac{\partial \overline{w}}{\partial t} + \frac{\partial \overline{q}_x}{\partial x} + \frac{C'}{\sqrt{t-t_0(x)}} = \frac{Q_0}{H}\delta(x),$$
$$\overline{q}_x = -\frac{1}{12H\mu}\frac{\partial p}{\partial x}\int_{-\frac{H}{2}}^{\frac{H}{2}} w^3 dy = -\frac{\overline{w}^3}{\pi^2 H\mu}\frac{\partial p}{\partial x}, \tag{11.2}$$

where μ is fluid viscosity, $\overline{q}_x$ is the average horizontal flux, $C' = 2C_l$ is the scaled Carter's leak-off coefficient [41], and $t_0(x)$ denotes the time instant at which the fracture front was located at the point x. Equations. (11.1) and (11.2) can be combined to yield

$$\frac{\partial \overline{w}}{\partial t} - \frac{E'}{2\pi^3 H\mu}\frac{\partial^2 \overline{w}^4}{\partial x^2} + \frac{C'}{\sqrt{t-t_0(x)}} = \frac{Q_0}{H}\delta(x), \tag{11.3}$$

which is the single governing equation for the PKN model.

To model lateral propagation condition that accounts for fracture toughness, the model from [39] is employed, namely

$$p(l) = \frac{2K_{Ic}}{\sqrt{\pi H}} \tag{11.4}$$

where K_{Ic} is fracture toughness and l is the fracture half-length. As shown in [40] this propagation condition is able to adequately capture the effect of toughness on lateral fracture growth, even though the near tip behavior is less accurate. Equation (11.4) can be rewritten using (11.1) as

$$\overline{w}(l) = \frac{\sqrt{\pi H}K_{Ic}}{E'} \tag{11.5}$$

This form of the propagation condition is used as a boundary condition for (11.3), which makes the whole problem formulated only in terms of $\overline{w}(x)$ and $l(t)$. Pressure, on the other hand, can be obtained from (11.1) if needed.

The evolution of length is governed by the volume balance at the tip, which requires that

$$\frac{dl}{dt} = \left.\frac{\overline{q}_x}{\overline{w}}\right|_{x=l} = -\left.\frac{2E'}{3\pi^3 H\mu}\frac{\partial \overline{w}^3}{\partial x}\right|_{x=l}, \tag{11.6}$$

which relates the spatial derivative of the width to the rate of fracture growth.

For completeness, global volume balance can be obtained by integrating (11.3) with respect to time and space as

$$\int_0^l \left[\overline{w}(x) + 2C'\sqrt{t-t_0}\,\right]dx = \frac{Q_0 t}{2H}. \tag{11.7}$$

In the above expression the first term signifies fracture volume, the second term is the total leaked volume, while the right hand side is the injected volume.

11.3 Tip Region

The problem of a tip region is considered first, which corresponds to the semi-infinite PKN fracture propagating steadily with the velocity V. In order to derive the relevant governing equations for the tip region from (11.3), a standard approach is taken, whereby a new moving coordinate is introduced $\hat{x} = Vt - x$ (see e.g. [3, 4]). In this case, the governing differential equation for the tip region becomes

$$\frac{E'}{2\pi^3\mu H}\frac{d\overline{w}^4}{d\hat{x}} = V\,\overline{w} + 2C'\sqrt{V\hat{x}}, \quad \overline{w}(0) = \frac{\sqrt{\pi H}K_{Ic}}{E'}. \tag{11.8}$$

Here, the second equation is simply the boundary condition (11.5).

Unfortunately, it is not possible to solve the differential Eq. (11.8) analytically. At the same time, one can obtain limiting solutions. In the limit of large toughness, the width is simply equal to the boundary condition in (11.8). To obtain the solution for zero toughness and zero leak-off (viscosity-dominated), both toughness and leak-off need to be set to zero in (11.8). This results in a solvable differential equation. In the limit of high leak-off and no toughness, the term "$V\overline{w}$" and toughness should be removed in (11.8), which again leads to the solvable differential equation. The summary of all three limiting or vertex solutions is

$$\overline{w}_k = \frac{\sqrt{\pi H}K_{Ic}}{E'}, \quad \overline{w}_m = \left(\frac{3\pi^3\mu HV}{2E'}\right)^{1/3}\hat{x}^{1/3}, \quad \overline{w}_{\tilde{m}} = \left(\frac{8\pi^3\mu HC'\sqrt{V}}{3E'}\right)^{1/4}\hat{x}^{3/8}, \tag{11.9}$$

where the first is the toughness solution, the second is the viscous solution, and the third is the leak-off solution. These results agree with the previous analysis in [34, 39].

It is also possible to compute the km and $k\tilde{m}$ solutions. Indeed, the differential Eq. (11.8) can be solved for these cases to obtain

$$\overline{w}_{km} = \left(\overline{w}_k^3 + \overline{w}_m^3\right)^{1/3}, \qquad \overline{w}_{k\tilde{m}} = \left(\overline{w}_k^4 + \overline{w}_{\tilde{m}}^4\right)^{1/4}, \tag{11.10}$$

Motivated by such a solution form, the following approximate solution is constructed

$$\overline{w}_{k\tilde{m}m} = \left(\overline{w}_k^p + \overline{w}_m^p + \overline{w}_{\tilde{m}}^p\right)^{1/p}, \qquad p = 3.4, \tag{11.11}$$

where the power p is tuned to minimize the maximum error with the numerical solution. This solution captures the limiting vertex cases precisely, while approximating the behavior in the transition regions. It is also possible to construct a more accurate solution:

$$\overline{w}_{k\tilde{m}m} = \left[\overline{w}_{km}\left(\overline{w}_{km}^4 + \overline{w}_{\tilde{m}}^4\right)^{1/4} + \overline{w}_{k\tilde{m}}\left(\overline{w}_{k\tilde{m}}^3 + \overline{w}_m^3\right)^{1/3}\right]\left[\overline{w}_{km} + \overline{w}_{k\tilde{m}}\right]^{-1}, \tag{11.12}$$

which in addition also respects the edge solutions outlined above.

In order to check accuracy of the suggested approximations and for simpler analysis, it is convenient to reformulate the problem (11.8) in the dimensionless form by introducing the following parameters:

$$\Omega = \frac{\overline{w}}{\overline{w}_k} = \frac{E'\overline{w}}{\sqrt{\pi H}K_{Ic}}, \quad \hat{\xi} = \frac{\pi^{\frac{3}{2}}\mu VE'^2\hat{x}}{2K_{Ic}^3H^{1/2}}, \quad \chi = \left(\frac{8C'^2K_{Ic}}{\pi^{\frac{5}{2}}\mu H^{\frac{1}{2}}V^2}\right)^{\frac{1}{2}}. \tag{11.13}$$

In this case, Eq. (11.8) reduces to

$$\frac{d\Omega}{d\hat{\xi}} = \frac{1}{\Omega^2} + \frac{\chi\hat{\xi}^{1/2}}{\Omega^3}, \qquad \Omega(0) = 1. \tag{11.14}$$

At the same time, the vertex solutions become

$$\Omega_k = 1, \quad \overline{\Omega}_m = \left(3\hat{\xi}\right)^{1/3}, \quad \overline{\Omega}_{\tilde{m}} = \left(\frac{8\chi}{3}\right)^{1/4}\hat{\xi}^{3/8}, \tag{11.15}$$

which also allows to recast the two suggested approximations (11.11) and (11.12).

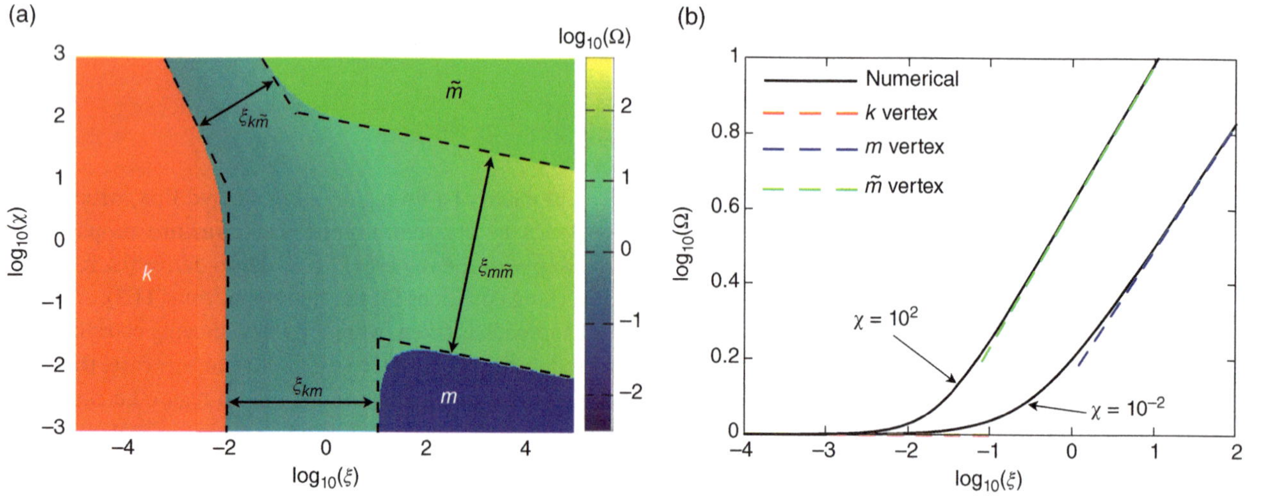

Figure 11.2 (a): Parametric space for the problem of tip region for PKN fracture. The colored regions indicate zones of applicability of the limiting vertex solutions (11.15) and they are also outlined by the dashed black lines defined by (11.16). (b): Numerical solution for the dimensionless opening versus dimensionless coordinate for two values of leak-off $\chi = 10^{-2}$ and $\chi = 10^{2}$ (solid black lines). The vertex solutions (11.15) are shown by the dashed colored lines.

Figure 11.2a plots the numerical solution of (11.14) (computed using Runge–Kutta fourth-order explicit integration of (11.14)) in the parametric space $(\hat{\xi}, \chi)$. Regions of applicability of the vertex solutions (11.15) are highlighted by the colored regions. These regions are defined as the zones, in which the relative difference between the global solution and the corresponding limiting solution does not exceed 1%. In particular, the region corresponding to the applicability of the toughness solution is marked by "k", the region representing the viscosity solution by "m", while the zone of applicability of the leak-off solution is marked by "$\tilde{m}$". There are three transition regions, namely from k to m, from k to $\tilde{m}$, and from m to $\tilde{m}$. These are quantified by the following parameters:

$$\hat{\xi}_{km} = \hat{\xi}, \quad \hat{\xi}_{km,1} \approx 0.010, \quad \hat{\xi}_{km,2} \approx 11,$$
$$\hat{\xi}_{k\tilde{m}} = \hat{\xi}\chi^{2/3}, \quad \hat{\xi}_{k\tilde{m},1} \approx 0.062, \quad \hat{\xi}_{k\tilde{m},2} \approx 4.4,$$
$$\hat{\xi}_{m\tilde{m}} = \hat{\xi}\chi^{6}, \quad \hat{\xi}_{m\tilde{m},1} \approx 1.85 \times 10^{-8}, \quad \hat{\xi}_{m\tilde{m},2} \approx 2.1 \times 10^{12}. \tag{11.16}$$

These parameters, alongside with their respective minimum and maximum values, quantitatively determine boundaries of applicability of the limiting vertex solutions and are plotted by the dashed black lines in Figure 11.2a. The simplest way to determine these transitional parameters is to equate the corresponding vertex $\Omega_k \sim \Omega_{\tilde{m}}$ or $1 \sim \chi^{1/4}\hat{\xi}^{3/8}$ which implies that $\hat{\xi}_{k\tilde{m}} = \hat{\xi}\chi^{2/3}$.

Figure 11.2b shows the linear plots of the numerical solution computed for $\chi = 10^{-2}$ and $\chi = 10^{2}$ versus $\hat{\xi}$ (black lines). The dashed colored lines indicate the vertex solutions (11.15). Gradual transitions from the toughness to the leak-off or viscosity solutions are observed. One peculiar observation is that the solution starts from the k vertex, then for some values of χ passes through the intermediate asymptote m, while $\tilde{m}$ is the far-field solution. This is in contrast to the classical solution for a semi-infinite plane strain hydraulic fracture [3], for which the far-field is always m solution, while $\tilde{m}$ is intermediate. Also, given the relatively gentle slope of the $m\tilde{m}$ transition in the parametric space Figure 11.2a, the transition from m to $\tilde{m}$ can be hard to achieve in practice.

In order to quantify accuracy of the approximations (11.11) and (11.12), Figure 11.3 plots the relative error between the approximate and numerical solutions. (a) gives the result for the simple solution (11.11), while (b) corresponds to the more comprehensive approximation (11.12). The maximum error within the entire parametric space for the first case is 3%, while it is just below 0.7% for the second, more accurate, case. The distribution of the error is different as well. For the simpler case (11.11), the error is distributed roughly uniformly within the transition regions, and decays toward the vertex limits. For the second case (11.12), the k vertex and the transitions from this vertex to m and $\tilde{m}$ are noticeably more accurate, while the error occurs mostly within the $m\tilde{m}$ transition. This is not surprising, since the exact solutions for km and $k\tilde{m}$ transitions are used to construct the approximation (11.12).

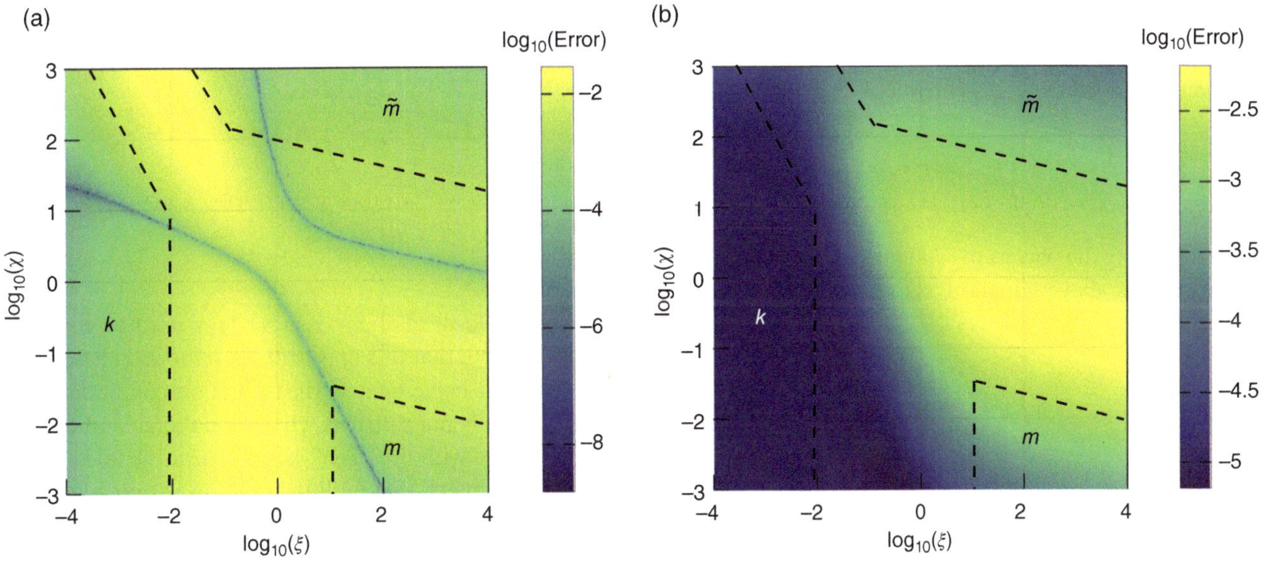

Figure 11.3 Relative error between the numerical solution and approximations (11.11) (a) and (11.12) (b).

11.4 Vertex Solutions

Similar to the problems of a plane strain and radial fracture, there are four limiting solutions for the PKN geometry. These are determined by the competition between toughness and viscosity, as well as storage of fluid within fracture and porous formation. In particular, let us define storage viscosity or M limit as the case corresponding to domination of viscosity and no leak-off, the storage toughness or K limit as the case of toughness domination and no leak-off, the leak-off viscosity or $\tilde{M}$ limit as the case of viscosity domination and high leak-off, and finally leak-off toughness or $\tilde{K}$ limit as the case of toughness domination and high leak-off, see e.g. the review paper [18].

11.4.1 Storage Viscosity

For the storage viscosity case, the governing Eq. (11.3), the boundary condition (11.5), and the volume balance (11.7) reduce to

$$\frac{\partial \overline{w}}{\partial t} - \frac{E'}{2\pi^3 H\mu}\frac{\partial^2 \overline{w}^4}{\partial x^2} = \frac{Q_0}{H}\delta(x),$$
$$\int_0^l \overline{w}(x)dx = \frac{Q_0 t}{2H}, \qquad \overline{w}(l) = 0. \tag{11.17}$$

By explicitly incorporating the viscosity dominated near-tip behavior (11.9) into the global solution and looking for the solution in the form

$$\overline{w} = \left(\frac{3\pi^3 \mu H}{2E'}\right)^{\frac{1}{3}} (\dot{l})^{\frac{1}{3}} l^{\frac{1}{3}} (1-\xi)^{\frac{1}{3}} f(\xi), \qquad \xi = \frac{x}{l(t)}, \tag{11.18}$$

the volume balance in (11.17) reduces to

$$\left(\frac{3\pi^3 \mu H}{2E'}\right)^{\frac{1}{3}} (\dot{l})^{\frac{1}{3}} l^{\frac{4}{3}} \int_0^1 (1-\xi)^{\frac{1}{3}} f(\xi) d\xi = \frac{Q_0 t}{2H}. \tag{11.19}$$

By balancing the powers of t and other dimensional parameters in the above equation, solution for length can be written as

$$l = c\left(\frac{5E' Q_0^3}{48\pi^3 \mu H^4}\right)^{1/5} t^{4/5}, \tag{11.20}$$

where c is a numeric constant and the volume balance equation becomes

$$c^{5/3} \int_0^1 (1-\xi)^{\frac{1}{3}} f(\xi) d\xi = 1. \tag{11.21}$$

With the above definition for width (11.18) and length (11.20), the lubrication equation in (11.17) transforms into

$$\frac{1}{4}(1-\xi)^{\frac{1}{3}} f(\xi) - \xi \frac{d(1-\xi)^{\frac{1}{3}} f(\xi)}{d\xi} - \frac{3}{4}\frac{d^2 (1-\xi)^{\frac{4}{3}} (f(\xi))^4}{d\xi^2} = 0, \tag{11.22}$$

where the boundary conditions become

$$-c^{\frac{5}{3}} \frac{3}{5} \frac{d(1-\xi)^{\frac{4}{3}} f(\xi)^4}{d\xi}\bigg|_{\xi=0} = 1, \qquad f|_{\xi=1} = 1. \tag{11.23}$$

Instead of solving the differential Eq. (11.22), the solution is sought in the form $f = 1 + a(1-\xi)$, i.e. expanding it near the tip. By substituting this result into (11.22) and balancing the powers of $(1-\xi)$, assuming that the latter is a small parameter, the result is $a = -1/96$. This demonstrates that the linear correction to the tip solution (i.e. $f = 1$) is on the order of 1%, even near the source. Therefore, it seems sufficient to consider only the linear term and write the approximate solution for f as

$$f(\xi) \approx 1 - \frac{1-\xi}{96}. \tag{11.24}$$

By substituting (11.24) into (11.21), the numeric coefficient c becomes

$$c = \left(\frac{224}{167}\right)^{3/5} \approx 1.193. \tag{11.25}$$

To summarize, the approximate M vertex solution is computed using (11.18), (11.20), (11.24), and (11.25) as

$$\overline{w}_M = 1.76\left(\frac{\mu Q_0^2}{E'H}\right)^{\frac{1}{5}} t^{\frac{1}{5}} (1-\xi)^{1/3}\left(1 - \frac{1-\xi}{96}\right),$$
$$p_M = 1.12\left(\frac{\mu E'^4 Q_0^2}{H^6}\right)^{\frac{1}{5}} t^{\frac{1}{5}} (1-\xi)^{1/3}\left(1 - \frac{1-\xi}{96}\right)$$
$$L_M = 0.38\left(\frac{E' Q_0^3}{\mu H^4}\right)^{\frac{1}{5}} t^{\frac{4}{5}} \tag{11.26}$$

Here the local elasticity Eq. (11.1) is used to compute pressure. Note that this approximate solution is based primarily on the behavior near the tip and the condition at the inlet (11.23) is not satisfied. Therefore, there can be some discrepancies with the "true" solution near the inlet. These expressions match that outlined in [1] to within a few percent error. This is an acceptable difference since approximations are also used to derive the expressions presented in [1].

11.4.2 Leak-off Viscosity

For the leak-off viscosity case, both toughness and fracture volume are negligibly small. As a result, the governing Eq. (11.3), the boundary condition (11.5), and the global volume balance (11.7) reduce to

$$-\frac{E'}{2\pi^3 H\mu}\frac{\partial^2 \overline{w}^4}{\partial x^2}+\frac{C'}{\sqrt{t-t_0(x)}}=\frac{Q_0}{H}\delta(x),$$
$$2C'\int_0^l \sqrt{t-t_0}\,dx=\frac{Q_0 t}{2H},\qquad \overline{w}(l)=0. \tag{11.27}$$

By taking the time dependence of length as $l \propto t^{\alpha}$, the exposure time needed for leak-off calculations becomes $t_0(x)=t\left(\frac{x}{l}\right)^{1/\alpha}$. In this case, the global volume balance reduces to

$$2C't^{1/2}l\int_0^1 \sqrt{1-\xi^{\frac{1}{\alpha}}}\,d\xi=\frac{Q_0 t}{2H}. \tag{11.28}$$

By balancing the powers of t in the above expression, one has $\alpha = 1/2$ and the length is

$$l=\frac{Q_0 t^{\frac{1}{2}}}{\pi C' H}. \tag{11.29}$$

Knowing the expression for $t_0(x)$, the lubrication equation in (11.27) reduces to

$$-\frac{E'}{2\pi^3 H\mu}\frac{\partial^2 \overline{w}^4}{\partial x^2}+\frac{C'}{\sqrt{t}\sqrt{1-\left(\frac{x}{l}\right)^2}}=\frac{Q_0}{H}\delta(x),\qquad \overline{w}(l)=0, \tag{11.30}$$

which can be solved to obtain

$$\overline{w}=\left(\frac{2\pi\mu Q_0^2}{E'C'H}\right)^{\frac{1}{4}} t^{\frac{1}{8}} g(\xi),$$
$$g(\xi)=\left[\xi\left(\sin^{-1}(\xi)-\frac{\pi}{2}\right)+\sqrt{1-\xi^2}\right]^{1/4},\qquad \xi=\frac{x}{l(t)}. \tag{11.31}$$

Finally, the solution for $\tilde{M}$ vertex can be summarized as

$$\overline{w}_{\tilde{M}}=\left(\frac{2\pi\mu Q_0^2}{E'C'H}\right)^{\frac{1}{4}} t^{\frac{1}{8}} g(\xi),$$
$$p_{\tilde{M}}=\left(\frac{32\mu E'^3 Q_0^2}{\pi^3 C' H^5}\right)^{\frac{1}{4}} t^{\frac{1}{8}} g(\xi), \tag{11.32}$$
$$l_{\tilde{M}}=\frac{Q_0 t^{\frac{1}{2}}}{\pi C' H}.$$

where the function $g(\xi)$ is defined in (11.31). This solution coincides precisely with the one in [1]. Note that the function $g(\xi)$ defined in (11.31) has the following asymptotic behavior

$$g(\xi)|_{\xi\to 1}=\left(\frac{8}{9}\right)^{1/8}(1-\xi)^{3/8},\qquad g(0)=1. \tag{11.33}$$

Given that $\left(\frac{8}{9}\right)^{1/8}\approx 0.985$, i.e. close to 1, the function g can be approximated via its asymptotic behavior as

$$g(\xi)\approx\left(\frac{8}{9}\right)^{\frac{1}{8}}(1-\xi)^{\frac{3}{8}}\left[1-\left(1-\left(\frac{9}{8}\right)^{\frac{1}{8}}\right)(1-\xi)\right]. \tag{11.34}$$

This form of the solution for fracture width is very similar to that for the viscosity storage solution (11.26). What is also common is that the multiplier in front of the "correction term" $1-\xi$ is relatively small, on the order of 1%.

11.4.3 Storage Toughness

In the limit of no viscosity and no leak-off, all the injected volume stays inside the fracture and there is no pressure gradient along the fracture. As a consequence of constant pressure and local elasticity, the width is constant throughout the fracture in this limit. The combination of the boundary condition (11.5), and the volume balance (11.7) leads to the solution in the form

$$\overline{w}_K=\frac{\sqrt{\pi H}K_{Ic}}{E'},$$
$$p_K=\frac{2K_{Ic}}{\sqrt{\pi H}}, \tag{11.35}$$
$$l_K=\frac{E'Q_0 t}{\sqrt{4\pi}\,K_{Ic}H^{3/2}}.$$

11.4.4 Leak-off Toughness

When toughness dominates, both the width and the pressure are constant along the whole fracture. Their respective values are determined solely by the boundary condition at the tip (11.5), similarly to the previous case of toughness storage limit. The difference comes from the volume balance, in which the injected fluid is now balanced by leak-off, namely

$$2C'\int_0^l \sqrt{t-t_0}\,dx=\frac{Q_0 t}{2H}. \tag{11.36}$$

Similar to the case of leak-off viscosity, by taking the time dependence of length as $l \propto t^{\alpha}$, one has $t_0(x) = t\left(\frac{x}{l}\right)^{1/\alpha}$, in which case the above integral reduces to

$$2C't^{1/2}l\int_0^1 \sqrt{1-\xi^{\frac{1}{\alpha}}}\,d\xi = \frac{Q_0 t}{2H}. \tag{11.37}$$

By balancing the powers of t, it is now clear that $\alpha = 1/2$. As a result, the above equation allows to find l, which, combined with the expressions for width and pressure stemming from the boundary condition, gives the following result

$$\begin{aligned} \overline{w}_{\tilde{K}} &= \frac{\sqrt{\pi H}K_{Ic}}{E'},\\ p_{\tilde{K}} &= \frac{2K_{Ic}}{\sqrt{\pi H}},\\ l_{\tilde{K}} &= \frac{Q_0 t^{\frac{1}{2}}}{\pi C' H}. \end{aligned} \tag{11.38}$$

It is interesting to observe that the difference of this result with the previous toughness storage solution is just in the fracture length, while the width and the pressure are exactly the same. The length, on the other hand is the same as for the viscosity leak-off limit.

11.5 Full Solution

This section aims to construct and analyze the full solution for the problem of PKN fracture, which includes the effects of toughness, viscosity, and leak-off. Two types of solutions are constructed: semi-analytical rapid approximate solution and fully numerical solution. The former allows to compute solution quickly and to investigate the whole parametric space, while the latter allows checking accuracy of the approximation.

As can be seen from the limiting vertex solutions (11.26), (11.32), (11.35), and (11.38), spatial variation of the fracture width is closely approximated by the corresponding tip asymptotes (11.9). There is some discrepancy for the storage viscosity and leak-off viscosity solutions, but it is on the order of 1% and is therefore ignored for simplicity. Motivated by this observation, the width of the global solution is taken in the form:

$$\overline{w} = \overline{w}_{m\tilde{m}k}\left(\hat{x} = l(1-\xi),\, V = \dot{l}\right), \qquad \xi = \frac{x}{l}, \tag{11.39}$$

where $\overline{w}_{m\tilde{m}k}$ is the approximate solution for the semi-infinite fracture (11.12), which also depends on E', K_{Ic}, μ, C', and H. Note that the approach for constructing the global solution from the tip asymptote was applied in [25] for a radial hydraulic fracture and [26] for a plane strain hydraulic fracture. One notable difference for PKN fracture is the fact that the tip asymptote is discontinuous at the tip and therefore it is not very accurate to approximate the spatial behavior by $(1-\xi)^{\delta}$ (with slowly varying function δ), as was done for the prior cases. In order to construct the global solution using (11.39), it is also necessary to consider global volume balance (11.7). By taking $l \propto t^{\alpha}$, where α is a slowly varying function of time, the leak-off exposure time becomes $t_0(x) = t\left(\frac{x}{l}\right)^{1/\alpha}$ and the volume balance (11.7) reduces to

$$l\int_0^1 \overline{w}_{m\tilde{m}k}(l(1-\xi), \alpha l/t)d\xi + 2C't^{1/2}l\int_0^1\sqrt{1-\xi^{\frac{1}{\alpha}}}\,d\xi = \frac{Q_0 t}{2H}. \tag{11.40}$$

The leak-off related integral in the above equation can be computed analytically, which results in the following system of equations

$$\begin{aligned} &l\int_0^1 \overline{w}_{m\tilde{m}k}(l(1-\xi), \alpha l/t)d\xi + \sqrt{\pi}C't^{1/2}l\frac{\Gamma(\alpha+1)}{\Gamma(\alpha+3/2)} = \frac{Q_0 t}{2H},\\ &\alpha = \frac{d\log(l)}{d\log(t)}, \end{aligned} \tag{11.41}$$

where $\Gamma(\cdot)$ denotes Gamma function. The integral of the fracture width in the above expression is computed numerically. Since α varies slowly, the first equation can be solved iteratively, e.g. using Newton's method for l. Once the length l computed for two-time instants, the value of α is updated. Such iterations are continued until convergence is reached. Once these quantities are computed, then the width solution is evaluated using (11.39), while pressure can always be computed with the help of (11.1).

To compute numerical solution for the problem under consideration, similar to the previous case, the moving coordinate $\xi = \frac{x}{l(t)}$ is introduced. To further simplify the problem, the following normalization is employed

$$\Omega = \frac{\overline{w}}{w_*}, \qquad \lambda = \frac{l}{l_*}, \qquad \tau = \frac{t}{t_*}, \tag{11.42}$$

where the width, length, and time scales are given by

$$w_* = \frac{(\pi H)^{\frac{1}{2}}K_{Ic}}{E'}, \qquad l_* = \frac{H^2K_{Ic}^4}{2\pi E'^3\mu Q_0}, \qquad t_* = \frac{H^{\frac{7}{2}}K_{Ic}^5}{2\pi^{1/2}E'^4\mu Q_0^2}. \tag{11.43}$$

With such a normalization, the governing Eqs. (11.3), (11.5), and (11.6) reduce to

$$\frac{\partial\Omega}{\partial\tau}-\xi\frac{\dot{\lambda}}{\lambda}\frac{\partial\Omega}{\partial\xi}-\frac{1}{\lambda^2}\frac{\partial^2\Omega^4}{\partial\xi^2}+\frac{\phi}{\sqrt{\tau-\tau_0(\xi)}}=\delta(\xi),$$
$$\Omega(1)=1,\qquad \frac{d\lambda}{dt}=-\frac{4}{3\lambda}\frac{\partial\Omega^3}{\partial\xi}\bigg|_{\xi=1} \tag{11.44}$$

Here the dimensionless parameter that describes leak-off is computed as

$$\phi=\left(\frac{\mathrm{H}^5\mathrm{K}_{Ic}^6C'^4}{4\pi^3\mathrm{E}'^4\mu^2\mathrm{Q}_0^4}\right)^{1/4}. \tag{11.45}$$

Solution for the above dimensionless problem (11.44) comprises of finding $\Omega(\tau,\phi)$ and $\lambda(\tau)$. Once the dimensionless problem is solved, the physical quantities can be restored using (11.42) and (11.43). This demonstrates that the parametric space for the problem is two-dimensional and consists of the dimensionless time τ, and dimensionless leak-off ϕ, which is somewhat similar to the case of radial hydraulic fracture [25]. Numerical solution for the problem (11.44) is computed using central difference to discretize spatial derivatives and backward difference for approximating the time derivative to ensure stability of the numerical scheme.

Figure 11.4 plots the parametric space for the problem that is computed using the fast approximate solution (11.41). In particular, the solution is plotted in the (τ,ϕ) space in terms of the ratio Ω/λ (see (11.42), (11.43), and (11.45)). This is done to ensure that the plotted quantity is different for all the limiting cases. Zones of applicability of the limiting vertex solutions are marked by K (storage toughness), M (storage viscosity), $\tilde{K}$ (leak-off toughness), and $\tilde{M}$ (leak-off viscosity) symbols. The latter are defined

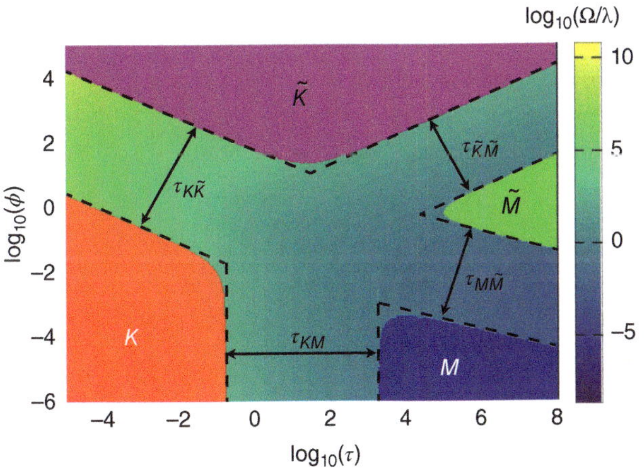

Figure 11.4 Parametric space for PKN hydraulic fracture. Zones of applicability of the limiting vertex solutions are highlighted by K (storage toughness), M (storage viscosity), $\tilde{K}$ (leak-off toughness), and $\tilde{M}$ (leak-off viscosity) symbols.

as the regions, in which the difference (of Ω/λ) with the corresponding vertex solution is below 1%. The K region corresponds to the storage toughness solution (11.35), the M region represents the storage viscosity limit (11.26), the $\tilde{K}$ region defines the leak-off toughness limit (11.38), while the $\tilde{M}$ region corresponds to the leak-off viscosity solution (11.32). The dashed lines in Figure 11.4 outline zones of applicability of the vertex solutions. To obtain the relevant parameter that determines the transition, one needs to equate either lengths or widths (whichever are different) between two regimes. For instance, the transition between K and M is determined by the parameter calculated from $l_M \sim l_K$ or $\overline{w}_M \sim \overline{w}_K$. To quantify the parameter for $K\tilde{K}$ transition, one should use $l_K \sim l_{\tilde{K}}$. The widths cannot be used since they are identical. Similarly, for the $\tilde{K}\tilde{M}$ transition one has to use widths $\overline{w}_{\tilde{K}} \sim \overline{w}_{\tilde{M}}$ since the lengths are identical. Finally, for the $M\tilde{M}$ either the widths or the lengths can be used, i.e. $l_M \sim l_{\tilde{M}}$ or $\overline{w}_M \sim \overline{w}_{\tilde{M}}$. The values of the transition parameters are determined numerically. Summary of the results is shown below:

$$\begin{aligned}
&\tau_{MK}=\tau, &&\tau_{MK,1}=0.11, &&\tau_{MK,2}=2.3\times10^3,\\
&\tau_{K\tilde{K}}=\tau\phi^2, &&\tau_{K\tilde{K},1}=5.7\times10^{-5}, &&\tau_{K\tilde{K},2}=3.1\times10^3,\\
&\tau_{\tilde{K}\tilde{M}}=\tau\phi^{-2}, &&\tau_{\tilde{K}\tilde{M},1}=0.18, &&\tau_{\tilde{K}\tilde{M},2}=6.5\times10^4,\\
&\tau_{M\tilde{M}}=\tau\phi^{10/3}, &&\tau_{M\tilde{M},1}=2.0\times10^{-7}, &&\tau_{M\tilde{M},2}=2.9\times10^3.
\end{aligned} \tag{11.46}$$

Note that the definitions of the applicability zones of the limiting solutions are somewhat arbitrary and the result can change if a different metric or different threshold is used. In addition, there can be some minor discrepancy since the approximate solution is used to construct the parametric map. At the same time, since the parametric space is plotted on a logarithmic scale, minor variations in the boundaries defined by (11.46) are not going to notably alter the result.

What is interesting about the obtained result is that the fracture evolves in time from the K vertex and ultimately reaches the $\tilde{M}$ vertex. It can pass through $\tilde{K}$, M, or none of those during the intermediate times. This is somewhat opposite to the radial fracture geometry, for which the solution always originates from the M vertex and then reaches the $\tilde{K}$ solution in the long term. It is also different from the plane strain fracture, for which the solution originates at the MK edge and eventually reaches $\tilde{M}\tilde{K}$ edge.

It is also instructive to verify that the solution constructed using the tip asymptote and volume balance accurately approximates the numerical solution. With the reference to Figure 11.4, three values of leak-off parameter are selected $\phi=\{10^{-4}, 1, 10^2\}$. These three values of ϕ allow to investigate transitions from the K solution to other three

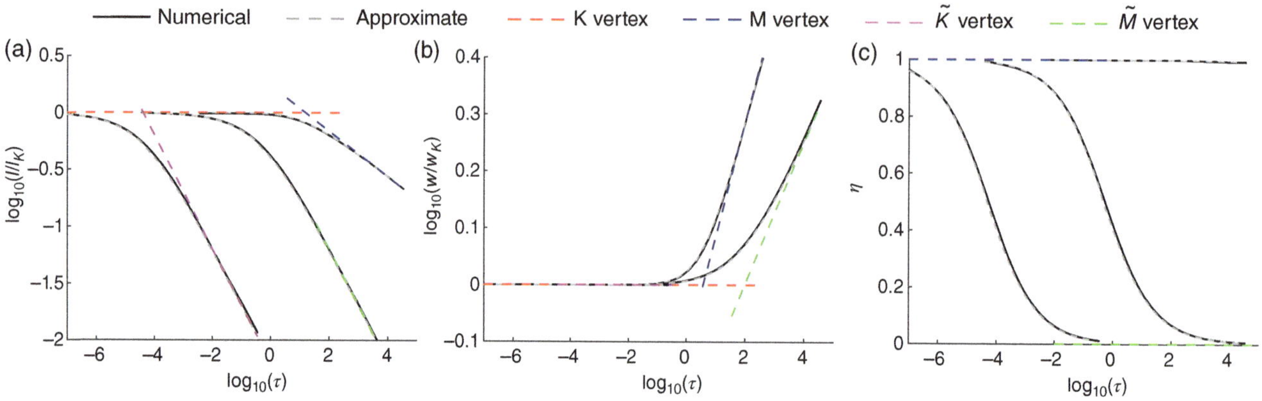

Figure 11.5 Comparison between the numerical, approximate, and vertex solutions for three values of the dimensionless leak-off $\phi = \{10^{-4}, 1, 10^2\}$. (a) shows the normalized fracture length versus time. (b) shows the normalized width at the wellbore versus time. (c) shows efficiency versus time.

vertices. Figure 11.5 shows the comparison between the numerical solution (solid lines) and the approximate solution (dashed lines) in terms of variation of the normalized fracture length l/l_K, width at the wellbore $\overline{w}/\overline{w}_K$, and efficiency η, where the latter is defined as the ratio between the fracture volume and the injected volume. The vertex solutions are indicated by the dashed straight lines. One can observe that the solution indeed transitions from the K limit to either M, $\tilde{M}$, or $\tilde{K}$ limit. Also, the approximate solution is visually nearly indistinguishable from the one computed numerically. To further compare the two solutions, Figure 11.6 plots the normalized width variation along the fracture length for $\phi = 10^{-4}$ and $\tau = \sqrt{4\pi}\{10^{-2}, 10, 10^3\}$, and $\phi = 1$ and $\tau = \sqrt{4\pi}\{0.5, 50, 5 \times 10^3\}$. Note that the case corresponding to $\phi = 10^2$ is trivial since the width is constant for this transition. Results again demonstrate that the difference between the numerical and approximate solutions is small. Thus, confirming that the parametric space, computed using the fast approximate solution, is constructed with a sufficient degree of accuracy.

11.6 Application Examples

To illustrate the developments, this section presents two examples of how the parametric map can be used in applications.

First, let us consider a field case with the parameters $H = 20$ m, $K_{Ic} = 1$ MPa$\cdot m^{\frac{1}{2}}$, $C' = 10^{-6}$ m/s$^{1/2}$, $E' = 30$ GPa, $\mu = 0.01$ Pa $\cdot$ s, $Q_0 = 0.01$ m^3/s, and $t = 1000$ s. With the reference to Eqs. (11.42), (11.43), and (11.45), the dimensionless parameters for this case are $\tau \approx 8 \cdot 10^4$ and $\phi \approx 4 \cdot 10^{-4}$. By looking at the parametric map in Figure 11.4, it can be

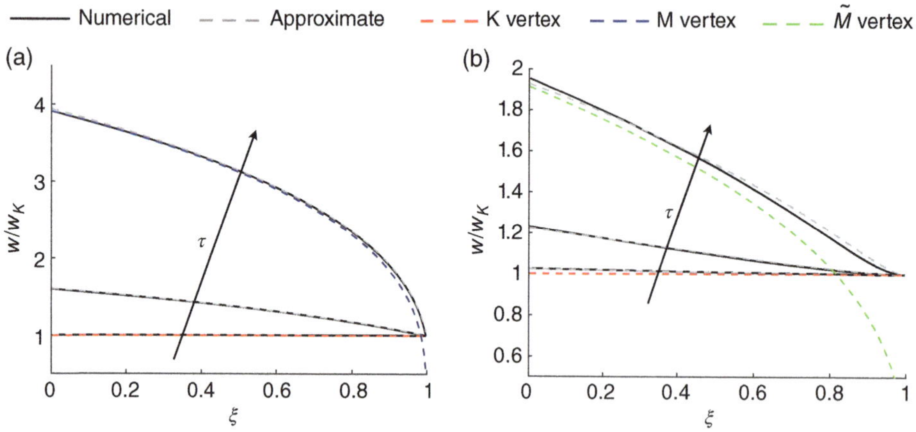

Figure 11.6 Comparison between the numerical, approximate, and vertex solutions in terms of the variation of width versus normalized spatial coordinate ξ. (a) corresponds to $\phi = 10^{-4}$ and $\tau = \sqrt{4\pi}\{10^{-2}, 10, 10^3\}$. (b) shows the results for $\phi = 1$ and $\tau = \sqrt{4\pi}\{0.5, 50, 5 \times 10^3\}$.

concluded that the solution follows the storage viscosity dominated case. Consequently, fracture length can be computed using Eq. (11.26) as $l \approx 170$ m.

Second, let us consider parameters used in [40] to evaluate performance of various numerical algorithms in the toughness regime. The relevant parameters are motivated by laboratory experiments and are given by $H = 0.05$ m, $K_{Ic} = 1.57$ MPa·$m^{\frac{1}{2}}$, $C' = 0$ m/s$^{1/2}$, $E' = 3.9$ GPa, $\mu = 30.2$ Pa · s, $Q_0 = 1.7$ mm^3/s, and $t = 604$ s. According to Eqs. (11.42), (11.43), and (11.45), the dimensionless parameters for this case are $\tau \approx 0.2$ and $\phi = 0$. These parameters fall into storage toughness regime in the parametric map shown in Figure 11.4. Thus, this confirms that in [40] different algorithms are indeed compared in the toughness regime. Finally, Eq. (11.35) can be used to calculate fracture length as 6.5 cm, which is also in agreement with the numerical results presented in [40].

11.7 Summary

The problem of a constant height or PKN hydraulic fracture is investigated. The combined effects of fracture toughness, fluid viscosity, and leak-off are considered. Also, to reduce complexity of the analysis, local elasticity is used in lieu of the more accurate nonlocal elasticity relation.

The analysis for the problem of a tip region is first outlined. It is shown that, similar to the classical semi-infinite plane strain hydraulic fracture, there are three limiting or vertex solutions related to domination of either toughness, viscosity, or leak-off. Parametric map of the solutions is obtained, in which zones of applicability of the limiting solutions are outlined. In addition, two approximate solutions for the problem are constructed and their accuracy is evaluated in the whole parametric space.

With respect to the solution for a finite PKN fracture, explicit expressions for the four limiting solutions are first obtained. The latter limits have similar meaning to those for other fracture geometries, such as plane strain and radial, and include storage viscosity, storage toughness, leak-off viscosity, and leak-off toughness limits. The full solution for a finite PKN fracture is computed using two approaches: numerical and fast approximate. The latter utilizes the developed tip asymptotic solution as an approximation for the spatial variation of width for the whole fracture. This fact, combined with the global volume balance, allowed constructing rapid solution for the problem. Accuracy of this approximation is evaluated by comparing its predictions to that from the numerical solution. Equipped with this quick solution, the whole parametric space for the problem is investigated and zones of applicability of the limiting solutions are outlined. One interesting observation is that the global solution evolves from early time storage toughness limit to the late time leak-off viscosity limit. This is different from the behavior for previously analyzed radial and plane strain geometries. The obtained results allow to readily estimate location inside the parametric space for any problem parameters for the constant height fracture, which can be used, for instance, to scale down field data to a laboratory experiment, or to use the closest limiting solution to estimate the fracture geometry.

References

1 Economides, M.J. and Nolte, K.G. (ed.) (2000). *Reservoir Stimulation*, 3rde. Chichester: Wiley.

2 Detournay, E. and Garagash, D. (2003). The tip region of a fluid-driven fracture in a permeable elastic solid. *Journal of Fluid Mechanics* 494: 1–32.

3 Garagash, D.I., Detournay, E., and Adachi, J.I. (2011). Multiscale tip asymptotics in hydraulic fracture with leak-off. *Journal of Fluid Mechanics* 669: 260–297.

4 Peirce, A. and Detournay, E. (2008). An implicit level set method for modeling hydraulically driven fractures. *Computer Methods in Applied Mechanics and Engineering* 197: 2858–2885.

5 Desroches, J., Detournay, E., Lenoach, B. et al. (1994). The crack tip region in hydraulic fracturing. *Proceedings of the Royal Society of London A* 447: 39–48.

6 Lenoach, B. (1995). The crack tip solution for hydraulic fracturing in a permeable solid. *Journal of the Mechanics and Physics of Solids* 43: 1025–1043.

7 Rice, J.R. (1968). Mathematical analysis in the mechanics of fracture. In: *Fracture: An Advanced Treatise*, vol. II, Chapter 3 (ed. H. Liebowitz), 191–311. New York, NY: Academic Press.

8 Dontsov, E. and Peirce, A. (2015). A non-singular integral equation formulation to analyze multiscale behaviour in semi-infinite hydraulic fractures. *Journal of Fluid Mechanics* 781: R1.

9 Dontsov, E.V. and Peirce, A.P. (2017). A multiscale implicit level set algorithm (ILSA) to model hydraulic fracture propagation incorporating combined viscous, toughness, and leak-off asymptotics. *Computer Methods in Applied Mechanics and Engineering* 313: 53–84.

10 Dontsov, E.V. and Kresse, O. (2018). A semi-infinite hydraulic fracture with leak-off driven by a power-law fluid. *Journal of Fluid Mechanics* 837: 210–229.
11 Bessmertnykh, A.O., Dontsov, E.V., and Ballarini, R. (2019). A semi-infinite hydraulic fracture driven by a Herschel-Bulkley fluid. *Journal of Applied Mechanics* 86: 121008.
12 Moukhtari, F.-E. and Lecampion, B. (2018). A semi-infinite hydraulic fracture driven by a shear-thinning fluid. *Journal of Fluid Mechanics* 838: 573–605.
13 Garagash, D. and Detournay, E. (2000). The tip region of a fluid-driven fracture in an elastic medium. *Journal of Applied Mechanics* 67: 183–192.
14 Dontsov, E.V. (2016). Tip region of a hydraulic fracture driven by a laminar-to-turbulent fluid flow. *Journal of Fluid Mechanics* 797: R2.
15 Garagash, D.I. (2019). Cohesive-zone effects in hydraulic fracture propagation. *Journal of the Mechanics and Physics of Solids* 133: 103727.
16 Bessmertnykh, A., Dontsov, E., and Ballarini, R. (2020). The effects of proppant on the near-front behavior of a hydraulic fracture. *Engineering Fracture Mechanics* 235: 107110.
17 Kanin, E.A., Garagash, D.I., and Osiptsov, A.A. (2020). The near-tip region of a hydraulic fracture with pressure-dependent leak-off and leak-in. *Journal of Fluid Mechanics* 892: A31.
18 Detournay, E. (2016). Mechanics of hydraulic fractures. *Annual Review of Fluid Mechanics* 48: 31139.
19 Savitski, A. and Detournay, E. (2001). Similarity solution of a penny-shaped fluid-driven fracture in a zero- toughness linear elastic solid. *Comptes Rendus- Academie Des Sciences Paris Serie 2B Mecanique* 329: 255–262.
20 Adachi, J.I. and Detournay, E. (2002). Self-similar solution of a plane-strain fracture driven by a power-law fluid. *International Journal for Numerical and Analytical Methods in Geomechanics* 26: 579–604.
21 Bunger, A., Detournay, E., and Garagash, D. (2005). Toughness-dominated hydraulic fracture with leak-off. *International Journal of Fracture* 134: 175–190.
22 Adachi, J.I. and Detournay, E. (2008). Plane-strain propagation of a hydraulic fracture in a permeable rock. *Engineering Fracture Mechanics* 75: 4666–4694.
23 Hu, J. and Garagash, D.I. (2010). Plane-strain propagation of a fluid-driven crack in a permeable rock with fracture toughness. *Journal of Engineering Mechanics* 136: 1152–1166.
24 Madyarova, M.V. (2003). Fluid-driven penny-shaped fracture in elastic medium. Master's thesis. University of Minnesota.
25 Dontsov, E.V. (2016). An approximate solution for a penny-shaped hydraulic fracture that accounts for fracture toughness, fluid viscosity, and leak-off. *Royal Society Open Science* 3: 160737.
26 Dontsov, E.V. (2017). An approximate solution for a plane strain hydraulic fracture that accounts for fracture toughness, fluid viscosity, and leak-off. *International Journal of Fracture* 205: 221–237.
27 Dontsov, E.V. (2019). Scaling laws for hydraulic fractures driven by a power-law fluid in homogeneous anisotropic rocks. *International Journal for Numerical and Analytical Methods in Geomechanics* 43: 519–529.
28 Dontsov, E., Bunger, A., Abell, B., and Suarez-Rivera, R. (2019). Ultrafast hydraulic fracturing model for optimizing cube development. *Proceedings of the Unconventional Resources Technology Conference,* https://doi.org/10.15530/urtec-2019-884.
29 Perkins, T.K. and Kern, L.R. (1961). Widths of hydraulic fractures. *Journal of Petroleum Technology* 13: 937–949.
30 Nordgren, R.P. (1972). Propagation of vertical hydraulic fractures. *Society of Petroleum Engineers. Journal* 12: 306–314.
31 Settari, A. and Cleary, M.P. (1986). Development and testing of a pseudo-three-dimensional model of hydraulic fracture geometry. *SPE Production Engineering*, 1, 449–466. SPE 10505-PA.
32 Adachi, J.I., Detournay, E., and Peirce, A.P. (2010). An analysis of classical pseudo-3D model for hydraulic fracture with equilibrium height growth across stress barriers. *International Journal of Rock Mechanics and Mining Sciences* 47: 625–639.
33 Dontsov, E. and Peirce, A. (2015). An enhanced pseudo-3D model for hydraulic fracturing accounting for viscous height growth, non-local elasticity, and lateral toughness. *Engineering Fracture Mechanics* 142: 116–139.
34 Kovalyshen, Y. and Detournay, E. (2010). A reexamination of the classical PKN model of hydraulic fracture. *Transport in Porous Media* 81: 317–339.
35 Xing, P., Yoshioka, K., Adachi, J. et al. (2017). Laboratory measurement of tip and global behavior for zero-toughness hydraulic fractures with circular and blade-shaped (PKN) geometry. *Journal of the Mechanics and Physics of Solids* 104: 172–186.
36 Zolfaghari, N., Meyer, C.R., and Bunger, A.P. (2017). Blade-shaped hydraulic fracture driven by a turbulent fluid in an impermeable rock. *Journal of Engineering Mechanics* 143 (11): 04017130.
37 Zia, H. and Lecampion, B. (2017). Propagation of a height contained hydraulic fracture in turbulent flow regimes.

International Journal of Solids and Structures 110–111: 265–278.

38 Nolte, K.G. (1991). Fracturing-pressure analysis for nonideal behavior. *Journal of Petroleum Technology* 43: 210–218.

39 Sarvaramini, E. and Garagash, D. (2015). Breakdown of a pressurized fingerlike crack in a permeable solid. *Journal of Applied Mechanics* 82: 061006.

40 Dontsov, E.V. and Peirce, A.P. (2016). Comparison of toughness propagation criteria for blade-like and pseudo-3d hydraulic fractures. *Engineering Fracture Mechanics* 160: 238–247.

41 Carter, E.D. (1957). Optimum fluid characteristics for fracture extension. In: *Drilling and Production Practices* (ed. G.C. Howard and C.R. Fast), 261–270.

12

Discrete Element Modeling of Hydraulic Fracturing

Mengli Li[1,2] *and Fengshou Zhang*[1,2]

[1] *Key Laboratory of Geotechnical and Underground Engineering of Ministry of Education, Tongji University, Shanghai, China*
[2] *Department of Geotechnical Engineering, College of Civil Engineering, Tongji University, Shanghai, China*

12.1 Introduction

Hydraulic fracturing refers to the process of fluid-driven fracture propagation in solids. The most common engineering application of hydraulic fracturing is the stimulation of oil and gas reservoirs by injecting high pressure fluid containing proppant particles into the rock formations in order to form high permeability paths for hydrocarbon [1]. The basic mechanics of hydraulic fracturing has been extensively studied in the past few decades [2–8]. The problem of hydraulic fracture propagation involves processes taking place on different time and length scales, created by the interplay of a few competing physical processes, including viscous fluid flow in the fracture, fracture propagation at the tip, fluid front lag behind the fracture tip, fluid leak-off into the reservoir, and elastic deformation of the rock matrix [9]. The interaction of these individual processes results in complexities in fracture growth, especially in heterogeneous lithology.

Both theoretical and experimental efforts have also been made previously to investigate the injection process in an unconsolidated medium. The unconsolidated nature of the material means that if an opening is created, the tip of the opening cannot sustain large stress contrast, and plastic yielding is expected. Influence of plasticity in hydraulic fracturing has been examined for both cohesive and cohesionless materials [10–14]. In general, one-dimensional leak-off model [15] is assumed for fluid flow from the fracture faces to the matrix in conventional LEFM-based models. To account for the large permeability typical of the unconsolidated formations, multi-dimensional fluid leak-off models have also been introduced [16–18]. Nevertheless, due to the highly nonlinear constitutive behaviors and the strong coupling between mechanical deformation and fluid flow, theoretical analysis of the injection process in the unconsolidated formations remains a great challenge.

The particle-based discrete element method (DEM) specializes in the micromechanical modeling of rock and soil behavior [19, 20]. The DEM code PFC2D is employed in this analysis. PFC2D solves Newton's second law of motion and force–displacement laws at the contact using an explicit finite difference method [21]. DEM coupled with the pore network model (PNM) has been applied to investigate hydromechanically coupled problems, e.g. two hydraulic fractures interaction in a cohesionless medium [22], stress-dependent permeability of sandstones [23, 24], hydraulic fracturing-induced microseismicity [25–27] and the transition from capillary fingering to capillary fracturing in gas invasion [28]. Compared with other DEM-based hydromechanical coupling schemes, e.g. the DEM-lattice Boltzmann method [29–31] and the DEM coupled with a fixed coarse grid fluid flow scheme ([32, 33]), the DEM-pore network coupling scheme is capable of resolving fluid flow at the particle scale and has the advantage of simplicity in the formulation.

In this work, the results of numerical analyses of a series of hydraulic fracturing problems in naturally fractured rock mass and dense granular media are presented. The goal is to demonstrate some attempts to address a few key coupled processes during hydraulic fracturing. The organization of the paper structure is as follows. First, we introduce the DEM-pore network coupling scheme of hydraulic fracturing. Then, a total of three case studies focusing on different aspects of hydraulic fracturing are presented. The first case study discusses the mechanism of hydraulic fracture interaction with natural fractures. The second case study is about the modeling of supercritical carbon dioxide (SC-CO_2) fracturing. In third case study, we address the

Mechanics of Hydraulic Fracturing: Experiment, Model, and Monitoring, First Edition.
Edited by Xi Zhang, Bisheng Wu, Diansen Yang, and Andrew Bunger.

challenges in the coupled displacement process of fluid injection into a dense granular medium.

12.2 Discrete Element Modeling of Hydraulic Fracturing

An additional pore network model based on the particle skeleton is constructed for the fluid flow coupling and hydraulic fracturing modeling [34–36]. Figure 12.1 shows the schematics of the pore network model. The pore domains are formed from close chains of bonded particles while the drag force F_{fluid} is calculated as a resultant from the pore pressure forces of surrounding pore domains.

Since the two dimensional nature of the model does not allow a physically open pore throat, a virtual flow pipe connecting two adjacent pore domains is assumed at each contact bond. The pipe flow is activated only if the bond fails, thus fluid leak-off into the rock matrix is not modeled although it can be activated if needed. Fluid flow inside each active pipe is driven by the pressure drop between the two connected pore domains, and the flow rate q is given by the Hagen–Poiseuille equation,

$$q = \frac{a^3}{12\mu}\frac{p_2 - p_1}{L_P} \tag{12.1}$$

where a is the aperture of the flow pipe, p_1 and p_2 are the fluid pressures in the two neighboring pores, and L_P is the length of the flow pipe. The flow pipe aperture is a function of model geometry and deformation, and is calculated differently depending on whether the contact is open (zero normal force) or closed. If the contact is open and has zero normal force, the aperture is calculated as,

$$a = \min\left(a_0 + a_{mech}, a_{\max}\right) \tag{12.2}$$

where a_0 is the initial hydraulic aperture and a_{mech} is the gap between the two particles, $a_{\max}$ is the maximum allowed pipe aperture that is introduced for the purpose of numerical stability.

If the contact is closed, the aperture depends on the magnitude of the normal force between the particles as,

$$a = \frac{a_0 F_0}{F + F_0} \tag{12.3}$$

where F is the current contact normal force, and F_0 is the normal force at which the hydraulic aperture reduces to 50% of the initial aperture a_0.

The increment of pore pressure due to flow and volumetric strain is:

$$\Delta p = \frac{K_f}{V}\left(\sum_{i=1}^{N} q\Delta t - \Delta V\right) \tag{12.4}$$

where K_f is the fluid bulk modulus; V is the current pore volume, N is the number of flow pipes around the pore, and Δt is fluid time step.

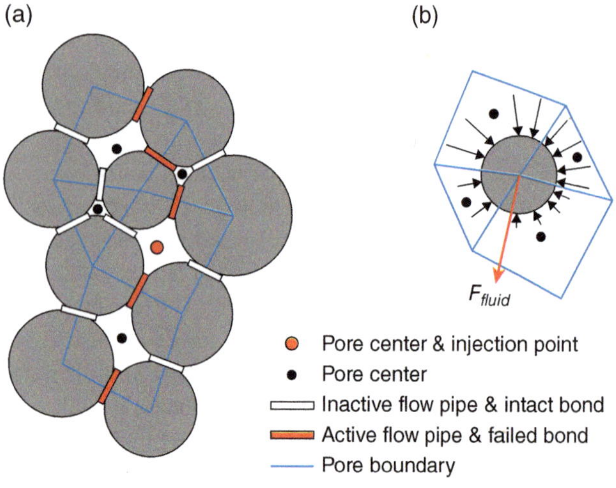

Figure 12.1 Schematics of the pore network model, (a) pore domains formed from close chains of bonded particles, and (b) the drag force on a particle F_{fluid} as a resultant from the pore pressure of surrounding pore domains (*Source:* Huang et al. [37]/ With permission of Elsevier).

12.3 Hydraulic Fracture Interacting with Natural Fractures

The behavior of a hydraulic fracture crossing a natural fracture is of fundamental importance for understanding the complexity of hydraulic fracturing. In this work, the propagation mechanism for a hydraulic fracture interacting with natural fractures is studied using a two dimensional fully coupled hybrid discrete-continuum method. The fully coupled hybrid scheme utilizes a discrete element model for an inner domain, within which the hydraulic fracture propagates and interacts with natural fractures. The inner domain is embedded in an outer continuum domain which is implemented to extend the length of the hydraulic fracture and to better approximate the boundary effects. The fracture is identified to propagate initially in the viscosity dominated regime and the numerical scheme is calibrated by using the theoretical plane strain hydraulic fracture solution. The simulation results for orthogonal crossing indicate three fundamental crossing scenarios, which occur for various stress ratios and friction coefficients of the natural fracture: (i) no crossing, i.e. the hydraulic fracture is arrested by the natural fracture and makes a T-shape intersection, (ii) offset crossing, i.e. the hydraulic fracture crosses the natural fracture with an offset, and (iii) direct crossing, i.e. the hydraulic fracture directly crosses the natural fracture without diversion. Each crossing scenario is associated with a distinct net pressure history. Additionally, the effects of strength contrast and stiffness contrast of rock materials, and intersection angle between the hydraulic fracture and the natural fracture are also investigated. The simulations also illustrate that the level of fracturing complexity increases as the number and extent of the natural fractures increase. As a result, we can conclude that complex hydraulic fracture propagation patterns occur due to complicated crossing behavior during the stimulation of naturally fractured reservoirs.

12.3.1 Hybrid Discrete-Continuum Method

A two-dimensional hybrid discrete/continuum modeling scheme is used in this study for the fully coupled hydro-mechanical simulations [38]. Two Itasca software codes, particle flow code (PFC) [21] and Fast Lagrangian Analysis of Continua (FLAC) [39], are employed for the discrete element modeling and continuum modeling, respectively. The DEM has a unique capability for modeling the particle-scale behavior of rock, but is computationally less efficient than continuum codes (e.g. finite difference or finite element). FLAC is a continuum modeling code based on an explicit finite difference scheme. To achieve an overall better computational efficiency, the hybrid discrete/continuum approach that is adopted for this study uses PFC in the relatively small core area, while FLAC is used to represent the rest of the domain.

The model setup is shown in Figure 12.2. The entire model domain in Figure 12.2a consists of an 8 m × 10 m rectangular FLAC domain and an embedded 1 m× 0.5 m PFC domain. The black solid lines represent the FLAC mesh while the yellow rectangle area represents the PFC domain. The hydraulic fracture initiates from the middle of the left boundary (the dot in Figure 12.2a shows the injection point), propagates horizontally along the predefined fracture path in the FLAC domain, and then enters the PFC domain. A natural fracture (or multiple fractures) is predefined in the PFC domain, as shown in Figure 12.2b. A roller boundary condition is set up on the left boundary of the outer FLAC domain and a zero-velocity boundary condition is set up on the other three boundaries. Due to symmetry, the model represents propagation of only the right wing of the hydraulic fracture. A two-dimensional Cartesian coordinate system is employed in the model, for which the origin is set at the center of the PFC domain, i.e. 3 m from the injection point on the left boundary and 5 m from the right boundary. The maximum horizontal stress is set along the *X* direction. The hydraulic fracture propagates in the PFC domain by breaking the particle bonds and does not have a predefined direction. Consequently, the model selects the fracture trajectory depending on stresses and interaction with the natural fractures within the PFC domain.

The PFC assembly is composed of 5173 disk-shape particles (two dimensional) with a radius ranging from 0.4 to 0.64 mm. The parallel-bond model is used to model the rock behavior at the particle scale [20]. The mechanical properties of the PFC assembly are summarized in Table 12.1 and the same mechanical properties are assigned to the FLAC domain.

The propagation of a hydraulic fracture in the PFC domain is achieved by progressively breaking the bonds

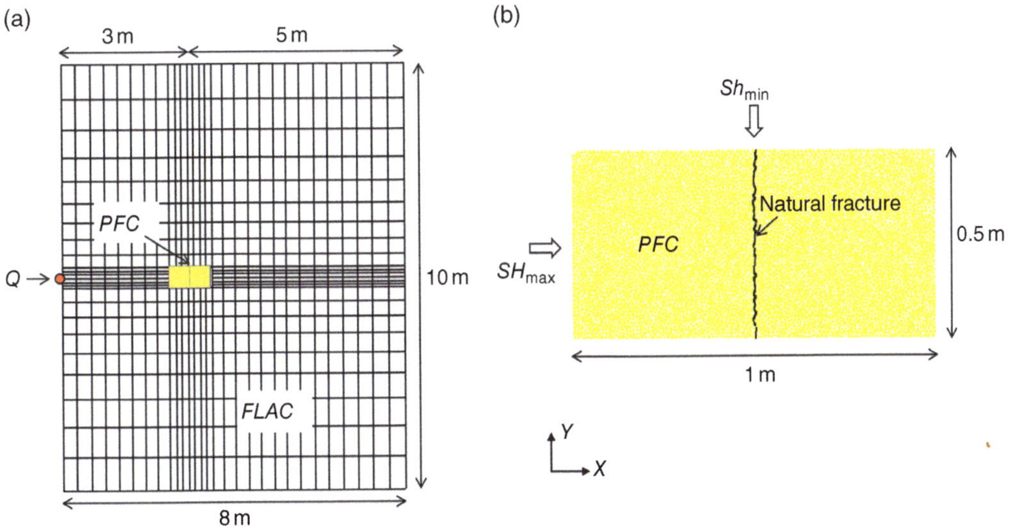

Figure 12.2 (a) Simulation domain for the discrete-continuum hybrid model. The yellow rectangle represents the PFC assembly while the black solid lines represent the FLAC mesh; (b) PFC assembly with a natural fracture vertically crossing the domain.

Table 12.1 Mechanical properties of the PFC assembly of rock material.

Mechanical property	Value
E (GPa)	19.4
ν	0.24
σ_t (MPa)	5.3
σ_c (MPa)	19.0
K_{IC} (MPa·m$^{0.5}$)	1.20

between the particles in either tensile or shear failure modes. For a bonded particle assembly, the mode-I fracture toughness K_{IC} is proportional to the tensile strength and particle size [20],

$$K_{IC} \sim \sigma_t \sqrt{\pi R} \tag{12.5}$$

where σ_t is the tensile strength of the rock and R is the particle radius. Thus, the toughness depends on the particle size. For cases with large particles, the toughness can be overestimated using the rock tensile strength. If that is the case, the local bond strength at the tip can be degraded in order to better match the fracture toughness.

12.3.2 Model Calibration for a Hydraulic Fracture in Intact Rock

In order to examine the accuracy of the hybrid model, simulation of hydraulic fracturing in intact rock (i.e. without any natural fractures in the PFC domain) is first performed. Mechanical properties of the PFC assembly are given in Table 12.1. The maximum and minimum horizontal stresses (SHmax and Shmin) are selected to be 22 MPa and 20 MPa, respectively. The initial pore pressure is set to 10 MPa, the injection rate is $Q = 0.0015\ \text{m}^2/\text{s}$ (2D), and the fluid viscosity is η=10 cp. The same values of the injection rate and fluid viscosity are used in all other cases in this paper. Note that the selected injection rate is specified for one fracture wing.

Figure 12.3 shows the resulting hydraulic fracture geometry in the PFC domain once the fracture has crossed the whole domain, which occurred at time t_0 = 1.7566 s. The red dots represent the fluid pressure magnitude, so that the trace of these dots corresponds to the hydraulic fracture. Although there is some local deviation of the fracture path, the primary direction of propagation lies along the SHmax direction, as expected.

The aperture along the hydraulic fracture for the test case without a natural fracture is plotted in Figure 12.4. The portion of the fracture in the FLAC domain is indicated by circular markers, while the portion in the PFC domain is shown by the square markers.

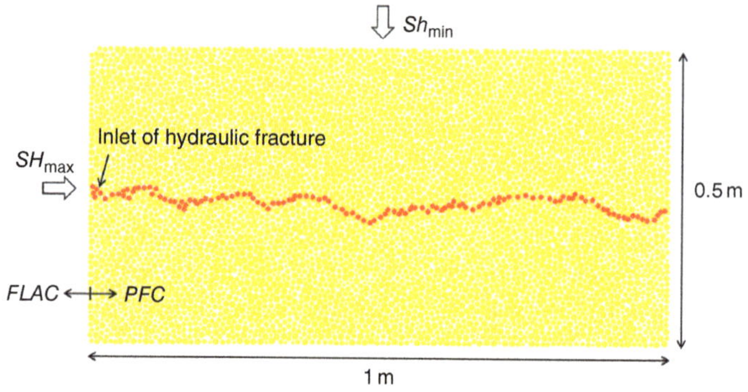

Figure 12.3 Hydraulic fracture geometry in the PFC domain for the case of no natural fractures.

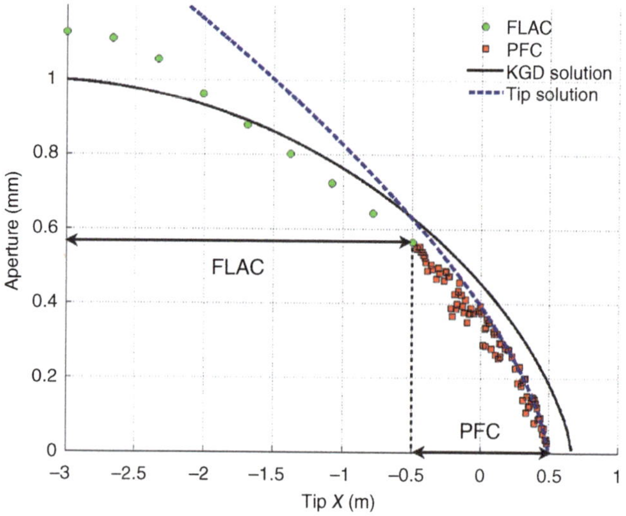

Figure 12.4 Aperture along the hydraulic fracture plane after crossing the PFC domain for the calibration case with no natural fractures.

To test the accuracy of the hybrid model, the results are compared to the reference solution for a plane strain hydraulic fracture, which is typically referred to as the Khristianovich–Zheltov–Geertsma–De Klerk (KGD) model. It is first necessary to determine the regime of propagation. As follows from [40, 41], the parameter that determines the regime of propagation for a plane strain crack in the absence of leak-off is,

$$K_m = \sqrt{\frac{32}{\pi}} \left(\frac{K_{IC}^4 (1-\nu^2)^3}{24\,\eta\, E^3 Q} \right)^{1/4} = 0.51 \tag{12.6}$$

As also shown in [40], if $K_m < 0.7$, then the fracture propagates in the viscosity-dominated regime (to within 1% error). For completeness, it is noted that this boundary is estimated as $K_m < 1.2$ in [41], but the tolerance is set to 5% and a different error measure was used.

The approximate solution for the viscosity-dominated regime is given in [40] as,

$$w(\xi, t) = 2.41 \left(\frac{\eta\, Q^3\,(1-\nu^2)}{E}\right)^{\frac{1}{6}} t^{\frac{1}{3}}\,(1-\xi)^{0.588}(1+\xi)^{\frac{2}{3}}, \quad \xi = \frac{x}{L} \tag{12.7}$$

$$L(t) = 0.576\left(\frac{Q^3 E}{\eta\,(1-\nu^2)}\right)^{\frac{1}{6}} t^{2/3} \tag{12.8}$$

where $w(\xi, t)$ is the fracture aperture, ξ is the normalized coordinate along the fracture, x is the coordinate that is measured from the injection point, and $L(t)$ is the fracture length. The KGD solution, calculated using the above expressions for $t_0 = 1.7566$ s, is shown in Figure 12.4 by a solid black line. Note that the coordinate system in Figure 12.4 is defined such that the origin coincides with the center of the PFC domain. The corresponding aperture at the injection point and the fracture length are,

$$w(0, t_0) = 1.00\,\text{mm}, \qquad L(t_0) = 3.66\,\text{m} \tag{12.9}$$

The aperture at the injection point and length that are calculated using the hybrid FLAC-PFC numerical model are 1.13 mm and the 3.5 m, respectively. In this case, the fracture width is overestimated by 13% and the fracture length is underestimated by 4.6%.

Even though the discrepancy is relatively small, it is clear that the fracture shape does not completely match. Comparison of the early time solution, for which the whole fracture was contained in the FLAC zone, revealed that the apparent value of fracture toughness is larger than the prescribed value. As a result, the fracture propagated in the toughness regime before entering the PFC zone, resulting in slightly larger aperture and shorter fracture length. Consequently, after encountering the PFC region, the fracture began to change its shape to accommodate the mismatch of fracture toughness. Thus, a possible explanation of the discrepancy between the KGD and FLAC-PFC solutions is that the fracture has not finished the transition after crossing the boundary between the FLAC and the PFC regions and therefore has not settled into the viscous solution at the time of sampling.

To further check the solution obtained by the hybrid scheme, the blue dashed line in Figure 12.4 shows the tip asymptotic solution that corresponds to the viscous regime of propagation [42, 43], which is calculated as,

$$w(s) = \beta_m\left(\frac{12\,\eta\,V\,(1-\nu^2)}{E}\right)^{\frac{1}{3}} s^{2/3}, \qquad \beta_m = 2^{1/3}3^{5/6} \tag{12.10}$$

where s is distance from the fracture tip and $V = 1.39$ m/s is the fracture tip velocity, where the latter is calculated by differentiating the fracture length solution $L(t)$. The FLAC-PFC result clearly follows the tip asymptotic solution in the tip region. This observation shows the ability of the hybrid scheme to accurately capture the viscous tip asymptotic solution, and allows us to conclude that the results presented in the remainder of this paper correspond to the case for which the initial fracture (i.e. before crossing) propagates in the viscosity dominated regime. It should also be noted here that fracture crossing is primarily affected by the processes occurring at the tip region. In view of having a good agreement in the tip region, the discrepancy between the hybrid numerical solution and the viscosity dominated solution for a KGD fracture, which is shown in Figure 12.4, is of less importance.

12.3.3 Orthogonal Crossing

12.3.3.1 Effects of Stress Ratio and Friction of Natural Fractures

Orthogonal crossing refers to the scenario in which a natural fracture is placed perpendicular to the *SH*max direction and vertically crosses the PFC domain as shown in Figure 12.2. The hydraulic fracture is expected to reach the natural fracture with 90° intersection angle. To investigate the effects of the stress ratio and the friction of the natural fracture, four stress ratios (i.e., *SH*max/*Sh*min = 20/20, 22/20, 26/20, and 30/20 MPa) and five coefficients of friction of the natural fracture (i.e. $\mu = 0.2$, 0.4, 0.5, 0.6, and 0.8) were selected for the study. A total of $4 \times 5 = 20$ simulation cases were carried out by varying these two parameters. As the pore pressure for all cases was fixed at 10 MPa, the corresponding ratios of the two effective principal stresses were 1.0, 1.2, 1.6, and 2.0.

Figure 12.5 shows the identified three types of intersection between the hydraulic fracture and the orthogonally aligned natural fracture from 20 simulation cases. The first crossing scenario is called "no crossing" and is shown in Figure 12.5a. The red dots represent the pressure increase at each pore domain along the fracture. The black line represents the natural fracture, which is initially impermeable. It can be seen that in this case the hydraulic fracture does not cross the natural fracture but forms a T-shape intersection. The hydraulic fracture, which is created by tensile micro-cracks, is arrested after intersecting the natural fracture. The latter slips as a result of the interaction with the hydraulic fracture and forms mainly shear micro-cracks. Note that although both tensile and shear failure modes are allowed for the particle contacts representing the intact rock and smooth joint contacts representing the natural fractures, there is only one primary failure mode for each contact type. The second crossing scenario is called "offset crossing" and is shown in Figure 12.5b. The hydraulic fracture forms a T-shape intersection initially, but eventually crosses the natural fracture with an offset. The failure map

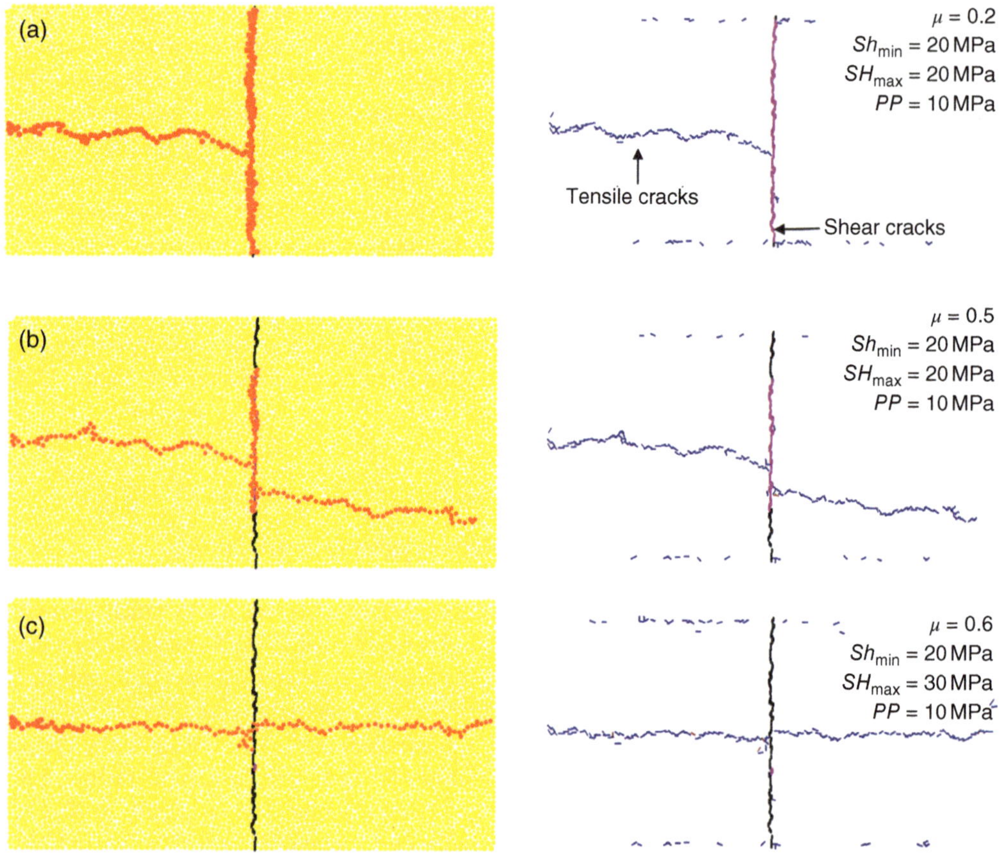

Figure 12.5 Three types of interaction between a hydraulic fracture and an orthogonal natural fracture for various stress ratios *SH*max/*Sh*min and the friction coefficient of the natural fracture: (a) no crossing, i.e. T-shape intersection, (b) offset crossing, i.e. crossing with an offset, and (c) direct crossing. The dots represent the pressure increase at each pore domain along the fracture; the vertical line represents the natural fracture. The segments represent micro-cracks in tensile or shear mode.

clearly shows the re-initiation of the hydraulic fracture at an offset distance along the natural fracture. Only a portion of the natural fracture fails in shear. The third crossing scenario is called "direct crossing" and is shown in Figure 12.5c. The hydraulic fracture crosses the natural fracture without any diversion into the natural fracture. The hydraulic fracture deforms due to elastic interactions with the natural fracture and there are almost no generated micro-cracks. Note that small tensile cracks at the boundary between the continuum and the discrete regions are caused by numerical noise at the interface. Due to their small size, the numerical results should not be affected.

Figure 12.6 summarizes the results of the 20 orthogonal crossing cases for various stress ratios and the friction coefficient of the natural fracture. Each case is marked with one of the three crossing scenarios mentioned above. The 20 cases form a 4 × 5 matrix in the figure. As the coefficient of friction or the stress ratio increase, the crossing pattern experiences the gradual transition from no crossing, to offset crossing, and then to direct crossing.

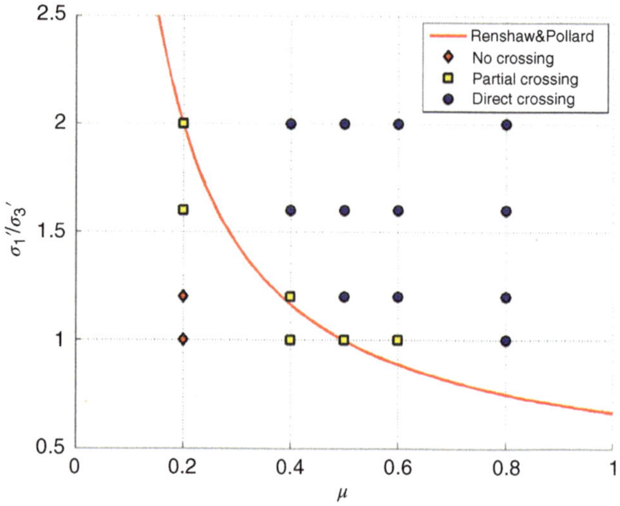

Figure 12.6 Results of the 20 orthogonal crossing cases for various the stress ratios and the friction coefficient of the natural fracture; the red line represents the crossing criterion reported in Renshaw and Pollard (*Source:* Adapted from Renshaw and Pollard [44]).

The simulation results were compared with the crossing criterion proposed in [44], in which the crossing is controlled by the magnitude of the applied stress, tensile strength of the rock and friction coefficient of the preexisting fracture. It can be expressed in the equation below and is plotted by the red line in,

$$\frac{-\sigma_1}{T_0 - \sigma_3} > \frac{0.35 + \dfrac{0.35}{\mu}}{1.06} \tag{12.11}$$

where σ_1 and σ_3 denote, respectively the applied maximum and minimum principal stresses, T_0 is the tensile strength of the rock, and μ is the friction coefficient of the preexisting fracture. The simulation results agree well with the predictions of the crossing criterion in Eq. (12.11).

Figure 12.7 shows the net pressure history at the inlet, or "receiver" pore, of the PFC domain versus the hydraulic fracture tip location for the three representative cases shown in Figure 12.5. Although the numerical results feature pressure oscillations, for each crossing scenario, the pressure behaves differently after the hydraulic fracture reaches the natural fracture and clear pressure patterns can be identified. For the no crossing case, the net pressure keeps building up after the hydraulic fracture is arrested by the natural fracture. For the offset crossing case, the net pressure increases first and then reduces because the hydraulic fracture eventually crosses the natural fracture with an offset. For the direct crossing case, the natural fracture has a minimal effect on the net pressure and is almost the same (about 1.5 MPa) before and after the intersection. This pressure behavior may suggest guidelines for interpreting field data for the injection pressure. A sharply increasing net pressure could indicate that the hydraulic

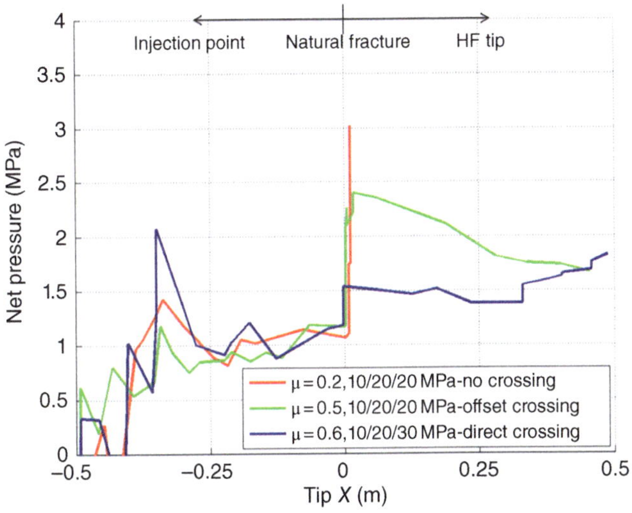

Figure 12.7 Net pressure at the inlet of the PFC domain versus the hydraulic fracture tip location for the three representative cases.

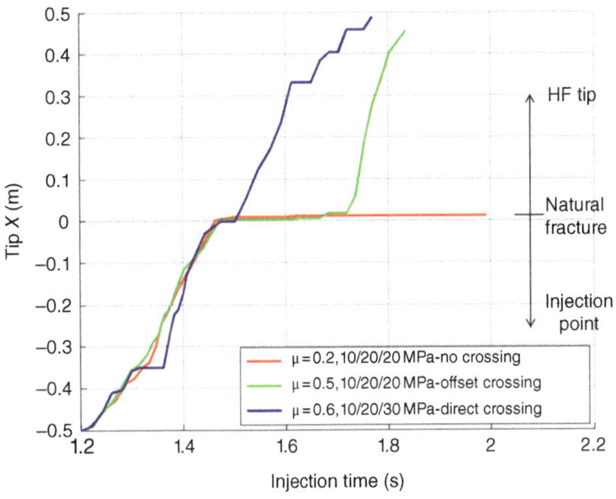

Figure 12.8 Hydraulic fracture tip location versus time in the PFC domain for the three representative hydraulic fracture intersection cases.

fracture is arrested by a natural fracture and the no crossing scenario is occurring. Further, sudden drops in net pressure might indicate a breakthrough crossing of a natural fracture.

Figure 12.8 plots the hydraulic fracture tip location versus time in the PFC domain for the three representative cases. For the no crossing case, the hydraulic fracture tip shows continuous growth before meeting the natural fracture but reaches a plateau after being arrested by the natural fracture. For the offset crossing case, the hydraulic fracture tip resumes growth after the hydraulic fracture crosses the natural fracture with an offset. For the direct crossing case, the hydraulic fracture tip shows nearly continuous growth during the whole crossing process.

Shear deformation during the propagation of a hydraulic fracture affects the hydraulic fracture aperture and therefore potentially affects the path for the proppant transport. Figure 12.9 shows the aperture of the hydraulic fracture and natural fracture for the "offset crossing" case shown in Figure 12.5b. Both the color scale and the thickness of the lines indicate the fracture width. The aperture of the failed natural fracture is smaller than the aperture of the hydraulic fracture section connected to the natural fracture. Thus, the existence of natural fractures could cause "pinch points" along the proppant transport path. Proppant forced into these "pinch points" could result in a local screenout and further increase of the fluid pressure. However, the pinch effect may be less critical in the real case due to the alternative path of fluid and proppant in the third dimension.

12.3.3.2 Effect of Strength (Toughness) Contrast

Three simulation tests were performed to investigate the behavior of a hydraulic fracture crossing a natural fracture, where the latter outlines an interface between two materials with different strength, as shown in Figure 12.10. The strength of the material on the left side of the natural fracture is fixed and equal to the value given in Table 12.1, i.e. $\sigma_t = \sigma_1$, while the strength of the material on the right side, $\sigma_t = \sigma_2$, was varied so that the left to the right strength ratio gradually decreased from 2 to 0.5. The tensile strength contrast is directly related to the fracture toughness ratio. The elastic properties such as elastic modulus and Poisson's ratio for the materials on both left and right sides were kept the same. As shown in Figure 12.10, increasing the strength of the material on the right side favors the initial arrest of the hydraulic fracture at the natural fracture followed by crossing with an offset, while reducing the strength of the material on the right side favors direct crossing of the hydraulic fracture through the natural fracture. The crossing scenario changes from direct crossing to offset crossing as the right-side material strength is increased. This result is expected since the increased strength introduces an additional resistance at the fracture tip on the other side of the natural fracture. It is also interesting to observe that the fracture does not necessarily propagate along the *SH*max direction after the crossing. One possible explanation for this is that the shearing along the interface results in a significant stress change, with magnitude comparable to the difference between the maximum and the minimum principal stresses.

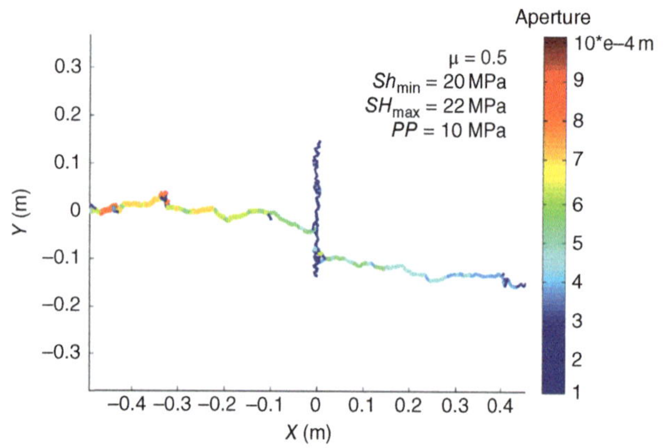

Figure 12.9 Aperture of the hydraulic fracture and natural fracture for the offset crossing case shown in Figure 12.5b. Both the color scale and the thickness of the lines indicate the fracture width.

Figure 12.11 plots the net pressure at the inlet of the PFC domain versus hydraulic fracture tip location for the three simulation cases with different strength ratios. For the low-to-high strength crossing case, the net pressure increases to more than 3.5 MPa (more than twice the net pressure for the case with a 1 : 1 strength ratio) before it drops as the hydraulic fracture enters the higher strength material. The high net pressure also causes a backward penetration of fracture into the low strength material, as shown in Figure 12.11. For the high-to-low strength crossing case, the net pressure is slightly smaller than the one for the case with 1 : 1 strength ratio, showing that the fracture experiences lower resistance after crossing the interface. Figure 12.12 plots hydraulic fracture tip location versus injection time in the PFC domain for the three simulation

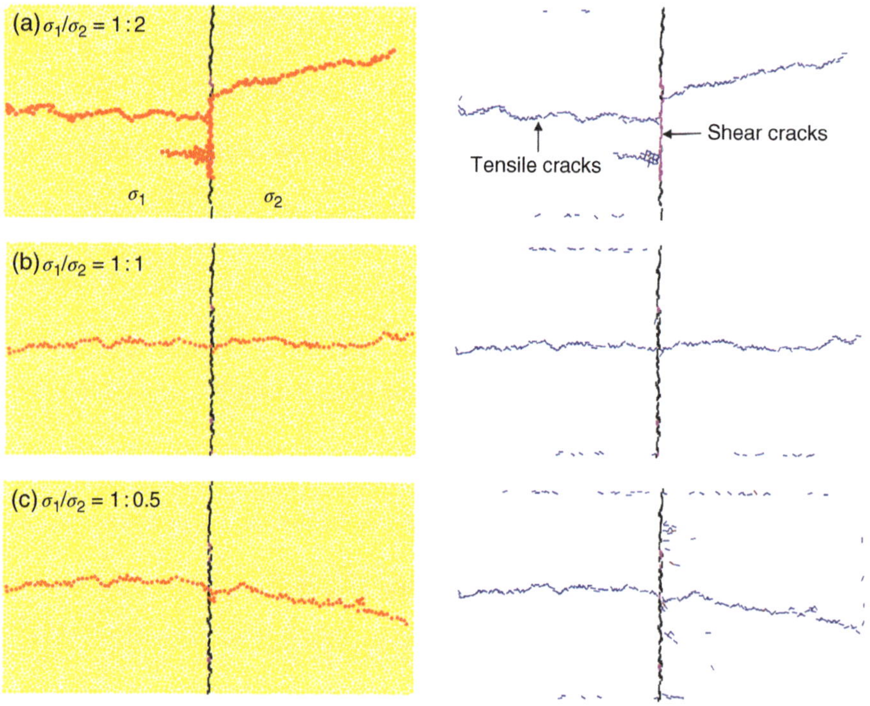

Figure 12.10 Profiles of a hydraulic fracture crossing a natural fracture between two materials with different strength ratios: (a) strength ratio 1 : 2, (b) strength ratio 1 : 1, and (c) strength ratio 1 : 0.5, for the conditions of μ = 0.5 and *SH*max/*Sh*min = 22/20 MPa.

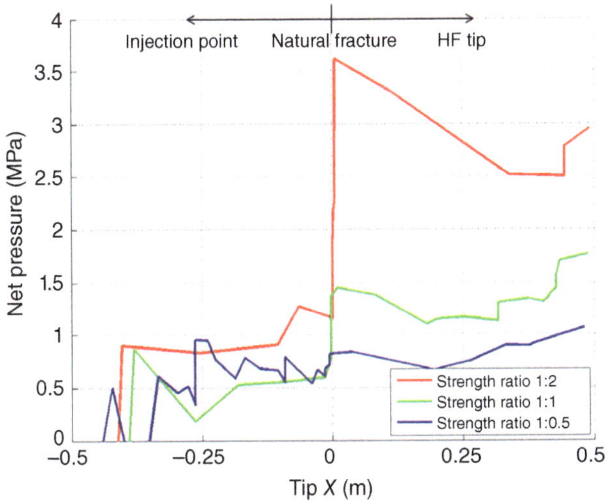

Figure 12.11 Net pressure at the inlet of the PFC domain versus hydraulic fracture tip location for the three simulation cases for different strength ratios.

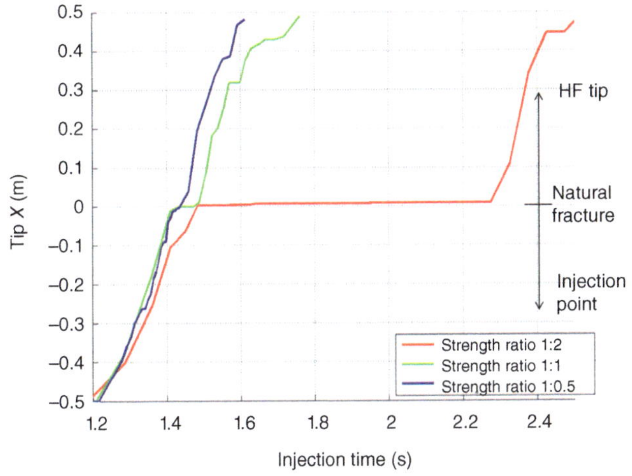

Figure 12.12 Hydraulic fracture tip location versus injection time in the PFC domain for the three simulation cases with different strength ratios.

cases with different strength ratios. Increasing the strength of the material on the right side significantly increases the hydraulic fracture crossing time, while reducing the strength of the material on the right side only slightly reduces the hydraulic fracture propagation time.

12.3.3.3 Effect of Stiffness (Modulus) Contrast

Three simulations were carried out to investigate the behavior of a hydraulic fracture crossing a natural fracture between two materials with different stiffnesses, as shown in Figure 12.13. The stiffness of the material on the left side was fixed and the elastic properties (Young's modulus and Poisson's ratio) were equal to the values listed in Table 12.1, i.e. $E = E_1$, while the stiffness of the material on the right side, $E = E_2$, was varied so that the left-to-right stiffness ratio gradually decreased from 2 to 0.5. The strength of the material on both the left and right sides was kept the same.

As shown in Figure 12.13, all cases show direct crossing of the hydraulic fracture though the natural fracture, regardless of the stiffness ratio. However, we can see from the plot of net pressure versus hydraulic fracture tip location in Figure 12.14 that for the high-to-low stiffness crossing case, the net pressure builds up as the hydraulic fracture reaches the natural fracture and then reduces as the hydraulic fracture enters into the lower stiffness material. For the low-to-high stiffness crossing cases, the net pressure behavior is similar to the case in which the stiffness ratio is one. The plot of hydraulic fracture tip location versus injection time in Figure 12.15 also shows that a substantial delay of hydraulic fracture propagation happens for the high-to-low stiffness crossing case. This may occur because the near tip fracture aperture is proportional to the ratio between the fracture toughness and the elastic modulus for the arrested fracture (toughness dominated propagation). Once the fracture encounters the softer material, the aforementioned ratio increases, which is equivalent to increasing the fracture toughness and keeping the same elastic modulus. As a result, the fracture is arrested until it becomes sufficiently wide to accommodate the change and to propagate further. Note that the above argument is applicable only in the absence of the natural fracture and before fracture starts penetrating the softer material.

The results illustrate that although the stiffness contrast (up to a four times difference) might not affect the overall crossing behavior of the hydraulic fracture, the pressure behavior can be significantly different. For example, for a hydraulic fracture entering a lower stiffness layer from a higher stiffness layer, the net pressure builds up before propagation resumes. However, the net pressure increase does not create a T-shaped intersection along the natural fracture, indicating that negligible slip or shear failure along the natural fracture occurs.

As mentioned in previous section, the initial fracture propagates in the viscosity-dominated regime. At the same time, Figures 12.8, 12.12, and 12.15 indicate that the fracture can be arrested by the natural fracture, in which case the tip behavior can be argued to be toughness-dominated. To avoid confusion, it should be noted that the regimes of propagation are defined only for fractures propagating in an intact material (which is not the case for Figures 12.8, 12.12, and 12.15). Therefore, the conclusion that viscosity dominates propagation is applicable only for the hydraulic fracture before crossing, and the propagation regime cannot be directly translated to fractures that interact with natural fractures.

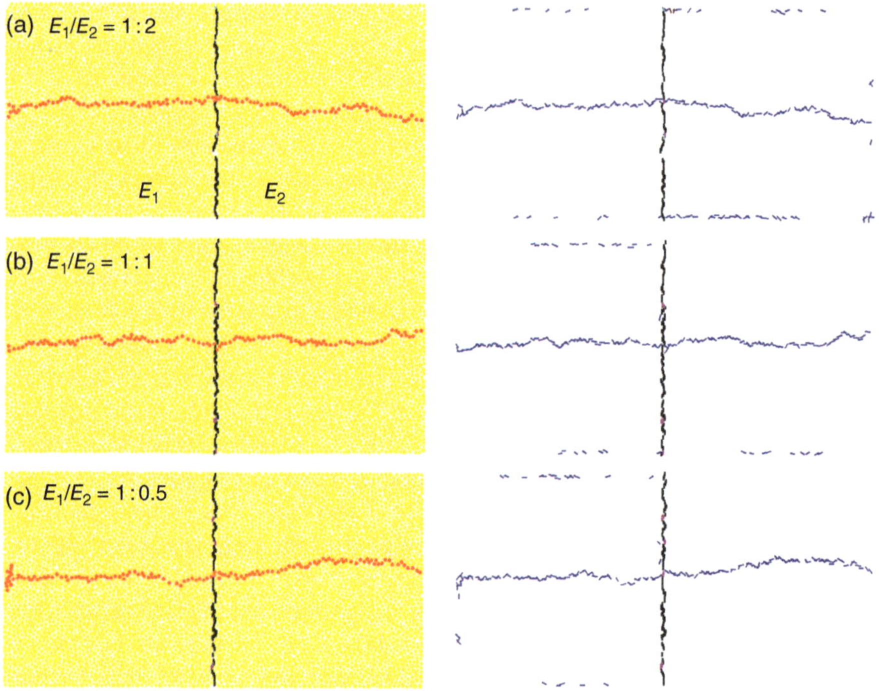

Figure 12.13 Profiles of hydraulic fracture crossing a natural fracture between two materials with different stiffnesses: (a) stiffness ratio 1 : 2, (b) stiffness ratio 1 : 1, and (c) stiffness ratio 1 : 0.5 for the conditions of $\mu = 0.5$ and *SH*max/*Sh*min = 22/20 MPa.

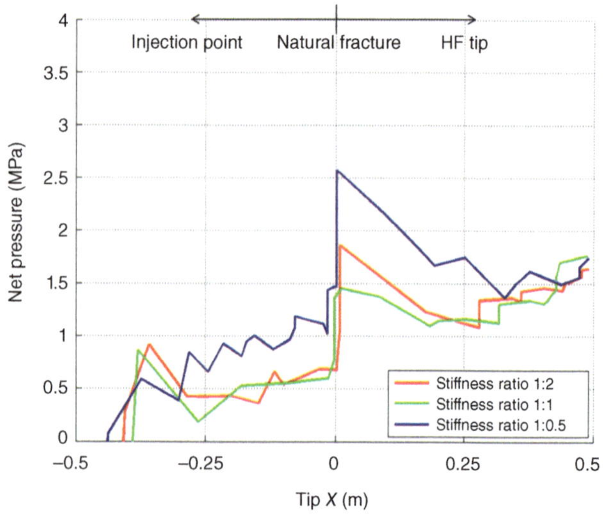

Figure 12.14 Net pressure at the inlet of the PFC domain versus the hydraulic fracture tip location for the three stiffness ratio cases.

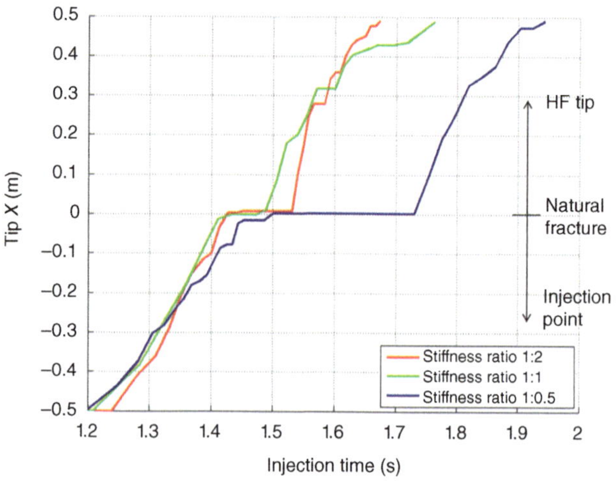

Figure 12.15 Hydraulic fracture tip location versus injection time in the PFC domain for the three stiffness ratio cases.

12.3.4 Non-Orthogonal Crossing

This section considers situations in which a natural fracture intersects the *SH*max direction less than 90° (i.e. non-orthogonal intersection). To investigate this case, two direct crossing cases from the crossing diagram in Figure 12.6 are chosen to test the effect of the intersection angle. The first case, with $\mu = 0.5$ and *SH*max/*Sh*min = 22/20 MPa, is located near the crossing/arresting border in Figure 12.6. The second case, with $\mu = 0.8$ and *SH*max/*Sh*min = 22/20 MPa, is relatively far away from the crossing/arresting border in Figure 12.6. The intersection angle β for these two cases was gradually decreased from 90°, with an interval of 15°, until the hydraulic fracture was arrested during crossing.

Figure 12.16 shows profiles of the hydraulic fracture crossing the natural fracture at 90° and 75° intersection angles for the first case. As the intersection angle decreases from 90° to 75°, the crossing scenario changes from direct crossing to no crossing while the fracture diverts into the lower part of the natural fracture and is

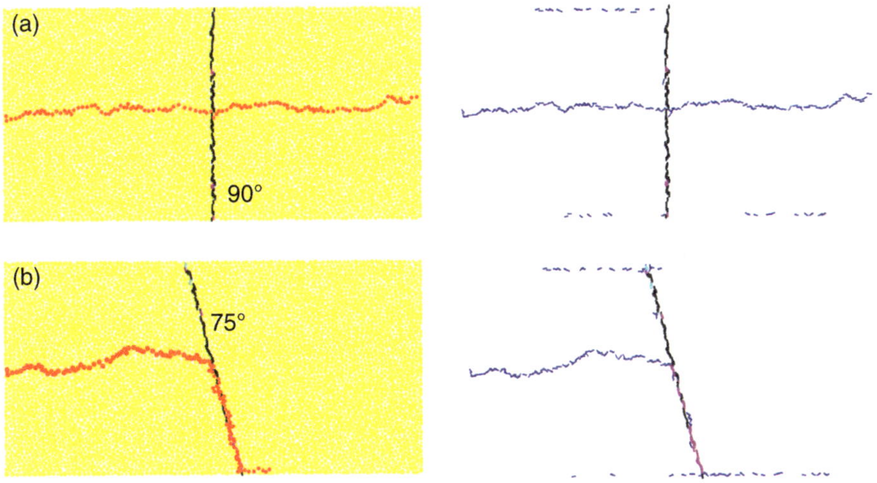

Figure 12.16 Profiles of a hydraulic fracture crossing a natural fracture at two intersection angles: (a) $\beta = 90°$ and (b) $\beta = 75°$, with the conditions of $\mu = 0.5$ and *SH*max/*Sh*min = 22/20 MPa.

arrested. The hydraulic fracture does not penetrate the upper part of the natural fracture because the upper part bears more compressive stress due to the hydraulic fracture propagation. A T-shaped intersection is less likely to occur at a non-orthogonal natural fracture, because the natural fracture offers less resistance in one direction than the other, so that the hydraulic fracture, if it diverts into the natural fracture, will propagate along the less resistance path and form an obtuse angle diversion.

Figure 12.17 shows the profiles of a hydraulic fracture crossing a natural fracture at five different intersection angles for the second case. It can be seen that as the intersection angle decreases, hydraulic fracture crossing behavior transitions from direct crossing when $\beta = 90°$ and 75°, to offset crossing when $\beta = 60°$ and 45°, and then to no crossing when $\beta = 30°$.

The results for the non-orthogonal crossing demonstrate that, if all other conditions are kept the same, decreasing the intersection angle is expected to impede direct crossing and favor arrest of the hydraulic fracture. Further, due to the non-orthogonal angle, slip along the natural fracture becomes more favorable and natural fracture failure occurs prior to the penetration of the hydraulic fracture to the opposite side of the natural fracture. It is worth mentioning that the non-orthogonal crossing criterion for any specific angle, although not shown in this paper, can be obtained if a sufficient number of simulation cases are executed. Compared with the crossing diagram for a hydraulic fracture with an orthogonally aligned natural fracture, the non-orthogonal crossing criterion simply shifts the crossing/arresting border further away from the origin in Figure 12.6, as shown in [45].

12.3.5 Fracturing Complexity

The existence of a natural fracture network may cause a number of complicated interaction situations to arise. Numerical simulations were performed to investigate the effect of two or more adjacent natural fractures on the hydraulic fracture crossing behavior. Figure 12.18 shows the crossing behavior when a hydraulic fracture crosses two natural fractures at different intersection angles: the first natural fracture intersects the *SH*max direction at 75° and the second natural fracture intersects it at 105°. The hydraulic fracture crosses the first natural fracture with an offset and propagates downward along the natural fracture, then crosses the second natural fracture with an offset but propagates upward along the natural fracture until the crossing. The propagation of the fracture always follows an obtuse angle diversion path because this is the path of least resistance. This case provides an illustrative example for the effect of a simple fracture network on hydraulic fracturing treatments. The hydraulic fracture may not be strictly planar as commonly assumed in hydraulic fracturing theory. In addition, this is also important for proppant transport, because narrower portions of the fracture at intersection points can cause accumulation of proppant and screen-out.

Figure 12.19 shows the crossing behavior when a hydraulic fracture crosses three natural fractures at intersection angles of 90°, 110°, and 85° from the *SH*max direction. The hydraulic fracture directly crosses the first natural fracture. After that, the hydraulic fracture propagates upward and reaches the edge of the second natural fracture and then propagates downward. Eventually, the hydraulic fracture breaks out from the upper edge of the second natural fracture and propagates horizontally to the third natural fracture, where it propagates downward until eventually crossing it. The simulation suggests that the complexity of trajectory of a propagating hydraulic fracture increases with the number and extent of the intersected natural fractures.

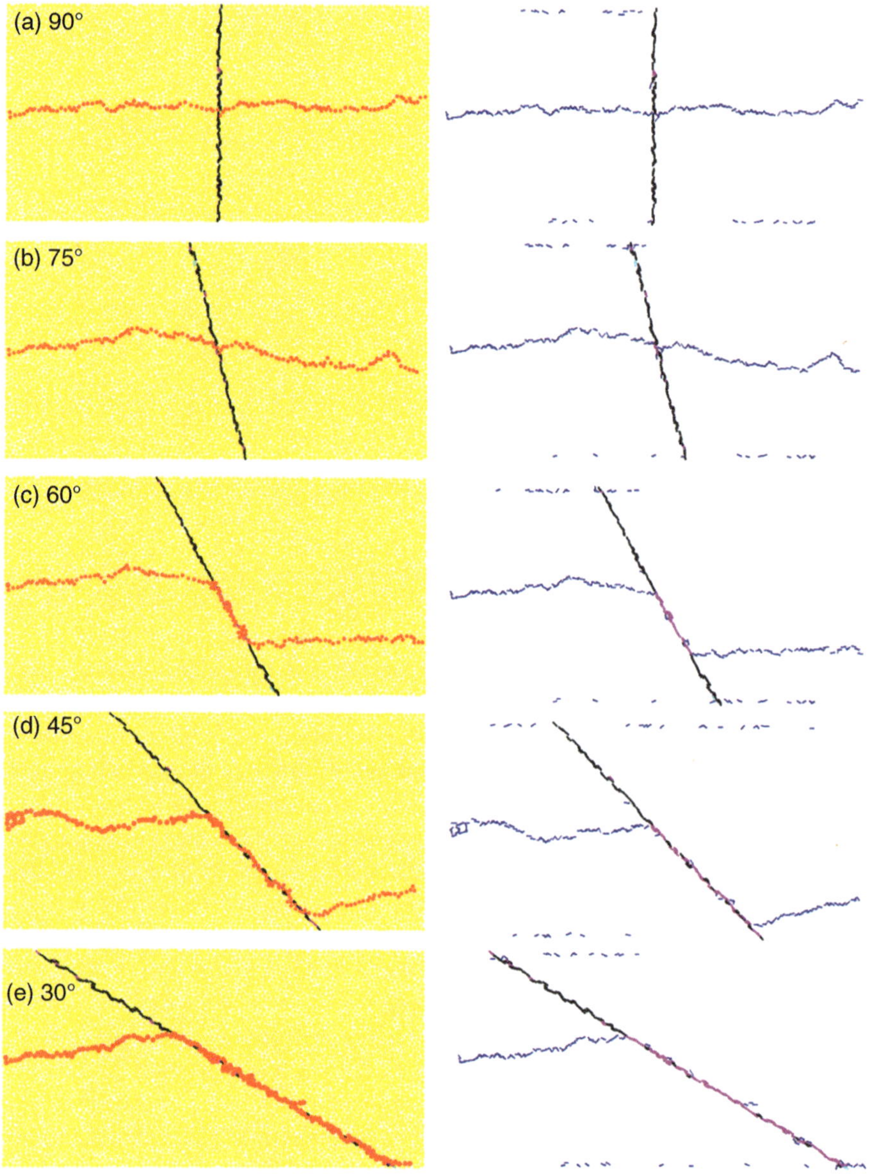

Figure 12.17 Profiles of a hydraulic fracture crossing a natural fracture at five different intersection angles: (a) $\beta = 90°$, (b) $\beta = 75°$, (c) $\beta = 60°$, (d) $\beta = 45°$, and (e) $\beta = 30°$, with the conditions of $\mu = 0.8$ and $SH\text{max}/Sh\text{min} = 22/20$ MPa.

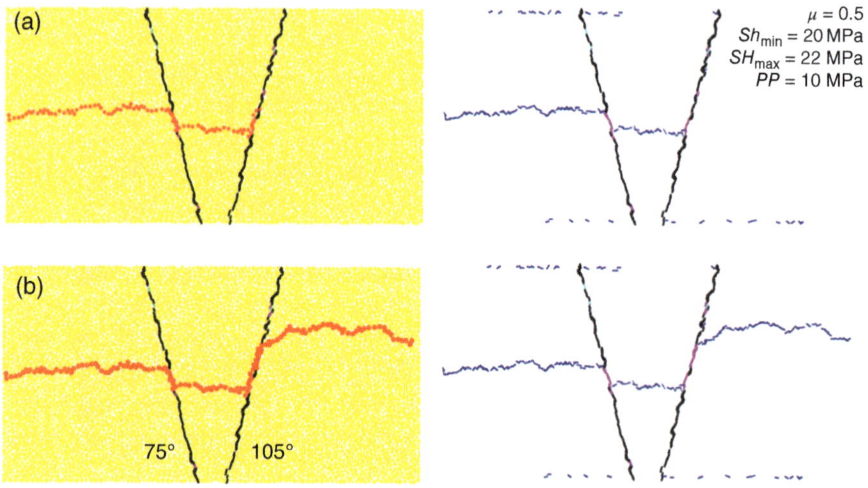

Figure 12.18 Profiles of a hydraulic fracture crossing two adjacent natural fractures at intersection angles of 75° and 105° with the SH_{max} direction at the injection times of (a) 1.52 s and (b) 1.83 s.

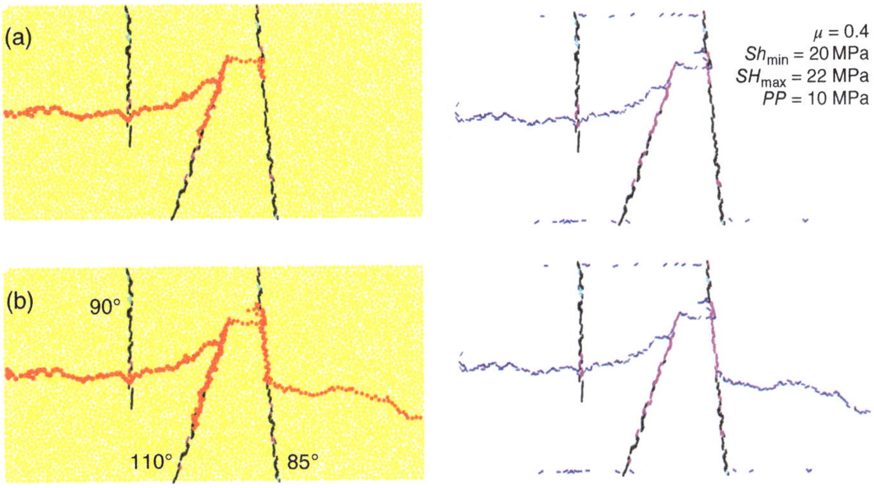

Figure 12.19 Profiles of a hydraulic fracture crossing three adjacent natural fractures at intersection angles of 90°, 110°, and 85° from the *SH*max direction at the injection times of (a) 1.59 s and (b) 2.13 s.

The results in Figures 12.18 and 12.19 are consistent with observations from field hydraulic fracture systems investigated through excavation of a mining tunnel through the fracture, or the so-called mineback experiments [46]. Multiple hydraulic fracture crossings and additional complexity of the hydraulic fracture path are expected to arise if more natural fractures are added. The existence of a complex discrete fracture network greatly influences the overall geometry of the hydraulic fracture propagation.

12.4 DEM Modeling of Supercritical Carbon Dioxide Fracturing

Conventional hydraulic fracturing with water-based fracturing fluid has some disadvantages, such as environmental impacts caused by the flow-back water and high cost of water usage [47, 48]. There is an urgent requirement to explore the feasibility of non-aqueous fracturing technologies to overcome these drawbacks. Supercritical carbon dioxide (SC-CO_2) is considered as an ideal non-aqueous fracturing fluid due to its superior properties of liquid-like density, gas-like viscosity, high compressibility, and diffusivity. This study aims to investigate the micromechanical behavior of SC-CO_2 fracturing in both intact and fractured rock samples by using a fluid–solid-coupled DEM model. A new numerical algorithm for hydraulic fracturing in the toughness-dominated regime is developed by assuming that the pressure in the whole fracture is uniform. This new numerical algorithm could achieve a much higher computational efficiency compared with the conventional hydro-mechanical scheme in DEM. Hydraulic fracturing cases using high viscosity fracturing fluid are also performed for comparison. The results indicate that the fracture propagation induced by SC-CO_2 tends to be less smooth and continuous, more asymmetric and tortuous compared to that induced by viscous fluid. Besides, the low viscosity fluid such as SC-CO_2 can lead to a lower breakdown pressure, and the fluid leak-off into the rock matrix can result in a lower breakdown pressure and higher fracture propagation pressure. The simulations also illustrate that SC-CO_2 fracturing tends to create a more complex and productive fracture network if the preexisting natural fractures are involved. As a result, we can conclude that SC-CO_2 could be an alternative fracturing fluid to induce a more effective fracture network for hydrocarbon production.

12.4.1 New Algorithm for the Toughness-Dominated Regime

The developed new algorithm for the toughness-dominated regime is shown in the flow chart in Figure 12.20 [49]. In the toughness-dominated regime, the energy is mainly spent in fracturing the rock, and energy dissipated by viscous flow is negligible. The pressure gradient along the fracture length is nearly zero. Thus, this algorithm assumes that the pressure in the whole fracture is uniform. The fluid volume inside the hydraulic fracture can be viewed as a whole, and the goal is to match the current opened fracture volume V_c with the injection fluid volume V_f. The change of fluid pressure is proportional to the volume difference as given in Eq. (12.12). Note that the compressibility of SC-CO_2 is not considered here, since the effect of fluid viscosity can dominate and mask the effect of compressibility [50, 51]. The fracture volume is calculated from the aperture width w and the length of flow channel l_p as shown in Eqs. (12.13) and (12.14). The apertures and current opened fracture volume are updated with the apparent fluid pressure acting on

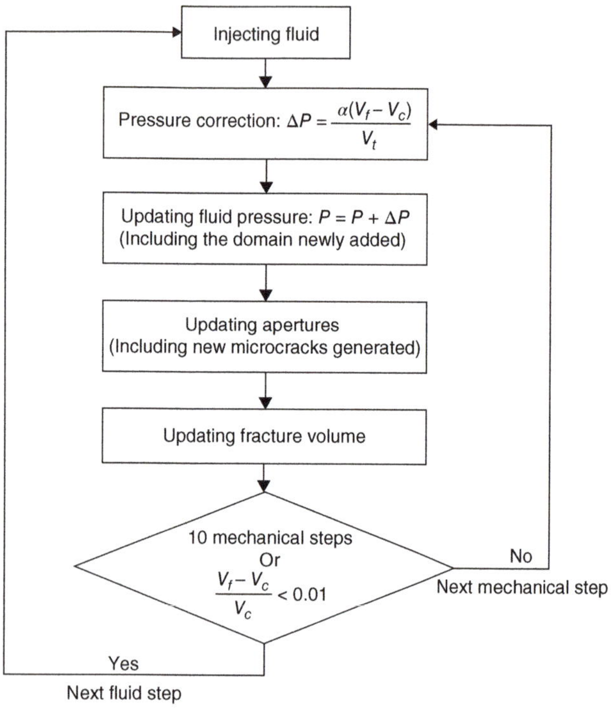

Figure 12.20 Proposed numerical algorithm for the toughness-dominated regime.

the surrounding particles. Not more than ten mechanical steps are taken between the fluid time steps to achieve the balance of the fluid volume and fracture volume.

$$\Delta P = \alpha(V_f - V_c)/V_t \tag{12.12}$$

$$V_c = \sum l_p(w - w_0) \tag{12.13}$$

$$V_t = \sum l_p w \tag{12.14}$$

where ΔP is the pressure increment at each mechanical time step, α is a coefficient proportional to the rock stiffness, V_t is the total fracture volume, w is the aperture width, w_0 is the assumed initial aperture width, and l_p is the length of the flow channel, which is the distance of two connecting domain centers.

In the conventional hydro-mechanical scheme, the fluid time step is proportional to the term of fluid viscosity over fracture aperture ($dt \propto \frac{\mu}{K_f w^2}$), so the fluid time step could be very small for low viscosity fluid such as SC-CO_2. In contrast, fluid viscosity and fracture aperture are irrelevant to the fluid time step for the toughness-dominated regime and it can therefore be two orders of magnitude larger than that for the conventional hydro-mechanical scheme. Consequently, a much higher computational efficiency can be achieved based on the new proposed algorithm for hydraulic fracturing in the toughness-dominated regime.

12.4.2 Numerical Model Setup

12.4.2.1 Model Description

A two-dimensional intact rock sample (1.0 m × 1.0 m) with an injection point at the center is generated in this study. The particle radius is randomly distributed between 3 and 5 mm by using a given random seed number. The sample is subjected to two types of bi-axial compressive stresses, one with 10 MPa in the x-direction and 12 MPa in the y-direction, and the other with 10 MPa in the x-direction and 15 MPa in the y-direction. The Young's modulus of the intact rock sample is 29.1 GPa and the Poisson's ratio is 0.30 based on the numerical simulation of unconfined compressive strength test. The uniaxial compressive strength and tensile strength of the intact rock sample are 22.3 MPa and 4.8 MPa, respectively. Through the biaxial compressive test, the intact rock friction angle and cohesion are 37.5° and 5.8 MPa, respectively. Besides the intact rock sample, a fractured rock sample which contains 18 natural fractures is also generated as shown in Figure 12.21. The natural fractures obey the SJM, which allows slip and separation at particle contacts [52]. The frictional stress on the weakly bonded natural fractures is assumed to obey the Coulomb's frictional law. The initial aperture is 0.1 mm, 17 times larger than the assumed residual aperture. The fracturing fluid is not able to invade the natural fracture until the corresponding bond breaks and connects to the hydraulic fracture. The algorithm for the toughness-dominated regime is applied to the modeling cases with SC-CO_2 (fluid viscosity 0.06 cp), while the traditional hydro-mechanical coupling scheme is applied to that with high viscosity of 100 cp. The input parameters for the numerical model and the microscale parameters for the contact models are summarized in Tables 12.2 and 12.3, respectively. Hydraulic fracturing simulation cases in both intact and fractured rock samples are listed in Table 12.4.

To test the accuracy of the new numerical algorithm for the toughness-dominated regime, the simulation results are compared with the reference solution for a plane strain KGD hydraulic fracture [53]. It is first necessary to determine the regime of propagation. There are four limiting regimes as shown in Figure 12.22: storage viscosity M [54], storage toughness K [55], leak-off viscosity $\tilde{M}$ [56], and leak-off toughness $\tilde{K}$ [57]. The fracture propagation regimes are determined by the dimensionless time τ and the toughness parameter K_m as follows:

$$\tau = \frac{E'C'^6}{\mu' Q^3} t \tag{12.15}$$

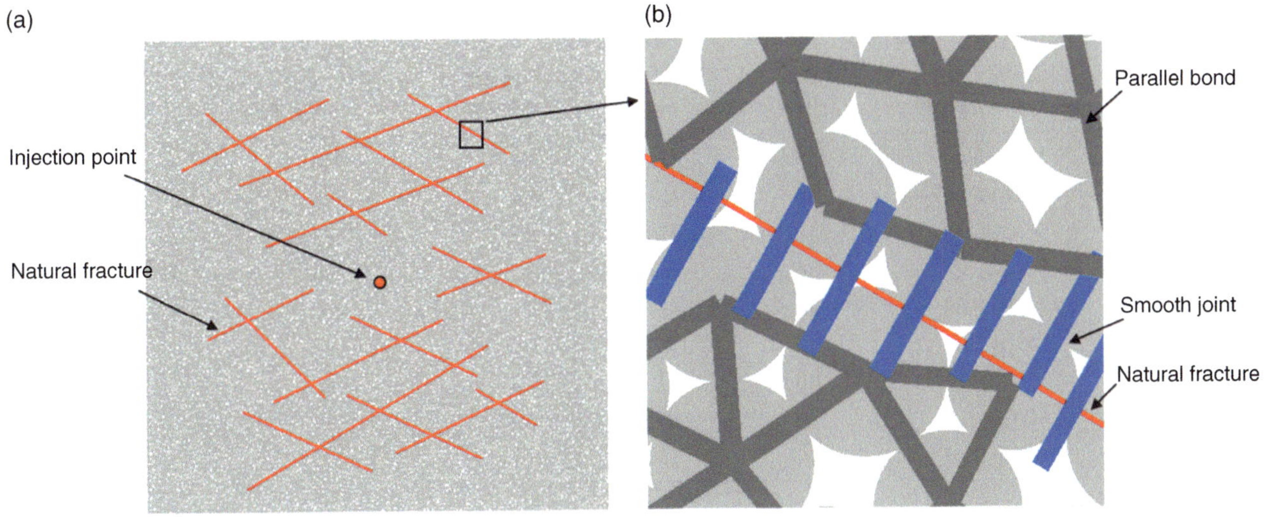

Figure 12.21 Preexisting natural fractures in rock sample, (a) 18 fractures inserted in the model sample, and (b) close-up view of one natural fracture. The thin lines in (a) represent pre-existing natural fractures, and the thick lines in (b) indicate the contact normal.

Table 12.2 The parameters for the numerical model.

Model properties	Values
Model size (m)	1.0 × 1.0
Number of particles	17 511
Particle radius (mm)	3–5
Particle density (kg/m^3)	3169
Compressive stress in x, $S_{\min}$ (MPa)	10
Compressive stress in y, $S_{\max}$ (MPa)	12/15
Young's modulus, E (GPa)	29.1
Poisson's ratio, v	0.26
Uniaxial compressive strength, σ_c (MPa)	23.5
Uniaxial tensile strength, σ_t (MPa)	3.6
Friction angle, φ (°)	37.9
Cohesion, c (MPa)	5.9
Bulk modulus of rock, K (GPa)	8.3
Proportionality coefficient, α (MPa)	0.2
Assumed residual aperture, w_0 (m)	6e−6
Assumed joint initial aperture, w_{0j} (m)	1e−4
Injection rate, Q (m^2/s)	0.5e−4
Injection fluid viscosity (cp)	0.06 and 100

$$K_m = \left(\frac{K'^4}{\mu' E'^3 Q}\right)^{1/4} \tag{12.16}$$

where Q is the fluid injection rate, t is the injection time, and μ', E', K', and C' are the material parameters defined as follows:

Table 12.3 The microscale parameters for the contact models.

Microscale input parameters	Values
Young's modulus of the particle (GPa)	17
Ratio of stiffness of the particle	2.5
Friction coefficient of the particle, μ	0.5
Young's modulus of the parallel bond (GPa)	17
Ratio of stiffness of the parallel bond	2.5
Tensile strength of the parallel bond (MPa)	8
Cohesion of the parallel bond (MPa)	15
Friction angle of the parallel bond (°)	30
Normal stiffness of SJM (GPa/m)	120
Shear stiffness of SJM (GPa/m)	60
Tensile strength of SJM (MPa)	0.1
Cohesion of SJM (MPa)	0.1
Friction coefficient of SJM	0.5

$$\mu' = 12\mu,\ E' = \frac{E}{1-v^2},\ K' = 4\left(\frac{2}{\pi}\right)^{1/2} K_{IC},\ C' = 2C_L \tag{12.17}$$

where μ is the fluid viscosity, E is the Young's modulus, v is the Poisson's ratio, K_{IC} is the mode-I fracture toughness of the rock, and C_L is the Carter's leak-off parameter.

In this study, K_m is the decisive parameter due to the absence of leak-off. As shown in [53], the hydraulic fracture propagates in the toughness-dominated regime for $K_m > 4.80$ and in the viscosity-dominated regime for $K_m < 0.70$. When the mode-I fracture toughness of the regular model is 2.66 Pa m$^{1/2}$, which is inversely calculated in the next

Table 12.4 Summary of hydraulic fracturing simulation cases.

Case ID	Fracturing fluid	S_{min}	S_{max}	Preexisting fractures	Leak-off
Case 1-S1	100 cp fluid	10	12	No	No
Case 1-S2	100 cp fluid	10	15	No	No
Case 2-S1	SC-CO_2	10	12	No	No
Case 2-S2	SC-CO_2	10	15	No	No
Case N1-S1	100 cp fluid	10	12	Yes	No
Case N1-S2	100 cp fluid	10	15	Yes	No
Case N2-S1	SC-CO_2	10	12	Yes	No
Case N2-S2	SC-CO_2	10	15	Yes	No
Case L1-S1	100 cp fluid	10	12	No	Yes

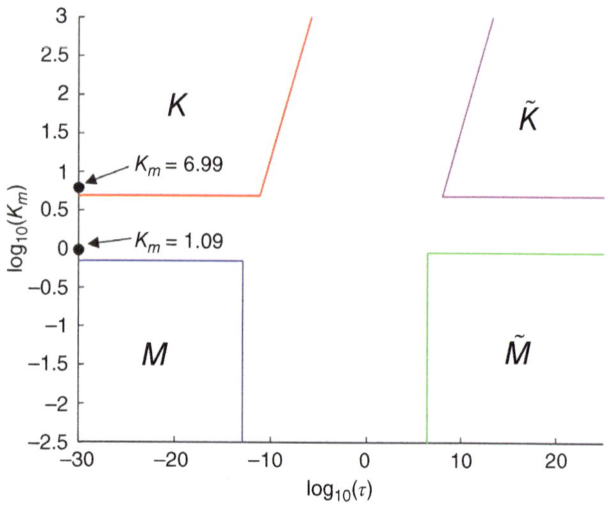

Figure 12.22 The phase diagram of the fracture propagation regimes versus dimensionless time τ and toughness parameter K_m (*Source:* Adapted from Dontsov [53]). M, K, $\tilde{M}$, and $\tilde{K}$ represent the storage viscosity regime, the storage toughness regime, the leak-off viscosity regime, and the leak-off toughness regime, respectively. The black dots indicate the fracture propagation regimes with the input parameters in Table 12.2.

section of this paper, the toughness parameter K_m calculated for the modeling cases with SC-CO_2 and 100 cp fluid are 6.99 and 1.09, respectively. Therefore, the fracture propagation driven by SC-CO_2 belongs to the toughness-dominated regime, while that induced by 100 cp fluid approaches the viscosity-dominated regime.

12.4.2.2 Model Verification

In order to verify the hydraulic fracturing model in the toughness-dominated regime, simulation of hydraulic fracturing using SC-CO_2 was first performed by using the cubic array particle assembly (Figure 12.23). The particle radius is 5 mm and the Young's modulus is 35.1 GPa. The two applied stresses are 10 MPa in the x direction and 12 MPa in the y direction. Other input parameters are the same as those in Table 12.2.

As shown in [53], the KGD solutions of fracture aperture w and single fracture wing length l for the toughness-dominated regime are given as follows:

$$w(\xi, t) = 0.6828\left(\frac{K'^2 Qt}{E'^2}\right)^{1/3}\left(1-\xi^2\right)^{1/2} \tag{12.18}$$

$$l(t) = 0.9324\left(\frac{E'Qt}{K'}\right)^{2/3} \tag{12.19}$$

where $\xi = x/l$ is the normalized coordinate along the fracture, and x is the coordinate measured from the injection point.

The evolutions of fracture aperture and fracture length at $t = 0.9$ s in the numerical model and KGD model are compared in Figure 12.24. Note that the input of mode-I fracture toughness of the regular model was 1.88 Pa m$^{1/2}$ in the KGD model in order to match numerical solution. The fracture length in the cubic array particle assembly is the distance between two fracture tips in y-direction. The fracture aperture at the injection point, and the length in the KGD model are as follows: $w(0, 0.9) = 7.52\text{e}{-5}$ m and 2 $l(0.9) = 0.763$ m, while that calculated in the numerical model are 7.24e−5 m and 0.770 m, respectively. The fracture width is underestimated by 3.9%, and the fracture length is overestimated by 1.1%. The small discrepancy can be attributed to the gradual increase of rock modulus in the DEM modeling due to the opening of the fracture. Nevertheless, the results in Figure 12.24 indicate that the numerical algorithm for the toughness-dominated regime is relatively accurate.

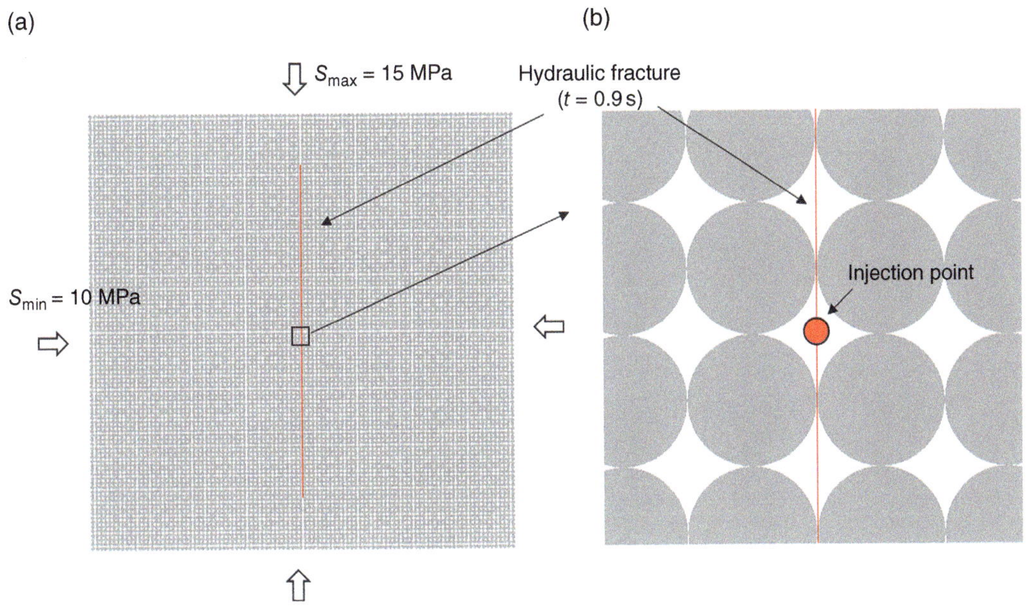

Figure 12.23 (a) Schematic of the hydraulic fracturing verification model at $t = 0.9$ s, and (b) close-up view of the cubic array particle assembly.

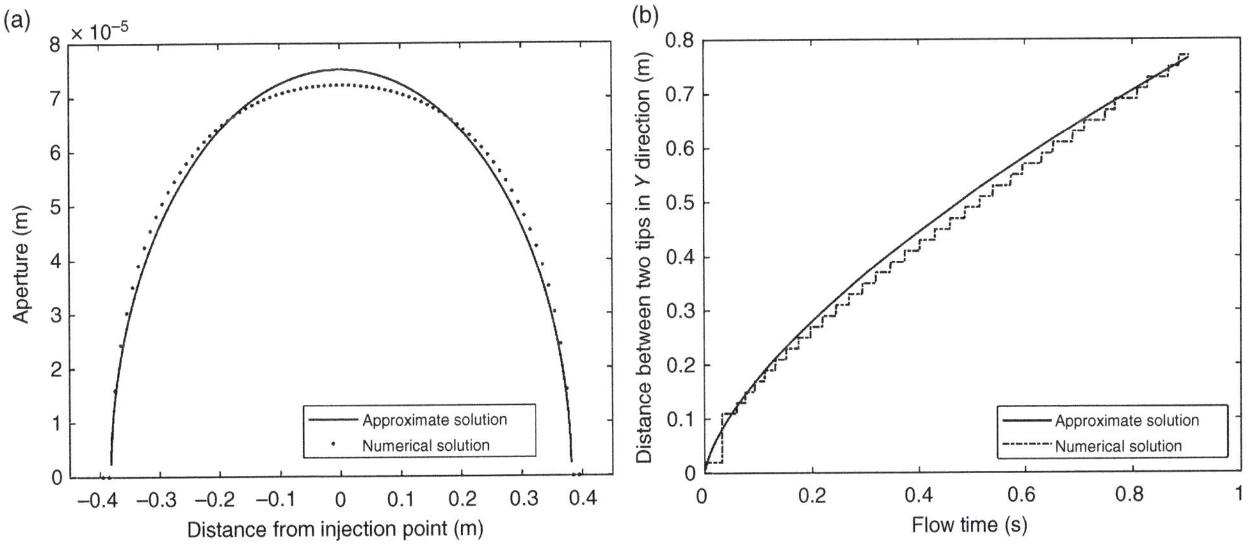

Figure 12.24 Simulation results of hydraulic fracturing verification model with the cubic array particle assembly, (a) comparison between the approximate solution and the numerical solution for fracture aperture, and (b) comparison between the approximate solution and the numerical solution for fracture length.

12.4.3 Hydraulic Fracturing in Intact Rock Sample

The cases of hydraulic fracturing with SC-CO_2 in the toughness-dominated regime and with 100 cp fracturing fluid approaching the viscosity-dominated regime were descripted and compared in this section. All cases were run with the random particle assembly to better mimic the rock materials.

In the cases for intact rock sample, the hydraulic fractures driven by the two types of fracturing fluid extend along the direction of maximum principal stress as shown in Figure 12.25. The hydraulic fracture is tortuous and the tortuosity can be used to evaluate fracture propagation pathways [58]. The tortuosity is defined as the total fracture length along a pathway (L_0) divided by the distance between the two ends of the fracture (L) in the reference

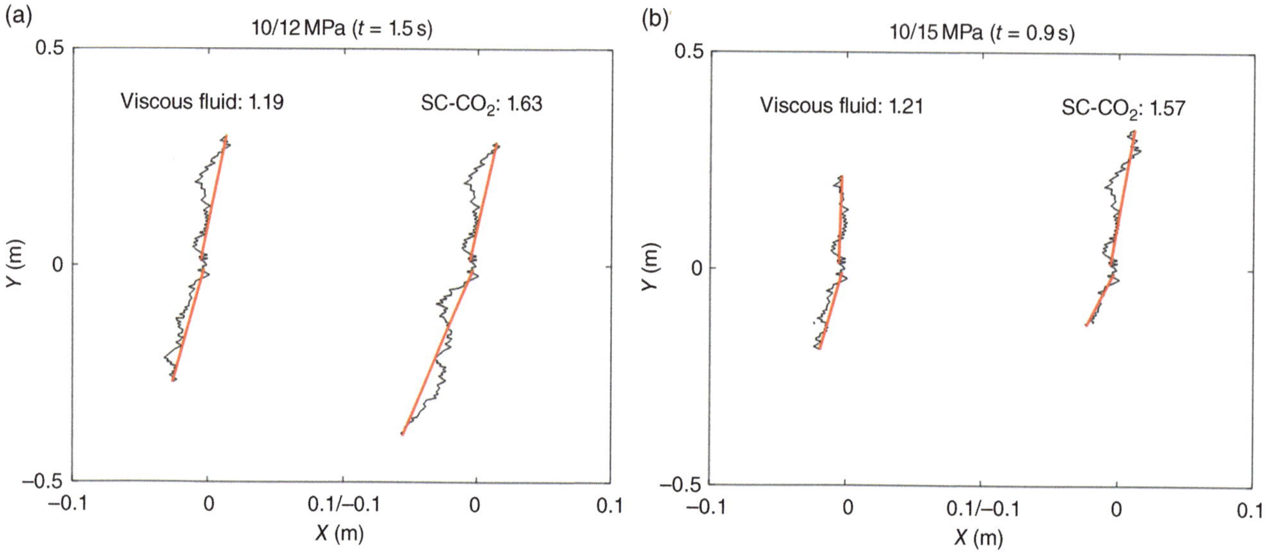

Figure 12.25 Fracture profiles induced by 100 cp fracturing fluid and SC-CO_2, (a) in Case 1-S1 and Case 2-S1 at $t = 1.5$ s, and (b) in Case 1-S2 and Case 2-S2 at $t = 0.9$ s. The black line represents the hydraulic fracture, and the red line shows the connection line of two ends. The numbers give the average tortuosity in the two directions for the fractures generated under different test conditions.

area, and the total fracture length is calculated as the sum of the length of all micro-cracks generated in the model. When the stress difference is 2 MPa in Case 1-S1 and Case 2-S1, the results show that the tortuosity is 1.63 for SC-CO_2 and 1.19 for 100 cp fluid. The tendency in laboratory experiments using granite and shale is similar but with different values [58–60]. In the experiments using the granite [58, 60], the fractures induced by SC-CO_2 have a tortuosity up to 1.109, while that induced by oil (~ 320 cp) have a tortuosity up to 1.062. In the experiments using shale [59], the measured total fracture length contains branches, and the tortuosity is 5.602 for L-CO_2 and 1.090 for oil (~ 270 cp). The cases with larger stress difference show similar results. The calculated tortuosity suggests that SC-CO_2 has potential to generate more complex fracture networks and is desirable to enhance the hydrocarbon production in the flied operations.

The net pressure and aperture along the hydraulic fractures are shown in Figures 12.26 and 12.27. When comparing at the same injection fluid volume, hydraulic fracturing using more viscous fluid tends to induce a thicker and shorter fracture than that using SC-CO_2 in

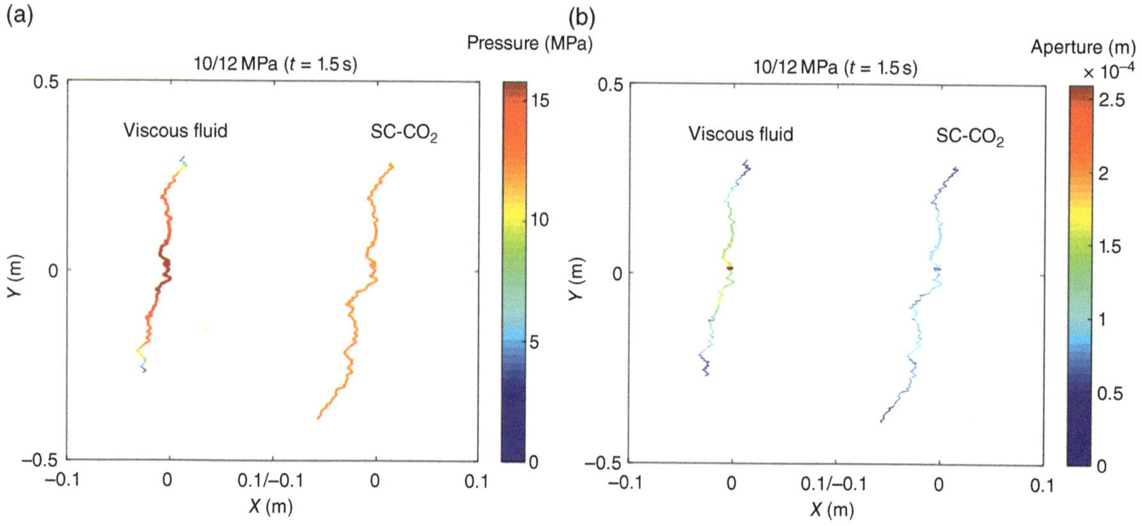

Figure 12.26 (a) The net pressure of hydraulic fractures in Case 1-S1 and Case 2-S1 at $t = 1.5$ s, and (b) the aperture of hydraulic fractures in Case 1-S1 and Case 2-S1 at $t = 1.5$ s. Both the gray scale and the thickness of the lines indicate the net pressure in (a) and fracture width in (b).

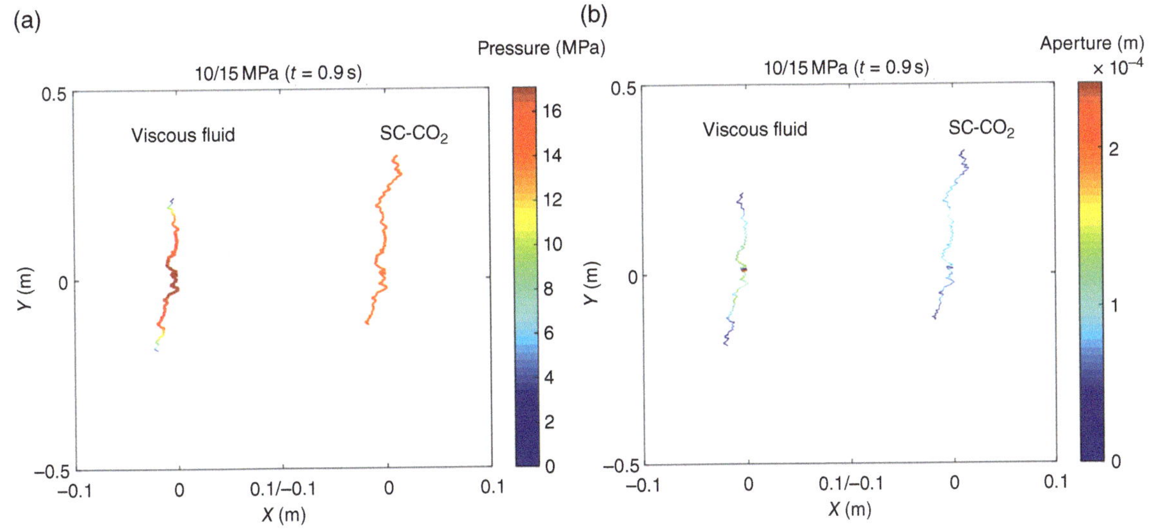

Figure 12.27 (a) The net pressure of hydraulic fractures in Case 1-S2 and Case 2-S2 at t = 0.9 s, and (b) the aperture of hydraulic fractures in Case 1-S2 and Case 2-S2 at t = 0.9 s. Both the color scale and the thickness of the lines indicate the net pressure in (a) and fracture width in (b).

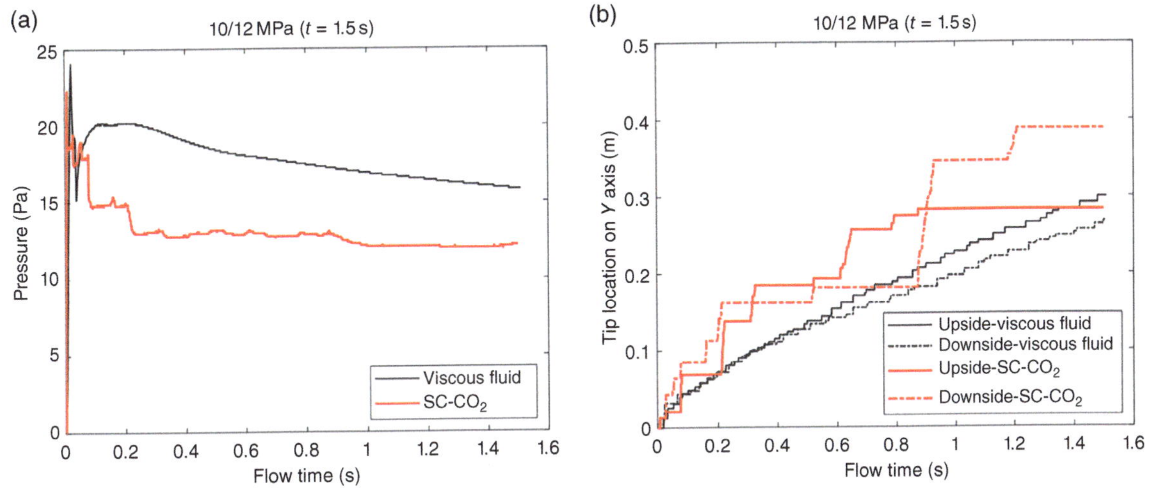

Figure 12.28 (a) The history of injection fluid pressure in Case 1-S1 and Case 2-S1 at t = 1.5 s, and (b) the evolution of the fractures in both sides of injection point in Case 1-S1 and Case 2-S1 at t = 1.5 s.

the toughness-dominated regime. The difference of fracture profiles can be explained by the distribution of fluid pressure in the fracture as shown in Figures 12.26a and 12.27a. Uniform fluid pressure is applied throughout the SC-CO_2 induced fracture surface, while large pressure gradient exists in the viscous fluid induced fracture and the closer it is to the tip, the greater the pressure gradient. The influence of fluid pressure distribution will be analyzed in the discussion part.

Evolution of injection fluid pressure and the tip location on Y-axis for the two types of fracturing fluid are compared in Figures 12.28 and 12.29. The fracture length growth with a series of jumps is observed in the cases using SC-CO_2, and the jumps cause the immediate drop of the injection fluid pressure in the initial state. Moreover, the fracture induced by SC-CO_2 is more asymmetric as shown in Figures 12.28b and 12.29b. The plateaus of the curve indicate that the fracture growth of one wing might be temporarily arrested due to the local inherent heterogeneity, which may shift the position of maximum aperture from the injection point (Figure 12.29b). By contrast, the hydraulic fracture driven by 100 cp fracturing fluid propagates more smoothly, continuously and symmetrically over the flow time. The results are in agreement with those observed in the hydraulic

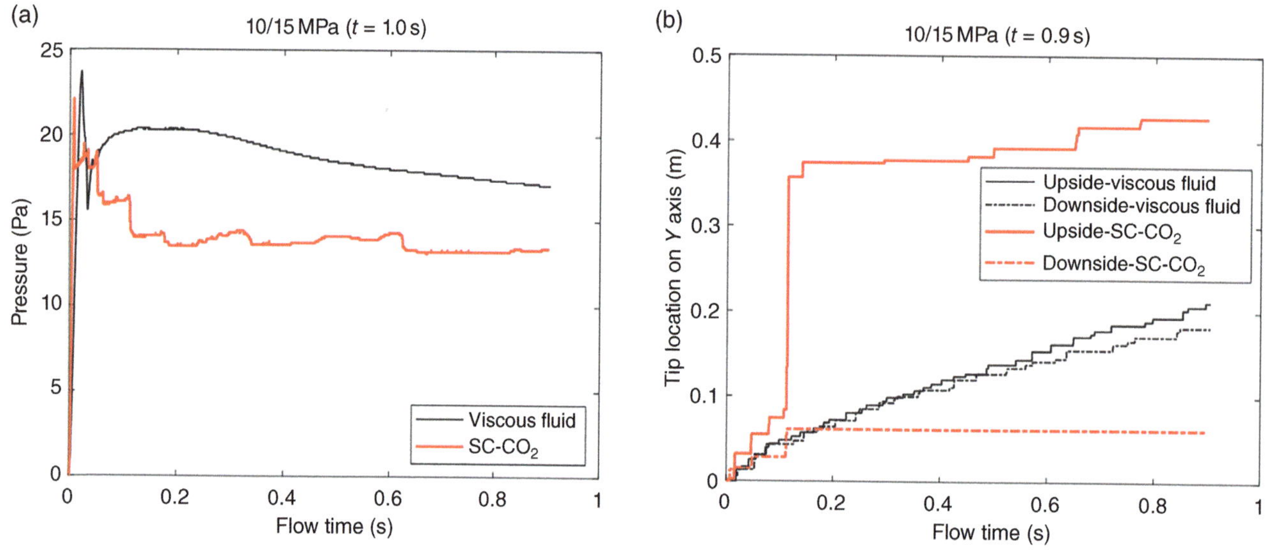

Figure 12.29 (a) The history of injection fluid pressure and the distance between the two tips in y direction in Case 1-S2 and Case 2-S2 at t = 0.9 s, and (b) the evolution of fractures in both sides of injection point in Case 1-S2 and Case 2-S2 at t = 0.9 s.

fracturing simulation in the toughness-dominated regime using traditional hydro-mechanical coupling scheme [37]. During the fracturing process, the ratio between the particle size and the accumulated fracture length gets smaller and smaller as the hydraulic fracture propagates, resulting in a stable injection fluid pressure with a decreasing fluctuation. Moreover, the propagation pressure for SC-CO_2 fracturing is much lower than that with high viscosity fluid, and the former is slightly larger than the minimum principal stress (10 MPa). It is also observed that SC-CO_2 fracturing can reach a lower breakdown pressure (22.3 MPa for Case 2-S1 and 22.1 MPa for Case 2-S2) than the viscous fluid fracturing (24.1 MPa for Case 1-S1 and 23.7 MPa for Case 1-S2). However, the difference of breakdown pressure is less profound than that observed in previous experimental studies [60, 61]. The discrepancy may be attributed to the absence of circular borehole and fluid leak-off in this study, which will be analyzed in the discussion part.

The numerical results of Case 2-S1 and the approximate solutions using KGD model at $t = 1.5$ s are shown in Figure 12.30. The fracture aperture at the injection point

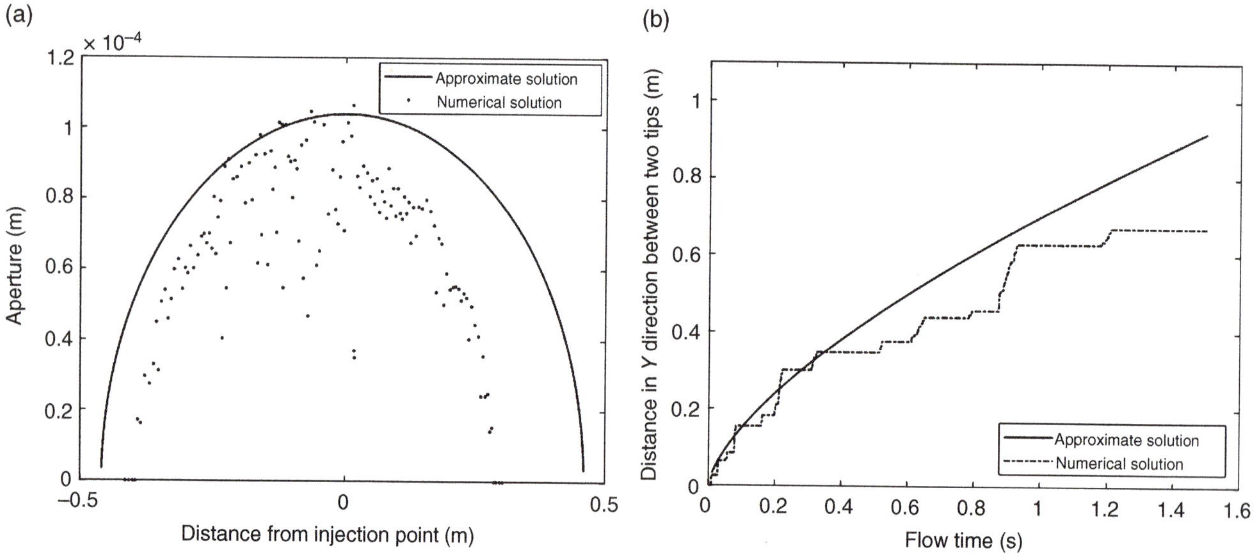

Figure 12.30 Simulation results of hydraulic fracturing for random models, (a) comparison between the approximate solution and the numerical solution for hydraulic fracture aperture, and (b) comparison between the approximate solution and the numerical solution for the distance in y direction between two fracture tips.

and the distance between two fracture tips in y-direction using KGD model were $w(0, 1.5) = 10.40\text{e}{-5}$ m and $2l(1.5) = 0.919$ m, while those calculated by the numerical model are 10.64e−5 m and 0.683 m, respectively. The fracture aperture at the injection point is overestimated by 2.3%, while the distance between the two tips in y-direction is underestimated by 25.7%. It is obvious that there is a relatively large discrepancy, though the input of mode-I fracture toughness (2.66 Pa m$^{1/2}$) is attempted to match the numerical solutions as well as possible. The apparent toughness K_{IC} scales with the tensile strength of the contact bond σ_t' and the particle radius R as shown in [20, 62]: $K_{IC} \propto \sigma_t'\sqrt{R}$. However, the mode-I fracture toughness in Case 2-S1 is larger than that in the regular model (1.88 Pa m$^{1/2}$) regardless of the negative effect of decreasing the particle size. The possible explanation is that the fracture propagation in the toughness-dominated is more sensitive to the local inherent heterogeneity of the sample, resulting in less smooth fracture growth, more asymmetric and tortuous fractures as aforementioned. In real rock mass, the contact between particles might not be aligned with the direction of the maximum or minimum principal stresses, and the maximum horizontal stress may make some contribution to the contact force along the path of the hydraulic fracture. It is harder to fracture the random solid medium and the hydraulic fracture propagation path is more tortuous. There is an increase of the effective toughness and a decrease of direct fracture length for the random model [37]. If the tortuosity is taken into consideration, the accumulated fracture total length is 1.099 m in Case 2-S1. Thus, the fracture in the numerical solution is longer, thinner and more tortuous, than the one predicted by the KGD theoretical solution.

12.4.4 Hydraulic Fracturing in Fractured Rock Sample

It is widely accepted that rock mass contains preexisting natural fractures and the propagation of hydraulic fracture could be greatly affected by those natural fractures or geological discontinuities [63, 64]. As a fluid driven fracture approaches the preexisting fractures, the stress state around the fracture tip is disturbed, and it may cause complex fracture interaction behavior [38, 44, 65]. In this study, four cases of hydraulic fracturing simulations were conducted to explore the fracture propagation behavior in rocks with preexisting natural fractures.

Figure 12.31 shows the evolution of injection fluid pressure using different fracturing fluids in the fractured rock sample at two horizontal stress differences. It is observed that the breakdown pressure and propagation pressure in SC-CO_2 fracturing are much lower compared to those during viscous fluid fracturing. This finding is similar to that observed in intact rock samples. However, the preexisting natural fractures cause the increase of breakdown pressure in fractured rock sample compared to that in intact rock sample. Note that the effect of natural fractures near wellbore on the breakdown pressure may be smaller in the real case because of the alternative path in the third dimension. Figures 12.32 and 12.33 display the hydraulic fracture profiles when the injection volumes of the two fracturing fluid are the same. Fracture propagation is deviated from the direction of the maximum confining stress through the

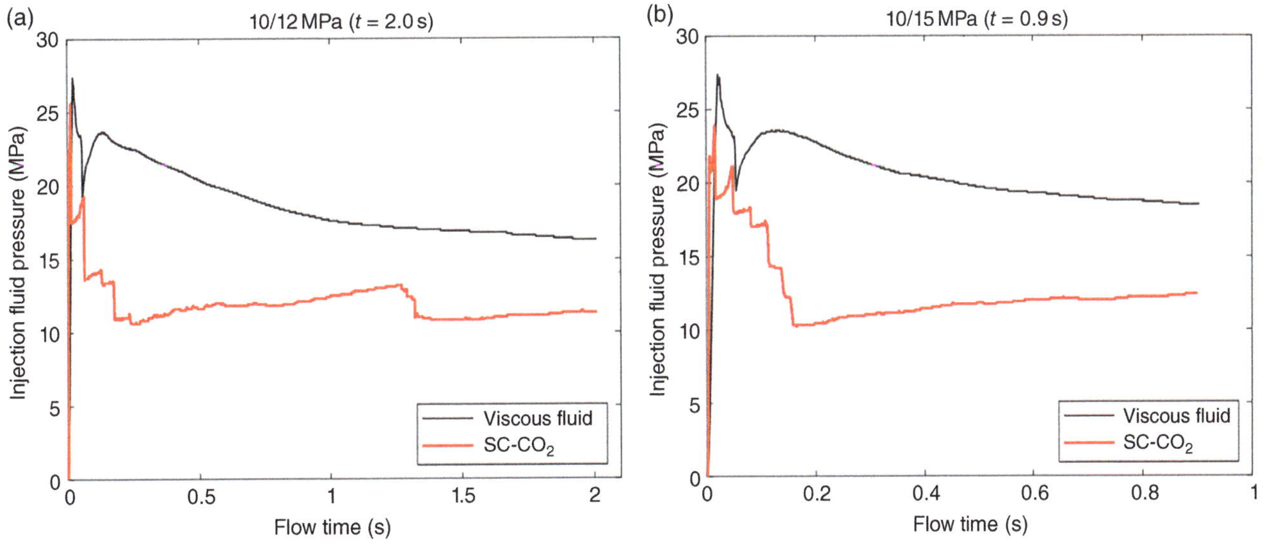

Figure 12.31 The history of injection fluid pressure of (a) Case N1-S1 and Case N2-S1, and (b) Case N1-S2 and Case N2-S2.

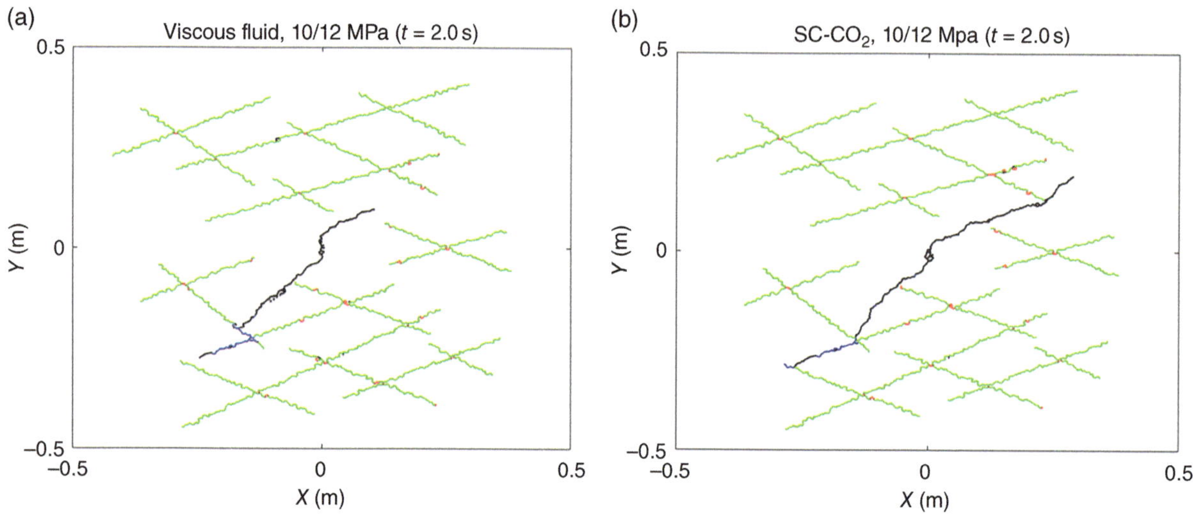

Figure 12.32 Fracture profiles at t = 2.0 s: (a) Case N1-S1, and (b) Case N2-S1.

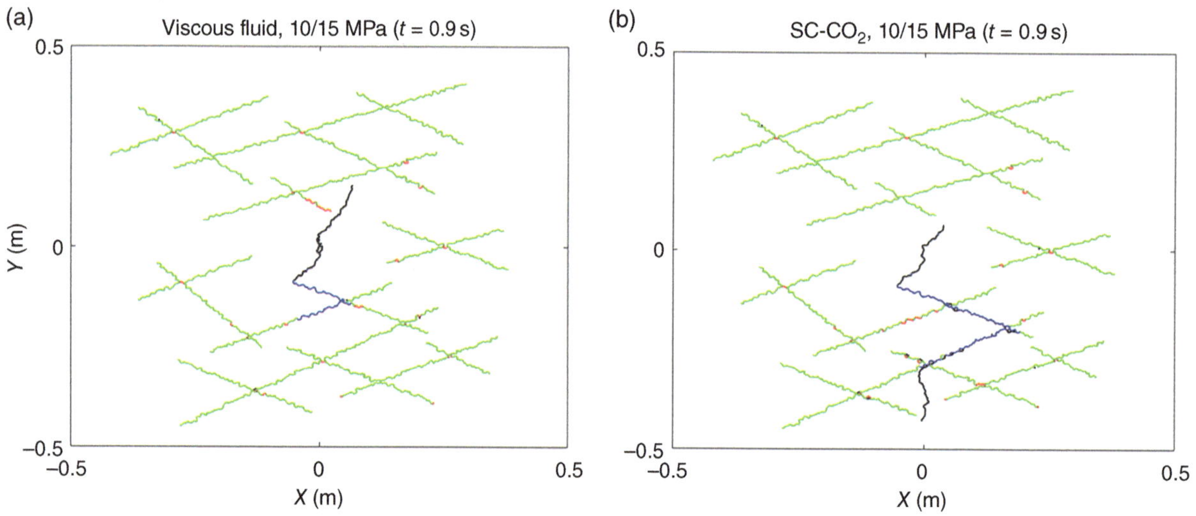

Figure 12.33 Fracture profiles at t = 0.9 s: (a) Case N1-S2, and (b) Case N2-S2.

interaction with natural fractures when the stress difference is low (see Figure 12.32). When the stress difference is high, the hydraulic fracture induced by viscous fluid tends toward the preexisting fractures and once it has reached them it propagates along the preexisting fractures, as shown in Figure 12.33a. Meanwhile, the asymmetric fractures driven by $SC\text{-}CO_2$ propagate along the preexisting natural fracture at first, and then branch off and turn to the direction of the maximum confining stress (see Figure 12.33b).

It is obvious that $SC\text{-}CO_2$ induces a much longer fracture path. The number of accumulated micro-cracks is obtained to evaluate the ability of stimulating natural fractures as listed in Table 12.5. Compared with viscous fracturing fluid, more micro-cracks corresponding to the broken parallel bond are generated using $SC\text{-}CO_2$ at a low and high stress difference. When the stress difference is relatively high, $SC\text{-}CO_2$ fracturing generates 168 micro-cracks in the preexisting natural fractures, and 117 of them are connected to the hydraulic fracture. However, in the case with viscous fluid, only 109 micro-cracks are generated and 57 of them are connected to the hydraulic fracture. The results indicate that $SC\text{-}CO_2$ fracturing tends to achieve a more complex and productive fracture network than fracturing with high viscosity fluid. The result is in agreement with the findings in the laboratory experiments [36, 66].

Table 12.5 Accumulated micro-crack number in the fractured rock samples.

Micro-cracks number	10/12 MPa (t = 2.0 s)		10/15 MPa (t = 0.9 s)	
	Case N1-S1	Case N2-S1	Case N1-S2	Case N2-S2
Broken parallel bonds	104	154	62	87
Broken smooth joints connected to main fracture	41	32	57	117
Broken smooth joints not connected to main fracture	51	55	52	51
Broken smooth joints	92	87	109	168
Sum	196	241	171	255

12.5 DEM Modeling of Fluid Injection into Dense Granular Media

In this work, the coupled displacement process of fluid injection into a dense granular medium is investigated numerically using DEM-PNM method. How a dense granular medium behaves in response to fluid injection is a subject of fundamental and applied research interests to better understand subsurface processes such as fluid or gas migration and formation of intrusive features as well as engineering applications such as hydraulic fracturing and geological storage in unconsolidated formations. The numerical analysis is performed with DEM executing the mechanical calculation and the network model solving the Hagen–Poiseuille equation between the pore spaces enclosed by chains of particles and contacts. Hydromechanical coupling is realized by data exchanging at predetermined time steps. The numerical results show that increase in the injection rate and the invading fluid viscosity and decrease in the modulus and permeability of the medium result in fluid flow behaviors displaying a transition from infiltration-governed to infiltration-limited and the granular medium responses evolving from that of a rigid porous medium to localized failure leading to the development of preferential paths. The transition in the fluid flow and granular medium behaviors is governed by the ratio between the characteristic times associated with fluid injection and hydromechanical coupling. The peak pressures at large injection rates when fluid leak-off is limited compare well with those from the injection experiments in triaxial cells in the literature. The numerical analysis also reveals intriguing tip kinematics field for the growth of a fluid channel, which may shed light on the occurrence of the apical inverted-conical features in sandstone and magma intrusion in unconsolidated formations.

12.5.1 Background and Experimental Motivation

Complex patterns arise when a granular medium is invaded by a fluid. Extensive literature [67–70] exists for the two limiting uncoupled cases when the granular medium either behaves as a rigid porous medium or is submerged in a fluid to behave as a dilute suspension. When the grain displacements become coupled with fluid flow, "granular fingering," an instability of Saffman–Taylor type, has been observed in experiments of air injection into granular media in Hele-Shaw cells [71–74]. The coupled fluid-grain displacement process is not only of fundamental scientific interest but also relevant to engineering applications such as ground improvement, environmental remediation, and reservoir stimulation.

In the experimental study, we study the coupled fluid-grain displacement process when a dry dense granular medium is invaded by an aqueous glycerin solution in a radial Hele–Shaw cell. We demonstrate that there exists a transition from the solid like to the fluidlike behaviors in the response of the dense granular medium. By varying the invading fluid viscosity via the weight concentration of glycerin, and by adjusting the normalized injection velocity via the injection rate and the gap size of the cell, we observe four distinct fluid-grain displacement regimes: (i) a simple radial flow regime, (ii) an infiltration-dominated regime, (iii) a grain displacement-dominated regime, and (iv) a viscous fingering-dominated regime.

The granular material used here is dry Ottawa F110 sand. The sand grain size is mostly in the range from 70 to 200 μm with the mean particle diameter by mass d_p = 110 μm. The radial Hele–Shaw cell is made from two 305 × 305 × 25.4 mm. Plexiglass plates bolted together at the four corners, with the gap size, b, controlled by the metal spacers in between the two plates. Sand is rained into the Hele–Shaw cell when held in an upright position. The granular medium is compacted first by manual vibration during the filling stage and then on a vibration table after the cell is filled. The grain mass fraction obtained through such a sample preparation procedure is about 0.63–0.65. The outer boundary of the Hele–Shaw cell is sealed in all the tests to prevent decompaction near the boundary prior to and during the tests. A summary of the test parameters is given in Table 12.6 for a series of 12 representative tests from a total of 58 tests. Tests A1–C4 are performed at four normalized injection velocities $v = Q/(\pi b D_i)$, where Q is the injection rate and D_i = 1.016 mm is the diameter of the injection inlet, while the remaining 46 tests are conducted at injection velocities selected from the range set by tests A1 and C4.

The images from tests A1-C4 before the fluid fronts reach the cell boundaries are summarized in Figure 12.34.

The light-colored areas are occupied by dry sands only. The dark areas indicate the fluid-only areas, and the medium gray areas are the fluid infiltrated areas. While fluid permeates through the granular medium with a nearly circular front and no fluid channels are created in test A1, in all other tests, fluid flow results in significant grain displacements, which in turn create channels occupied by fluid only. We may interpret the term "granular finger" in a broad sense and use it here to describe the fluid channels. As the invading fluid viscosity and the injection velocity increases from test A1 to C4, we can identify a transition in the fluid flow behaviors from infiltration dominated to infiltration limited. Meanwhile, the response of the granular medium also changes from that of a rigid porous medium in test A1 to fluidlike in test C4 where the ramified finger morphology resembles the viscous fluid–fluid displacement pattern. In tests A2, A3, and B1, the permeation fronts remain nearly circular and are affected very little by the development of the granular fingers. In tests A4, B2, B3, C1, and C2, the permeation fronts reflect the propagation of the granular fingers, but not yet the development of

Table 12.6 The invading fluid viscosity η and the normalized injection velocity v in tests A1-C4.

Test parameters	η = 5 cp	176 cp	942 cp
v = 8.22 mm/s	A1	B1	C1
82.90 mm/s	A2	B2	C2
165.80 mm/s	A3	B3	C3
829.52 mm/s	A4	B4	C4

The injection velocity v is adjusted through the injection rate Q and the gap size b of the Hele-Shaw cell. Tests A1–C1: Q = 5 ml/min, b = 3.175 mm; tests A2–C2: Q = 25 ml/min, b = 1.575 mm; tests A3–C3: Q = 50 ml/min, b = 1.575 mm; and tests A4–C4: Q = 125 ml/min, b = 0.787 mm.

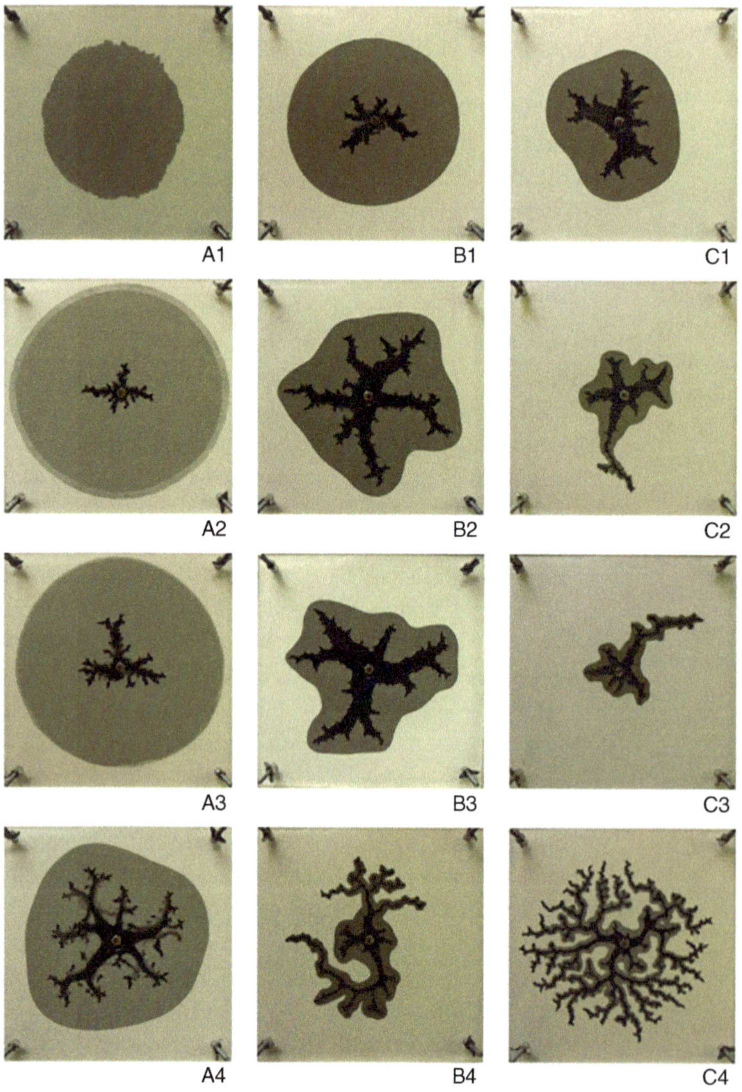

Figure 12.34 Displacement patterns from the injection experiments with dry Ottawa F110 sand and aqueous glycerin solutions (*Source:* Huang et al. [75], American Physical Society).

individual granular fingers. The fingers grow much wider than those intestsA2, A3, and B1. In test C4, where the fluid viscosity and the injection velocity are the largest, the permeation front now closely follows individual granular fingers.

Two dimensionless times are defined to classify the four displacement regimes (Huang et al. [75], American Physical Society) i.e.

$$\tau_1 = \eta v l / E k \tag{12.20}$$

$$\tau_2 = \eta' v / l E \tag{12.21}$$

where l is a characteristic length of the injection process, E and k are the small strain Young's modulus and the intrinsic permeability, and η' is the apparent viscosity of the granular mixture, assuming the medium can be treated as a nonlinear viscoelastoplastic solid. The two dimensionless times are obtained from the ratios of characteristic times associated with the injection process, hydromechanical coupling and the retardation time of the granular medium. While the dimensionless time τ_1 determines whether the granular medium response is infiltration-governed or deformation-governed, the time τ_2 characterizes the viscoelastic response of the granular medium. As τ_2 increases, the material behaviors evolve from solid-like to becoming fluid-like. For a dense granular medium, it is expected that $\tau_1 < \tau_2$. A t large τ_2, it is the fluid-like response of the granular medium that results in the viscous fingering type of morphology in Figure 12.34, C4.

Take the injection inlet diameter D_i as the characteristic process length l and the small strain modulus $E = 66.2$ MPa and permeability $k = 749$ mD as the index properties for the dense packing of dry Ottawa F110 sand. The ratio of the area of the fingers (the dark area in Figure 12.34) over the area enclosed by the infiltration front (the medium gray and the dark areas in Figure 12.34) at late time of injection is used as an index to indirectly measure the injection efficiency in creating open space. Variation of this area ratio λ as a function of the dimensionless time τ_1 is shown in Figure 12.35 from a total of 58 tests. The area ratio λ is close to zero when $\tau_1 \leq 6 \times 10^{-3}$ and increases nearly linearly with τ_1 when $\tau_1 \leq 0.1$. When $0.1 \leq \tau_1 \leq 3$, the increase in λ becomes nonlinear. When $\tau_1 \geq 3$, the area ratio reaches a plateau around $\lambda \approx 0.45$. For this granular medium system, $\tau_1 \approx 6 \times 10^{-3}$ may be used to define the transition from simple radial flow to the infiltration dominated regime; $\tau_1 \approx 0.1$ defines the transition from the infiltration-dominated regime to the grain displacement dominated regime; and $\tau_1 \approx 3$ (or $\tau_2 \geq 3k\eta'/\eta l^2$, if the apparent viscosity η' is known) corresponds to the transition from the grain displacement-dominated regime to the viscous fingering-dominated regime.

12.5.2 Model Setup

Fluid injection from a circular wellbore into an initially dry granular medium is modeled here. The capillary effect is assumed to be negligible. The injection simulations are performed using a hollow circular domain with an outer diameter $D_0 = 8$ mm and an inner diameter $D_i = 160$ mm. The particle configuration at the wellbore vicinity with the pore network added on and the overview of the numerical domain are shown in Figure 12.36. The assembly consists of 15 605 particles with radii ranging from 0.5 to 0.7 mm ($D_0/\overline{R} \cong 267$ and $D_i/\overline{R} = 13$). The default microscale parameters specified for the particles in the injection tests are: $K_n = K_s = 83.3$ MN/m^2, $\mu = 0.577$ and $\rho = 2650$ kg/m^3.

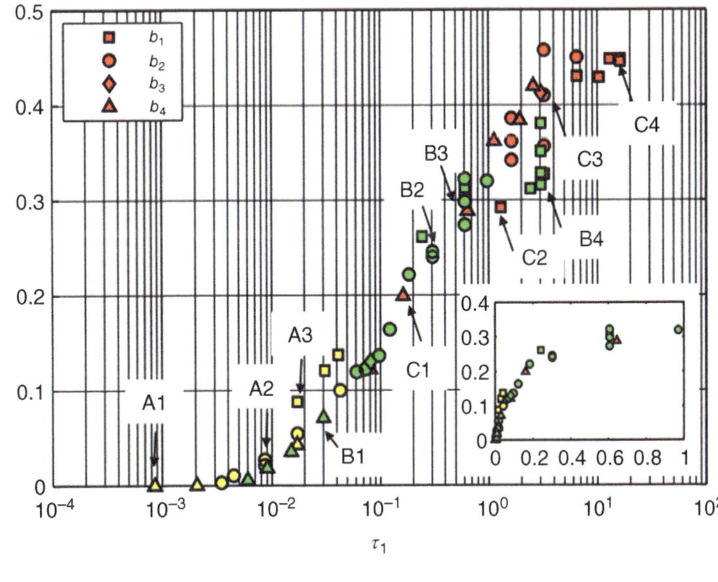

Figure 12.35 Variation of the area ratio λ as a function of the dimensionless time τ_1 from a total of 58 tests; the marker shape corresponds to the Hele–Shaw cell gap size, $b_1 = 0.787$ mm, $b_2 = 1.575$ mm, $b_3 = 2.362$ mm, and $b_4 = 3.175$ mm; the marker grayscale indicates the fluid type: hollow, corresponding to the 50% c.w. aqueous glycerin solution ($\eta = 5$ cp), light, 90% ($\eta = 176$ cp), and dark, 100% ($\eta = 942$ cp); see linear scale plot in the inset (*Source:* Huang et al. [75]/American Physical Society).

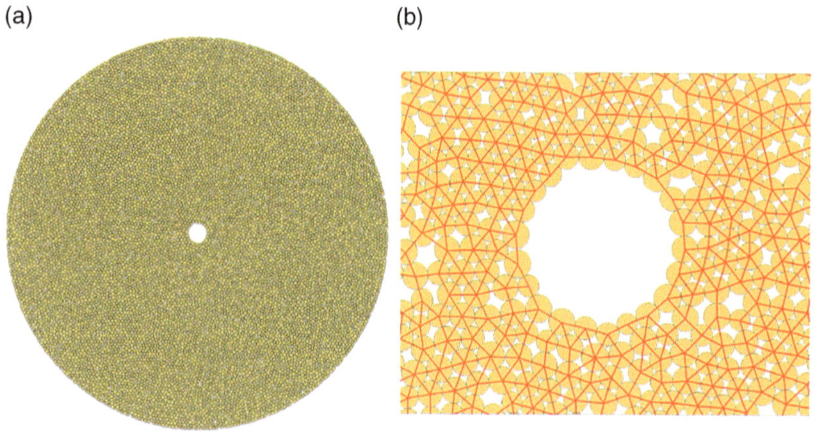

Figure 12.36 (a) Overview of the simulation domain and (b) wellbore vicinity with the pore network added on.

Two frictionless circular walls are placed at the inner and outer boundaries. The wall stiffnesses are set to be the same as those of the particles. The assembly is generated with an initial radial confining stress $\sigma_0 = 0.5$ MPa. At this level of confinement, the effective material properties are as follows: $E'_{50} = 34.13$ MPa , $\nu'_{50} = 0.222$, $\varphi'_{50} = 25.7°$, and $k = 0.849 \times 10^{-9}\text{m}^2$.

Pore pressures are applied as the boundary conditions in the fluid flow calculation. However, a constant injection rate condition is implemented by considering the wellbore as one pore space with the wellbore pressure change Δp_w for a given fluid time step determined from

$$\Delta p_w = K_f \left(\frac{Q \Delta t_f - \Delta V'_w}{V_w} \right) \tag{12.22}$$

where K_f is the fluid bulk modulus, V_w is the volume of the pore space that represents the wellbore, and $\Delta V'_w$ is the volume increment due to wellbore expansion. If any localized fluid channel develops during the simulation, the opening is merged with the wellbore pore space. However, constant pressure is assumed for the wellbore and any preferential path connected to the wellbore. In other words, no fluid flow is simulated along the preferential paths. The wellbore volume and the pressure are updated at every five mechanical steps during the mechanical calculation.

12.5.3 Effect of the Injection Rate

Effect of the injection rate on the grain displacement and fluid flow patterns is first examined. A series of eight tests are carried out by increasing the injection flow rate from $Q = 0.0–0.16$ m^2/s, see Table 12.7 for the test number and parameters for each test. The corresponding range of the normalized injection velocity is $v = \dfrac{Q}{\pi D_i} = 0.398 - 6.366\text{m/s}$. To maintain numerical stability, the fluid time step is decreased as the injection rate increases.

Table 12.7 Injection rate, fluid time step, and total simulation time for test series I to investigate the effect of the injection rate.

Test no.	Injection rate (m^2/s)	Fluid time step (s)	Simulation time (s)
I1	0.01	1.2×10^{-7}	0.048
I2	0.02	1×10^{-7}	0.04
I3	0.04	5×10^{-8}	0.024
I4	0.05	4×10^{-8}	0.0144
I5	0.08	2×10^{-8}	0.008
I6	0.1	2×10^{-8}	0.0056
I7	0.12	1.5×10^{-8}	0.0036
I8	0.16	1×10^{-8}	0.0012

Fluid viscosity η = Pa · s and far field confinement $\sigma_0 = 0.5$ MPa are prescribed for the tests in this series.

The displacement patterns at the end of the simulation times for the eight tests are shown in Figure 12.37. The black circle indicates the initial borehole position. The gray hollow circles represent particles associated with dry pore spaces and the filled circles represent those associated with saturated pore spaces. The filled color indicates the magnitude of the locally averaged pore pressure at the positions of the particles. From red to blue, the pore fluid pressure decreases. The interface between the blue disks and the gray hollow circles can therefore be interpreted as the infiltration front. The grain displacement and fluid infiltration patterns illustrated in Figure 12.37 are consistent with the experimental results in Figure 12.34. As the injection rate increases, fluid flow shows a transition from infiltration-governed to infiltration-limited behaviors while the granular medium response evolves from that of a fixed bed to localized failure leading to the growth of fluid channels.

At a relatively low injection rate $Q = 0.01$ m^2/s (Test I1), the particle displacements are mostly negligible, and

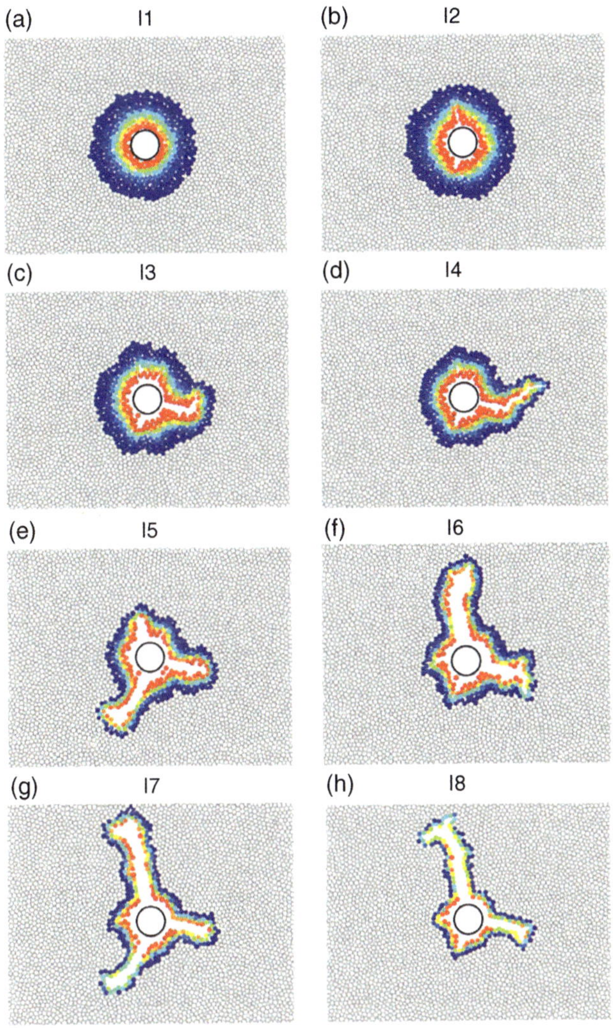

Figure 12.37 Displacement patterns from Test Series I at the end of the simulations: (a) Test I1: $Q = 0.01\ m^2/s$, (b) Test I2: $Q = 0.02\ m^2/s$ (c) Test I3: $Q = 0.04\ m^2/s$ (d) Test I4: $Q = 0.05\ m^2/s$, (e) Test I5: $Q = 0.08\ m^2/s$, (f) Test I6: $Q = 0.1\ m^2/s$, (g) Test I7: $Q = 0.12\ m^2/s$, (h) Test I8: $Q = 0.16\ m^2/s$. The black circle indicates the initial borehole position. The gray hollow circles represent particles associated with dry pore spaces and the filled circles represent those associated with saturated pore spaces. The filled color indicates the magnitude of the locally averaged pore pressure at the positions of particles. From red to blue, the pore fluid pressure decreases.

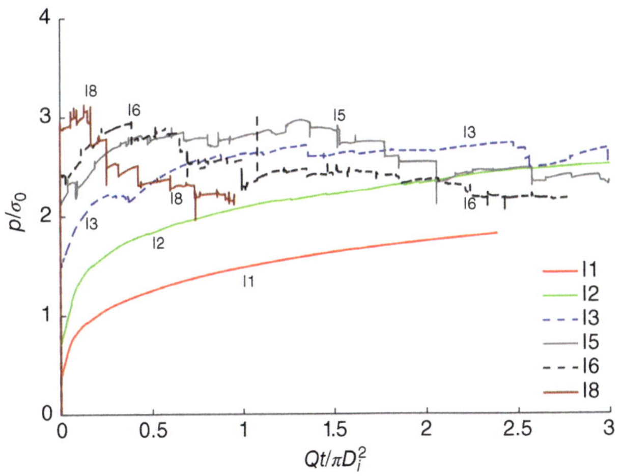

Figure 12.38 Wellbore pressure histories at injection rates $Q = 0.01, 0.02, 0.04, 0.08, 0.1$ and $0.16\ m^2/s$ from Test Series I.

the pressure profile is nearly axisymmetric indicating a radial flow pattern. Consequently, the borehole pressure increases continuously with time, see Figure 12.38. Axisymmetric fluid flow also results in the particle contact force chains to become radially aligned in the near wellbore area. This case is therefore similar to Test A1 in Figure 12.34a in the simple radial flow regime.

As the injection rate increases to $Q = 0.02\ m^2/s$ (Test I2), the infiltration pattern remains circular. Particle displacements at the early time of injection are negligible. However, as fluid injection continues, the wellbore starts to expand slightly with fluid channels of a few particle diameters in length developed near the wellbore, see Figure 12.34b. Nevertheless, the wellbore pressure increases continuously with almost no local drops. The continuously increasing trend in the wellbore pressure history concurs with the pore pressure distribution in Figure 12.34b that the fluid flow pattern remains radial. This test case indeed displays behaviors similar to Test A2 in Figure 12.34b in the infiltration-dominated regime.

As similarly done in the experimental analysis [75], an index λ_t, which indirectly measures the efficiency of fluid injection in creating open space, may be defined as the ratio between the area of the opening and the infiltrated area at a given time. The area of the opening now includes the area from wellbore expansion and the area of the fluid channels. Meanwhile, the area enclosed by the boundary of fully saturated pores minus the initial wellbore area may be considered as the infiltrated area. The area ratio λ_t is nearly zero at $Q = 0.01\ m^2/s$ and is less than 4% at $Q = 0.02\ m^2/s$. The small values of λ_t in these two cases reflect that the efficiency is rather low when the injection rate is small.

The shape of the infiltration front becomes slightly affected by the growth of the fluid channels when $Q = 0.04\ m^2/s$ (Test I3). The abrupt local drops and the subsequent gradual increases in the pressure curve are the signatures of sudden fluid channel growth events and the arrest periods when fluid flow results in limited grain displacements. Extension of the fluid channels is indeed in spurts. It is interesting to note, however, that in an average sense, the pressure curve as well as the area ratio λ_t reach plateaus after $Qt/\pi D_i^2 > 1$ ($Qt/\pi D_i^2$ may be interpreted as a scaled injection time or the normalized injection volume), see Figures 12.38 and 12.39. After the initial increase periods, $p/\sigma_0 \cong 2.69$ and $\lambda_t \cong 10\%$.

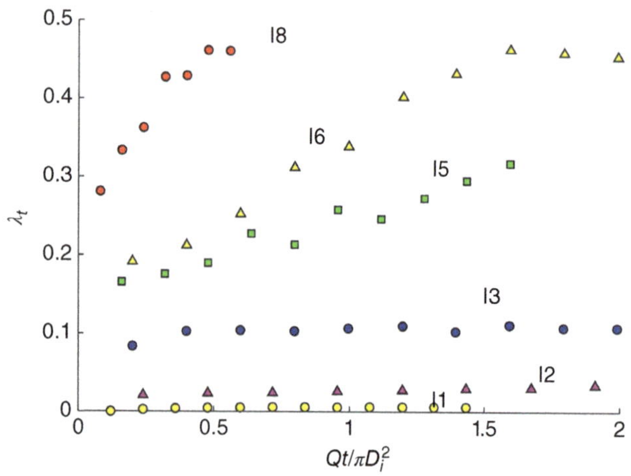

Figure 12.39 Variations of the area ratio λ_t with the scaled time $Qt/\pi D_i^2$ at injection rates Q = 0.01, 0.02, 0.04, 0.08, 0.1 and 0.16 m^2/s from Test Series I.

As the injection rate increases further, the infiltration fronts follow the channel profiles closely. The depth of infiltration decreases with the injection rate. At $Q = 0.12$ (Test I7) and 0.16 m^2/s (Test I8), the depth of infiltration is limited to 2–3 particle diameters. The wellbore pressure histories now exhibit a global trend of an initial increase followed by a decline after reaching a peak around p/σ_0~3. The pressure decline is fairly gradual at $Q = 0.08$ (Test I5) and 0.10 m^2/s (Test I6), but relatively steep at $Q = 0.16\ m^2/s$. The scaled time, corresponding to the peak pressure, decreases as the injection rate increases. In these cases, the area ratio λ_t increases with time. When $Q > 0.1\ m^2/s$, the area ratio λ_t seems to reach a plateau around 50% near the end of the simulations. Since the medium is assumed to be elastic friction (although damping is present in the numerical calculation for stability), these simulations are not expected to be in the viscous fingering-dominated regime.

12.5.4 Dimensionless Time Scaling

The experimental evidence in [75] suggests that the injection process in a dense granular medium is governed by the dimensionless time $\tau_1 = \eta v l/Ek$. If the wellbore diameter is chosen as the characteristic length, i.e. $l = D_i$, $\tau_1 = Q\eta/\pi Ek$. Here, we intend to verify the scaling of the material properties in τ_1 by varying the fluid viscosity, modulus, and permeability of the medium while keeping the dimensionless time τ_1 constant. The effective modulus of the medium can be adjusted proportionally at the particle scale through the particle stiffness K_n while keeping the stiffness ratio K_s/K_n and the ratio σ_0/K_n constant. The latter ensures that the particle configuration and, consequently, the porosity and permeability of the assembly remains unchanged. Meanwhile, permeability of the particle assembly can be changed through the mean particle diameter d_p and the porosity φ according to the Kozeny–Carman correlation [67].

$$k = \frac{1}{180}\frac{d_p^2\phi^3}{(1-\phi)^2} \tag{12.23}$$

Four additional series of tests, Test Series II–IV, are carried out at an injection rate $Q = 0.08\ m^2/s$. The variables in these tests are summarized in Table 12.7. Fluid viscosity is varied in Series II while the particle stiffness (or the elastic modulus) and the particle sizes (or the permeability) are varied in Series III and IV. The particle assembly in Series II and III is identical to the baseline configuration in Test Series I. In order to understand whether variation in the assembly configuration has a significant effect, another series (Series IIIb) of tests is conducted with the same microscale parameters as those in Series III, but with a different assembly configuration. In Series IV, the particle radii are uniformly distributed within the ranges of 0.4–0.6 mm (Test IV1) and 0.9–1.1 mm (Test IV2). Since $(R_{\max} - R_{\min})/\overline{R}$ (and consequently the porosity) is not kept constant in these cases, the effective permeability is affected by not only the change in the mean particle radius but also the porosity of the assembly. The numerically measured permeability values in Series IV are $k = 0.604 \times 10^{-9}\ m^2$ for $\overline{R} = 0.5$ mm and $k = 2.779 \times 10^{-9}\ m^2$ for $\overline{R} = 1.0$ mm. Change in the particle size and distribution also affects the effective modulus slightly.

The wellbore pressure histories from Series II are compared with those from Test I3 ($Q = 0.04\ m^2/s$) and Test I8 ($Q = 0.16\ m^2/s$) in Figure 12.40. The pressure histories

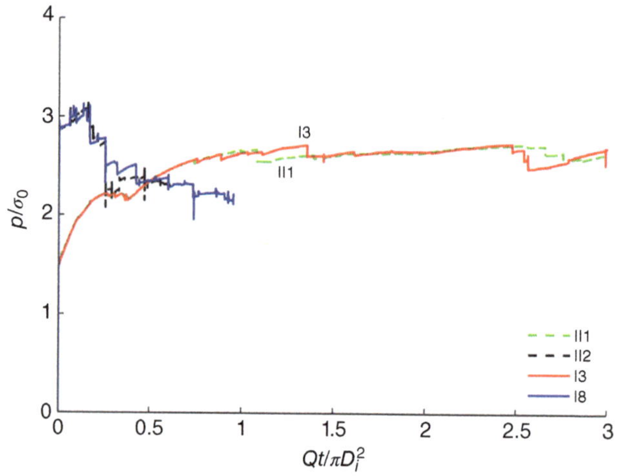

Figure 12.40 Comparison of the wellbore pressure histories from Series I and II; Test II1: $Q = 0.08\ m^2/s$, $\eta = 0.5$ Pa · s, Test II2: $Q = 0.08\ m^2/s$, $\eta = 2$ Pa · s, Test I3: $Q = 0.04\ m^2/s$, $\eta = 1$ Pa · s, and Test I8: $Q = 0.16\ m^2/s$, $\eta = 1$ Pa · s.

plotted as a function of the normalized time $Qt/\pi D_i^2$ are nearly identical for the cases with the same $Q\eta$. Excellent agreement is also obtained for the comparison among Series III, IIIb and Tests I3 and I8, for the same ratio of Q/E. Although the tests in Series I and IV do not yield identical Q/k (based on the permeability values in Series IV, the injection rates in Series I need to be $Q = 0.0244$ and $0.112\ m^2/s$ for the same Q/k), the pressure histories from Series IV are in reasonable agreement with Test I2 ($Q = 0.02\ m^2/s$) and Test I6 ($Q = 0.10\ m^2/s$).

The grain displacement and fluid flow patterns from Series II and III also display nearly identical morphology to those cases having the same $Q\eta$ or Q/E in Series I, see the comparison among Tests II1, III2, and III2b in Figure 12.41. As expected, with a different assembly configuration, the locations of fluid channels in Series IIIb differ from those in Series I–III. The displacement patterns from Series IV are shown in Figure 12.42. It is interesting to note that Test IV1 with a smaller mean particle radius in fact produces an opening similar to the bi-wing planar fracture.

Denote λ as the area ratio at a normalized reference time, i.e., $\lambda = \lambda_t(t_0)$. A reference time $t_0 = 2.4$ ms is chosen for Test I5 ($Q_0 = 0.08\ m^2/s$). For all the other cases, the simulation time to calculate the area ratio λ is determined from $t = t_0(Q_0\eta_0 Ek/Q\eta E_0 k_0)$. An interpolated value is used, if time t falls in between two output times. Variation of the area ratio λ as a function of the dimensionless time $\tau_1 = Q\eta/\pi Ek$ is shown in Figure 12.43. The agreement among the cases with the same τ_1 is indeed remarkable. Furthermore, although the numerical model is rather different from the experimental setup (e.g. the grain size, the inlet configuration, and the material properties), the numerically obtained threshold value t_d, which defines the transition from the infiltration-dominated to the grain displacement-dominated regime ($t_d = 0.44$ assuming Test I3 as the transition case), compares exceptionally well with the experimental value of $t_d = 0.1$ determined from Figure 12.35. In both the experiments and the numerical simulations, λ reaches around 50% at late time when $\tau_1 \sim O(1)$.

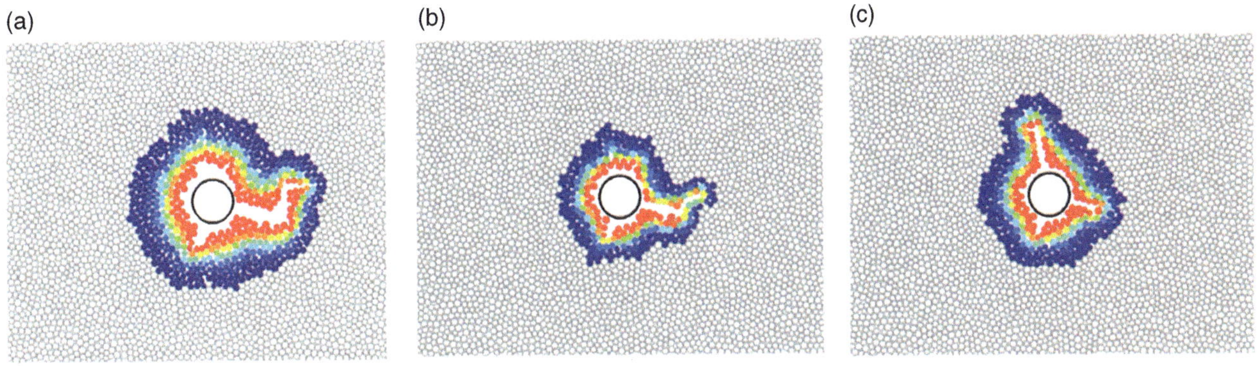

Figure 12.41 Displacement patterns from Tests (a) II1, (b) III2, and (c) III2b at $Qt/\pi D_i^2 = 4.77$. The parameters that differ from Test I3 are the following: (a) II1: $Q = 0.08\ m^2/s$, $\eta = 0.5$ Pa • s, (b) III2: $Q = 0.08\ m^2/s$, $E'_{50} = 68.37$ MPa, (c) III2b: $Q = 0.08\ m^2/s$, $E'_{50} = 68.37$ MPa and different particle configurations. The black circle indicates the initial borehole position. The gray hollow circles represent particles associated with dry pore spaces and the filled circles represent those associated with saturated pore spaces. The filled color indicates the magnitude of the locally averaged pore pressure at the positions of particles. From red to blue, the pore fluid pressure decreases.

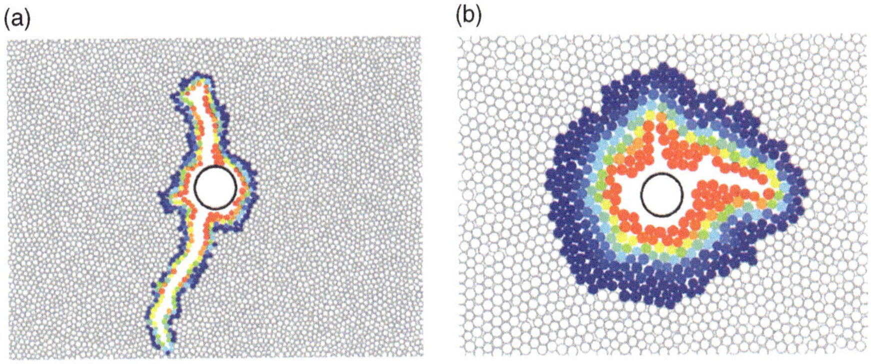

Figure 12.42 Displacement patterns from Series IV: (a) IV1: $R = 0.5$ mm at $t = 0.0032$ s and (b) IV2: $R = 1$ mm at $t = 0.025$ s. The black circle indicates the initial borehole position. The gray hollow circles represent particles associated with dry pore spaces and the filled circles represent those associated with saturated pore spaces. The filled color indicates the magnitude of the locally averaged pore pressure at the positions of particles. From red to blue, the pore fluid pressure decreases.

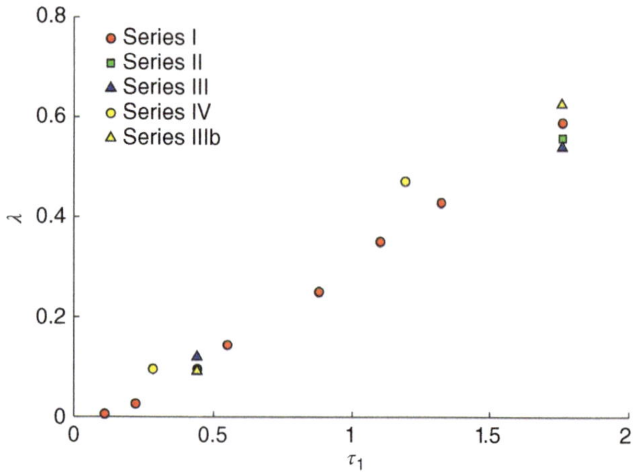

Figure 12.43 The area ratio λ at a normalized reference time as a function of the dimensionless time τ_1.

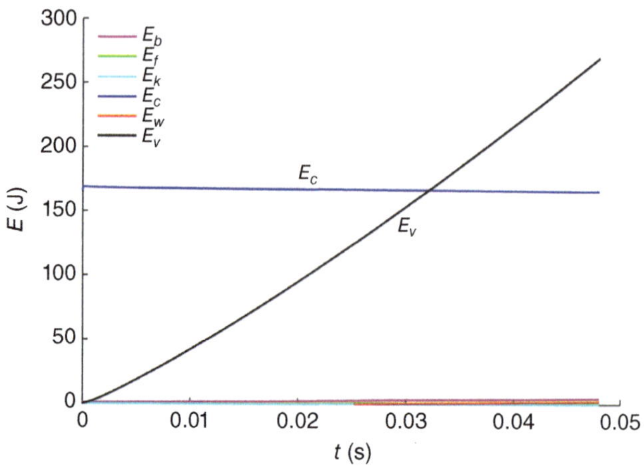

Figure 12.44 Histories of the energy components in Test I1 ($Q = 0.01\ \mathrm{m^2/s}$).

12.5.5 Energy Partition

In this numerical model, the energy components in the particle assembly include the following: (i) the body force work E_b, which in this case comes from the work done by the resultant drag forces exerted onto the particles, (ii) the work done by the boundary E_w, (iii) frictional work E_f as a result of intergranular slip, (iv) kinetic energy of the particles E_k, including the translational and rotational motion, and (v) strain energy E_c, which can be determined from the energy stored at all the contacts. Finally for fluid flow along the flow paths at the contacts, viscous dissipation E_v can be determined from

$$\Delta E_v = \sum_i q_p^{(i)} \Delta t_f \left[p_2^{(i)} - p_1^{(i)} \right] \tag{12.24}$$

Since constant pressure is assumed for the fluid channels connected to the wellbore, there is no viscous dissipation from fluid flow along the channels.

Evolution of the energy components from Tests I1, I5, and I8 are presented in Figures 12.44–12.46. In all these tests, since the numerical model is subjected to a constant far field confining stress prior to fluid injection, the total strain energy is nonzero initially and remains more or less constant during the simulations. In the case of $Q = 0.01\ \mathrm{m^2/s}$, the input energy is mostly dissipated when fluid flows through the flow paths at the contacts. Since fluid flow results in negligible particle displacements, increments in all other forms of energy are nearly zero. When the injection rate increases to $Q = 0.08\ \mathrm{m^2/s}$, although viscous dissipation E_v is still the largest term, the work done by the drag forces E_b and by the boundary E_w as well as the frictional work E_f gradually increase with time. The work E_b and E_f are the two components of energy dissipation

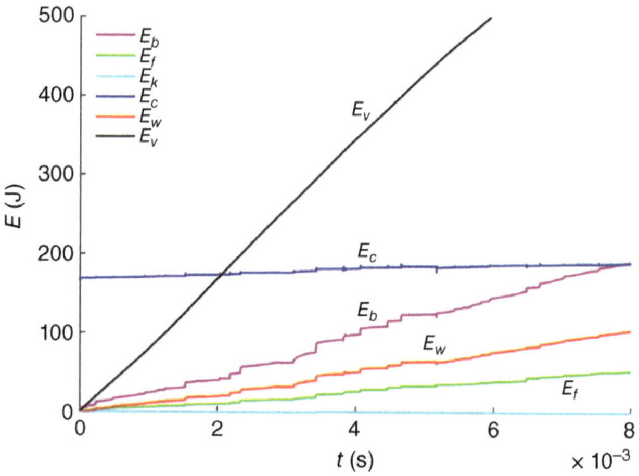

Figure 12.45 Histories of the energy components in Test I5 ($Q = 0.08\ \mathrm{m^2/s}$).

associated with grain displacements. The fact that the frictional work is relatively small indicates that intergranular slip is likely localized (most logically near the tip of the opening). At $Q = 0.16\ \mathrm{m^2/s}$, the work E_b done by the drag forces is comparable with the viscous dissipation E_v, and the sum $E_b + E_f$ is slightly larger than E_v, suggesting that the dissipation associated with grain displacements is becoming the main mechanism of energy dissipation. The intermittent propagation of the fluid channels during fluid injection is also reflected by the step increase in the energy components in Figure 12.46.

The energy analysis presented here supports our argument that the fluid-displacement regimes in the injection process essentially emerge as a result of the competition among various forms of energy dissipation mechanisms. The fluid-grain displacement regime is infiltration-governed

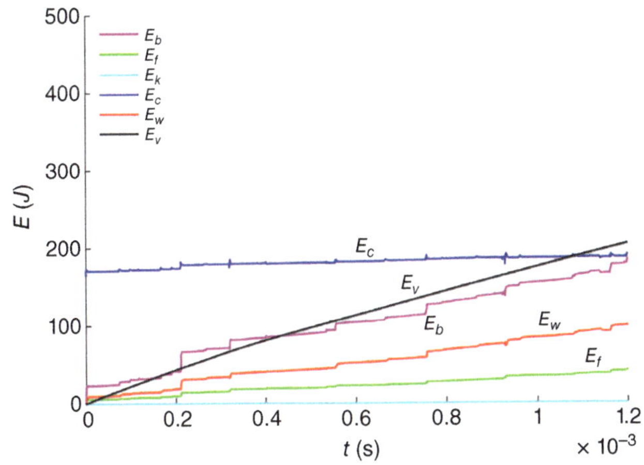

Figure 12.46 Histories of the energy components in Test I8 (Q = 0.16 m^2/s).

when the input energy is mainly dissipated through flow in the pore spaces. In the grain displacement-dominated regime, the energy dissipated due to grain displacements in the viscous fluid and intergranular slip is comparable to the viscous dissipation from flow in the pore spaces. Although the viscous dissipation from fluid flow along the channels is not counted here, it is expected that the percentage of this portion would increase as the injection rate increases.

12.6 Discussion

Hydraulic fracturing in naturally fractured rock mass and dense granular media involves strong hydro-mechanical coupling, and could result in complex fracture systems. In this work, we carried out fully coupled numerical analyses using DEM coupled with the pore network model to demonstrate some attempts to address a few key coupled processes during hydraulic fracturing. The first case study discusses the underlying mechanisms of hydraulic fracture interaction with natural fractures, which are critical to optimize the recovery of unconventional resources. The second case study develops a new numerical algorithm for the toughness-dominated regime to investigate the micromechanical behavior of SC-CO_2 fracturing, concluding that SC-CO_2 could be an alternative fracturing fluid to induce a more effective fracture network for hydrocarbon production. The third case study addresses the challenges in the coupled displacement process of fluid injection into an initially dry dense granular medium, which is most relevant to geological storage in a depleted reservoir. Through those case studies, we demonstrate that with an advanced numerical modeling tool, the complex fracturing associated with hydraulic fracturing in naturally fractured rock mass and dense granular media can be quantitatively analyzed and the extent of various uncertainties can be assessed.

It should be mentioned that there are some limitations in those case studies and future work is required. The first case study focuses on the case in which the initial (i.e. before crossing) hydraulic fracture propagates in the viscosity-dominated regime. The interaction results of hydraulic fracture with natural fractures for other regimes of propagation, such as the toughness-dominated regime, require further study. In the second case study, the improvement of the new algorithm for further investigation on the fluid compressibility and fluid infiltration of SC-CO_2 is required in the future work. Besides, the high injection rate and low viscosity of SC-CO_2 can lead to high Reynolds numbers and potential turbulence in the fracture [76], which could be also considered. The analysis in the third case study has limitations in assuming isotropic confining stresses and neglecting the capillary effect, which may become important as fluid migrates further away from the wellbore. In addition, the numerical results of the three case studies are subject to the restrictions of being only 2D. In order to address more realistic fracturing complexity, a fully coupled three-dimensional geomechanical model is required.

It is also worth mentioning that although numerical analysis is probably the most powerful tool to analyze complex problems such as hydraulic fracturing in naturally fracture rock masses, it is still an approximation of the real physical process. With different assumptions, even for the same engineering problem, different numerical schemes might not give the exact same results. Given the levels of complexity, the numerical analysis, hence, should not be taken as the absolute answer of those challenging problems discussed in this paper, but it provides important insights of the controlling behaviors.

12.7 Conclusions

In this work, the fully coupled numerical analyses using DEM coupled with the pore network model is performed to investigate some key coupled processes of hydraulic fracturing in naturally fractured rock mass and dense granular media. We introduce a total of three case studies, including the interaction of hydraulic fracture and natural fractures, supercritical carbon dioxide fracturing, and the coupled displacement process of fluid injection into dense granular media. The numerical analysis shows that the methodology is capable of reproducing phenomena consistent with those observed in laboratory injection experiments and in engineering applications. The main conclusions of this work are as follows:

1) Crossing behavior is affected by a combination of multiple factors. Nevertheless, three fundamental crossing scenarios can be identified from the investigations of orthogonal crossing: (i) no crossing, that is, the hydraulic fracture is arrested by the natural fracture and forms a T-shape intersection, (ii) offset crossing, that is, the hydraulic fracture crosses the natural fracture with an offset, and (iii) direct crossing, that is, the hydraulic fracture directly crosses the natural fracture without diversion into the natural fracture. Each crossing scenario is associated with a distinct net pressure history.
2) The intersection angle between a hydraulic fracture and a natural fracture can significantly affect the crossing behavior (both the propagation pattern and injection pressure). Decreasing the intersection angle impedes direct crossing and favors the arrest of the hydraulic fracture.
3) Material inhomogeneity (strength contrast and stiffness contrast) can also greatly affect crossing behavior. Crossing is impeded for the low-to-high strength crossing and high-to-low stiffness crossing.
4) The level of fracturing complexity increases as the number and extent of the natural fractures increase. This suggests that complex hydraulic fracture propagation patterns will occur because of complicated crossing behavior during hydraulic fracturing in naturally fractured reservoir systems.
5) The hydraulic fracture propagation in intact rock driven by SC-CO_2 in the toughness-dominated regime tends to be less smooth and continuous, more asymmetric and tortuous than that driven by high viscosity fluid.
6) In the fractured rock mass, the fracture propagation could deviate from the direction of the maximum confining stress under the low stress difference. Compared with viscous fluid, SC-CO_2 tends to cause a more complex fracture network with a longer main fracture in the rock matrix and more activated micro-cracks in the preexisting natural fractures.
7) The increase of the injection rate causes the fluid flow behaviors to change from infiltration-governed to infiltration-limited and the granular medium response to evolve from that of a rigid porous medium to localized failure leading to the development of preferential paths. The transition in the fluid flow and granular medium behaviors is governed by the ratio between the characteristic times associated with fluid injection and hydromechanical coupling, which is in reasonable agreement with that from injection experiments in a Hele–Shaw cell [75].

References

1 Economides, M., Nolte, K., and Ahmed, U. (2000). *Reservoir Stimulation – References*, (3rd ed.).e. Wiley http://do.rulitru.ru/docs/4/3081/conv_1/file1.pdf.
2 Barenblatt, G.I. (1962). The mathematical theory of equilibrium cracks in brittle fracture. *Advances in Applied Mechanics* 7 (C): 55–129. https://doi.org/10.1016/S0065-2156(08)70121-2.
3 Geertsma, J. and Haafkens, R. (1979). A Comparison of the theories for predicting width and extent of vertical hydraulically induced fractures. *Journal of Energy Resources Technology* 101 (1): 8–19. https://doi.org/10.1115/1.3446866.
4 Howard, G. and Fast, C.R. (1957). Optimum fluid characteristics for fracture extension. *Proceedings of the American Petroleum Institute* 261–270. https://doi.org/API-57-261.
5 Hubbert, M.K. and Willis, D.G. (1957). Mechanics of hydraulic fracturing. *Society of Petroleum Engineers* 210: 153–168.
6 Khristianovic, S.A. and Zheltov, Y.P. (1955). Formation of vertical fractures by means of highly viscous liquid. In: *4th World Petroleum Congress*, 579–586. Rome: World Petroleum Congress.
7 Nordgren, R.P. (1972). Propagation of a vertical hydraulic fracture. *Society of Petroleum Engineers Journal* https://doi.org/10.2118/3009-pa.
8 Perkins, T.K. and Kern, L.R. (1961). Widths of hydraulic fractures. *Journal of Petroleum Technology* 13 (9): https://doi.org/10.2118/89-PA.
9 Detournay, E. (2016). Mechanics of hydraulic fractures. *Annual Review of Fluid Mechanics* 48 (1): 311–339. https://doi.org/10.1146/annurev-fluid-010814-014736.
10 Abou-Sayed, A., Zaki, K., Wang, G. et al. (2004). Fracture propagation and formation disturbance during injection and frac-pack operations in soft compacting rocks. *Proceedings – SPE Annual Technical Conference and Exhibition*, pp. 3453–3464. Society of Petroleum Engineers (SPE). https://doi.org/10.2118/90656-ms

11 van Dam, D.B., Papanastasiou, P., and de Pater, C.J. (2002). Impact of rock plasticity on hydraulic fracture propagation and closure. *SPE Production and Facilities* 17: 149–159. SPE. https://doi.org/10.2118/78812-PA.

12 Papanastasiou, P. and Thiercelin, M. (1993). Influence of inelastic rock behaviour in hydraulic fracturing. *International Journal of Rock Mechanics and Mining Sciences and Geomechanics Abstracts* 30 (7): 1241–1247. https://doi.org/10.1016/0148-9062(93)90102-J.

13 Papanastasiou, P. and Atkinson, C. (2006). Representation of crack-tip plasticity in pressure sensitive geomaterials: large scale yielding. *International Journal of Fracture* 139 (1): 137–144. https://doi.org/10.1007/s10704-006-8371-8.

14 Wu, R. (2006). *Some Fundamental Mechanisms of Hydraulic Fracturing*. Georgia Institute of Technology.

15 Carter, E.D. (1957). Optimum fluid characteristics for fracture extension. In: *Drilling and Production Practices* (ed. G.C. Howard and C.R. Fast), 261–270. OnePetro.

16 Entov, V. M., Chekhonin, E. M., Detournay, E., and Thiercelin, M. (2007). Fracture propagation in high-permeability rocks: the key influence of fracture tip behavior. *SPE – Hydraulic Fracturing Technology Conference 2007* Vol. 2007, pp. 352–362. Society of Petroleum Engineers. https://doi.org/10.2118/106225-ms

17 Van den Hoek, P.J. (2002). A simple and accurate description of nonlinear fluid leakoff, in high-permeability fracturing. *SPE Journal* 7 (1): 14–23. https://doi.org/10.2118/77183-PA.

18 Mathias, S.A. and van Reeuwijk, M. (2009). Hydraulic fracture propagation with 3-D leak-off. *Transport in Porous Media* 80 (3): 499–518. https://doi.org/10.1007/s11242-009-9375-4.

19 Cundall, P.A. and Strack, O.D.L. (1979). A discrete numerical model for granular assemblies. *Géotechnique* 29 (1): 47–65. https://doi.org/10.1680/geot.1979.29.1.47.

20 Potyondy, D.O. and Cundall, P.A. (2004). A bonded-particle model for rock. *International Journal of Rock Mechanics and Mining Sciences* 41 (8): 1329–1364. https://doi.org/10.1016/j.ijrmms.2004.09.011.

21 Itasca Consulting Group (2014). PFC2D – Particle flow code in 2 dimensions, version 5.0. Minneapolis, MN.

22 Thallak, S., Rothenburg, L., and Dusseault, M. (1991). Simulation of multiple hydraulic fractures in a discrete element system. *32nd U.S. Symposium on Rock Mechanics*, USRMS 1991, pp. 271–280. Rotterdam: Balkema. https://doi.org/10.1016/0148-9062(92)93766-d

23 Bruno, M.S. (1994). Micromechanics of stress-induced permeability anisotropy and damage in sedimentary rock. *Mechanics of Materials* 18 (1): 31–48. https://doi.org/10.1016/0167-6636(94)90004-3.

24 Li, L. and Holt, R.M. (2001). Simulation of flow in sandstone with fluid coupled particle model. *DC Rocks 2001 – 38th U.S. Symposium on Rock Mechanics (USRMS)*. OnePetro.

25 Al-Busaidi, A., Hazzard, J.F., and Young, R.P. (2005). Distinct element modeling of hydraulically fractured Lac du Bonnet granite. *Journal of Geophysical Research: Solid Earth* 110 (6): 1–14. https://doi.org/10.1029/2004JB003297.

26 Shimizu, H., Murata, S., and Ishida, T. (2011). The distinct element analysis for hydraulic fracturing in hard rock considering fluid viscosity and particle size distribution. *International Journal of Rock Mechanics and Mining Sciences* 48 (5): 712–727. https://doi.org/10.1016/j.ijrmms.2011.04.013.

27 Zhao, X. and Paul Young, R. (2011). Numerical modeling of seismicity induced by fluid injection in naturally fractured reservoirs. *Geophysics* 76 (6): 167–180. https://doi.org/10.1190/geo2011-0025.1.

28 Jain, A.K. and Juanes, R. (2009). Preferential mode of gas invasion in sediments: grain-scale mechanistic model of coupled multiphase fluid flow and sediment mechanics. *Journal of Geophysical Research* 114 (B8): B08101. https://doi.org/10.1029/2008JB006002.

29 Boutt, D.F., Cook, B.K., McPherson, B.J.O.L., and Williams, J.R. (2007). Direct simulation of fluid-solid mechanics in porous media using the discrete element and lattice-Boltzmann methods. *Journal of Geophysical Research* 112 (B10): B10209. https://doi.org/10.1029/2004JB003213.

30 Cook, B. K., Noble, D. R., & Williams, J. R. (2004). A direct simulation method for particle-fluid systems. *Engineering Computations* (Swansea, Wales) (Vol. 21, pp. 151–168). Emerald Group Publishing Limited. https://doi.org/10.1108/02644400410519721

31 Han, Y. and Cundall, P.A. (2013). LBM-DEM modeling of fluid-solid interaction in porous media. *International Journal for Numerical and Analytical Methods in Geomechanics* 37 (10): 1391–1407. https://doi.org/10.1002/nag.2096.

32 Shimizu, Y. (2006). Three-dimensional simulation using fixed coarse-grid thermal-fluid scheme and conduction heat transfer scheme in distinct element method. *Powder Technology* 165 (3): 140–152. https://doi.org/10.1016/j.powtec.2006.04.003.

33 Tsuji, Y., Kawaguchi, T., and Tanaka, T. (1993). Discrete particle simulation of two-dimensional fluidized bed. *Powder Technology* 77 (1): 79–87. https://doi.org/10.1016/0032-5910(93)85010-7.

34 Fatt, I. (1956). The network model of porous media. *Transactions of the AIME* 207 (1): 144–181. https://doi.org/10.2118/574-g.

35 Zhang, F., Damjanac, B., and Huang, H. (2013). Coupled discrete element modeling of fluid injection into dense granular media. *Journal of Geophysical Research: Solid*

Earth 118 (6): 2703–2722. https://doi.org/10.1002/jgrb.50204.

36 Zhang, X., Lu, Y., Tang, J. et al. (2017). Experimental study on fracture initiation and propagation in shale using supercritical carbon dioxide fracturing. *Fuel* 190: 370–378. https://doi.org/10.1016/j.fuel.2016.10.120.

37 Huang, L., Liu, J., Zhang, F. et al. (2019). Exploring the influence of rock inherent heterogeneity and grain size on hydraulic fracturing using discrete element modeling. *International Journal of Solids and Structures* 176–177: 207–220. https://doi.org/10.1016/j.ijsolstr.2019.06.018.

38 Zhang, F., Dontsov, E., and Mack, M. (2017). Fully coupled simulation of a hydraulic fracture interacting with natural fractures with a hybrid discrete-continuum method. *International Journal for Numerical and Analytical Methods in Geomechanics* 41 (13): 1430–1452. https://doi.org/10.1002/nag.2682.

39 Itasca Consulting Group (2009). FLAC2D – Fast lagrangian analysis of continua in two dimensions. Minneapolis, MN.

40 Dontsov, E.V. (2016). An approximate solution for a penny-shaped hydraulic fracture that accounts for fracture toughness, fluid viscosity and leak-off. *Royal Society Open Science* 3 (12): 160737. https://doi.org/10.1098/rsos.160737.

41 Hu, J. and Garagash, D.I. (2010). Plane-strain propagation of a fluid-driven crack in a permeable rock with fracture toughness. *Journal of Engineering Mechanics* 136 (9): 1152–1166. https://doi.org/10.1061/(ASCE)EM.1943-7889.0000169.

42 Dontsov, E.V. and Peirce, A.P. (2015). A non-singular integral equation formulation to analyse multiscale behaviour in semi-infinite hydraulic fractures. *Journal of Fluid Mechanics.* https://doi.org/10.1017/jfm.2015.451.

43 Garagash, D.I., Detournay, E., and Adachi, J.I. (2011). Multiscale tip asymptotics in hydraulic fracture with leak-off. *Journal of Fluid Mechanics* https://doi.org/10.1017/S002211201000501X.

44 Renshaw, C.E. and Pollard, D.D. (1995). An experimentally verified criterion for propagation across unbounded frictional interfaces in brittle, linear elastic materials. *International Journal of Rock Mechanics and Mining Sciences & Geomechanics Abstracts* 32 (3): 237–249. https://doi.org/10.1016/0148-9062(94)00037-4.

45 Gu, H., Weng, X., Lund, J.B. et al. (2012). Hydraulic fracture crossing natural fracture at nonorthogonal angles: a criterion and its validation. *SPE Production & Operations* 27: https://doi.org/10.2118/139984-PA.

46 Warpinski, N.R. and Teufel, L.W. (1987). Influence of geologic discontinuities on hydraulic fracture propagation. *Journal of Petroleum Technology* 39 (2): 209–220. https://doi.org/10.2118/13224-PA.

47 Annevelink, M.P.J.A., Meesters, J.A.J., and Hendriks, A.J. (2016). Environmental contamination due to shale gas development. *Science of the Total Environment* 550 (August): 431–438. https://doi.org/10.1016/j.scitotenv.2016.01.131.

48 Middleton, R.S., Carey, J.W., Currier, R.P. et al. (2015). Shale gas and non-aqueous fracturing fluids: opportunities and challenges for supercritical CO_2. *Applied Energy* 147: 500–509. https://doi.org/10.1016/j.apenergy.2015.03.023.

49 Li, M., Zhang, F., Zhuang, L. et al. (2020). Micromechanical analysis of hydraulic fracturing in the toughness-dominated regime: implications to supercritical carbon dioxide fracturing. *Computational Geosciences* 24 (5): 1815–1831. https://doi.org/10.1007/s10596-019-09925-5.

50 Wang, D., Chen, M., Jin, Y., and Bunger, A.P. (2018). Impact of fluid compressibility for plane strain hydraulic fractures. *Computers and Geotechnics* 97(September 2017): 20–26. https://doi.org/10.1016/j.compgeo.2017.12.009.

51 Zhou, X. and Burbey, T.J. (2014). Fluid effect on hydraulic fracture propagation behavior: A comparison between water and supercritical CO_2-like fluid. *Geofluids* 14 (2): 174–188. https://doi.org/10.1111/gfl.12061.

52 Mas Ivars, D., Pierce, M.E., Darcel, C. et al. (2011). The synthetic rock mass approach for jointed rock mass modelling. *International Journal of Rock Mechanics and Mining Sciences* 48 (2): 219–244. https://doi.org/10.1016/j.ijrmms.2010.11.014.

53 Dontsov, E.V. (2017). An approximate solution for a plane strain hydraulic fracture that accounts for fracture toughness, fluid viscosity, and leak-off. *International Journal of Fracture* 205 (2): 221–237. https://doi.org/10.1007/s10704-017-0192-4.

54 Savitski, A.A. and Detournay, E. (2002). Propagation of a penny-shaped fluid-driven fracture in an impermeable rock: asymptotic solutions. *International Journal of Solids and Structures* 39 (26): 6311–6337. https://doi.org/10.1016/S0020-7683(02)00492-4.

55 Garagash, D.I. (2006). Plane-strain propagation of a fluid-driven fracture during injection and shut-in: asymptotics of large toughness. *Engineering Fracture Mechanics* 73 (4): 456–481. https://doi.org/10.1016/j.engfracmech.2005.07.012.

56 Adachi, J.I. and Detournay, E. (2008). Plane strain propagation of a hydraulic fracture in a permeable rock. *Engineering Fracture Mechanics* 75 (16): 4666–4694. https://doi.org/10.1016/j.engfracmech.2008.04.006.

57 Bunger, A.P., Detournay, E., and Garagash, D.I. (2005). Toughness-dominated hydraulic fracture with leak-off. *International Journal of Fracture* 134 (2): 175–190. https://doi.org/10.1007/s10704-005-0154-0.

58 Chen, Y., Nagaya, Y., and Ishida, T. (2015). Observations of fractures induced by hydraulic fracturing in anisotropic granite. *Rock Mechanics and Rock Engineering* 48 (4): 1455–1461. https://doi.org/10.1007/s00603-015-0727-9.

59 Bennour, Z., Watanabe, S., Chen, Y. et al. (2018). Evaluation of stimulated reservoir volume in laboratory hydraulic fracturing with oil, water and liquid carbon dioxide under microscopy using the fluorescence method. *Geomechanics and Geophysics for Geo-Energy and Geo-Resources* 4 (1): 39–50. https://doi.org/10.1007/s40948-017-0073-3.

60 Ishida, T., Chen, Y., Bennour, Z. et al. (2016). Features of CO_2 fracturing deduced from acoustic emission and microscopy in laboratory experiments. *Journal of Geophysical Research: Solid Earth* 121 (11): 8080–8098. https://doi.org/10.1002/2016JB013365.

61 Ishida, T., Aoyagi, K., Niwa, T. et al. (2012). Acoustic emission monitoring of hydraulic fracturing laboratory experiment with supercritical and liquid CO_2. *Geophysical Research Letters* 39 (16): L16309. https://doi.org/10.1029/2012GL052788.

62 Huang, H. (1999). Discrete element modeling of tool-rock interaction. ProQuest dissertations and theses. University of Minnesota, Minneapolis.

63 Hou, B., Chang, Z., Fu, W. et al. (2019). Fracture initiation and propagation in a deep shale gas reservoir subject to an alternating-fluid-injection hydraulic-fracturing treatment. *SPE Journal* 24 (4): 1839–1855. https://doi.org/10.2118/195571-PA.

64 Zhang, F., Damjanac, B., and Maxwell, S. (2019). Investigating hydraulic fracturing complexity in naturally fractured rock masses using fully coupled multiscale numerical modeling. *Rock Mechanics and Rock Engineering* 52 (12): 5137–5160. https://doi.org/10.1007/s00603-019-01851-3.

65 Tang, J., Wu, K., Zuo, L. et al. (2019). Investigation of rupture and slip mechanisms of hydraulic fracture in multiple-layered formations. *Society of Petroleum Engineers Journal* https://doi.org/10.2118/197054-PA.

66 Zou, Y., Li, N., Ma, X. et al. (2018). Experimental study on the growth behavior of supercritical CO_2-induced fractures in a layered tight sandstone formation. *Journal of Natural Gas Science and Engineering* 49(June 2017): 145–156. https://doi.org/10.1016/j.jngse.2017.11.005.

67 Bear, J. (1972). *Dynamics of Fluids in Porous Media*. New York: American Elsevier Publishing.

68 Homsy, G.M. (1987). Viscous fingering in porous media. *Annual Review of Fluid Mechanics* 19 (1): 271–311. https://doi.org/10.1146/annurev.fl.19.010187.001415.

69 Lenormand, R. (1989). Flow through porous media: limits of fractal patterns. *Proceedings of the Royal Society of London. A. Mathematical and Physical Sciences* 423 (1864): 159–168. https://doi.org/10.1098/rspa.1989.0048.

70 Saffman, P.G. and Taylor, G.I. (1958). The penetration of a fluid into a porous medium or Hele-Shaw cell containing a more viscous liquid. *Proceedings of the Royal Society of London. Series A. Mathematical and Physical Sciences* 245 (1242): 312–329. https://doi.org/10.1098/rspa.1958.0085.

71 Cheng, X., Xu, L., Patterson, A. et al. (2008). Towards the zero-surface-tension limit in granular fingering instability. *Nature Physics* 4 (3): 234–237. https://doi.org/10.1038/nphys834.

72 Johnsen, Ø., Toussaint, R., Måløy, K.J., and Flekkøy, E.G. (2006). Pattern formation during air injection into granular materials confined in a circular Hele-Shaw cell. *Physical Review E - Statistical, Nonlinear, and Soft Matter Physics* 74 (1): 011301. https://doi.org/10.1103/PhysRevE.74.011301.

73 Johnsen, Ø., Toussaint, R., Måløy, K.J. et al. (2008). Coupled air/granular flow in a linear Hele-Shaw cell. *Physical Review E - Statistical, Nonlinear, and Soft Matter Physics* 77 (1): 011301. https://doi.org/10.1103/PhysRevE.77.011301.

74 Sandnes, B., Flekkøy, E.G., Knudsen, H.A. et al. (2011). Patterns and flow in frictional fluid dynamics. *Nature Communications* 2 (1): 1–8. https://doi.org/10.1038/ncomms1289.

75 Huang, H., Zhang, F., Callahan, P., and Ayoub, J. (2012). Granular fingering in fluid injection into dense granular media in a Hele-Shaw cell. *Physical Review Letters* 108 (25): 1–4. https://doi.org/10.1103/PhysRevLett.108.258001.

76 Zia, H. and Lecampion, B. (2017). Propagation of a height contained hydraulic fracture in turbulent flow regimes. *International Journal of Solids and Structures* 110–111: 265–278. https://doi.org/10.1016/j.ijsolstr.2016.12.029.

13

Interaction of a Hydraulic Fracture with Natural Fractures of Lesser Height and Weak Bedding Interfaces as a Possible Mechanism for Fracture Swarms

Xiaowei Weng[1] *and Olga Kresse*[2]

[1] *retired, Houston, TX, USA*
[2] *Schlumberger, Houston, TX, USA*

13.1 Introduction

Over the last two decades, horizontal wells with multi-stage hydraulic fractures had become the standard choice of completion for unconventional shale reservoirs, which resulted in exponential increase in the hydrocarbon production in North America. Economic production from these reservoirs depends greatly on the effectiveness of hydraulic fracturing stimulation treatment. Microseismic monitoring of fracturing treatments routinely showed laterally diffusive events cloud, in contrast to the signature for planar fractures with more tightly grouped events along the fracture orientation in conventional tight sand formations, suggesting creation of complex fracture networks. This was especially evident in the early unconventional reservoir development, such as in Barnett Shale, where the spacings between stages and perforation clusters were large and/or when vertical wells were stimulated (Maxwell et al. 2002; Fisher et al. 2002; Warpinski et al. 2005). In some cases, drastically different microseismic patterns, e.g. both tightly packed and highly diffusive events clouds were observed in different stages in the same well (Daniels et al. 2007). The created fracture complexity is believed to be strongly influenced by the preexisting natural fractures and in situ stress anisotropy in the formation. Microseismic as well as production data also presented major challenges to the conventional planar fracture models in terms of being able to simulate the fracture dimensions that were consistent with the microseismic observations, as well as matching treating pressure response and production data (Cipolla et al. 2011).

To find better agreement with microseismic observations and address the impact of natural fractures on the induced hydraulic fracture networks geometry, complex fracture models had been developed (see, for example, Weng (2014) for a review of the complex fracture models). Complex fracture models explicitly account for the interaction of hydraulic fractures with the existing natural fractures that lead to potential hydraulic fracture arrest and branching at the natural fractures and consequently the creation of complex fractures. The model by Weng et al. (2011) was able to explain the different microseismic patterns in the same wells resulting from variation of preexisting natural fracture distribution based on the geophysical interpretation (Cipolla et al. 2011), or from different treatments using slick water vs. cross linked gel as shown in Figure 13.1 (Kresse et al. 2013).

The model was successfully applied to field-scale simulations, production matching, refrac and infill studies, and completion and well spacing optimization in several unconventional reservoirs (Cipolla and Wallace 2014; Ejofodomi et al. 2015; Malpani et al. 2015; Marongiu-Porcu et al. 2015; Wallace et al. 2016; Gakhar et al. 2017, Xiong et al. 2018, 2019). When simulating field-scale problems, only relatively large natural fractures (in the order of tens of feet and above) are typically considered. The shale formations may also contain abundant small natural fractures, most of which are filled with minerals. It would be challenging to account for these small fractures since it requires very high numerical grid resolution which renders the problem impractical to solve. Therefore, they are generally lumped into the rock "matrix."

As the completion technology evolved in the unconventional plays in North America over the last decade, there had been a trend towards a steady increase in the number of perforation clusters in each stage, and a decrease in perforation clusters spacing from 75 to 100 feet in the early vintage of completions to as low as 20 feet or less in recent years. The reduced spacing is aimed to maximize the number of hydraulic fractures created within a stage and hence the contact area with reservoir. With increasing fracture density, the differences between planar and complex fractures become less pronounced in terms of fracture surfaces

Mechanics of Hydraulic Fracturing: Experiment, Model, and Monitoring, First Edition.
Edited by Xi Zhang, Bisheng Wu, Diansen Yang, and Andrew Bunger.

(a)

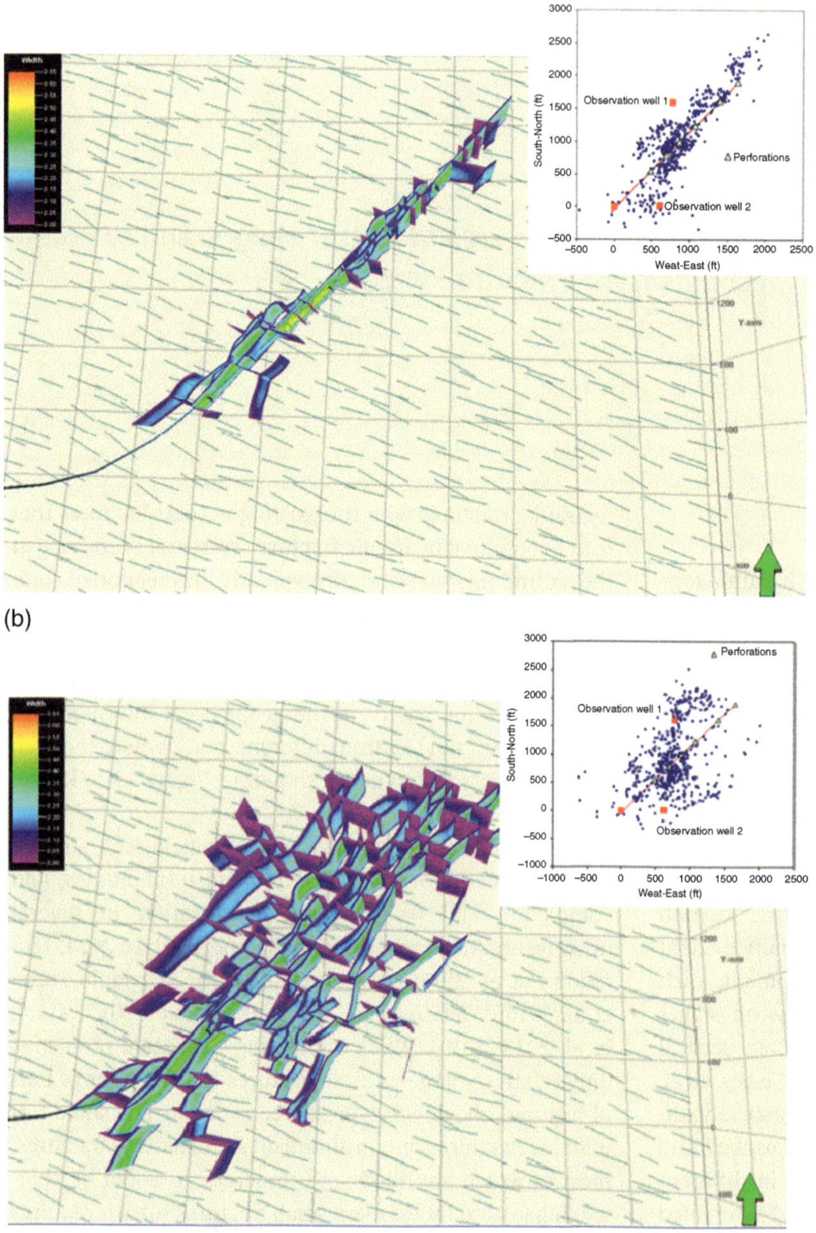

Figure 13.1 Hydraulic fracture network simulated using a complex fracture model by Kresse et al. (2013) for a Barnett shale case with microseismic monitoring (the insets) reported by Warpinski et al. (2005): (a) the initial stimulation with cross linked gel; (b) the slick water refrac of the same interval. The thin blue lines indicate the traces of the natural fractures on a horizontal plane. (*Source:* Adapted from Kresse et al. (2013) and Warpinski et al. (2005)).

generated, and microseismic monitoring does not have sufficient resolution to make the distinction. Nevertheless, the complex fracture models are still often applied since the conventional planar models tend to overpredict the fracture length without drastic increase of fluid leakoff, either artificially or requiring calibration against time consuming injection tests, especially for slick water treatments. One of the reasons for overprediction of fracture length by planar models is that they do not take into account fracture complexity, either tortuosity in the fracture path or multiple fracture strands, which lead to increased net pressure and fracture width. With an artificially increased leakoff, a planar model can obtain a shorter fracture length, but fails to properly account for the fluid storage inside the fractures, leading to other side effects such as difficulty in matching the pressure decline during shut-in or erroneous occurrence of tip-screen-out during the treatment that is not observed in the actual treating pressure response.

To obtain better understanding of fracture geometries and complexity in order to further optimize the recovery from unconventional plays, several large-scale field experiments were conducted in the industry in recent years. In field experiments in Eagle Ford (Raterman et al. 2017) and hydraulic fracturing test site (HFTS) in Wolfcamp formation of Midland Basin (Ciezobka et al. 2018), the slanted cores taken through the fractured zones revealed many closely spaced hydraulically induced fractures, referred to as fracture swarms. These fractures are subparallel and

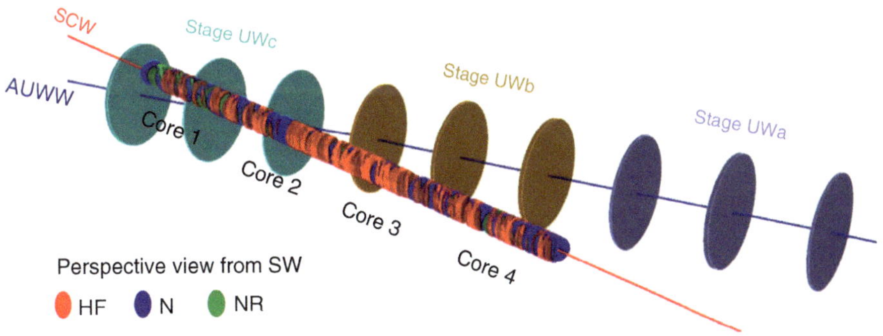

Figure 13.2 Perspective view of fractures in cores 1–4 from the southwest. Large circles are centered at locations of perforation clusters from stages UWb and UWc of well AUWW (*Source:* Shrivastava et al. (2018)/ Society of Petroleum Engineers).

oriented in the dominant fracture direction, and many have spacings less than 1 foot. Figure 13.2 shows a perspective view of the hydraulic and natural fractures identified in the first sections of the cores (Shrivastava et al. 2018). The number of the hydraulically induced fractures identified in the cores (Gale et al. 2018) far exceeded the number of the perforation clusters in the well that was stimulated.

In this paper, we will first discuss some of potential mechanisms that may cause fracture bifurcation or branching, a necessary precursor to the formation of fracture swarms. We will then examine the interference (stress shadow) between closely spaced subparallel hydraulic fractures and the associated strong instability that has tendency to generally prevent and/or suppress simultaneous propagation of parallel fractures in close proximity. And lastly, we will discuss some possible mechanisms to cause closely spaced parallel fractures being created sequentially in time, focusing specifically on the interaction of a hydraulic fracture with a natural fracture of lesser height or a weak bedding interface that could potentially lead to fracture swarms.

13.2 Possible Mechanisms for Fracture Bifurcation

Some possible causes for multiple fractures and fracture swarms had previously been postulated by various authors (Germanovich et al. 1997a; Raterman et al. 2017, Gale et al. 2018; Sesetty and Ghassemi, 2019; Fu et al. 2020; McClure et al. 2020). In order for hydraulic fracture swarms to form, there needs to be mechanisms for either initiation of multiple closely offsetting hydraulic fractures at the wellbore, or fracture bifurcation or branching at a propagating fracture front. Below, we discuss some of these mechanisms, with illustrations shown in Figure 13.3.

1) Initiation of Multiple Fractures from Spiral Perforations

Plug and perf completion is most commonly used in horizontal wells with multi-stage fracturing. Perforations clusters are shot in a spiral pattern around the wellbore in short clusters, with a limited number of holes per cluster to achieve sufficiently large perforation friction to effectively break down and propagate multiple fractures from the perforation clusters. The perforations are shot spirally at different angles around the well, with an offset along the well axis between adjacent shots, due to the physical size constraint of the shape charges which are difficult to fit in the same plane to facilitate the creation of a transverse fracture. Consequently, multiple starter fractures may initiate from individual perforations, with each fracture slightly offsetting from and partially overlapping with the neighboring fractures, forming a flower petal-like structure. This has been previously observed in laboratory block experiments as shown in Figure 13.3a (Weijers 1995; van de Ketterij and de Pater 1999). In many cases the fracture branches originating from different perforations may stay disconnected due to the offset at the wellbore, especially when the stress anisotropy (the difference between the stresses normal and parallel to the fracture) is high, and they only overlap over a small distance along the edges of the adjacent fractures due to the fracture interference (i.e. stress shadow).

2) Microcracks in the Process Zone Ahead of Fracture Tip

Long before the recent field observations of multi-stranded closely spaced hydraulic fractures in shales, similar observations were made in the cores drilled through the hydraulically fracture originated from a vertical well at the Multi-Well Site in one of the largest field experiments conducted by Gas Research Institute and DOE in tight gas sand (Warpinski et al. 1993). In discussion of the possible causes for the creation of these multiple fractures, Nolte (1993) attributed it to the overpressure of the formation and existence of extensive microcracks in the formation being fractured and considered it to be an atypical situation. Figure 13.3b illustrates a propagating hydraulic fracture tip in a medium with preexisting microcracks. Ahead of the fracture tip, a tensile stress field is induced on top of the initially compressive in-situ stress. In a small region very close to the fracture tip, referred to as process zone,

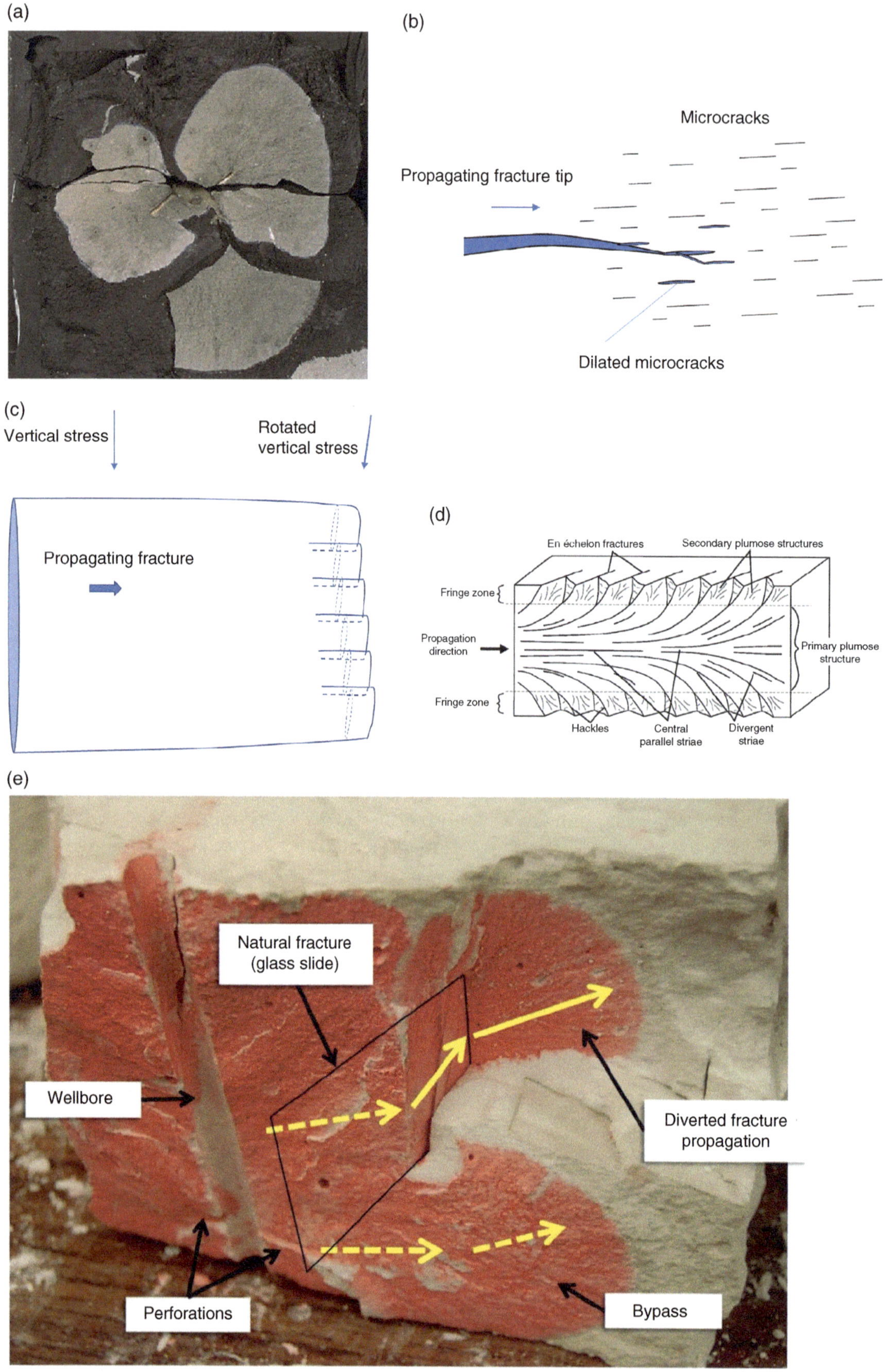

Figure 13.3 Diagrams showing different potential mechanisms that may cause multiple fractures and fracture branching/bifurcation. (a) Multiple fractures from perforations (*Source:* van de Ketterij and de Pater (1999), Society of Petroleum Engineers); (b) Microcracks ahead of fracture tip; (c) Fracture front splitting due to Mode III; (d) Front splitting at fracture top/bottom – twist hackles (*Source:* Cobain et al. (2015)/Geological Society of America/CC BY-4.0); (e) Laboratory experiment showing a hydraulic fracture being diverted by a finite-height artificial natural fracture, while the remainder of the fracture not blocked by the natural fracture continuing along the original plane (*Source:* Bahorich et al. (2012), Society of Petroleum Engineers).

the absolute stress is tensile and highly concentrated near the tip due to the fracture tip singularity, and therein the rock is subjected to plastic deformation and failure. The process zone contributes to the total resistance to fracture tip propagation and is characterized by the apparent fracture toughness. As pointed out by Nolte, the overpressure condition lead to low effective confining stress and consequently a larger process zone ahead of the tip. If the process zone is sufficiently large (greater than the length scale of the microcracks), it may cause microcracks in the vicinity to dilate or shear, and connect with the main fracture tip, causing it to zig-zag, or potentially create branched tips. Fracture tip splitting had also been observed in smaller scale in laboratory experiments and presented in Germanovich et al. (1997a).

3) Fracture Lateral Front Splitting

For a complex geological formation with tilting and undulating layers, and/or under the influences of structural features such as faults, one of the local principal stresses may not be perfectly vertical and its orientation may change slightly as one moves laterally in the formation. When a "vertical" hydraulic fracture is created, its plane is aligned with the local vertical stress near the well. As it propagates away, the fracture front, which was aligned with the original vertical stress direction, may no longer be aligned with the in situ vertical stress as it moves further and further from the well. Consequently, the fracture front may be subjected to both Mode I and Mode III (shear along the front) loadings, potentially leading to the front splitting into multiple en echelon fractures, as illustrated in Figure 13.3c. This had been demonstrated in laboratory experiments with transparent specimens under combined Mode I and III loadings (Wu et al. 2009), and also been observed in geological features such as magmatic dikes (Germanovich et al. 1997a).

Germanovich et al. (1997a) also suggested that fracture front may also split into multiple subparallel segments as a result of material heterogeneity in a layered formation, as a segment of the vertical fracture front within a layer may deviate slightly from the segments in the layers above or below, leading to the split.

4) Fracture Top/Bottom Front Splitting When Growing into Layers with Rotated Horizontal Stress

Similar to the lateral front splitting if the vertical stress is rotated, fracture top and bottom edges can also split if the maximum horizontal stress orientation is rotated slightly from one layer to the next. This front splitting again results in en echelon fractures, as observed in magmatic dikes (see the extensive discussion in Germanovich et al. (1997a) and the references therein) or as "twist hackles" and plumose structures on fracture surfaces (Cobain et al. 2015, Gale et al. 2018), as illustrated in Figure 13.3d (from Cobain et al. 2015). On a small scale, twist hackles were directly observed on fracture faces in some of the cores taken at HFTS by Gale et al. (2018) and they gave an indication of the direction of fracture propagation. On a larger scale, fracture front splitting into en echelon fractures could conceivably lead to larger subparallel fractures as they continue to extend vertically and laterally. This phenomenon had manifested in geological features and was previously described by Helgeson and Aydin (1991) from their observations of the joints near surface interfaces in outcrops.

5) Hydraulic Fracture Front Branching When Intersecting A Natural Fracture of Lesser Height

When a hydraulic fracture intersects a weak interface such as a natural fracture (NF), the fracture may cross straight through the NF, or cross with an offset along the NF, or be completely arrested and later open along the NF itself, depending on many factors such as in situ stress anisotropy, approach angle, friction coefficient, cohesion, permeability of the interface, fluid viscosity, etc. These interaction characteristics had been observed extensively in many fracture mine-back experiments at shallower depths (Warpinski and Teufel, 1987; Jeffrey et al. 1994; Jeffrey et al. 2009a; Warpinski et al. 1993). The crossing behaviors had also been studied extensively both experimentally and via analytical and numerical modeling (Blanton 1982, 1986; Renshaw and Pollard 1995; Gu and Weng 2010; Gu et al. 2011, Zhang and Jeffrey 2006, 2008; Thiercelin and Makkhyu 2007; Zhang et al. 2007a, 2007b; Jeffrey et al. 2009b; Chuprakov et al. 2010; Dahi-Taleghani and Olson 2011; Sesetty and Ghassemi 2012; Chuprakov et al. 2013; Jeffrey et al. 2015; Kear et al. 2017).

When a hydraulic fracture intersects a natural fracture of a lesser height, only a part of the fracture front interacts with the NF, which may cause it to be arrested or diverted along the NF or cross it with an offset. The remaining part of the fracture front above and/or below the NF continues to propagate along the original path. This leads to bifurcation of the fracture front into multiple branches. This phenomenon had been demonstrated in laboratory experiments by Bahorich et al. (2012) as shown in Figure 13.3e. It had also been replicated in numerical simulations (Fu et al. 2015; Cohen et al. 2017).

The fracture branching as discussed here not only can occur for a hydraulic fracture intersecting a NF of lesser height, it can also occur in the situation where the hydraulic fracture is of similar or lesser height than the NF initially which causes the hydraulic fracture to be arrested/diverted. As the fracturing pressure increases, the hydraulic fracture can grow in height and may propagate around the natural fracture along the original fracture plane, leading to a delayed branching. Furthermore, for the diverted fracture along the NF, it can grow in height as well when the pressure increases. Since the NF is at an angle with the maximum horizontal stress and is hence subjected to combined Mode I and III loading at the top/bottom edges, they can split into en-echelon fractures as discussed above.

6) Fracture Branching at Bedding Interfaces

In the field experiment in Eagle Ford by Raterman et al. (2017), fracture offset/stepping at bedding interfaces was observed. The authors stipulated that fracture deflection, offset and branching at bedding surfaces and other naturally occurring heterogeneities as possible causes for the observed fracture swarms.

Hydraulic stepping and branching at bedding interfaces had long been observed in geological features (Helgeson and Aydin 1991, Cooke and Underwood 2001) and in fracture mine-back experiments (e.g. Warpinski and Teufel 1987). While branching at the bedding interface could be caused by the Mode III loading as a result of stress rotation as already discussed above in 4), fracture stepping occurring at a local region could also be a potential source of branching. This is not much different from branching at a natural fracture discussed in 5) and illustrated in Figure 13.3e, except the fracture front is propagating vertically and the weak interface is horizontal. If fracture stepping occurs only along a section of the fracture plane, it leads to the side-stepped part of the fracture potentially growing laterally to overlap with the un-altered part of the fracture, and vice versa.

13.3 Interaction of Closely Spaced Parallel Fractures

Even though the mechanisms discussed above can possibly lead to fracture bifurcation and branching, multiple-branched fractures may not develop into parallel fracture swarms as those observed in the field experiments. In fact, for most of the mechanisms listed, most direct observations in the labs, outcrops, or mine-back experiments showed en echelon fractures with only partial overlap of the fractures. This is due to the strong mechanical interference, or stress shadow, between closely spaced subparallel fractures, which not only suppresses the fracture tip propagation into the "shadowed" area of the adjacent fracture, but also creates strong instability to cause initially overlapping parallel hydraulic fractures to self-readjust to avoid each other to minimize the flow resistance. We will elaborate these points in more details below.

13.3.1 Fracture Tip Extension in Overlapped Region

For multiple en echelon fractures to develop into subparallel fractures with significant overlap, i.e. forming twin, triplets or a swarm of parallel fractures as observed in the cores, the stress intensity factor at the fracture tip that is growing into the "shadow" of its neighboring fracture needs to be sufficiently large to overcome fracture toughness to allow it to propagate. To assess this condition, we examine a simple case of two en echelon fractures subjected to the same net pressure $P_1 = P_2 = P_0$ as shown in Figure 13.4. The total distance between the two outer tips is H, the length of each fracture is h, the horizontal fracture offset is d, the length of the overlap region is δ, and $\delta = 2h - H$.

Figure 13.5 shows the computed fracture tip stress intensity factor, normalized by the nominal stress intensity factor for a single fracture $K_{10} = P_0\sqrt{\pi h/2}$, as a function of δ/h, for various normalized offset d/H. The stress intensity factor is computed using 2D displacement discontinuity method (DDM) (Crouch and Starfield 1983).

As shown in Figure 13.5, when the overlapped length increases, the tip stress intensity factor decreases rapidly. This drop in stress intensity factor becomes more drastic when the fracture offset decreases. For closely spaced fractures, after there is a relatively small amount of fracture overlap, the tip stress intensity factor can drop to below

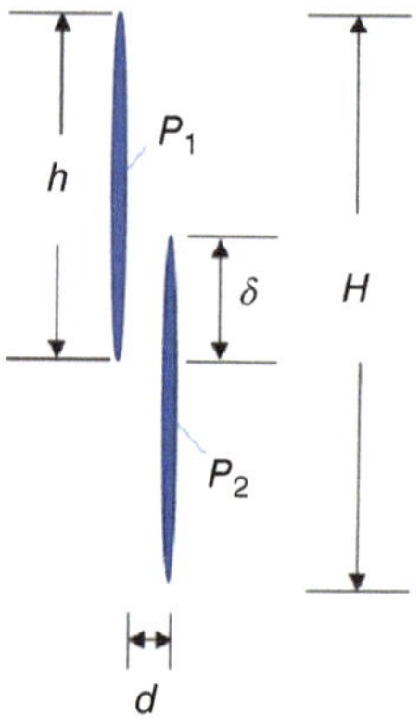

Figure 13.4 Diagram of two en echelon fractures.

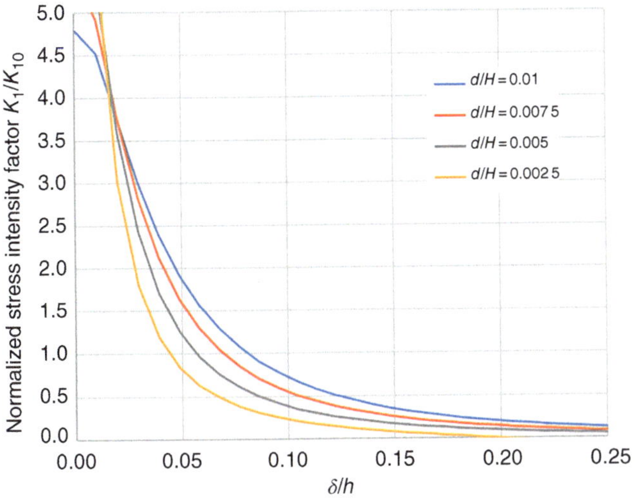

Figure 13.5 Stress intensity factor at the overlapped tips normalized by that of a single fracture of the same dimension and net pressure.

the fracture toughness, preventing the tip from further extension.

Consider an example with the fracture parameters given as

$$H = 40\text{ m}, P_0 = 1.38\text{ MPa}, \text{d} = 0.2\text{ m}, K_{1c} = 1\text{ MPa} - \text{m}^{1/2},$$

which corresponds to the curve of $d/H = 0.005$ in Figure 13.5. For reference, at the given net pressure, the nominal stress intensity factor K_{10} for a fracture of half of the height H is 7.73 MPa-m$^{1/2}$, which is much greater than the fracture toughness. This can be the case if the hydraulic fracture propagation is viscosity-dominated, or the outer tips are contained by stress barriers. Even so, the stress intensity factor at the inner tips drops quickly as the amount of fracture overlap increases. When δ/h exceeds 0.175, the tip stress intensity factor falls below the toughness in this case, and the inner tips will stop growing and the en echelon fractures will not develop into parallel fractures even at a net pressure significantly higher than what is normally sufficient to extend a single fracture. It is plausible, however, that at a very large net pressure, e.g. due to containment of outer tips by stress barriers, the stress intensity factor at the inner tips can still overcome the toughness, leading to the fractures fully extending their heights within the stress layer to become parallel fractures.

13.3.2 Instability of Closely Spaced Parallel Hydraulic Fractures – Shared Inlet

Consider an ideal case of two closely spaced infinitely long parallel fractures with the same height i.e. $\delta = h = H$. They are subjected to a uniform stress and the same internal net pressure P_0. The width of each fracture will be identical and nearly equal to a half of the width of a single fracture of the same height, i.e.

$$w_1 = w_2 \approx \frac{1}{2} w_0 = \frac{hP_0}{E'} \tag{13.1}$$

where w_1 and w_2 are the maximum width of each fracture, respectively, w_0 is the maximum width of an identical single fracture subjected to the same net pressure, and E' is the elastic plane-strain modulus. The reason is that, due to a very small fracture spacing, the incremental compression of the rock in between the fractures (from its in situ condition) is nearly negligible (equal to $d{*}P_0/E$, and $d<<h$). Therefore, the total combined width of both fractures created by the net pressure is resulted from the deformation of the outer half spaces, which is the same as a single fracture.

For the widths of both fractures to be equal, the net pressures in the fractures need to be equal. Now let us examine what happens if the pressures are slightly unbalanced, with $P_1 = P_0$, and $P_2 = P_0 + \delta P$, where $\delta P << P_0$. This small pressure increase does not cause a significant increase in deformation of the rock in the outer half spaces and hence little increase in the total width. However, the rock in between the two fractures is equivalent to a thin plate, and it can bend easily even with a small pressure difference. This would result in a much drastic increase in w_2, at the expense of w_1. The amount of deflection at the midpoint, δw, can be calculated using the equation for deformation of a plate clamped at both ends given in the following (refer to, for example, Gere and Goodno, 2013),

$$\delta w = \frac{\delta P h^4}{384EI} = \frac{\delta P h^4}{32Ed^3} \tag{13.2}$$

where I is the area moment of inertia with $I = \frac{d^3}{12}$. Considering the limiting case where the deflection of the rock between the fractures is equal to the initial width, equating Eq. (13.2) with Eq. (13.1) leads to

$$\frac{\delta P}{P_0} = 32\left(1 - \nu^2\right)\left(\frac{d}{h}\right)^3 \tag{13.3}$$

where ν is the Poisson's ratio. Since $d << h$, it takes a differential pressure of only a very small fraction of the net pressure to completely close the fracture with the lesser pressure. For example, assume both hydraulic fractures are of a height of 40 m, with an offset of 0.4 m, and subjected to a net pressure of 3.45 MPa, and the rock's Poisson's ratio is 0.25. It will only require a tiny differential pressure of 0.001 MPa to completely close one of the fractures. Therefore, for all practical purposes, in order for closely spaced hydraulic fractures to co-exist, fluid pressures inside the fractures must be nearly identical.

Now, let us examine the pressure gradient in the hydraulic fractures. It follows the Poiseuille's law and is given as follows for a Newtonian fluid.

$$\frac{dP}{dx} = -\frac{12\mu}{\overline{w}^3}\frac{q}{h} \quad (13.4)$$

where μ is fluid viscosity, q is flow rate, and $\overline{w}$ is average width.

Let us consider two initially identical parallel fractures, with the same inlet pressure. This is equivalent to a case of two fractures branched from a single fracture. In the event there is a tiny perturbation, e.g. a reduction in fracture width, for one of the fractures, it will result in a much more drastic increase in the pressure gradient if the flow rate remains the same and hence a decrease in the pressure downstream. As discussed above, even a slight pressure underbalance would cause the fracture to be closed. However, this cannot happen instantly since the volume in the fracture needs to be conserved. Consequently, the pressures need to be rebalanced among the fractures to keep them both open, which would require the flow rate to decrease, by a much larger amount per Eq. (13.4), to keep the pressure gradients the same. However, with reduction in the inflow rate, the fracture will gradually close with diminishing width due to fluid leakoff and/or its length growth will slow or stop. This suggests that branched parallel hydraulic fractures are highly unstable and will not be able to sustain continued growth over a large distance.

13.3.3 Instability of Closely Spaced Parallel Hydraulic Fractures – Separate Inlets

If the inflow rates at the inlets of the fractures are independently prescribed or regulated by a strong choke mechanism such as in the case of limited-entry perforating, the pressure balance between parallel fractures can indeed be self-regulated to satisfy the mass balance dictated by the flow rate distribution mechanism. For example, in the case of multiple fractures from limited-entry perforation clusters, the net pressure in the middle fracture would be inherently higher than the outer fractures to counter-act the higher stress shadow (Wu et al. 2012; Sesetty and Ghassemi 2019), at the expense of a small reduction in flow rate which leads to a reduction in perforation friction to offset the increase of pressure in the middle fracture. As such, multiple fractures can propagate simultaneously with nearly equal flow rates, enabled by the limited entry condition with a very high perforation entry friction.

However, while the separate inlet rates or the limited entry condition force the multiple fractures to propagate simultaneously, it does not mean that they will stay in parallel. The strong fracture interference as discussed in the previous section can also manifest itself in another type of instability that causes the initially parallel fractures to evolve into an en echelon arrangement along the height direction. As previously shown by Germanovich et al. (1997b), a set of en echelon fractures with small offset have nearly the same width as a single fracture with the same total height as the en echelon fractures. To further illustrate this point, Figure 13.6a shows two parallel fractures occupying a total height of H and their width profiles, and Figure 13.6b shows two fractures of approximately half the height in en echelon configuration. Figure 13.7 shows the computed width profiles of the en echelon fractures, corresponding to Figure 13.6b, from 2D DDM, in comparison to that of the parallel fractures (Figure 13.6a) at the same internal net pressure, for $d/H = 0.005$ and $\delta/H = 0.05$. It shows that for $d << H$, the fracture width of the en echelon fractures is very close to the width of a single fracture, which is nearly twice the width of the parallel fractures, even though each fracture is only half as tall.

Referring again to Eq. (13.4), with approximately twice the width and half of the height, the pressure gradient in the en echelon fractures is only one-fourth of that of the corresponding parallel fractures. Therefore, the en echelon configuration provides much less flow resistance than parallel fractures with nearly the same cross-sectional area. For two initially identical parallel fractures, a small perturbation (e.g. a reduction in pressure) in the upper or lower half of one of the fractures would lead to pressure imbalance with the neighboring fracture and consequently width reduction, which would in turn increase the flow resistance and divert a greater portion of the flow to the other half of the height, leading to an asymmetric width profile. The

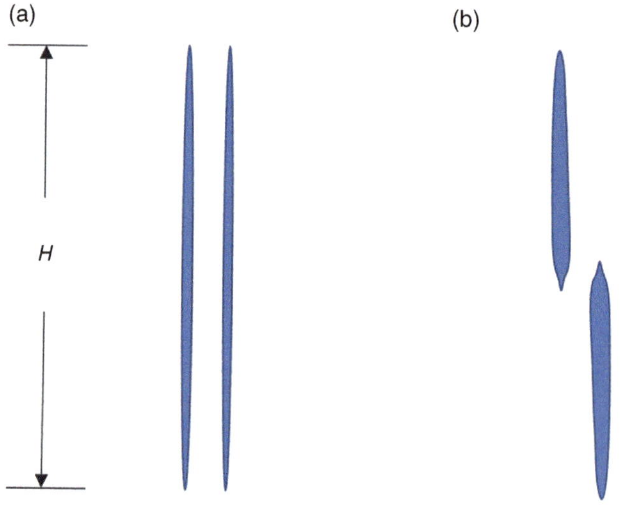

Figure 13.6 Illustration of two parallel fractures and their width profiles and that of two en echelon fractures occupying the same height.

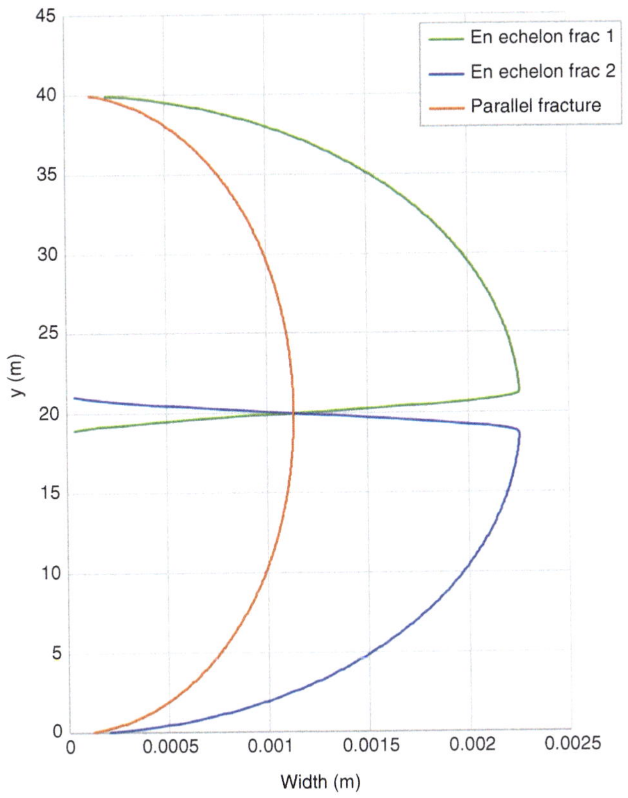

Figure 13.7 Comparison of the computed width profiles for two en echelon fractures (with offset d = 0.2 m, fracture height h = 21.1 m, overlap δ = 2.2 m and P_0 = 1.38 MPa) to that of two parallel fractures with the same horizontal offset and total height (H = 40 m).

width profile is opposite for the companion fracture. Consequently, the pressure in the wider half of the fracture would increase due to the reduced pressure gradient, leading to further pressure imbalance between the two fractures. This positive feedback loop eventually leads to one fracture only occupying a half of the total height and the other fracture occupying the other half, similar to what is shown in Figure 13.6b.

This instability is also what causes multiple fractures initiated from spiral perforations to form a flower petal like structure as shown in Figure 13.3a, instead of creating multiple overlapping parallel radial fractures. It can manifest itself in 3D numerical simulation of multiple parallel hydraulic fractures. For instance, Castonguay et al. (2013) simulated multiple transverse fractures at fixed flow rates into the fractures using a fully 3D BEM-based model. While the fractures initiated from the wellbore as parallel transverse radial fractures, as the fractures became larger, they started to interact more strongly and eventually evolved into asymmetrical fractures growing in different directions, likely caused by the instability as discussed above, to reduce overlapping to attain wider fracture width and less flow resistance.

13.4 Possible Mechanisms for Creating Fracture Swarms

Due to the strong instability associated with closely spaced parallel hydraulic fractures as discussed in the previous section, we hold a view that simultaneously propagating parallel fractures in close proximity are highly unlikely. We believe that the fracture swarms as observed in the field experiments are more likely to have occurred sequentially, rather than simultaneously. Sequential parallel fracturing can happen if one or more of the initially nonoverlapping or partially overlapping fractures in en echelon arrangement grow into the "shadowed" region overlapping with other fractures at a later time. In order for a later fracture to penetrate into the stress shadowed region of a fracture already in place, one could expect either the stress shadow exerted by the earlier fracture has already dissipated, for example due to fracture closure resulting from flow blockage by proppant bridging, or the pressure in the later fracture is greater than the pressure in the earlier created fracture.

One such scenario could be possible with the flower petal-like fractures as shown in Figure 13.3a, with each fracture occupying an angular section of the transverse fracture plane and connected to a limited number of perforations. As the overall fracture system grows in size radially, the fractures growing in the upward direction may see drop in the in-situ stress, which is common in a shale formation. This leads to generally upward height growth, and the pressure inside these fractures drops accordingly. Consequently, the pressures inside the side or downward propagating fractures may become greater than the upward propagating fractures, causing the former to "encroach" into the area originally occupied by the upper fractures as they migrate further up. Since these fractures were originally offset along the well axis due to the spiral perforations, this upward migration of the fractures may lead to the appearance of much greater areal overlap between adjacent fractures, even though the actual overlap at any given time is very limited along the edges of the fractures due to the stress shadow/instability.

In the remainder of this work, however, we will focus on the specific scenario of hydraulic fracture front branching when it intersects a natural fracture of smaller height, as observed in the lab experiment as shown in Figure 13.3e and illustrated here in the diagram in Figure 13.8, as a possible mechanism for creating sequential parallel fractures. More specifically, we look at the bypass fracture branch above or below the natural fracture as it grows around

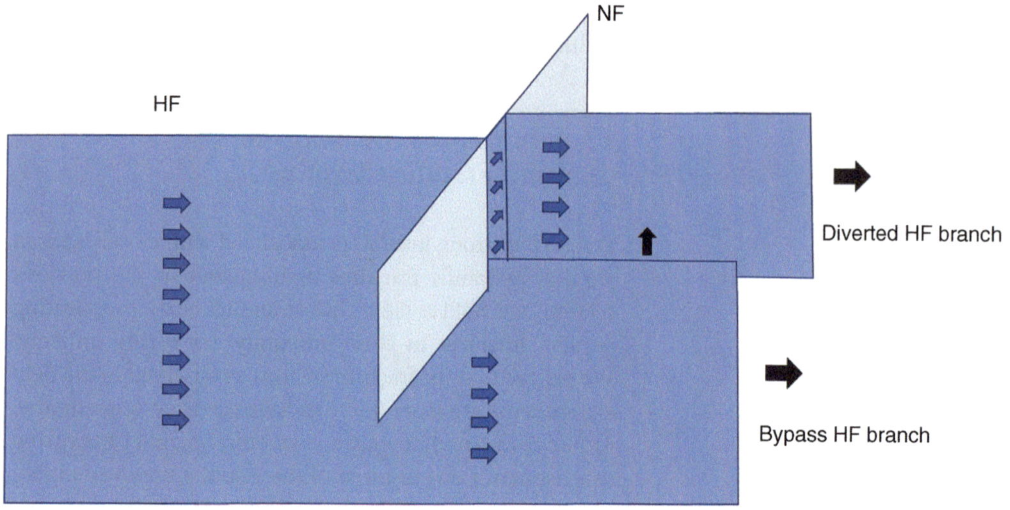

Figure 13.8 Schematic of a hydraulic fracture (HF) that crosses a natural fracture (NF) of smaller height with an offset, creating a diverted branch; the unobstructed part of the HF bypasses NF along the original plane and propagates both laterally and vertically to overlap with the diverted branch.

the NF to become overlapped with the diverted branch. When the hydraulic fracture first reaches the NF, the part of the fracture front intersecting the NF within its height interval may cross the NF with an offset (Thiercelin and Makkhyu 2007; Chuprakov et al. 2010; Kear et al. 2017), or get diverted along the NF and subsequently intersect another NF or a defect, or have reached the end of the NF to get redirected back into the main fracture direction. This diverted fracture branch is created immediately after the intersection. The bypass branch is formed from the portion of the original fracture front that is unobstructed by the NF and continues along the original fracture path. As the bypass branch goes beyond the NF, it can possibly grow its height into the area behind the natural fracture, and become overlapped with the already created diverted branch. We will examine whether the bypass branch can grow in height in the presence of the stress shadow from the diverted branch.

Due to the width restriction in the deviated section along the NF, there is an added pressure drop along this section. As a result, the pressure inside the diverted branch would be slightly less than the pressure in the bypass branch (assuming the pressure loss associated with the flow around the natural fracture is negligible). Assuming the width and length of the deviated section is w_d and L_d, then the pressure drop across the section ΔP_{div} is given as

$$\Delta P_{div} = \frac{12\mu L_d}{{w_d}^3} \frac{q_{DF}}{h_{DF}} \tag{13.5}$$

where q_{DF} and h_{DF} are the flow rate and the height of the diverted fracture branch. If the width of the deviated section along the NF is severely restricted, it could cause a significant pressure drop.

To study whether the bypass branch can grow height to overlap with the diverted branch, we again use 2D DDM to compute the stress intensity factor at the tip of the bypass branch. Considering a vertical cross section of the two fractures beyond the NF, they are en echelon fractures similar to that shown in Figure 13.4, with the diverted branch as fracture 1 and the bypass branch as fracture 2. Unlike in the previous case, the pressures in two fractures are now not equal, with $P_1 = P_0$, and P_2 greater than P_1 by an amount of ΔP_{div}. Without loss of generality, we consider two fractures with initially equal height $h = H/2$, and horizontal offset $d/H = 0.005$. Figure 13.9 shows the computed stress intensity factor at the tip of the bypass fracture for varying amount of overlap δ and differential pressure ΔP_{div}, normalized by the stress intensity factor for a single fracture of the same dimension and net pressure P_0. When the overlap between the two fractures is small, the stress intensity factor at the tip is much greater than that of a single fracture due to the interaction of the two fracture tips, and is dominated by P_0, since it is much greater than the differential pressure. The stress intensity factor decays quickly as the fracture overlap increases, similar to the corresponding curve without pressure difference as shown previously in Figure 13.5. However, after certain overlap distance, the stress intensity factor stops decreasing and starts to increase with the amount of fracture overlap. This increase is attributed to the contribution of ΔP_{div} to the tip stress intensity factor.

Figure 13.10 shows the computed stress intensity factor at the tip of the bypass fracture due to ΔP_{div} alone, normalized by the nominal stress intensity factor for a single fracture subjected to only ΔP_{div}, i. e. $K_{1div} = \Delta P_{div}\sqrt{\pi h/2}$, for various normalized fracture offsets d/H. It shows that the

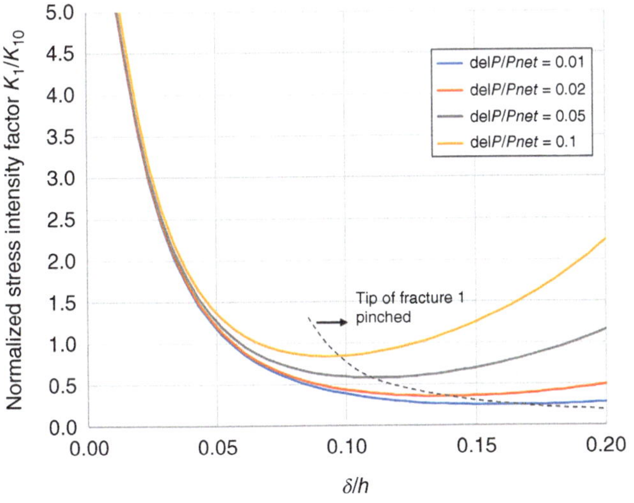

Figure 13.9 Normalized stress intensity factor at the upper tip of the bypass branch for $d/H = 0.005$ and different values of $\Delta P_{div}/P_O$.

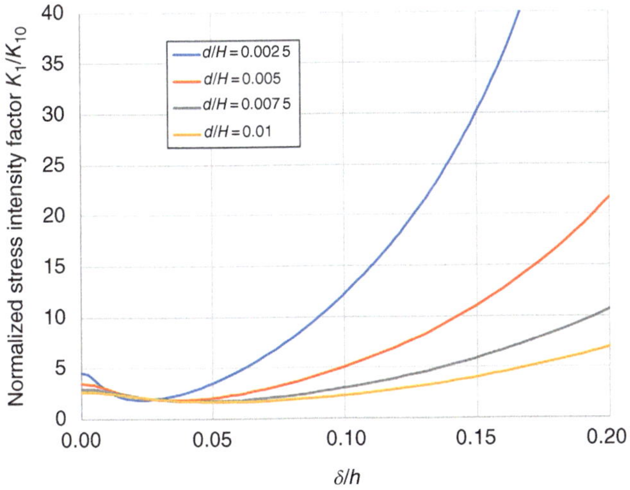

Figure 13.10 Contribution of the differential pressure to stress intensity factor at the upper tip of the bypass branch (normalized by K_{1div}) for various values of horizontal offsets d/H.

differential pressure induced tip stress intensity factor increases as the fracture overlap increases, as well as when the fracture offset decreases. At a large overlap and/or small offset, the K_1/K_{1div} is much greater than 1, due to the large bending of the rock in between the fractures, which acts as a thin plate, leading to the large tip stress intensity factor.

The actual stress intensity factor at the tip of the bypass branch is the sum of that due to P_O as given in Figure 13.5 and that from ΔP_{div} as shown in Figure 13.10. If the stress intensity factor maintains above the fracture toughness as the height of the bypass branch extends into the shadowed region of the diverted fracture branch, it will continue to grow in height and eventually can become fully overlapped with the diverted branch.

It should also be noted that due to the high compliance of the thin plate that is the rock in between the fractures in the overlapped region, the differential pressure will cause the fracture 1 to partially close off near the tip region when the overlap is sufficiently large, as indicated by the region to the right of the dash line in Figure 13.9. The DDM computation given in Figure 13.9 does not take into consideration the partial contact of the faces of fracture 1 near the tip region which limits further opening of the fracture 2. Therefore, the computed stress intensity factor to the right of the dash line may be overestimated.

Let us consider again the previous example, with $H = 40$ m, $P_O = 1.38$ MPa, $d = 0.2$ m, and $K_{Ic} = 1$ MPa-m$^{1/2}$. The normalized fracture toughness, K_{Ic}/K_{10}, is 0.129. The normalized stress intensity factor for the bypass fracture tip in this case is as given in Figure 13.9. The normalized toughness is well below all the curves for $\Delta P_{div}/P_O$ from 0.01 to 0.1 and for all values of fracture overlap. It means that even a small differential pressure of only 0.0138 MPa (2 psi) is more than enough to extend the height of the bypass fracture for it to become fully overlapped with the diverted fracture.

Referring to Eq. 13.5, and with a very simplified estimation of P_O, we have the following approximation

$$\frac{\Delta P_{div}}{P_0} \approx \frac{L_d}{L_{DF}} \left(\frac{w_{DF}}{w_d} \right)^3 \tag{13.6}$$

where L_{DF} and w_{DF} are the length and average width of the diverted branch, respectively. Even for a small L_d, which is proportional to the horizontal offset d, a moderate width reduction in the diverted section can create sufficient pressure drop $\Delta P_{div}/P_O$ to cause the bypass fracture to grow in height, overcoming the stress shadow from the diverted branch to become significantly overlapped.

It should be mentioned that the analyses of the tip stress intensity factor presented above are over-simplified with assumption of constant net pressure and differential pressure. However, these quantities change dynamically in reality as the fracture branches grow and become overlapped with each other. Once sufficient fracture overlap occurs, a part of the diverted fracture branch gets squeezed by the bypass branch or even closes, leading to a reduction in the flow rate into the diverted branch. This dynamic will help reduce the pressure difference between the diverted branch and the bypass branch, but the reduction in the inflow rate will likely cause the diverted branch to slow down or stop its growth in length and be eventually overtaken by the bypass fracture branch. Therefore, the parallel branches may only exist over some lateral distance before becoming a single fracture again. However, if the fractures intersect more natural fractures along the way and shift at these intersections, it is possible that multiple closely spaced parallel branches can be created, forming fracture swarms within the formation layers containing the natural fractures.

The analyses and discussions above focus on the interaction of the bypass branch and the diverted branch that has an offset when the lateral front of a propagating hydraulic fracture intersects a natural fracture. But they can also be applicable to the vertically propagating upper/lower front of the fracture that intersects weak bedding planes or interfaces which can potentially cause the fracture to have offsets/steps at the interfaces. However, the offset may not occur uniformly along the entire fracture front due to rock heterogeneity, leading to the front potentially breaking into sections that do experience an offset at the interface (the diverted branches) and others that do not (the bypass branches). The lateral edges of a bypass branch can propagate in the horizontal direction and become overlapped with the neighboring diverted branches. Similar to the interaction of a lateral fracture front with a natural fracture discussed above, the diverted branch has a severe width restriction where it has a step along the interface, which causes a significant pressure drop for the fluid to flow through. This leads to more favorable lateral growth of the bypass branch into the area next to the diverted branch due to the positive differential pressure. The bypass branch will eventually dominate over the diverted branch, causing the latter to terminate its further upward/downward growth or be completely closed off. The parallel fracture branches created in this manner are likely localized at where the fracture jogs at a weak bedding interface. However, they could be more abundant if many weak bedding interfaces exist that cause the fractures to jog frequently as they propagate vertically through these interfaces.

13.5 Conclusions

To understand the fracture swarms observed in unconventional shale formations in recent field experiments, we presented some of the mechanisms that can cause multiple fractures to initiate from a perforation cluster and for the hydraulic fracture front to bifurcate or branch, as possible precursor to forming fracture swarms. However, most of these branching mechanisms create offsetting fractures that are in en echelon fashion without significant fracture overlap per observations in laboratory experiments and from geological features. To investigate whether these initially mildly overlapping fracture branches can evolve into subparallel fractures with substantial area of overlap, we evaluated the stress intensity factor at the tips of the partially overlapping fractures. For en echelon fractures subjected to the same pressure, the numerical results show that the tip stress intensity factor drops rapidly when fracture overlap increases, which could prohibit further tip extension to attain substantial overlap if the stress intensity factor falls below the fracture toughness. For the fractures to evolve into parallel fractures, a net pressure drastically greater than the pressure corresponding to the toughness-dominated propagation will be required.

We also analyzed the instability associated with propagation of parallel fractures. It is shown that in order for closely spaced parallel fractures to propagate simultaneously, the pressures in the fractures must be nearly equal. This is due to extremely high flexibility of the rock in between the fractures that acts as a thin plate, which can bend easily with even a tiny differential pressure to completely close off the fracture with lower pressure. Consequently, for parallel fractures that share the same inflow source, their propagation is highly unstable. A very small disturbance in a fracture with slightly wider width among otherwise identical parallel fractures would result in a less flow resistance and positive differential pressure relative to its neighbors, and consequently trigger a positive feedback loop of rapidly increasing width and eventual dominance over the others.

For fractures hydraulically fed through independent sources of inflow or regulated by a strong flow distribution mechanism such as limited-entry perforations, the fractures are forced to be kept open because of the constant inflow. However, the instability still manifests itself by forcing the fractures to propagate in different nonoverlapping regions of the fracture plane to minimize fracture overlap, often in the form of en echelon fractures along the height for vertically contained fractures or flower petal-like branches if growing radially. The main driver for this behavior is the fact that multiple en echelon fractures with small offset attain almost the same width as a single fracture of the same overall height. In contrast, the width of fully parallel fractures at the same pressure is nearly equal to the single fracture width divided by the number of fractures. As a result, the flow resistance in the en echelon system, despite each fracture only occupying a fraction of the total height, is much smaller than in the parallel fractures. This causes the fractures to propagate in favor of the en echelon pattern rather than in parallel to minimize the energy.

Due to the strong instability associated with parallel fractures, we postulate that the individual fractures in the fracture swarms as observed in the field experiments are most likely formed sequentially in time, rather than simultaneously. A subsequent fracture can propagate in parallel to an existing open fracture if it has an internal pressure greater than that in the existing fracture. Using 2D DDM, it is shown numerically that unlike in the case of fractures with equal pressures, even a small differential pressure in the newer fracture can create a relatively large stress intensity factor at the fracture tip, and it increases as the overlap increases. It allows the stress intensity factor at the tip of the newer fracture to stay above the toughness as the overlap increases to continue the propagation in parallel to the existing fracture.

One of the scenarios where sequential parallel propagation may occur is in the case of a hydraulic fracture front intersecting a natural fracture of lesser height, branching into a diverted branch with an offset at the natural fracture, and bypass branches above or below the natural fracture that continue on the original path. The fracture width reduction in the deviated section along the natural fracture creates an added pressure drop in the diverted branch which may be sufficient to cause a favorable differential pressure in the bypass branch to extend its height to become parallel to the diverted branch. Even though the instability will eventually lead to the bypass branch dominating over the diverted branch and leaving again only one dominant fracture some distance beyond the branching point, multiple intersections with natural fractures along the way can potentially lead to more fracture branching and formation of fracture swarms. Similar fracture branching and sequential overlapping growth mechanism may occur at the vertically propagating fracture front when it intersects weak bedding interfaces. Fracture front may jog along the weak interfaces, but may not do so along the whole front due to rock heterogeneity, creating diverted branches (with a step along the interface) and bypass branches (straight through the interface). Similarly, the diverted branches are subjected to an additional pressure drop due to the width restriction, leading to subsequent favorable lateral growth of the bypass branches to overlap with and dominate over the diverted branches.

The analyses presented in this study are based on the results under static conditions with constant pressures in the fractures. However, the actual process of unstable hydraulic fracture interaction and propagation is highly dynamic and complex. Further research of this dynamic process will be needed to fully understand the fracture swarming process.

References

Bahorich, B., Olson, J.E., and Holder, J. (2012). Examining the effect of cemented natural fractures on hydraulic fracture propagation in hydrostone block experiments. *Paper SPE 160197, Presented at SPE Annual Technical Conference and Exhibition*, San Antonio, TX (8–10 October).

Blanton, T.L. (1982). An experimental study of interaction between hydraulically induced and pre-existing fractures. *SPE 10847, Presented at the SPE/DOE Unconventional Gas Recovery Symposium*, Pittsburgh, PA (16–18 May).

Blanton, T.L. (1986). Propagation of hydraulically and dynamically induced fractures in naturally fractured reservoirs. *SPE Unconventional Gas Technology Symposium*, Louisville, KY (1 January 1986).

Castonguay, S.T., Mear, M.E., Dean, R.H., and Schmidt, J.H. (2013). Prediction of the growth of multiple interacting hydraulic fractures in three dimensions. *Paper SPE 166259, SPE Annual Technical Conference and Exhibition*, New Orleans, LA (30 September–2 October).

Chuprakov, D.S., Akulich, A.V., Siebrits, E., and Thiercelin, M. (2010). Hydraulic fracture propagation in a naturally fractured reservoir. *SPE 128715, Presented at the SPE Oil and Gas India Conference and Exhibition held in Mumbai* (20–22 January).

Chuprakov, D., Melchaeva, O., and Prioul, R. (2013). Injection-sensitive mechanics of hydraulic fracture interaction with discontinuities. *ARMA Symposium*, San Francisco, CA.

Ciezobka, J., Courtier, J., and Wicker, J. (2018). Hydraulic fracturing test site (HFTS) – project overview and summary of results. *SPE/AAPG/SEG Unconventional Resources Technology Conference*, Houston, TX (23–25 July).

Cipolla, C.L., Weng, X., Mack, M. et al. (2011). Integrating microseismic mapping and complex fracture modeling to characterize fracture complexity. *SPE 140185, SPE Hydraulic Fracturing Technology Conference and Exhibition*, The Woodlands, TX (24–26 January).

Cipolla, C. and Wallace, J. (2014). Stimulated reservoir volume: a misapplied concept? *Paper SPE 168596, Presented at Hydraulic Fracturing Technology Conference*, The Woodlands, TX (4–6 February).

Cohen, C. Kresse, O., and Weng, X. (2017). Stacked height growth model to improve fracture height growth prediction, and simulate interactions with multi-layer DFNs and ledges at weak interfaces. *Paper SPE 184876, Presented at SPE Hydraulic Fracturing Technology Conference and Exhibition*, Woodlands, TX (24–26 January).

Cobain, S.L., Peakall, J., and Hodgson, D.M. (2015). Indicators of propagation direction and relative depth in clastic injectites: Implications for laminar versus turbulent flow processes. *Geological Society of America Bulletin* 127: 1816–1830.

Cooke, M.L. and Underwood, C.A. (2001). Fracture termination and step-over at bedding interfaces due to frictional slip and interface opening. *Journal of Structural Geology* 23: 223–238.

Crouch, S.L. and Starfield, A.M. (1983). *Boundary Element Methods in Solid Mechanics*, firste. London: George Allen & Unwin Ltd.

Dahi-Taleghani, A. and Olson, J.E. (2011). Numerical modeling of multistrand hydraulic-fracture propagation: accounting for the interaction between induced and natural fractures. *SPE Journal* 16 (2011): 575–581.

Daniels, J., Waters, G., LeCalvez, J. et al. (2007). Contacting more of the Barnett Shale through an integration of real-time microseismic monitoring, petrophysics, and hydraulic fracture design. *Paper SPE 110562 Presented at the 2007 SPE Annual Technical Conference and Exhibition*, Anaheim, CA (12–14 October).

Ejofodomi, E.A., Baihly, J.D., and Silva, F. (2015). Using a calibrated 3D fracturing simulator to optimize completions of future wells in the eagle ford shale. *Paper SPE 178680/ URTeC 2172668, Presented at Unconventional Resources Technology Conference*, San Antonio, TX (20–22 July).

Fisher, M.K., Davidson, B.M., Goodwin, A.K. et al. (2002). Integrating fracture mapping technologies to optimize stimulations in the Barnett Shale. *SPE 77411, SPE Annual Technical Conference and Exhibition*, San Antonio, TX (29 September–2 October).

Fu, P., Cruz, L., Moos, D. et al. (2015). Numerical investigation of a hydraulic fracture bypassing a natural fracture in 3D. *ARMA 15-0671, Presented at the 49th US Rock Mechanics/ Geomechanics Symposium*, San Francisco, CA (28 June – 1 July).

Fu, W., Morris, J., Fu, P. et al (2020). Developing upscaling approachh for swarming hydraulic fractures observed at hydraulic fracturing test site through multiscale simulations. *Paper SPE 199689, Presented at SPE Hydraulic Fracturing Technology Conference and Exhibition*, The Woodlands, TX (4–6 February).

Gakhar, K., Rodinov, Y., Defeu, C. et al. (2017). Engineering an effective completion and stimulation strategy for in-fill wells. *Paper SPE 184835, Presented at SPE Hydraulic Fracturing Technology Conference and Exhibition*, The Woodlands, TX (24–26 January).

Gale, J.F.W., Elliott, S.J., and Laubach, S.E. (2018). Hydraulic fractures in core from stimulated reservoirs: core fracture description of hfts slant core, Midland Basin, West Texas. *URTeC 2902624, Presented at SPE/AAPG/SEG Unconventional Resources Technology Conference*, Houston, TX (23–25 July).

Gere, J.M. and Goodno, B.J. (2013). *Mechanics of Materials*, 8the. Cengage Learning.

Germanovich, L.N., Astakhov, D.K., Mayerhofer, M.J. et al. (1997a). Hydraulic fracture with multiple segments I. Observations and model formulation. *International Journal of Rock Mechanics and Mining Sciences* 34 (3–4): 1–97.

Germanovich, L.N., Ring, L.M., Astakhov, D.K. et al. (1997b). Hydraulic fracture with multiple segments ii. modeling. *International Journal of Rock Mechanics and Mining Sciences* 34 (3–4): 1–98.

Gu, H. and Weng, X. (2010). Criterion for fractures crossing frictional interfaces at non-orthogonal angles. *44th US Rock Mechanics Symposium and 5th US-Canada Rock Mechanics Symposium*, Salt Lake City, Utah (27 – 30 June).

Gu, H., Weng, X., Lund, J.B. et al. (2011). Hydraulic fracture crossing natural fracture at non-orthogonal angles, a criterion, its validation and applications. *SPE 139984, Hydraulic Fracturing Conference and Exhibition*, The Woodlands, TX (24–26 January).

Helgeson, D.E. and Aydin, A. (1991). Characteristics of joint propagation across layer interfaces in sedimentary rocks. *Journal of Structural Geology* 13 (8): 897–911.

Jeffrey, R. G., Bunger, A., Lecampion, B. et al. (2009a). Measuring hydraulic fracture growth in naturally fractured rock. *SPE 124919, SPE Annual Technical Conference and Exhibition*, New Orleans, LA (4–7 October).

Jeffrey, R.G., Weber, C.R., Vlahovic, W., and Enever, J.R. (1994). Hydraulic fracturing experiments in the great northern coal seam. *SPE 28779, SPE Asia Pacific Oil & Gas Conference*, Melbourne (7–10 November).

Jeffrey, R.G., Zhang, X., and Thiercelin, M. (2009b). Hydraulic fracture offsetting in naturally fractured reservoirs: quantifying a long-recognized process. *SPE 119351, SPE Hydraulic Fracturing Technology Conference*, The Woodlands, TX (19–21 January).

Jeffrey, R.G., Kear, J., Kasperczyk, D., and Zhang, X. (2015). A 2d experimental method with results for hydraulic fractures crossing discontinuities. *ARMA 15–439, Presented at 49th US Rock Mechanics/Geomechanics Symposium*, San Francisco, CA (28 June – 1 July).

Kear, J., Kasperczyk, D., Zhang, X. et al. (2017). 2D experimental and numerical results for hydraulic fractures interacting with orthogonal and inclined discontinuities. *ARMA 17-404, Presented at 51st US Rock Mechanics/ Geomechanics Symposium*, San Francisco, CA (25–28 June).

Kresse, O., Weng, X., Chuprakov, D. et al. (2013). Effect of flow rate and viscosity on complex fracture development in UFM model. *International Conference for Effective and Sustainable Hydraulic Fracturing*, Brisbane, (20–22 May).

Malpani, R., Sinha, S., Charry, L. et al. (2015). Improving hydrocarbon recovery of horizontal shale wells throughrefracturing. *Paper SPE 175920, Presented at Unconventional Resources Conference*, Calgary, Alberta (20–22 October).

Marongiu-Porcu, M., Lee, D., Shan, D., and Morales, A. (2015). Advanced modeling of interwell fracturing interference: an eagle ford shale oil study. *Paper SPE 174902, Presented at Annual Technical Conference and Exhibition*, Houston, TX (28–30 September).

Maxwell, S.C., Urbancic, T.I., Steinsberger, N.P., and Zinno, R., 2002. Microseismic imaging of hydraulic fracture complexity in the Barnett Shale. *SPE 77440, SPE Annual Technical Conference and Exhibition*, San Antonio, TX, USA, (29 September–2 October).

McClure, M., Picone, M., Fowler, G. et al. (2020). Nuances and frequently asked questions in field-scale hydraulic fracture modeling. *Paper SPE 199726, Presented at SPE Hydraulic Fracturing Technology Conference and Exhibition*, The Woodlands, TX (4–6 February).

Nolte, K. (1993). *Discussion of Examination of a Cored Hydraulic Fracture in a Deep Gas Well.* SPE Prod. Facil: Society of Petroleum Engineers.

Raterman, K.T., Farrell, H.E., Mora, O.S. et al. (2017). Sampling a stimulated rock volume: an eagle ford example. *Presented at Unconventional Resources Technology Conference*, Austin, TX (24–26 July 2017).

Renshaw, C.E. and Pollard, D.D. (1995). An experimentally verified criterion for propagation across unbounded frictional interfaces in Brittle, linear elastic-materials. *International Journal of Rock Mechanics and Mining Sciences & Geomechanics Abstracts* 32 (3): 237–249.

Sesetty, V. and Ghassemi, A. (2012). Simulation of hydraulic fractures and their interactions with natural fractures. *ARMA 12-331 Presented at 46th US Rock Mechanics/ Geomechanics Symposium*, Chicago, IL (24-27 June).

Sesetty, V. and Ghassemi, A. (2019). Simulation and analysis of fracture swarms observed in the eagle ford field experiment. *Paper SPE 194328. Presented at SPE Hydraulic Fracturing Technology Conference and Exhibition*, The Woodlands, TX (5–7 February).

Shrivastava, K., Hwang, J., and Sharma, M. (2018). Formation of complex fracture networks in the wolfcamp shale: calibrating model predictions with core measurements from the hydraulic fracturing test site. *Paper SPE 191630, Presented at SPE Annual Technical Conference and Exihibition*, Dallas, TX (24–26 September).

Thiercelin, M. and Makkhyu, E. (2007). Stress field in the vicinity of a natural fault activated by the propagation of an induced hydraulic fracture. *Proceedings of the 1st Canada-US Rock Mechanics Symposium* 2: 1617–1624.

Van de Ketterij, R.G. and de Pater, C.J. (1999). Impact of perforations on hydraulic fracture tortuosity. *SPE Production & Facilities* 14 (2): 131–138.

Wallace, K.J., Reyes Aquirre, P., Jinks, E. et al. (2016). Understanding completion performance in Niobrara-Codell reservoirs through the use of innovative software-guided workflows and models. *Paper SPE 179172, Presented at Hydraulic Fracturing Technology Conference*, The Woodlands, TX (9–11 February).

Warpinski, N.R., Kramm, R.C., Heinze, J.R., and Waltman, C. K. (2005). Comparison of single- and dual-array microseismic mapping techniques in the Barnett Shale. *SPE 95568, SPE Annual Technical Conference and Exhibition*, Dallas, TX (9–12 October).

Warpinski, N.R., Lorenz, J.C., Branagan, P.T. et al. (1993). Examination of a cored hydraulic fracture in a deep gas well. *SPE Production & Facilities* 8: 150–158.

Warpinski, N.R. and Teufel, L.W. (1987). Influence of geologic discontinuities on hydraulic fracture propagation (includes associated papers 17011 and 17074). *Journal of Petroleum Technology* 39 (2): 209–220.

Weijers, L. (1995). The near-wellbore geometry of hydraulic fractures intiated from horizontal and deviated wells. Ph.D dissertation. Delft University of Technology, Delft University Press.

Weng, X., Kresse, O., Cohen, C. et al. (2011). Modeling of hydraulic fracture network propagation in a naturally fractured formation. *SPE 140253, SPE Hydraulic Fracturing Conference and Exhibition*, The Woodlands, TX (24–26 January).

Wu, R., Germanovich, L.N., and Hurt, R.S. (2009). Experimental and theoretical study of mixed-mode I+III crack propagation and segmentation. *43rd U.S. Rock Mechanics Symposium & 4th U.S. - Canada Rock Mechanics Symposium*, Asheville, North Carolina (28 June–1 July).

Wu, R., Kresse, O., Weng, X. et al. (2012). Modeling of Interaction of Hydraulic Fractures in Complex Fracture Networks. *SPE Paper 152052, SPE Hydraulic Fracturing Technology Conference*, The Woodlands, TX (6–8 February).

Xiong, H., Wu, W., and Gao, S. (2018). Optimizing well completion design and well spacing with integration of advanced multi-stage fracture modeling & reservoir simulation - a permian basin case study. *Paper SPE 189855 Presented at the SPE Hydraulic Fracturing Technology Conference and Exhibition*, The Woodlands, TX (23–25 January).

Xiong, H., Liu, S., Feng, F. et al. (2019). Optimize completion design and well spacing with the latest complex fracture modeling & reservoir simulation technologies–a permian basin case study with seven wells. *Paper SPE 194367,*

Presented at the SPE Hydraulic fracturing Technology Conference and Exhibition, Woodlands, TX (5–7 February).

Zhang, X. and Jeffrey, R.G. (2006). The role of friction and secondary flaws on deflection and re-initiation of hydraulic fractures at orthogonal pre-existing fractures. *Geophysical Journal International* 166 (3): 1454–1465.

Zhang, X. and Jeffrey, R.G. (2008). reinitiation or termination of fluid-driven fractures at frictional bedding interfaces. *Journal of Geophysical Research Atmospheres* 113 (B8): B08416.

Zhang, X., Jeffrey, R.G., and Thiercelin, M. (2007a). Effects of frictional geological discontinuities on hydraulic fracture propagation. *SPE 106111, SPE Hydraulic Fracturing Technology Conference*, College Station, TX (29–31 January).

Zhang, X., Jeffrey, R.G., and Thiercelin, M. (2007b). Deflection and propagation of fluid-driven fractures as frictional bedding interfaces: a numerical investigation. *Journal of Structural Geology* 29: 390–410.

14

Hydraulic Fracturing Mechanisms Leading to Self-Organization Within Dyke Swarms

Andrew P. Bunger[1,2], D. Gunaydin[1], S. T. Thiele[3], and A. R. Cruden[4]

[1] Department of Civil and Environmental Engineering, University of Pittsburgh, Pittsburgh, PA, USA
[2] Department of Chemical and Petroleum Engineering, University of Pittsburgh, Pittsburgh, PA, USA
[3] Helmholtz Institute Freiberg for Resource Technology, Helmholtz-Zentrum Dresden-Rossendorf, Freiberg, Germany
[4] School of Earth Atmosphere & Environment, Monash University, Clayton, Victoria, Australia

14.1 Introduction

Dykes are one of the most widespread mechanisms of magma transport in the brittle crust [1]. Some reach the surface to cause eruptions, but many also propagate laterally over large distances without breaching the surface. Among the most striking and widespread examples of these are giant continental dyke swarms, thought to originate from mantle-plume driven large igneous provinces [2–4]. Individual swarms contain hundreds to thousands of individual dykes that apparently grew laterally from a common source to attain lengths on the order of hundreds to thousands of kilometers. More than 100 of these dyke swarms are known on Earth and when combined, occur more than 300 times on Earth, Venus, and Mars [5]. Structure and geochemistry have been extensively studied for both giant dyke swarms (e.g. [4–12]) and smaller-scale swarms associated with local magma chambers and volcanic centers (e.g. [9, 13–17]).

It has been observed that spacing among dykes within swarms often follows a negative exponential or log normal distribution [14, 15], with the spacing in the distal part of the swarm (farthest from the source) converging to a value similar to the dyke height [18]. Meanwhile, geochemical data makes it clear that swarms are generated by multiple discrete magma injection events, or pulses ([6, 19]). Cross-cutting relationships indicate that each pulse possesses the potential to generate several dykes and/or dyke strands ([20]), and, at least in some cases, with an initial large intrusion event followed by a series of progressively smaller ones [21]. However, cross-cutting relationships cannot elucidate the timing between spatially separate dykes, and where dykes do crosscut give only relative age, making it difficult to assess if (i) dykes grow sequentially or in pulses of contemporaneous intrusions, and (ii) how long it takes to form giant dyke swarms.

Our understanding of giant dyke swarms is also limited to observation of ancient systems on Earth, Venus, and Mars. Direct seismological observations of present-day dyke growth are typically confined to cases of <60 km long, <10 km high, and 1–8 m wide single dykes propagating laterally from shallow magma reservoirs (e.g., [22–24]). While these cases provide potentially useful analogues to giant swarms, the length scales, time scales, and tectonic conditions of such observations make it necessary to be cautious and to avoid unequivocal and/or universal inferences based only upon analogy to small-scale, present-day swarms.

In this contribution, we discuss the components of swarm theory as they apply to dyke swarm formation. In this way, we develop a framework for understanding populations of mechanically interacting dykes based on swarm theory. Such a model is then illustrated via a minimalist dyke swarm simulator, focusing on a particular illustrative case for the Mackenzie Giant Dyke Swarm, Ontario, Canada.

14.2 Swarm Morphology and Fundamental Drivers

Mechanisms leading to swarming behavior have been recognized since at least the 1950s [25, 26], and were eventually studied in a mathematical framework in the 1980s, with the purpose of generating realistic looking groups of animals in computer graphics [27]. The basic result is to demonstrate that swarming morphologies emerge from systems involving members that move in a manner that

Mechanics of Hydraulic Fracturing: Experiment, Model, and Monitoring, First Edition.
Edited by Xi Zhang, Bisheng Wu, Diansen Yang, and Andrew Bunger.

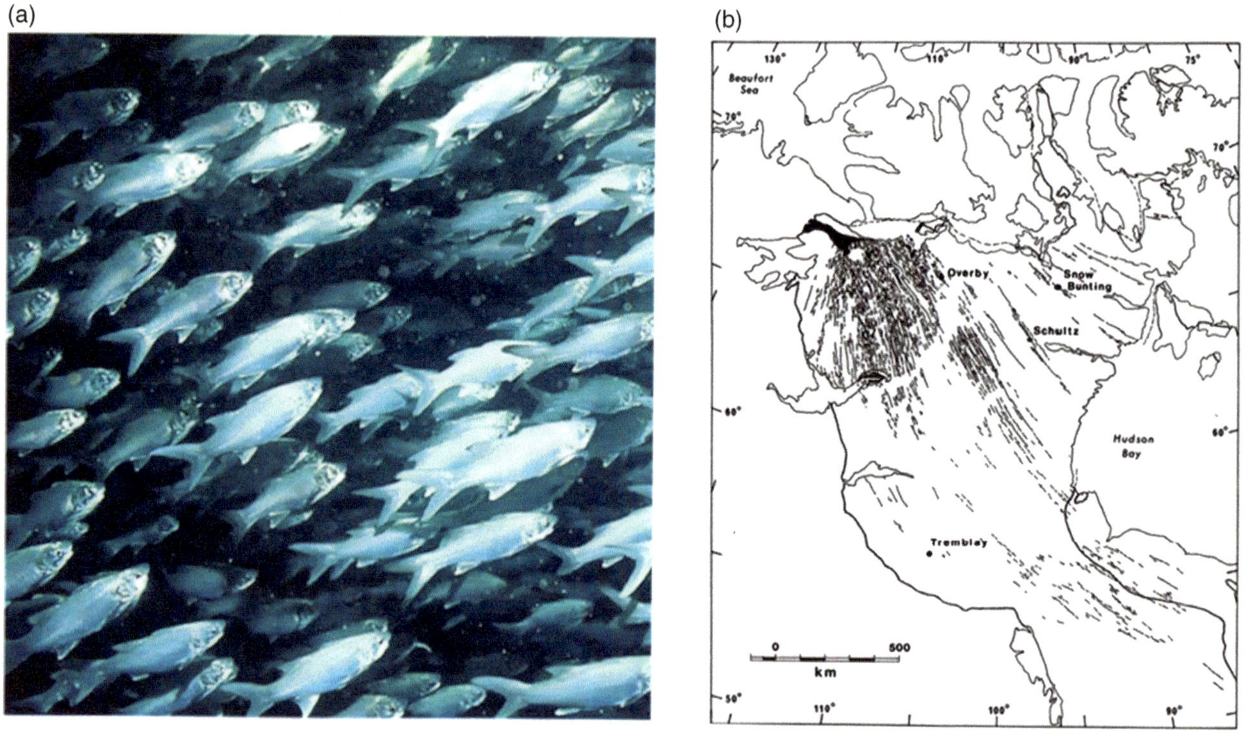

Figure 14.1 Observed swarm morphologies characterized by aligned members with a finite spacing among neighbors in (a) *Polydactylus sexfilis* fish (*Source:* Photo from NOAA Photo Library), and (b) Mesoproterozoic Mackenzie Dyke Swarm, Canada (*Source:* LeCheminant and Heaman [12] in [18], used with permission).

is governed by three fundamental drivers (as summarized in [28]):

1) Alignment: move in the same direction as neighboring members
2) Avoidance: do not run into other members
3) Attraction: stay close to other members

Swarm modeling comprises one of the early demonstrations of complex, life-like behavior emerging from computer-simulated systems governed by only a few simple rules. It also provides the basis for the development of subsequent fields such as particle swarm optimization [29], and swarm intelligence in control applications such as mobilization of multiple interacting robots [30, 31].

This biologically inspired paradigm provides a useful framework for developing mechanical models of dyke swarms. Like other swarming morphologies in nature, dyke swarms comprise multiple aligned members separated by a finite characteristic spacing. Indeed, such striking behavior in fish (as illustrated by Figure 14.1a) first led Breder [26] to propose that swarm morphology arises from interplay of attraction and avoidance. Drawing on insights from these past studies of biological systems, we propose that the geometry of giant dyke swarms (as illustrated by the Mesoproterozoic Mackenzie and Paleoproterozoic Matachewan Dyke Swarms, Canada, Figures 14.1b and 14.2, respectively) results from the interplay of the same three fundamental drivers. Alignment has been well studied in giant dyke swarms [5], as it is generally thought to reflect the external tectonic stresses during e.g. rifting or supercontinent breakup. However, the mechanical interactions between contemporaneous or consecutive dyking events have received far less attention. We suggest that these interactions determine the attraction and avoidance drivers in a swarm, which are critical to determine its emergent properties. The purpose of this paper is thus to elucidate these drivers and combine them in a minimalist way in order to simulate emergent spacing in dyke swarms.

14.3 Dyke Swarm Model and Energetics

Like all swarming behavior in nature, the self-organization expressed by the emergent geometry should be considered its most fundamental characteristic [26, 27]. Earth science examples of self-organizing swarms with geometries defined by a characteristic spacing include the formation of joint sets ([34–36]), fracture sets induced by thermal and/or pore pressure changes ([37–39]), and shrinkage

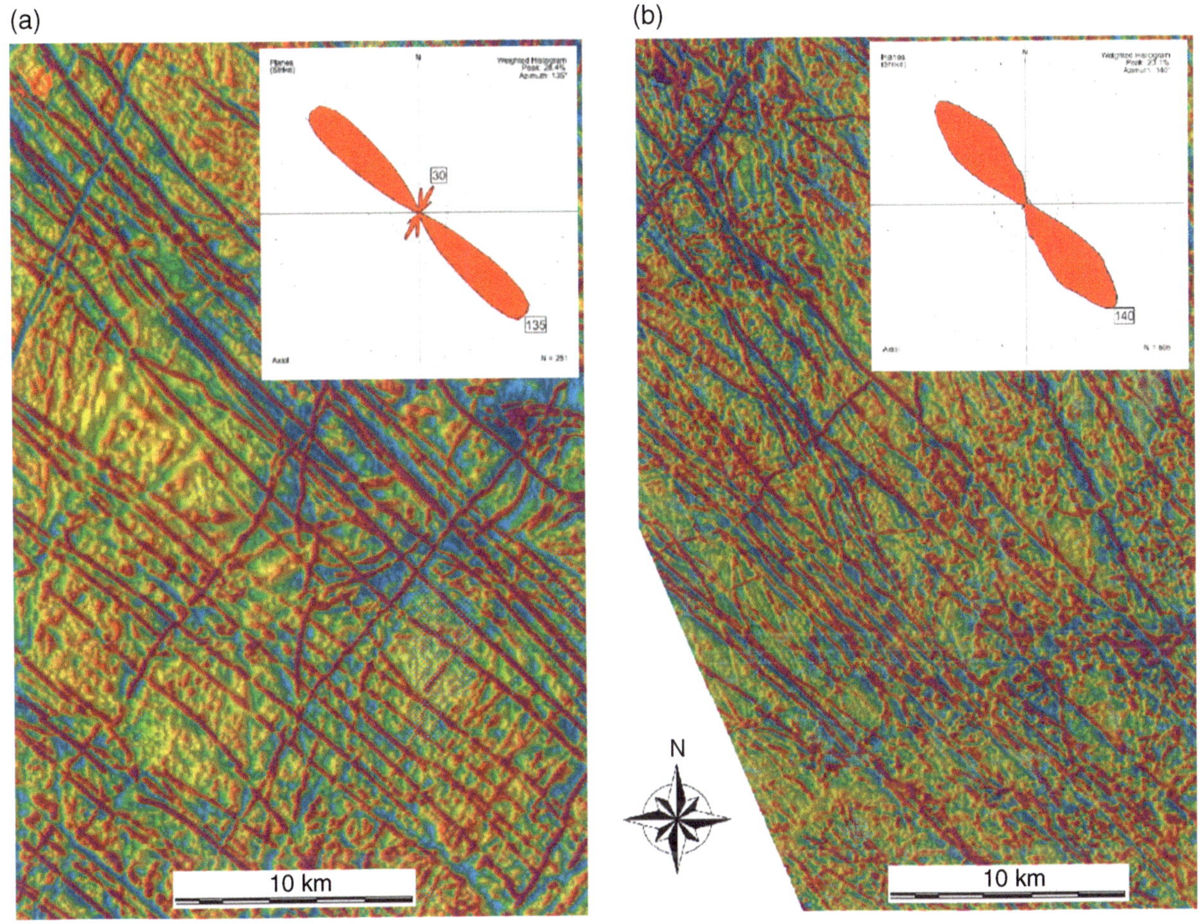

Figure 14.2 Aeromagnetic images (first vertical derivatives of reduced to the pole total magnetic field surveys) of, (a) an area near White River (WR) area, north of Lake Superior, Ontario, Canada (*Source:* Adapted from NWMO [32]); and (b) an area near Elliot Lake and Blind River, North of Lake Huron (NH), Ontario, Canada (*Source:* Adapted from NWMO [33]). Purple/red colors are high intensity positive magnetic gradient values (200–100 nT/km) and yellow-green-blue are increasingly negative gradient values (maximum = −0630 nT/km). Thin red, linear magnetic anomalies are mafic dykes: the northwest trending sets are the ca. 2450 Ma Matachewan diabase dykes; the north-east trending set are ca. 2167 Ma Biscotasing diabase dykes. Insets are rose diagrams (Gaussian smoothed spherical histograms, weighted by trace length) plotting the orientations of all dyke-related lineaments in each area. Also note that the dykes (i) exhibit generally regular spacing, and (ii) can be divided into distinct parallel and more closely spaced sub-clusters.

cracks associated with cooling of initially molten rock ([40, 41]) or dehydration of clay-bearing sediments ([42, 43]). These previously studied systems have a common assumption that is vital to the theories used to explain them, namely, that fracturing is the predominant energy dissipation mechanism. In contrast, fluid flow (not fracturing) is almost always the dominant mechanism of energy dissipation for growing dykes and hydraulic fractures ([1, 18]). Previous analyses of swarming fractures are therefore not directly applicable to fluid-driven hydraulic fracture swarms because they neglect the most important contributor to the energy balance of the system. In other words, dyke swarms, along with industrial hydraulic fractures and perhaps some fluid-driven sub-glacial fractures [44], form a family of fractures that, although different in some important ways, are related in that: (i) they can generate swarms, and (ii) the mechanics of fluid flow dominates the energetics of the system.

When considering the energetics of magma driven fracturing, it is useful to define the power, P, required to drive dyke growth. This input power is provided by the magma source and, fundamentally, the bulk volume change occurring during melting. During dyke propagation, this power is expended to: (i) increase the strain energy of deformed country rock (P_u) surrounding the dyke, (ii) overcome work done on each dyke by the compressive stresses induced by neighboring dykes (P_{int}), (iii) overcome preexisting rock stresses such as far-field/tectonic stresses and stress disturbance from the magma chamber that feeds the dykes (P_o), (iv) break rock at the fracture tip (P_c), (v) cause viscous flow within the dyke (P_f). Hence (after [18])

$$P_{growth} = P_u + P_{int} + P_o + P_c + P_f \quad (14.1)$$

All these terms are readily, though not trivially, quantified from classical continuum mechanics definitions of elastic strain energy, work performed on a body by a surface traction, and dissipation of energy due to fracturing and fluid flow. The symbolically intensive details for linear arrays of fluid-driven cracks are already established by Bunger [18] and Bunger et al. [18], and only the results necessary to the central arguments of the present paper will be reproduced in the sections below.

14.4 Alignment

The energy balance argument (Eq. 14.1) is based on the premise that magma does work on the host rock. This work must be equal to the difference between the total energy input from the magma source minus the dissipated energy due to viscous fluid flow and rock breakage. One of the terms in this resulting energy balance is the rate of work done by the magma in order to overcome the negative work being done on the dike by the compressive in situ stress, including both far field stresses and stress disturbance by the magma chamber, given by

$$P_o = -\sigma_n Q \quad (14.2)$$

where σ_n is the normal stress acting on the dyke, resisting its opening, and Q is the volumetric magma influx rate. As is well established, this negative work is minimized when the dyke is oriented in a direction such that its opening is resisted by the least compressive in-situ stress. In the context of a dyke swarm model, this impetus of the dykes to orient preferentially in a given stress field is equivalent to alignment.

Assuming that the minimum principal compressive stress is not vertical (as otherwise a sill complex would form rather than a dyke swarm), we can propose that alignment may be stronger or weaker depending upon the difference between the minimum and maximum horizontal in situ stresses. If these two stresses are equal, there is effectively no driver of alignment because it is energetically equivalent for the dykes to grow in any (vertical) orientation. On the other hand, if these two stresses are vastly different, dykes pay a much larger energy penalty for growing in any orientation other than that which causes their opening to be resisted by the least compressive stress. It is reasonable, then, to propose that a certain dimensionless parameter embodying the difference between the two horizontal principal stresses should govern whether a dyke swarm is strongly aligned or weakly aligned. If the magma pressure is well defined and equal to p_{source}, then the relevant parameter governing whether the stress difference is sufficient to cause strong alignment should compare the stress difference to the magma overpressure ($p_{over} = p_{source} - \sigma_{hmin}$), that is

$$\chi = \frac{\sigma_{Hmax} - \sigma_{hmin}}{p_{source} - \sigma_{hmin}} \quad (14.3)$$

Here, σ_{Hmax} and σ_{hmin} are the maximum and minimum horizontal stresses, respectively. A number of numerical simulations for fluid-driven cracks support this proposed alignment parameter ([45–47]). Note that we are neglecting the disturbance from the magma chamber, which may also be important in some settings.

Such an approach enables a comparison between well-aligned and poorly aligned dyke swarms. Firstly, however, we must state an important caveat that dyke swarms clearly form from multiple pulses. Hence, adjacent dykes are expected to typically display cross-cutting relationships indicative of sequential growth and, as far as alignment is concerned, we will limit the discussion only to overall trends that we propose to be associated with the tendency of large contrasts in the pre-existing stresses (numerator of Eq. 14.3) to overcome local stress variations, including principal stress rotations, induced by other simultaneously growing and/or previously-emplaced dykes (denominator of Eq. 14.3).

As an illustrative example, the Mackenzie Swarm in Ontario, Canada (Figure 14.1) is a well-aligned swarm in which most dykes grew nearly parallel to their neighbors and with a predominant NNW–SSE strike. This ca. 1267 Ma swarm was emplaced into Archean and Paleoproterozoic crystalline host rocks. Given that these host rocks will have had $a > 1000$ Ma long pre-dyking cooling, uplift and erosion history, we ascertain that they could have sustained high differential stresses. It is also understood that the Mackenzie swarm was intruded during a time of rifting and plume-induced uplift such that the tectonic conditions of the time of emplacement would have been conducive to generation of high differential stresses. It is well known that maximum horizontal stress increases with depth at a rate exceeding that for the minimum horizontal stress [48]. For the sake of argument, we estimate the differential stress ($\sigma_{Hmax} - \sigma_{hmin}$) to increase with depth in the order of 10 MPa/km. We also estimate the available magma overpressure to be around $p_{over} \sim 70$ MPa at 15 km depth. Hence, the alignment parameter (Eq. 14.3) takes on a value of about $\chi = 2$ at 15 km depth.

To further illustrate the stress control on alignment we compare the relative alignment of dykes in the ca. 2452 Ma Matachewan swarm, Ontario, Canada (Figure 14.2). The Matachewan dyke swarm radiates to the north and northwest from an unexposed point source located approximately under the southeast tip of Manitoulin

Island, Lake Huron, Ontario [49,50]. The subvertical dykes of the swarm intrude ca. 2700 Ma granitic rocks and greenstones of the Superior Province. Three sub-swarms have been identified within the fan-like structure [50]; the north-trending M1 sub-swarm in the east, the north-northwest-trending M2 sub-swarm further west, and the westernmost northwest-trending M3 sub-swarm. The dykes shown in Figure 14.2 are part of the M3 sub-swarm, which can be traced over ~700 km from the southern Abitibi greenstone belt in the southeast to the Sachigo province in the northwest. It is also transected by the Kapuskasing structural zone (KSZ), which exposes deeper crustal rocks in its hangingwall to the northwest [50, 51].

High resolution aeromagnetic images (first vertical derivatives of the reduced to the pole total magnetic field) show two sections of the M3 swarm, one from the White River (WR) area of the Wawa greenstone belt (Figure 14.2a) and the other from the southern Abitibi greenstone belt, north of Lake Huron (NH) (Nuclear Waste Management Organization [NWMO], [33, 36]). The NH area dykes are closer to the inferred source of the swarm (~230 km) and the present level of erosion represent emplacement into granitic rocks in an upper crustal setting (5–10 km). The WR area dykes occur northwest of the KSZ and represent emplacement at deeper paleodepths in the crust (10–15 km). They are located ~450 km from the inferred source. The WR area is located 220 km northwest of the NH area, such that the dykes in both images are effectively along strike of each other (Figure 14.2).

Matachewan dykes in the NH area range in peak trends from 140° to 155° (Figure 14.2b, inset). The 155° set can be seen to bifurcate from the 140° set in the assumed northwest propagation direction and then rejoin other 140° dykes further along strike. Matachewan dykes in the WR area are very strongly aligned with one peak trend at 135° (Figure 14.2a, inset). Some bifurcation and reconnection of dykes can still be observed but it is much less common than in the NH area.

Here we see an example of a dyke swarm where dykes emplaced at shallower depth and closer to the inferred magma source (NH) are relatively poorly aligned (15° dispersion) compared to dykes emplaced at greater depth and further from the source (WR). In both cases, we consider it reasonable to assume: (i) the host rocks had sufficient cooling time so as to be brittle at the relatively rapid strain rates associated with dyking, and (ii) the differential stress gradient and magma overpressure were probably similar to the Mackenzie swarm. Hence, for the well-aligned, deep, far-from-source WR dykes, it is reasonable to estimate a similar value of the alignment parameter as for the Mackenzie dykes, that is, $\chi = 2$. However, for the less-aligned, shallower, closer-to-source NH dykes, it is estimated that the alignment parameter is around half the value compared to the WR dykes due to emplacement at half of the depth. Additionally, the stress disturbance due to the proximity of the source is expected to further reduce the differential stress. While modeling this disturbance in detail is beyond the current scope, it is reasonable to take the depth-related reduction in differential stress as an upper bound so that the alignment parameter for the NH dykes is bounded by $\chi < 1$. Hence, the difference in alignment between these two parts of the Matachewan Swarm is proposed to stem from variation in the differential stress across the swarm.

Through this analysis and comparison between field examples of well- and poorly aligned dyke swarms, we are thus proposing that the driver of alignment is mainly associated with minimizing energy associated with overcoming the far field stress. Next, we consider potential avoidance and attraction drivers that lead to the characteristic spacing emerging within the swarms (e.g. Figure 14.2).

14.5 Avoidance

14.6 Stress Shadow

Avoidance has been previously related to the stresses induced by dyke intrusion, leading to a tendency for fluid-driven cracks such as dykes to avoid growing in close proximity to one another (e.g. [52–55]). This phenomenon can be understood by first realizing that dyke growth is accompanied by accommodation of magma volume through deformation of the host rock. The process is embodied in the change in elastic strain energy (Pu, recalling Eq. 14.1). Such deformation of the host rock is accompanied by a local increase in the compressive stress surrounding each dyke. This compressive stress acts on each of the dyke's neighbors in a manner that is counteractive to the magma pressure driving dyke propagation. Hence, for each dyke, work by the magma on the host rock is required to counteract the negative work being performed by its neighbors. This work of interaction is embodied by P_{int} (again recalling Eq. 14.1), which is shown for a uniform array of blade shaped dykes to have the form

$$P_{int} \propto \frac{1}{s^2} \qquad (14.4)$$

Because of its inverse proportionality to the separation between the dikes, s, we can see that this term comprises a driver of avoidance. That is, minimizing this negative work drives dykes to grow as far apart as possible.

The interaction stress is not the only driver of avoidance within these systems. Looking in detail at the energy dissipated in association with rock breakage (P_c, Eq. 14.1), one finds that this term also is minimized when fewer dykes propagate [56], which would result in longer but more widely spaced dykes. Specifically, the energy dissipated by rock breakage scales with the surface area of the cracks that are created, which is minimized when the fluid is concentrated to one dyke rather than many. It is important to note, however, that the energy dissipation associated with rock fracturing can be substantially smaller than the energy dissipation associated with fluid flow. In such cases, the contribution of the avoidance force from rock fracturing is insignificant because its contribution to the overall energy balance is negligible.

The consequences of these drivers of avoidance are often referred to as a stress shadow effect (after e.g. [57]). Their most recognized manifestation is the suppression of some hydraulic fractures by others in industrial horizontal well stimulations, as illustrated by Peirce and Bunger [58] (Figure 14.3a). Similarly, how some dykes suppress the growth of others through the stress shadow effect has also been explored, for example, by Mukhamediev et al. [61].

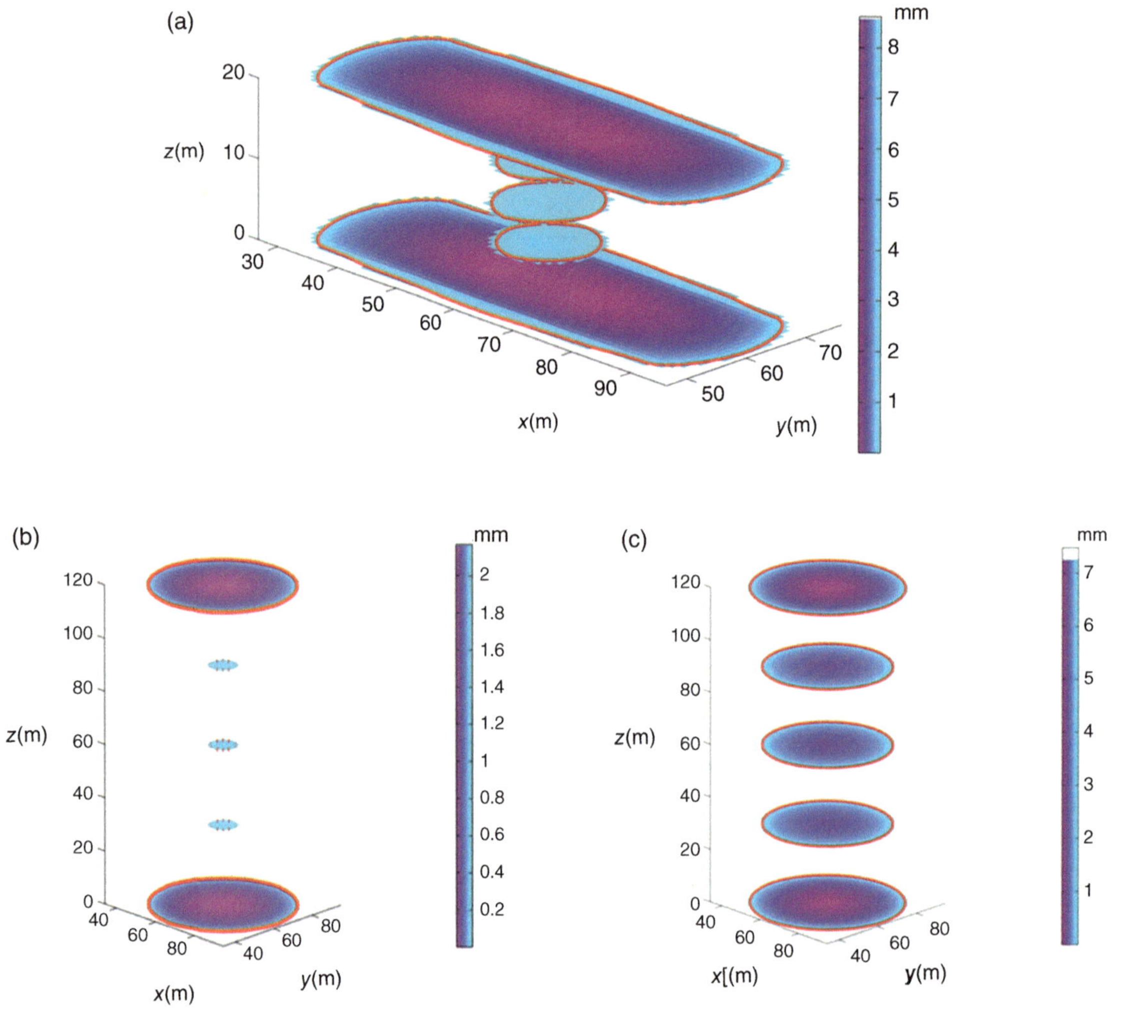

Figure 14.3 Numerical simulations of fluid-driven cracks using a 3D simulator accounting for full stress interaction among the fractures, but under the limiting assumption that all cracks grow in pre-defined, parallel planes. These illustrate: (a) Example of the so-called "stress shadow" phenomenon leading to suppression of inner fractures in an array of five blade-shaped fluid-driven cracks (*Source:* Peirce and Bunger [59], with permission), (b) Array of five radial fluid-driven cracks showing instability leading to immediate suppression of central fractures under conditions where rock breakage associated with rock fracture toughness comprises the dominant energy dissipation mechanism, i.e. in the "toughness dominated" regime (*Source:* Bunger and Peirce [60], with permission), and (c) Array of five radial fluid-driven cracks showing greater tendency for swarm-like growth of all fractures when viscous fluid flow is the dominant energy dissipation mechanism, i.e. in the "viscosity dominated" regime (*Source:* Bunger and Peirce [60], with permission). Case (b) can thus be considered to show the consequences of a stronger ratio of avoidance to attraction compared to Case (c). Here, the color scale refers to the fracture opening.

14.7 Stress Plugs

The stress-shadow concept can also be extended by considering the sensitivity of fracture propagation to stress. All fluid driven fractures, dykes included, act to reduce the overall differential stress through elastic strain perpendicular to the fracture surface, which is optimally perpendicular to the least compressive stress direction. This reduction in differential stress, or increase in least compressive stress, has two important effects: it increases the work done during fracture opening (as discussed above), increasing the viscous drag during magma flow, and also moves the host-rock farther from failure, which increases the energy needed for rock breakage by subsequent fractures. In large systems this change in stress can be sufficient to deflect propagating dykes away from previously intruded areas by favoring propagation along regions of the 3-D fracture front (leading edge) that is farthest from high stress regions caused by previous intrusions. These "stress plugs" have been proposed to explain the migration of eruptive activity at ocean island volcanoes [62], and they represent a subtly different avoidance mechanism to the well-established stress-shadow effect.

It is important to note that the influence of both avoidance effects (stress shadows and stress plugs) will decay with time, as inelastic deformation and/or tectonic extension dissipates strain energy stored in the surrounding rock mass. Although insignificant for engineered hydraulic fractures, this dissipation could be important for natural dyke swarms that form over longer timescales. Such time-dependence adds important complexity to the balance between competing drivers controlling dyke swarm growth.

14.8 Attraction

The model element that has been most elusive over decades of study of dyke swarms has been a basic accounting for the driver(s) of attraction in these systems. Observations of spatial and temporal clustering within dyke swarms (e.g. Figure 14.2) suggest that such a driver of attraction is necessary (as clustering cannot be explained by avoidance and alignment alone), but it is difficult to imagine why fluid-driven cracks could have any mechanical impetus to grow near one another in the first place.

A simple attractive element can be included if fractures are required to nucleate from a localized source, such as a magma chamber or sill. By starting in the same location, propagating fractures must follow some tradeoff between propagation in the regionally favorable direction (alignment), and separation from their neighbors (avoidance). However, the influence of this tradeoff would be expected to decrease rapidly away from the common source of the dykes, meaning it cannot explain distal observations of regular spacing and clustering, such as observed in Figure 14.2.

More subtly, the viscous energy dissipation accompanying dyke growth (embodied by P_F in Eq. 14.1) could also promote attraction. It turns out that splitting the magma among as many dykes as possible minimizes this dissipation [18, 56]. At the other extreme, pushing all of the magma through a single dyke maximizes the viscous dissipation associated with expulsion of an equivalent volume of magma from the source. However, it is not obvious that this should be the case. It is true that splitting the magma as many ways as possible diminishes the volumetric flow rate for each individual dyke, thereby diminishing the shear strain rate and in turn minimizing the viscous dissipation (which scales with the strain rate, see e.g. [63]). But, as previously discussed, splitting magma between many dykes maximizes the surface area of the dykes, and the quantity used to compute the fluid flow dissipation ends up being integrated over the domain of these fractures (see details in [56], as adapted from [63]). So, on one hand we have a splitting of magma that drives the fluid dissipation to be minimized through the creation of many dykes, while on the other hand minimizing the creation of surface area is achieved through growth of fewer dykes.

To discern which of these drivers is stronger, one must do the analysis in detail. Upon carrying out such analysis for plane strain, radial, and blade shaped fluid-driven cracks, Bunger [18] shows that the energy is minimized through splitting of the magma into as many fluid-driven cracks as possible for radial and blade-shaped geometries. On the other hand, for plane strain the energy is minimized by concentrating the fluid flow into a single dominant fracture/dyke. From this we can conclude two things. First, for three-dimensional, or at least simple three-dimensional planar geometries (e.g. radial and blade-shaped), minimizing viscous fluid dissipation provides a driver for attraction within dyke swarms. Second, this prior analysis finds that the role the viscous dissipation term depends on the geometry of the model. In particular, plane strain models have no obvious source for attraction and therefore do not approximate the energetics of a three-dimensional system.

Besides the need for a simulator to account for three-dimensional geometry (even in a simplified manner), the origin of this attractive force within the fluid flow equations requires that magma flow in a dyke must be modeled explicitly, rather than taking a simplified fracture mechanics approach (i.e. assuming uniform pressure within the dykes). To illustrate this point, consider an array of five fluid-driven cracks. Bunger and Peirce [60] used a fully coupled hydraulic fracture simulator to show that the tendency to sustain simultaneous growth of multiple fractures is profoundly different between two limiting regimes. In the first regime ("Toughness Dominated"), the fracture toughness of the rock is large enough and the fluid viscosity small enough that viscous dissipation can be neglected. In this case, the fluid pressure is nearly uniform inside each fracture. This results in an almost immediate suppression of the internal fractures within the array (Figure 14.3b), indicating little impact of attraction force within the system. On the other hand, in the second limiting regime ("Viscosity Dominated"), the fracture toughness of the rock is taken to be negligibly small, such that the dominant dissipation mechanism is viscous flow. In this case, multiple fluid-driven cracks grow stably until the fracture radius exceeds the fracture spacing, at which point suppression of the inner fractures begins due to stress interaction (avoidance force). This extended period of stable, nearly uniform partitioning of the fluid among the fractures (Figure 14.3c) is a consequence of the drive to minimize energy dissipation through fluid partitioning.

14.9 Emergent Spacing

The self-organization expressed by the emergence of a natural spacing among a swarm's members should be considered its most fundamental characteristic, and grasping the mechanics governing the swarming behavior is synonymous with grasping the drivers that lead to the emergent spacing [27]. Emergent spacing results from a competition between attraction and avoidance forces, and so quantifying both is necessary to predict spacing. Also, it has been previously shown that:

- Plane strain (2D) geometry never leads to a well-defined, emergent spacing due to the lack of a driver of attraction (see above and [56], and [64]).
- Blade-shaped geometry gives rise to a stable, emergent spacing among many growing dykes/fluid-driven cracks. This emergent spacing scales with the dyke height and depends on the inlet boundary conditions.
- Constant source pressure leads to an emergent spacing that is predicted to be equal to the dyke height [18].
- Constant volumetric flow rate in the source leads to an emergent spacing that is predicted to be 1.5–3 times the dyke height, where the lower end of this range corresponds to cases where there is a large pressure drop associated with the magma entering the dyke (perforation friction in the hydraulic fracturing analogue system, [64]) and the upper end corresponds to cases where such pressure drop (inlet friction) is negligible [56].

Such predictions provide testable hypotheses for comparison to field data. A compilation of available data, both from dyke swarms and industrial hydraulic fractures, is presented in Table 14.1. Note that for inclusion it is required that the distal spacing has been measured and the dyke height can be estimated. Hence, there are very few examples in spite of a relatively large number of investigations of dyke and dyke swarm structure in the literature. Consistent with theoretical prediction, in all of the cases in Table 14.1 we observe aligned swarms with a spacing ranging from 1 to 2 times the dyke/fracture height.

Table 14.1 Dyke and hydraulic fracture swarm observed geometry.

Name	Location	Approx. maximum length	Approx. mean height	Approx. mean spacing	Source type	Reference(s) and notes
Mackenzie Swarm	Canada	3000 km	30 km	30 km	Constant pressure	Distal spacing presented by Bunger et al. [18]. Height inferred from crustal thickness.
Alfafjordur Swarm	Iceland	30 km	2.8 km	2.5 km	Constant pressure	Height inferred from Krafla episode and mean spacing presented by Bunger et al. [18]
Cotton Valley	Texas, USA	300 m	100 m	150 m	Constant rate	For industrial hydraulic fracturing of horizontal wells, Fisher et al. [53] observe microseismic data indicating ~100 m hydraulic fracture height and "regularly spaced cross-cutting fracture that tend to appear at ~150 m intervals regardless of perforation cluster location." Further discussion by Bunger et al. [64].
Barnett Shale	Texas, USA	300 m	90 m	90 m	Constant rate	Hydraulic fracturing from vertical wells results, in 4 of 6 cases, in multiply-branched network with spacing between aligned, dominant lineaments with spacing around 90 m, as evidenced by microseismic data ([65]; [66]).

14.10 Simulating Dyke Swarm Assembly

A simple numerical model can be devised in order to capture the alignment, avoidance, and attraction forces in a minimalist way. Although details are omitted for the sake of brevity, this model accounts for swarm assembly drivers as follows:

- Alignment is obtained by brute force by assuming parallel propagation in a direction determined by the regional horizontal principal stress direction.
- Avoidance is obtained by accounting for the elastic stress interaction among fractures using a functional interpolation between near and far field interaction stresses exerted on each dyke by the summation of its neighbors, where the near and far field expressions of this interaction stress are presented by Bunger et al. [18].
- Attraction is obtained by accounting for laminar flow of a viscous fluid according to Poiseuille equation.

The model therefore assumes parallel planar fractures that are limited in height and grow laterally with a blade shape ([67]). Coupling among fracture length, height, pressure, and width (aperture) is provided by elasticity. Fracture stress interaction is also provided by elasticity. Fluid flow is governed by local volume balance of an incompressible fluid as well as Poiseuille equation providing a relationship between fluid flux and the pressure gradient for laminar flow of a Newtonian fluid ([67, 68]). Propagation assumes rock fracturing consumes negligible energy compared to viscous fluid flow and therefore takes on a form that ensures satisfying fluid flow and elasticity equations in the near tip region. The magma source is modeled as a pressure boundary condition acting equally on the inlets of all dykes that evolves with time, assuming a fixed volume and compressibility magma chamber recharging at a fixed volumetric rate.

Dykes are taken to initiate according to a classical kinetic fracture theory (e.g. [69] as adapted by [58]) whereby the time to fracturing is a power law of the ratio of the applied tensile stress (generated around the magma source due to the magma overpressure) and the nominal tensile strength of the material. By assuming stochastic variability of the tensile strength of the rock, each dyke initiates at a different time depending on tensile strength variability and how each dyke is impacted by prior growth of nearby dykes (which reduce or suppress tensile stresses). Dykes are taken to solidify according to a 1D chill layer expansion model (Carslaw and Jaeger [70] as applied to solidifying dykes by Lister [71], Bolchover and Lister [72], and Chanceaux and Menand [73]). Hence, the model does not make an a priori assumption about simultaneous versus sequential dyke growth. Both limits are captured, with simultaneous growth occurring when rock strength is relatively uniform, overpressure is relatively large, and solidification is slow (i.e. due to relatively warm host rock). Sequential emplacement occurs when rock strength is highly variable, magma overpressure is relatively low, and/or solidification is rapid (i.e. due to relatively cool host rock). Both limits are physically realistic according to this model, with typical cases falling between these two limits having several dykes growing simultaneously and the swarm being constructed as this process repeats over time.

The interplay among the governing parameters is complex and a detailed parametric study is beyond the scope of this discussion. However, a key outcome is the emergent spacing, which is qualitatively unchanged by parameter combinations favoring emplacement that is sequential, simultaneous, or mixed. Figure 14.4 shows a typical result,

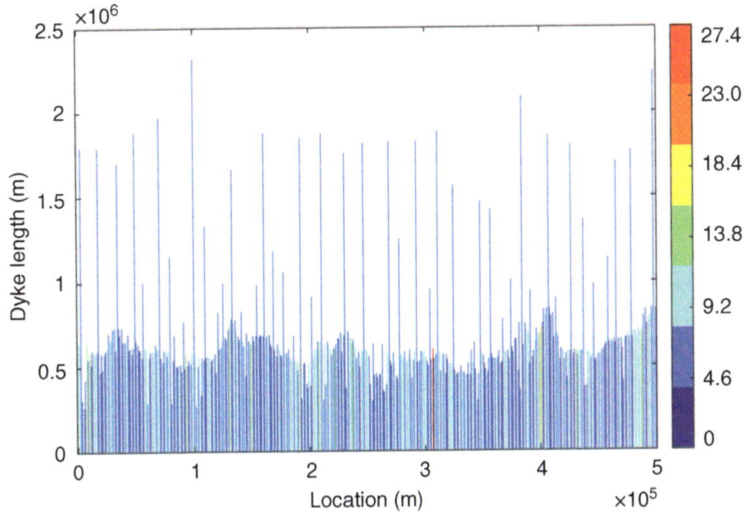

Figure 14.4 Dyke length as a function of location for simulation with 35 km dyke height, where the color scale is in years from time of first dyke propagation.

making use of input parameters intended to represent the aforementioned Mackenzie dyke swarm, including a dyke height of 35 km. Here it is seen that the longest dykes are observed to grow a few at a time and predominantly within a period of a couple of years, with the propagation time for each dyke on the order of days. The hundreds of very short dykes should be considered to have been trial locations where dyke growth was attempted but quickly arrested owing to the large stresses (i.e. stress-plug type limiting of growth). Hence, the longest dykes that emerge are the focus of the subsequent discussion.

The final length of the longest dykes is around 2000 km and average thickness is around 20 m, both of which are similar to observations, with measurements from a variety of sources that are reviewed in Bunger et al. [18]. Additionally, the spacing attains a value similar to the 25 km average distal spacing from measurements (Figure 14.5). Consistent with this mean distal spacing, the frequency spectrum of the dyke lengths (Figure 14.6) indicates a dominant frequency of periodicity emerges around 28 km, and therefore similar to the fracture height and consistent, again, with measurements. Repeated numerical experiments show that this emergent spacing varies as different random distributions of rock strength (and hence initiation times of the dykes) are used. However, the emergent spacing stays within the range of about 15–50 km and therefore present a clear tendency to scale with the dyke height. Repeated numerical experiments with other dyke heights confirms that emergent spacing is similar to dyke height, a result predicted by Bunger et al. [18] as a spacing that is energetically advantageous because it maximizes the volumetric discharge rate from a source under constant pressure. Hence, this simple model that brings together basic embodiments of alignment, attraction, and avoidance forces is shown to give promisingly similar predictions to observations and theory.

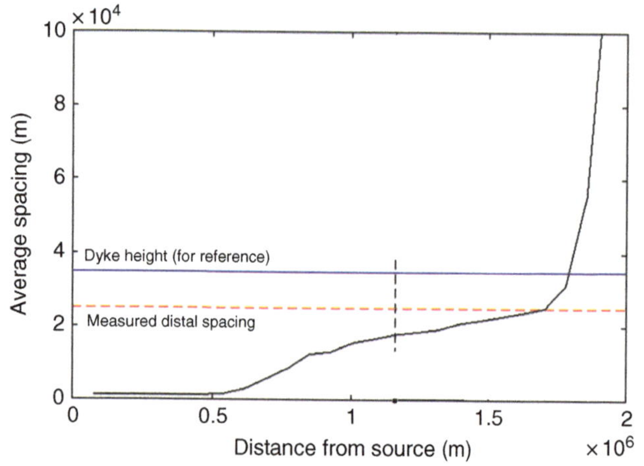

Figure 14.5 Intradyke spacing as a function of distance from source for simulations of the Mackenzie dyke swarm, with mean distal spacing as reported in Bunger et al. [18] (*Source:* Adapted from Bunger et al. [18]).

14.11 Conclusions

Dyke swarm morphologies are consistent with a system driven by the interplay between alignment, avoidance, and attraction forces. Models suitable for the purpose of predicting emergent spacing and alignment must couple together the influence of all three of these drivers. On the other hand, incomplete models that do not account for all three drivers, or that reduce the dimensionality of the model in such a way that the forces are fundamentally altered, are not appropriate. Using our examination of the mechanical origins of swarming within systems of dykes, we conclude with a basic set of criteria, which can be used to evaluate the suitability of a given model for capturing the three fundamental

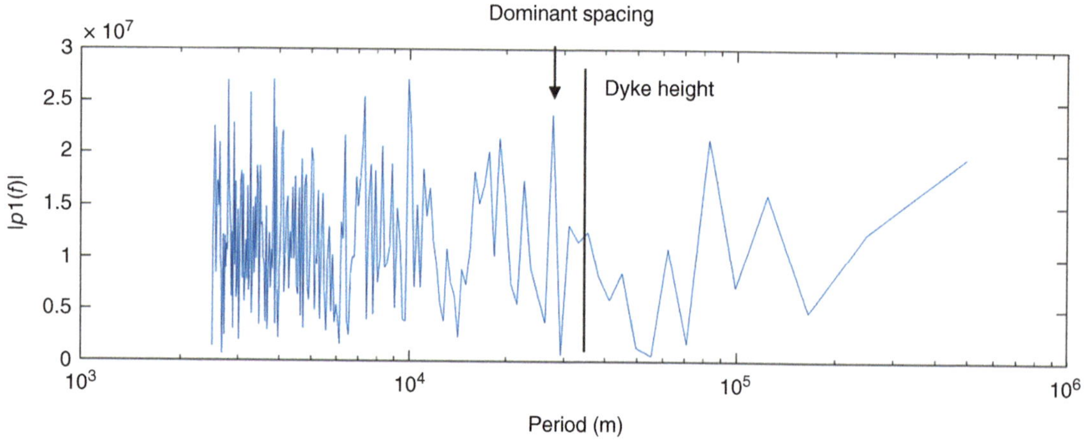

Figure 14.6 Spectrum of dyke lengths indicating a dominant spacing around 28 km.

drivers of swarming. Based on these, a suitable simulator must model:

1) The presence of in situ stresses. Without these the simulator does not account for the overall driver of alignment.
2) Interaction stresses among various dykes. Without these the simulator does not account for the main driver of avoidance.
3) Energy dissipation associated with fluid flow. Without accounting for fluid flow, the simulator does not include the basic ingredients required to capture the driver of attraction. There is, of course, a certain temptation to use fracture mechanics approaches that neglect coupled fluid flow in an effort to obtain a qualitative understanding of the system. However, such efforts will mostly be misguided because of the fundamental difference in the energetics of a system that accounts for fluid flow versus one that does not.
4) Three-dimensional (3D), or at least a simplified 3D geometry such as radial or blade-shaped growth. Attraction arising from fluid dissipation does not occur in a plane strain model. Hence, while there is a certain appeal to using 2D models in order to obtain a qualitative picture of how dyke swarms grow, such efforts will generally give misleading results because the energetics of the 2D system fundamentally differs from the energetics of a 3D system. Specifically, the 2D simulated system will lack the presence of the attraction force (from the fluid flow equations, as previously discussed), and therefore will overestimate the tendency of the system to localize fluid flow into just one or a few dykes.

These guidelines have several implications, which are not obvious at the outset of a modeling program. For example, a relatively simple 3D model might be appropriate while an advanced, multi-physics 2D model is not. Additionally, a model that accounts in a basic way for the impact of fluid dissipation may be suitable, while a state-of-the-art fracture mechanics model that fails to account for fluid flow is not. Indeed, it is shown that even a minimal complexity model that accounts for these combined model features is able to provide promising similarity to length, thickness, and intra-dyke spacing observed in a field example based on the Mackenzie Dyke Swarm. Hence, the swarm theory paradigm provides an overarching framework for model selection to ensure simulators are fundamentally capable of predicting alignment and emergent spacing within dyke swarms and, by extension, other fluid-driven systems of cracks.

Acknowledgments

This material is based upon work supported by the National Science Foundation under Grant No. 1645246 to APB and Australian Research Council Discovery Grant DP190102422 to ARC and APB. We also thank Nuclear Waste Management Organization, Toronto, Canada, for permission to use the aeromagnetic and structural data in Figure 14.2.

References

1 Lister, J.R. (1990). Buoyancy-driven fluid fracture: similarity solutions for the horizontal and vertical propagation of fluid-filled cracks. *Journal of Fluid Mechanics* 217: 213–239.
2 Ernst, R.E. and Buchan, K. (1997). Giant radiating dyke swarms: their use in identifying pre-mesozoic large igneous provinces and mantle plumes. In: *Large Igneous Provinces: Continental, Oceanic, and Planetary Flood Volcanism*, Geophysical Monograph Series, vol. 100 (ed. J.J. Mahoney and M.F. Coffin), 297–333. Washington, DC: American Geophysical Union.
3 Ernst, R.E. and Buchan, K. (2003). Recognizing mantle plumes in the geological record. *Annual Review of Earth and Planetary Sciences* 31 (1): 459–523.
4 Halls, H.C. (1982). The importance and potential of mafic dyke swarms in studies of geodynamic processes. *Geoscience Canada* 9 (3): 145–154.
5 Ernst, R.E., Head, J.W., Parfitt, E. et al. (1995). Giant radiating dyke swarms on earth and Venus. *Earth-Science Reviews* 39 (1–2): 1–58.
6 Baragar, W.R.A., Ernst, R.E., Hulbert, L., and Peterson, T. (1996). Longitudinal petrochemical variation in the Mackenzie dyke swarm, northwestern Canadian shield. *Journal of Petrology* 37 (2): 317–359.
7 Buchan, K.L. and Ernst, R.E. (2013). *Diabase Dyke Swarms of Nunavut, Northwest Territories, and Yukon, Canada*. Geological Survey of Canada Open File 7464.
8 Ernst, R.E. (1991). *The Near-Surface Geometry of the Great Abitibi Dyke, Superior Province, Canada*. Geological Society of America Map & Chart Series MCH-071.
9 Fahrig, W.F. and Halls, H.C. (ed.) (1987). *Mafic Dyke Swarms*. Geological Association of Canada.

10 Halls, H.C. (1991). The Matachewan Dyke Swarm, Canada: an early Proterozoic magnetic field reversal. *Earth and Planetary Science Letters* 105 (1–3): 279–292.

11 Hou, G., Kusky, T.M., Wang, C., and Wang, Y. (2010). Mechanics of the giant radiating Mackenzie dyke swarm: a paleostress field modeling. *Journal of Geophysical Research: Solid Earth* 115: B02402:https://doi.org/10.1029/2007JB005475.

12 LeCheminant, A.N. and Heaman, L.M. (1989). Mackenzie igneous events, Canada: middle Proterozoic hotspot magmatism associated with ocean opening. *Earth and Planetary Science Letters* 96 (1–2): 38–48.

13 Gudmundsson, A. (1995). Infrastructure and mechanics of volcanic systems in Iceland. *Journal of Volcanology and Geothermal Research* 64: 1–22.

14 Jolly, R.J.H. and Sanderson, D.J. (1995). Variation in the form and distribution of dykes in the Mull swarm, Scotland. *Journal of Structural Geology* 17 (11): 1543–1557.

15 Jolly, R.J.H., Cosgrove, J.W., and Dewhurst, D.N. (1998). Thickness and spatial distributions of clastic dykes, northwest Sacramento Valley, California. *Journal of Structural Geology* 20 (12): 1663–1672.

16 Paquet, F., Dauteuil, O., Hallot, E., and Moreau, F. (2007). Tectonics and magma dynamics coupling in a dyke swarm of Iceland. *Journal of Structural Geology* 29 (9): 1477–1493.

17 Walker, G.P.L. (1986). Koolau dike complex, Oahu: intensity and origin of a sheeted-dike complex high in a Hawaiian volcanic edifice. *Geology* 14: 310–313.

18 Bunger, A.P., Menand, T., Cruden, A. et al. (2013). Analytical predictions for a natural spacing within dyke swarms. *Earth and Planetary Science Letters* 375: 270–279.

19 Halls, H.C. (1986). Paleomagnetism, structure, and longitudinal correlation of middle Precambrian dykes from northwestern Ontario and Minnesota. *Canadian Journal of Earth Sciences* 23 (2): 142–157.

20 Burchardt, S., Tanner, D.C., Troll, V.R. et al. (2011). Three-dimensional geometry of concentric intrusive sheet swarms in the Geitafell and the Dyrfjöll volcanoes, eastern Iceland. *Geochemistry, Geophysics, Geosystems* 12 (7): 1–21.

21 Passarelli, L., Rivalta, E., and Shuler, A. (2014). Dike intrusions during rifting episodes obey scaling relationships similar to earthquakes. *Scientific Reports* 4: 3886.

22 Ayele, A., Keir, D., Ebinger, C. et al. (2009). September 2005 mega-dike emplacement in the Manda-Harraro nascent oceanic rift (Afar depression). *Geophysical Research Letters* 36: L20306:https://doi.org/10.1029/2009GL039605.

23 Eibl, E.P.S., Bean, C.J., Vogfjörd, K.S. et al. (2017). Tremor-rich shallow dyke formation followed by silent magma flow at Bárđarbunga in Iceland. *Nature Geoscience* 10: https://doi.org/10.1038/NGEO2906.

24 Sigmundsson, F., Hooper, A., Hreinsdóttir, S. et al. (2015). Segmented lateral dyke growth in a rifting event at Bárðarbunga volcanic system, Iceland. *Nature* 517: 191–195.

25 Breder, C.M. (1951). Studies on the structure of the fish school. *Bulletin of the American Museum of Natural History* 98 (1): 1–28.

26 Breder, C.M. (1954). Equations descriptive of fish schools and other animal aggregations. *Ecology* 35 (3): 361–370.

27 Reynolds, C.W. (1987). Flocks, herds and schools: a distributed behavioral model. *Computer Graphics* 21 (4): 25–34.

28 Fisher, L. (2009). *The Perfect Swarm: The Science of Complexity in Everyday Life*. Basic Books.

29 Kennedy, J. and Eberhart, R.C. (1995). Particle swarm optimization. *Proceedings of the IEEE International Conference on Neural Networks IV*. pp. 1942–1948. Piscataway, NJ: IEEE.

30 Beni, G. (1988). The concept of cellular robotic system. *Proceedings IEEE International Symposium on Intelligent Control*. pp. 57–62. (24–26 August 1988). Arlington, VA. IEEE: New York.

31 Fukuda, T. and Nakagawa, S. (1988). Approach to the dynamically reconfigurable robotic system. *Journal of Intelligent and Robotic Systems* 1 (1): 55–72.

32 Nuclear Waste Management Organization (2017a). Phase 2 Geoscientific Preliminary Assessment Initial Findings: Township of White River and Area, ON. *Report APM-REP-01332-0212*. 63pp. (www.nwmo.ca)

33 Nuclear Waste Management Organization (2017b). Phase 2 Geoscientific Preliminary Assessment Initial Findings: Town of Blind River and City of Elliot Lake and Area, ON. *Report APM-REP-01332-0220*. 45pp. (www.nwmo.ca).

34 Bai, T., Pollard, D.D., and Gao, H. (2000). Explanation for fracture spacing in layered materials. *Nature* 403 (6771): 753.

35 Hobbs, D.W. (1967). The formation of tension joints in sedimentary rocks: an explanation. *Geological Magazine* 104 (6): 550–556.

36 Olson, J.E. (2004). Predicting fracture swarms – the influence of subcritical crack growth and the crack-tip process zone on joint spacing in rock. In: *The Initiation, Propagation, and Arrest of Joints and Other Fractures*, vol. 231 (ed. J.W. Cosgrove and T. Engelder), 73–87. London: Geological Society.

37 Bažant, Z.P. and Cedolin, L. (2010). *Stability of Structures: Elastic, Inelastic, Fracture and Damage Theories*. Singapore: World Scientific.

38 Nemat-Nasser, S. and Oranratnachai, A. (1979). Minimum spacing of thermally induced cracks in brittle solids. *Journal of Energy Resources Technology* 101 (1): 34–40.

39 Townsend, M., Pollard, D.D., Johnson, K., and Culha, C. (2015). Jointing around magmatic dikes as a precursor to the development of volcanic plugs. *Bulletin of Volcanology* 77 (10): 92.

40 Grossenbacher, K.A. and McDuffie, S.M. (1995). Conductive cooling of lava: columnar joint diameter and stria width as functions of cooling rate and thermal gradient. *Journal of Volcanology and Geothermal Research* 69 (1–2): 95–103.

41 Spry, A. (1962). The origin of columnar jointing, particularly in basalt flows. *Journal of the Geological Society of Australia* 8 (2): 191–216.

42 Burst, J.F. (1965). Subaqueously formed shrinkage cracks in clay. *Journal of Sedimentary Research* 35 (2): 348–353.

43 Plummer, P.S. and Gostin, V.A. (1981). Shrinkage cracks: desiccation or synaeresis? *Journal of Sedimentary Research* 51 (4): 1147–1156.

44 Tsai, V.C. and Rice, J.R. (2010). A model for turbulent hydraulic fracture and application to crack propagation at glacier beds. *Journal of Geophysical Research* 115: F03007: https://doi.org/10.1029/2009JF001474.

45 Bunger, A.P., Zhang, X., and Jeffrey, R.G. (2012). Parameters affecting the interaction among closely spaced hydraulic fractures. *Society of Petroleum Engineers Journal* 17 (01): 292–306.

46 Kresse, O., Weng, X., Gu, H., and Wu, R. (2013). Numerical modeling of hydraulic fractures interaction in complex naturally fractured formations. *Rock Mechanics and Rock Engineering* 46 (3): 555–568.

47 Olson, J.E. and Taleghani, A.D. (2009). Modeling simultaneous growth of multiple hydraulic fractures and their interaction with natural fractures. *SPE Hydraulic Fracturing Technology Conference* (19–21 January), The Woodlands, TX, Society of Petroleum Engineers. SPE 119739.

48 McGarr, A. and Gay, N.C. (1978). State of stress in the Earth's crust. *Annual Review of Earth and Planetary Sciences* 6 (1): 405–436.

49 Bates, M.P. and Halls, H.C. (1991). Broad-scale Proterozoic deformation of the central Superior Province revealed by paleomagnetism of the 2.45 Ga Matachewan dyke swarm. Canadian. *Journal of Earth Science* 28: 1780–1796.

50 West, G.F. and Ernst, R.E. (1991). Evidence from aeromagnetics on the configuration of Matachewan dykes and the tectonic evolution of the Kapuskasing structural zone, Ontario, Canada. *Canadian Journal of Earth Science* 28: 1997–1811.

51 Halls, H.C. and Zhang, B. (1998). Uplift structure of the southern Kapuscasing zone from 2.45 Ga dike swarm displacement. *Geology* 26 (1): 67–70.

52 Abass, H.H., Soliman, M.Y., Tahini, A.M. et al. (2009). Oriented fracturing: a new technique to hydraulically fracture an openhole horizontal well. *Proceedings SPE Annual Technical Conference and Exhibition*, New Orleans, LA. SPE 124483.

53 Fisher, M.K., Heinze, J.R., Harris, C.D. et al. 2004. Optimizing horizontal completion techniques in the Barnett shale using microseismic fracture mapping. *Proceedings SPE Annual Technology Conference and Exhibition*, Houston, TX (29–29 September 2004). SPE 90051. Society of Petroleum Engineers: Richardson, TX.

54 Meyer, B. and Bazan, L. (2011). A discrete fracture network model for hydraulically induced fractures-theory, parametric and case studies. *Proceedings SPE Hydraulic Fracturing Technology Conference and Exhibition*, The Woodlands, TX. SPE 140514.

55 Germanovich, L.N., Ring, L.M., Astakhov, D.K. et al. (1997). Hydraulic fracture with multiple segments II. *Modeling. International Journal of Rock Mechanics and Mining Sciences* 34 (3-4): 98–e1.

56 Bunger, A.P. (2013). Analysis of the power input needed to propagate multiple hydraulic fractures. *International Journal of Solids and Structures* 50 (10): 1538–1549.

57 Meehan, D.N. (1995). Stimulation results in the Giddings (Austin chalk) field. *SPE Production & Facilities* 10 (02): 96–102.

58 Bunger, A.P., Kear, J., Dyskin, A.V., and Pasternak, E. (2015). Sustained acoustic emissions following tensile crack propagation in a crystalline rock. *International Journal of Fracture* 193 (1): 87–98.

59 Peirce, A.P. and Bunger, A.P. (2015). Interference fracturing: nonuniform distributions of perforation clusters that promote simultaneous growth of multiple hydraulic fractures. *Society of Petroleum Engineers Journal* 20 (02): 384–395.

60 Bunger, A.P. and Peirce A.P. (2014). Numerical simulation of simultaneous growth of multiple interacting hydraulic fractures from horizontal wells. *Proceedings ASCE Shale Energy Engineering Conference*, Pittsburgh, PA (21–23 July 2014). pp. 201–210. American Society of Civil Engineers: Reston, VA.

61 Mukhamediev, S.A., Galybin, A.N., and Morozov, Y.A. (2017). The geometry of a dyke swarm as a result of dyke interaction with each other and with external stresses. *Doklady Earth Sciences* 473 (2): 406–410.

62 Thiele, S., Cruden, A., Micklethwaite, S. et al. (2020). Dyke apertures record stress accumulation during sustained volcanism. *Scientific Reports* 10 (1): 17335–17335.

63 Landau, L.D. and Lifshitz, E.M. (1987). *Fluid Mechanics*, 6, Course of Theoretical Physics, 2e, vol. 6. Oxford: Pergamon (repr. 2000).

64 Bunger, A.P., Jeffrey, R.G., and Zhang, X. (2014). Constraints on simultaneous growth of hydraulic fractures

from multiple perforation clusters in horizontal wells. *Society of Petroleum Engineers Journal* 19 (04): 608–620.

65 Urbancic, T.I. and Maxwell, S.C. (2002). Microseismic imaging of fracture behavior in naturally fractured reservoirs. *SPE/ISRM Rock Mechanics Conference* (20–23 October), Irving, TX. SPE 78229.

66 Maxwell, S.C., Urbancic, T.I., Steinsberger, N., and Zinno, R. (2002). Microseismic imaging of hydraulic fracture complexity in the Barnett shale. *Proceedings SPE Annual Technical Conference and Exhibition*, San Antonio, TX (29 September–2 October 2002). SPE 77440.

67 Rubin, A.M. (1995). Propagation of magma-filled cracks. *Annual Review of Earth and Planetary Sciences* 23 (1): 287–336.

68 Lister, J.R. (1995). Fluid-mechanical models of the interaction between solidification and flow in dykes. In: *Physics and Chemistry of Dykes* (ed. G. Baer and A. Heimann), 115–124. Rotterdam: Balkema.

69 Zhurkov, S.N. (1965). Kinetic concept of the strength of solids. *International Journal of Fracture Mechanics* 1 (4): 311–323.

70 Carslaw, H.S. and Jaeger, J.C. (1959). *Conduction of Heat in Solids*. Oxford University Press.

71 Lister, J.R. (1994). The solidification of buoyancy-driven flow in a flexible-walled channel. Part 1. Constant-volume release. *Journal of Fluid Mechanics* 272: 21–44.

72 Bolchover, P. and Lister, J.R. (1999). The effect of solidification on fluid–driven fracture, with application to bladed dykes. *Proceedings of the Royal Society of London A* 455: 2389–2409.

73 Chanceaux, L. and Menand, T. (2016). The effects of solidification on sill propagation dynamics and morphology. *Earth and Planetary Science Letters* 442: 39–50.

15

Numerical Simulation of Thermal Fracturing During Heat Extraction from a Closed-Loop Circulation Enhanced Geothermal System

Z. Lei[1], Bisheng Wu[1], and Z.R. Chen[2]

[1] Department of Hydraulic Engineering, Tsinghua University, Beijing, China
[2] CSIRO Energy, Private Bag 10, Clayton South, Victoria, Australia

KEY POINTS

- A thermoelastic model for predicting fracture propagation during heat extraction from an EGS is developed.
- Intensive cooling and fluid pressure significantly facilitate fracture propagation.
- Thermal effects on fracture propagation during fluid circulation should be considered.

15.1 Introduction

Geothermal energy has been considered as one of the most promising renewable energy sources due to its huge reserves and small environmental impact compared to other carbon-based fuel energies. Hot dry rock (HDR), one form of rock medium containing no or a small amount of water and having a temperature higher than 200 °C, stores a large amount of geothermal heat [1]. It is reported that, if 2% of HDR heat resources at a depth of 3–10 km could be exploited in China, it would provide 4000 times the total energy consumption of China in 2017 [2].

Enhanced geothermal system (EGS) is the main approach to extract geothermal energy from the deep HDR for power generation [3]. The basic principle of EGS is to increase the permeability of rock matrix by well stimulation, normally hydraulic fracturing, through which a closed loop forms for fluid circulation between the injection and production well. After the low-temperature fluid, including water, nitrogen, or supercritical carbon dioxide [4, 5], is injected into a fracture channel which is generated by hydraulic fracturing, it is heated by the surrounding rock with higher temperature and then pumped out at the production well for power generation or other uses. The cold fluid generated from the heat exchanger is then reinjected into the injection well. Two critical technical challenges include how to create effective fracture networks between wells in rock through hydraulic fracturing and how to ensure reservoir fracture stability during long-time cold fluid circulation [6].

A great deal of mathematical modeling has been carried out for the production performance, such as outlet temperature and power capacity, of an EGS. Analytical or semi-analytical model is an efficient tool to determine the thermal behavior of the fluid and reservoir. For example, Wu et al. [7] developed a semi-analytical model for an EGS with dipole wells considering the areal flow by using velocity potential and streamline functions. Compared with dipole-well configuration, multiple wells and multiple fractures provide a more effective way to improve the production performance of the EGS [8]. Wu et al. [9] derived a semi-analytical solution for optimization design of a multiple-fracture EGS, revealing the relationship between the number of fractures, fracture spacing and output temperature. Numerical method is another efficient way to predict the heat extraction from a geothermal reservoir. Xu et al. [10] and Lei et al. [11] investigated the production temperature and electricity generation in a fractured reservoir by using the open source code TOUGH2, which is based on finite difference method (FDM). Other widely used commercial software that can deal with heat transfer in a geothermal reservoir includes COMSOL Multiphysics [12] and ANSYS [13] based on finite element method (FEM), and Fluent [14] based on finite volume method (FVM). However, most of the above studies focus on thermal energy output and few pays attention to the rock deformation and failure during the long-time heat extraction.

The effect of temperature change on mechanical behavior of rock is significant. Take granite for example. Given a

Mechanics of Hydraulic Fracturing: Experiment, Model, and Monitoring, First Edition.
Edited by Xi Zhang, Bisheng Wu, Diansen Yang, and Andrew Bunger.

temperature change $\Delta T = 100\,°C$, thermal expansion coefficient $\alpha = 7.8 \times 10^{-6}$/K, Young's modulus $E = 35$ GPa and Poisson's ratio $\nu = 0.23$, the thermal induced tensile stress for a temperature change of 100 °C is calculated as $\sigma = E\alpha\Delta T/(1-\nu) = 35.5$ MPa. Obviously, such a large stress cannot be ignored when studying the fracture behaviors of rocks. Since the notable temperature change is easy to be achieved for an EGS with high temperature, it is important to consider the thermal effects and induced fracturing behaviors when the geothermal reservoir is stimulated for a time longer than tens of years.

Numerical simulations of hydraulic fracture initiation and propagation have received considerable attention in the area of EGS and petroleum engineering. The finite element method (FEM) has widely been used to solve the problems related to fracture mechanics [15, 16]. However, as the FEM model needs remeshing after crack propagation and the mesh must conform to the geometry, the computational cost is very high [17]. The boundary element method (BEM) and its variants such as displacement discontinuity method (DDM) [18] provide effective approaches to analyze the propagation of hydraulic fractures in an infinite or semi-infinite elastic medium. Since BEMs only implement discretization on the domain boundaries, a much lower demand on memory is required. However, the application of BEM in modeling fracture propagation in nonhomogeneous and inelastic rocks subjected to complex boundary conditions is limited due to the difficulty in finding the corresponding Green's functions.

The extended finite element method (XFEM) [19, 20] overcomes the inherent drawbacks associated with remeshing when dealing with fracture propagation problems. Besides, the XFEM can be used to model fracture propagation in the rock with nonlinear mechanical properties. Therefore, the XFEM has been widely used for predicting rock deformation and fracture propagation. The finite element software ABAQUS includes XFEM package for crack growth, and so it has extensively been applied in hydraulic fracturing models [21–23]. In this chapter, the XFEM is adopted to solve the fracture propagation problem that involves thermal stress changes.

Most of the EGS models do not consider the thermal effects on fracture propagation by assuming the reservoir perimeter is not changed during heat extraction processes and thus the thermal effects on fracture propagation are not accounted for. For an EGS, a large temperature change is easy to be achieved after long time circulation of fluid, inducing large thermal stresses that have a significant influence on fracture growth and stability of the geothermal reservoir. To the best of the authors' knowledge, no analytical or numerical solutions are available for studying the influence of induced thermal stresses on the fracture propagation in an EGS during long time fluid circulation. The motivation of the present work is to establish a thermo-elastic model to predict EGS efficiency.

This chapter is organized as follows. In Section 2, the model description for the EGS of penny-shaped fractures is given, and then the governing equations for fluid flow, heat transfer, and XFEM-based fracture propagation are described. It is followed by the solution methodology and computational procedures in Section 3. Next, numerical results conducted by using ABAQUS are given in Section 4, and conclusions are drawn in Section 5.

15.2 Mathematical Formulation

15.2.1 Problem Description

The geometry of the model studied is shown in Figure 15.1. At a depth of H_0, two horizontal wells (one for injection and the other for production) are drilled in an isotropic and homogenous reservoir and intercept one or more parallel vertical penny-shaped fractures with an identical radius of R. The fractures based on the distance to the inlet and outlet are numbered 1, 2 ... N from left to right. The origin of the cylindrical coordinate system is at the center of the first fracture and the positive z-axis points upwards. The initial temperature of the reservoir is denoted by $T_0(x, y, z, t = 0) = A_0(H_0 - z) + B_0$, where B_0 is the ground temperature and A_0 the geothermal gradient. The two wells and fractures form a closed-loop multiple-fracture doublet system. In order to obtain the maximum heat exchange efficiency, the injection and production wells are drilled through the uppermost and lowermost points, respectively, of the fracture boundary. At $t > 0$, a cold fluid is injected into the injection well, absorbs heat from the surrounding rock with higher temperature and then is pumped out from the production well. The total injection rate, Q_t, and injection temperature, T_{in}, are assumed to be constant. In addition, the in-situ stresses, including the vertical principal stress, σ_v, maximum horizontal principal stress, σ_H, and minimum horizontal principal stress, σ_h, are kept the same.

Some assumptions are made for the present model to be tractable:

a) The rock formation is isotropic, homogenous, impermeable, and linearly elastic.
b) The fluid is incompressible and Newtonian, and the fluid flow is single-phase laminar flow.
c) The mechanical and thermal properties of the fluid and rock are temperature-independent.

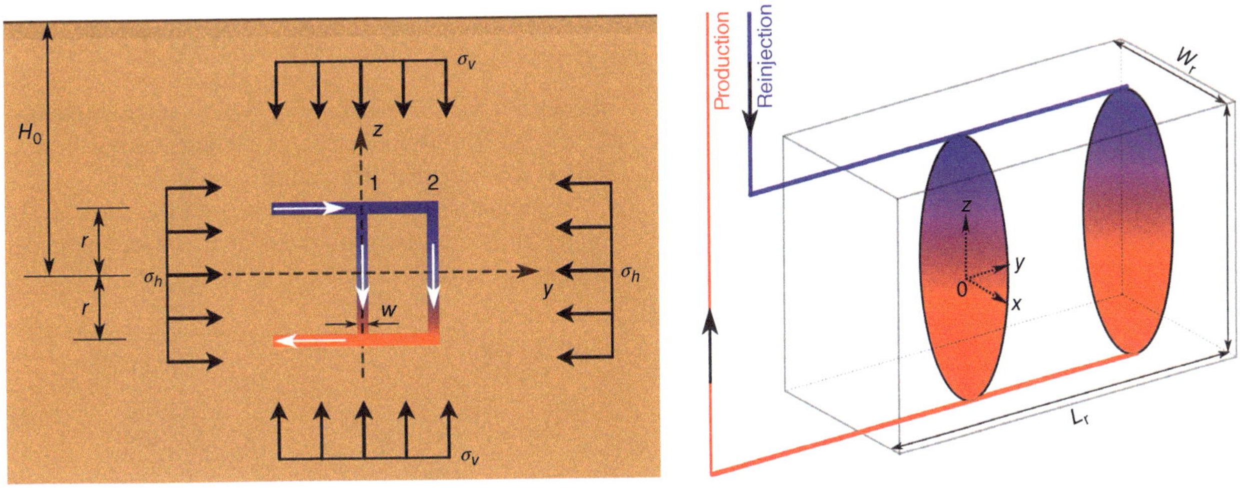

Figure 15.1 Geometry of the model containing two penny-shaped fractures.

15.2.2 Governing Equations of Coupled Thermoelastic Model

Heat production in a closed-loop circulation EGS can be considered as a multiphysics system, involving: (i) fluid flow in fractures, (ii) rock deformation and fracture propagation induced by pressure and thermal effects, and (iii) thermal transport through fluid flow and heat diffusion in rock matrix. In this section, the governing equations for the 3D thermoelastic model are described as follows.

15.2.2.1 Fluid Flow

As the fracture size is much larger than fracture width, the local fluid flux within the fracture is governed by Poiseuille's law [24], given as

$$\mathbf{q}_f = -\frac{w_n^3}{12\mu}\nabla P_f, \tag{15.1}$$

where $\mathbf{q}_f$, the fluid flux at a point within the fracture, is equal to the fracture aperture, w_n, times the specific discharge, $\mathbf{v}_f$ (i.e. $\mathbf{q}_f = w_n\mathbf{v}_f$); μ is the fluid dynamic viscosity, and ∇P_f is the fluid pressure gradient. The conductivity k is equal to $w_n^3/12$.

According to the mass conservation law, the governing equation for fluid flow within the fracture can be simplified as

$$\nabla\cdot\mathbf{q}_f + \frac{\partial w_n}{\partial t} + \delta(z-z_1^\ell)Q_{in}^\ell - \delta(z-z_2^\ell)Q_{out}^\ell = 0, \tag{15.2}$$

where $\nabla\cdot$ is the divergence operator, and δ denotes the Dirac delta function; z_1^l and z_2^l denote the location of injection well and production well, respectively, in z-axis direction for lth fracture, as shown in Figure 15.1. Q_{in}^l and Q_{out}^l are the injection and production rates, respectively, for lth fracture. It should be noted that Q_{in}^l and Q_{out}^l are defined as positive.

15.2.2.2 Rock Deformation

According to the theory of solid mechanics [25], the stress field, $\boldsymbol{\sigma}$, inside the domain of the thermoelastic model, is related to the external prescribed tractions, $\mathbf{t}$, thermal stress, $\boldsymbol{\sigma}_T$, and fluid pressure, P_f. The equilibrium equations are

$$\begin{aligned}
&\nabla\cdot\boldsymbol{\sigma} = 0 \quad \text{on } \Omega,\\
&\boldsymbol{\sigma}\cdot\mathbf{n} = \mathbf{t} \quad \text{on } S_F,\\
&\boldsymbol{\sigma}\cdot\mathbf{n}^- = -\boldsymbol{\sigma}\cdot\mathbf{n}^+ = -P_f\mathbf{n}^+ = P_f\mathbf{n}^- \quad \text{on } S_c,
\end{aligned} \tag{15.3}$$

where the domain of rock medium is denoted by Ω, with unit normal vector $\mathbf{n}$ at the boundary, external tractions acting area S_t, and fracture faces S_c. The superscripts "+" and "−" refer to the left and right fracture faces, respectively, for the fracture propagating upwards, as shown in Figure 15.2.

The linear elastic constitutive law for rocks is given below

$$\boldsymbol{\sigma} = \mathbf{D} : \boldsymbol{\varepsilon}, \tag{15.4}$$

where $\mathbf{D}$ is Hooke's tensor, and $\boldsymbol{\varepsilon}$ is the strain tensor which includes the elastic part and thermal part.

The Kinematic equations contain the strain–displacement relationship, the prescribed displacement boundary conditions and the fracture surface separation as follows [26]

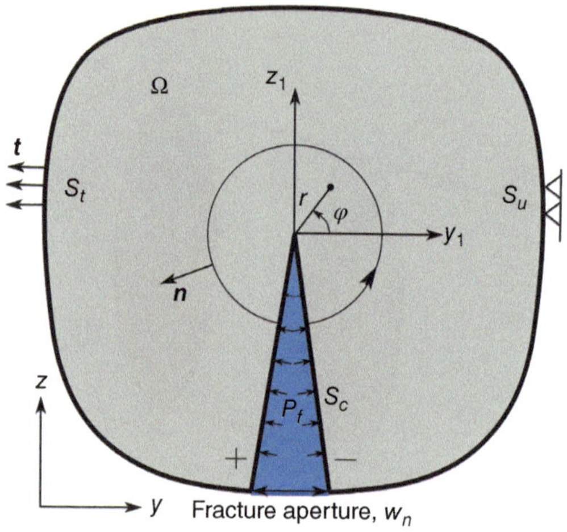

Figure 15.2 Fracture propagating upward in a rock medium in the y-z plane.

$$
\begin{aligned}
&\boldsymbol{\varepsilon} = \boldsymbol{\varepsilon}_e + \boldsymbol{\varepsilon}_T \quad \text{on } \Omega,\\
&\boldsymbol{\varepsilon}_e = \frac{1}{2}\left[\nabla \mathbf{u} + [\nabla \mathbf{u}]^{\mathrm{T}}\right] \quad \text{on } \Omega,\\
&\boldsymbol{\varepsilon}_T = \alpha_T \Delta T \boldsymbol{I} \quad \text{on } \Omega,\\
&\mathbf{u} = 0 \quad \text{on } S_u,\\
&\boldsymbol{w} = \mathbf{u}^+ - \mathbf{u}^- \quad \text{on } S_c,
\end{aligned}
\tag{15.5}
$$

where $\mathbf{u}$ is the displacement vector; $\mathbf{w}$ is the displacement discontinuity vector, consisting of the normal displacement (fracture aperture w_n) and shear displacement, w_s. $\boldsymbol{\varepsilon}_e$ and $\boldsymbol{\varepsilon}_T$ are the elastic part and thermal part, respectively, of the strain. ΔT is the change of temperature, and α_T is the thermal expansion coefficient of rocks. $\mathbf{I}$ is the second-order identity tensor. S_u is the prescribed displacement boundary.

15.2.2.3 Fracture Initiation and Propagation

The cohesive law is adopted to define the criteria of fracture initiation and propagation. The initiation of fractures is governed by the traction-separation relation, which is described by the maximum principal stress (MPS) in ABAQUS [27]

$$
\hat{f}_c = \frac{\langle \sigma_{\max} \rangle}{\sigma_T}, \tag{15.6}
$$

where σ_T is the tensile strength of rock, $\sigma_{\max}$ is the allowable MPS, and the Macaulay bracket operator, < >, is used to avoid the compressive loading effects.

The fracture plane is perpendicular to the direction of the MPS, and the fractures initiate when the MPS reaches the critical value (i.e. $\hat{f}_c = 1$). Once the fracture initiates, the fracture propagation is governed by the Benzeggagh–Kenane fracture criterion [28] in terms of fracture energy theory, which is defined as

$$
G_C = G_n^C + \left(G_s^C - G_n^C\right)\left(\frac{G_s + G_t}{G_s + G_t + G_n}\right)^{\eta}, \tag{15.7}
$$

where η is a material parameter; G_C is the cohesive energy, G_n, G_s, and G_t are fracture energies in the normal, the first and the second tangential directions, respectively.

The normal stress component, t_n, and the shear stress components, i.e. t_s and t_t, of the traction-separation model affected by damage are determined as

$$
\begin{aligned}
&t_n = \begin{cases} \left(1-\hat{D}_v\right)T_n & (T_n \geq 0) \\ T_n & (T_n < 0) \end{cases},\\
&t_s = \left(1-\hat{D}_v\right)T_s,\\
&t_t = \left(1-\hat{D}_v\right)T_t,
\end{aligned}
\tag{15.8}
$$

where T_n, T_s, and T_t are the normal and the shear stress components predicted by the elastic traction–separation behavior without damage [22]; $\hat{D}_v$ is a scalar damage variable.

The elastic traction–separation model assumes a linear elastic behavior after the initiation of damage. As shown in Figure 15.3, the area under the softening part of the traction–separation curve denotes the cohesive energy G_C [22]. The scalar damage variable $\hat{D}_v$ is given as

$$
\hat{D}_v = \frac{w_{n,m}(w_n - w_{n,0})}{w_n\,(w_{n,m} - w_{n,0})}, \tag{15.9}
$$

where $w_{n,0}$ and $w_{n,m}$ are the fracture aperture at fracture initiation and complete failure, respectively.

In the current study, only Mode-I fracture growth is considered. Therefore, Eq. (15.7) can be simplified to be

$$
G_C = G_n^C. \tag{15.10}
$$

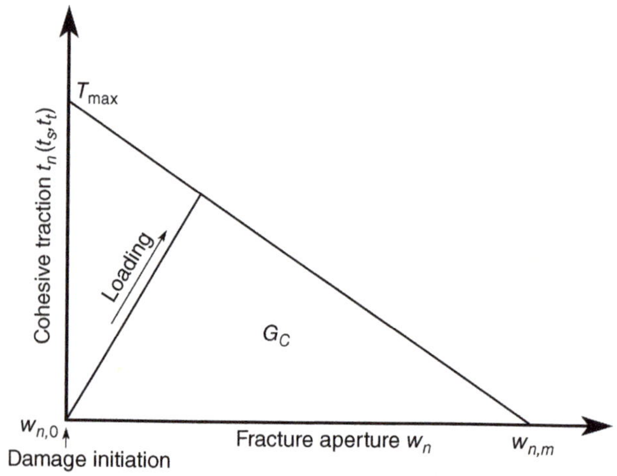

Figure 15.3 Cohesive traction–separation law.

15.2.2.4 Thermal Transport Through Fluid Flow

The thermal transport through fluid flow has four components, including heat storage, advection, dispersion and conduction from the aquifer walls [29], and can be written as

$$\frac{\partial T_f}{\partial t} + \mathbf{v}_f \cdot \nabla T_f - D_L \nabla^2 T_f - \frac{\dot{q}}{\rho_f c_f w_n} = 0, \tag{15.11}$$

where T_f is the fluid temperature, ρ_f and c_f are the fluid density and fluid specific heat, respectively. D_L is the fluid dispersion coefficient. In addition, $\partial/\partial t$ is called the local derivative, which is physically the time rate of change at a fixed point; ∇^2 denotes the Laplace operator; $\dot{q}$ denotes the heat flux from rock to fluid.

Since the fracture aperture is far less than the fracture radius (i.e. $w_n << R$), it is rational to assume that the heat flux in the rock is only along the y-direction, although this assumption may be approximately valid near the fracture fringe. Therefore, along the y-direction, the rock is separated by the fractures. For fracture ℓ, there are two contacting surfaces between the fluid and rock, and thus, the heat flux from rock to fluid, $\dot{q}$, is calculated as

$$\dot{q} = -k_r\left[\frac{\partial T_{\mathrm{r}}^{\ell}}{\partial y} - \frac{\partial T_{\mathrm{r}}^{\ell+1}}{\partial y}\right], \tag{15.12}$$

where k_r is the rock thermal conductivity; the subscript ℓ also denotes the sequence number of a rock layer in between fracture ℓ−1 and fracture ℓ.

As Cheng et al. [29] have demonstrated the dispersion effect during the heat transport in fractures is so small that can be neglected, Eq. (15.11) combining with Eq. (15.12) is simplified to be

$$\frac{\partial T_f}{\partial t} + \mathbf{v}_f \cdot \nabla T_f + \frac{k_r}{\rho_f c_f w_n}\left[\frac{\partial T_r^{\ell+1}}{\partial y} - \frac{\partial T_r^{\ell}}{\partial y}\right] = 0. \tag{15.13}$$

15.2.2.5 Heat Transfer in Rock Matrix

Since most EGS reservoirs are composed of granite that is impermeable to fluid flow and the contribution of rock deformation to temperature change is trivial, the heat transfer in the granite matrix is described by the Fourier's law [30]

$$\rho_r c_r \frac{\partial T_r}{\partial t} - k_r \nabla^2 T_r = 0, \tag{15.14}$$

where T_r is the rock temperature; ρ_r and c_r are density and specific heat, respectively, of the rock.

15.2.3 Boundary and Initial Conditions

Q_{in}^{l} and Q_{out}^{l} are assumed to be identical as follows

$$Q_{in}^{\ell} = Q_{out}^{\ell} = \frac{Q_t}{N}, \tag{15.15}$$

where N is number of fractures. For the purpose of simplicity, the maximum of N is chosen to be 2 in the present work.

To avoid the difficulty of calculating the heat transfer, it is reasonable to assume that the temperature of the fluid and the surfaces of the rock formation are equal [31, 32], i.e.

$$T_f = T_r^2 = T_r^1, \quad \text{at the same depth } z. \tag{15.16}$$

The fluid temperature at each injection point is fixed

$$T_{f,in} = T_{\mathrm{in},j}(t), \quad \text{at } \left(x_j, y_j, z_1\right)\ (j = 1, ..N), \tag{15.17}$$

where z_1 denotes the vertical coordinate of the injection wells.

The initial rock temperature is defined as follows

$$T_r(x, y, z, t = 0) = A_0(H_0 - z) + B_0, \tag{15.18}$$

Apart from the contacts between the rock and fluid, there is no heat flux at all exterior sides of the rock

$$\mathbf{n} \cdot \nabla T_r = 0. \tag{15.19}$$

15.3 Solution Methodology and Computational Procedures

The solution methods and computational procedures for the 3D sequentially coupled thermoelastic model are provided in this section. In the model, the coupled fluid-fracture problem and the coupled fluid-thermal problem are solved sequentially. Thes coupled fluid-fracture solver means the governing equations for fluid flow in fracture and for fracture propagation are solved simultaneously. Besides, the heat transfer models for fluid flow and rock matrix are decoupled from the coupled fracture-thermal model, which means the temperature variables are solved independently. During a specific time step Δt, the coupled fluid-fracture problem is first solved by the XFEM approximation of the displacement and pressure field as described below, and then, the thermal transport models are solved by FEM at the same time step.

15.3.1 Coupled Fluid-Fracture Solver

15.3.1.1 Weak Form and FEM Discretization

The discretized weak form of equilibrium equation is written as

$$\int_{\Omega} \delta\boldsymbol{\varepsilon}^T \boldsymbol{\sigma} d\Omega - \int_{S_t} \delta\mathbf{u}^T \boldsymbol{t} dS - \left(-\int_{S_c^+} \delta\mathbf{u}^{+T} P_f \mathbf{n}^+ dS + \int_{S_c^-} \delta\mathbf{u}^{-T} P_f \mathbf{n}^- dS\right) = 0, \tag{15.20}$$

where $\delta\mathbf{u}$ is the arbitrary virtual displacement and $\delta\boldsymbol{\varepsilon}$ is the corresponding virtual strain.

The fluid pressure acting on the fracture faces, **P**, and the fracture opening displacement, $\mathbf{w}_n$, are defined as

$$\mathbf{P} = -P_f\mathbf{n}^+ = P_f\mathbf{n}^-, \tag{15.21}$$

$$w_n = \mathbf{n}(\mathbf{u}^+ - \mathbf{u}^-), \text{ or } \mathbf{w}_n = \mathbf{n}(\mathbf{u}^+ - \mathbf{u}^-)\mathbf{n}. \tag{15.22}$$

Considering Eq. (15.21) and Eq. (15.22), Eq. (15.20) can be expressed in a compact form as

$$\int_\Omega \delta\boldsymbol{\varepsilon}^T\boldsymbol{\sigma}d\Omega - \int_{S_t}\delta\mathbf{u}^T\mathbf{t}dS - \int_{S_c}\delta\mathbf{w}_n^T\mathbf{P}dS = 0. \tag{15.23}$$

The weak form of Eq. (15.2) (i.e. fluid flow in the fracture) is expressed as

$$\int_{S_c}\delta\mathbf{P}^T\left(\nabla\cdot\mathbf{q}_f + \frac{\partial\mathbf{w}_n}{\partial t} + \delta(z - z_1^\ell)Q_{in}^\ell - \delta(z - z_2^\ell)Q_{out}^\ell\right)dS = 0, \tag{15.24}$$

which, after applying boundary conditions and integration by parts, takes the form

$$\int_{S_c}\delta\mathbf{P}^T\frac{\partial\mathbf{w}_n}{\partial t}dS + \int_{S_c}\nabla^T(\delta\mathbf{P})\mathbf{k}\nabla P_f dS + \int_{S_c}\delta\mathbf{P}(\delta(z - z_1^\ell)Q_{in}^\ell - \delta(z - z_2^\ell)Q_{out}^\ell)dS = 0. \tag{15.25}$$

The standard nodal displacement vector, **u**, the fluid pressure P_f, and the displacement discontinuity vector, **w**, are approximated as

$$\mathbf{u} \approx \hat{\mathbf{u}} = \sum_{i=1}^{n} N_i^u\mathbf{u}_i = \mathbf{N}^u\tilde{\mathbf{u}}, \quad \delta\mathbf{u} \approx \mathbf{N}^u\delta\tilde{\mathbf{u}},$$
$$P_f \approx \hat{P}_f = \sum_{i=1}^{n} N_i^P P_{f,i} = \mathbf{N}^P\tilde{\mathbf{P}}, \quad \delta P_f \approx \mathbf{N}^P\delta\tilde{\mathbf{P}},$$
$$\mathbf{w} \approx \hat{\mathbf{w}} = \sum_{i=1}^{n} N_i^w\mathbf{u}_i = \mathbf{N}^w\tilde{\mathbf{u}}, \quad \delta\mathbf{w} \approx \mathbf{N}^w\delta\tilde{\mathbf{u}}, \tag{15.26}$$

where N_i^u and N_i^P denote the shape functions of nodal displacement $\mathbf{u}_i$ and fluid pressure $P_{f,\,i}$, respectively, at node i. N_i^w denotes the crack displacement discontinuity shape function.

A system of algebraic equations of the discrete solid problem are obtained by substituting Eqs. (15.26) into the constitutive equation (i.e. Eq. (15.4))

$$\mathbf{K}\tilde{\mathbf{u}} - \mathbf{Q}\tilde{\mathbf{P}} - \mathbf{f}^u = \mathbf{0}, \tag{15.27}$$

where **K**, **Q** and $\mathbf{f}^u$ are the structural stiffness matrix, coupling term and equivalent nodal force vector, respectively, and are defined as

$$\boldsymbol{K} = \int_\Omega \boldsymbol{B}^T\boldsymbol{D}\boldsymbol{B}d\Omega, \quad \boldsymbol{Q} = \int_{S_c}(\boldsymbol{N}^w)^T\boldsymbol{n}\boldsymbol{N}^P dS,$$
$$\boldsymbol{f}^u = \int_{St}(\boldsymbol{N}^u)^T\boldsymbol{t}dS,$$

where **B** is the strain–displacement matrix.

A system of algebraic equations for the fluid flow are obtained by substituting Eq. (15.26) into the weak form of the governing equation for the fluid flow in the fracture (i.e. Eq. (15.25))

$$\mathbf{C}\tilde{\mathbf{u}} - \mathbf{H}\tilde{\mathbf{P}} - \mathbf{f}^P = \mathbf{0}, \tag{15.28}$$

where **C**, **H**, and $\mathbf{f}^P$ are defined as

$$\mathbf{C} = \mathbf{Q}^T = \int_{S_c}(\mathbf{N}^P)^T\mathbf{n}\mathbf{N}^w dS, \quad \mathbf{H} = \int_{S_c}(\nabla\mathbf{N}^P)^T\mathbf{k}\nabla\mathbf{N}^P dS, \mathbf{f}^P = \int_{Sc}(\mathbf{N}^P)^T(\delta(z - z_1^\ell)Q_{in}^\ell - \delta(z - z_2^\ell)Q_{out}^\ell)dS.$$

Therefore, a system of algebraic equations for the coupled fluid-fracture problem can be given in matrix form

$$\begin{bmatrix} \boldsymbol{0} & \boldsymbol{0} \\ \mathbf{C} & \boldsymbol{0} \end{bmatrix}\begin{pmatrix} \dot{\tilde{\mathbf{u}}} \\ \dot{\tilde{\mathbf{P}}} \end{pmatrix} + \begin{bmatrix} \mathbf{K} & -\mathbf{Q} \\ \boldsymbol{0} & \mathbf{H} \end{bmatrix}\begin{pmatrix} \tilde{\mathbf{u}} \\ \tilde{\mathbf{P}} \end{pmatrix} = \begin{pmatrix} \mathbf{f}^u \\ \mathbf{f}^P \end{pmatrix}, \tag{15.29}$$

where $\dot{\tilde{\mathbf{u}}}$ and $\dot{\tilde{P}}_f$ are the time derivative of displacement and fluid pressure. By using the finite difference technique, the time derivative of a variable Ξ is defined as

$$\frac{\partial X}{\partial t} = \frac{X^{t+\Delta t} - X^t}{dt}, \tag{15.30}$$

where $\Xi^{t+\Delta t}$ and Ξ^t are the values of Ξ at time $t + \Delta t$ and t, respectively.

15.3.1.2 Extended Finite Element Approximation

The formulation of XFEM involves using the Heaviside step function, $H(x)$, and the enrichment tip functions, $\psi_\alpha(x)$, to represent the crack displacement [19]. In this chapter, ABAQUS is used for XFEM discretization, and the enriched displacement approximation of the point $\mathbf{x}_p$, $\mathbf{u}^h(\mathbf{x}_p)$, is given as follows [33]

$$\mathbf{u}^h(\mathbf{x}_p) = \sum_{i\in I}\mathbf{N}_i(\mathbf{x}_p)\mathbf{u}_i + \sum_{j\in J}\mathbf{N}_j(\mathbf{x}_p)H(\mathbf{x}_p)\mathbf{a}_j + \sum_{k\in K}\sum_{\alpha=1}^{4}\mathbf{N}_k(\mathbf{x}_p)\psi_\alpha(\mathbf{x}_p)\mathbf{b}_k^\alpha, \tag{15.31}$$

where $\mathbf{u}_i$ are the standard nodal displacements; $\mathbf{a}_j$ and $\mathbf{b}_k^{\alpha}$ are the enrich degrees of freedom (DOF); $N_i(\mathbf{x}_p)$ denotes the shape function at node i; I represents the number of all nodes in the discretization, J is the set of nodes in the element entirely split by fractures and K is the set of nodes in the element containing the crack tip; The Heaviside step function $H(\mathbf{x}_p)$ is assumed to be 1 at one side of the fracture and −1 at the other side. The enrichment tip functions, $\psi_{\alpha}(\mathbf{x}_p)$, are defined as [34]

$$\psi_{\alpha}(\mathbf{x}_p) = \psi_{\alpha}(r, \varphi) = \sqrt{r}\left\{\sin\left(\frac{\varphi}{2}\right), \sin\left(\frac{\varphi}{2}\right)\sin(\varphi), \cos\left(\frac{\varphi}{2}\right), \cos\left(\frac{\varphi}{2}\right)\sin(\varphi)\right\}, \tag{15.32}$$

where r and φ are the local polar coordinates around the fracture tip.

The enriched fluid pressure approximation of the point $\mathbf{x}_p$ within the crack, $P_f^h(\mathbf{x}_p)$, is given as follows

$$P_f^h(\mathbf{x}_p) = \sum_{j \in J} N_j^P(\mathbf{x}_p) P_{f,j}, \tag{15.33}$$

where $N_j^P(\mathbf{x}_p)$ denotes the fluid pressure shape function for the set of nodes in the element entirely split by fractures.

For the coupled fluid-fracture solver, a three-dimensional eight-node hexahedral element is used and a coupled fluid/stress analysis is performed within ABAQUS. Each node has the standard displacement degrees of freedom $\mathbf{u}_I$, the Heaviside enriched degrees of freedom $\mathbf{a}_I$ and $\mathbf{b}_I^{\alpha}$, and the fluid pressure degrees of freedom $P_{f,\ I}$.

15.3.2 Coupled Fluid-Thermal Solver

The weak form of the governing equation for thermal transport through fluid flow is given as

$$\int_{S_c} \delta\mathbf{T}_f^T \frac{\partial T_f}{\partial t} dS + \int_{S_c} \delta\mathbf{T}_f^T \mathbf{v}_f \cdot \nabla T_f dS - \int_{S_c} \delta\mathbf{T}_f^T \frac{k_r}{\rho_f c_f w_n}\left[\frac{\partial T_r^{\ell+1}}{\partial y} - \frac{\partial T_r^{\ell}}{\partial y}\right] = 0, \tag{15.34}$$

where $\delta\mathbf{T}_f$ is the virtual fluid temperature.

The weak form of the governing equation for heat diffusion in rock matrix is given as

$$\int_{\Omega} \delta\mathbf{T}_r^T \rho_r c_r \frac{\partial T_r}{\partial t} d\Omega - \int_{\Omega} (\nabla\delta\mathbf{T}_r)^T \cdot (k_r \nabla T_r) d\Omega = 0, \tag{15.35}$$

where $\delta\mathbf{T}_r$ is the virtual rock temperature.

The nodal fluid temperature T_f and the nodal rock temperature T_r are approximated as

$$\begin{aligned} T_f \approx \hat{T}_f &= \sum_{i=1}^{n} N_{f,i}^T \widetilde{\mathbf{T}}_{f,i} = \mathbf{N}_f^T \widetilde{\mathbf{T}}_f, \quad \delta T_f \approx \mathbf{N}^T \delta\widetilde{\mathbf{T}}_f, \\ T_r \approx \hat{T}_r &= \sum_{i=1}^{n} N_{r,i}^T \widetilde{\mathbf{T}}_{r,i} = \mathbf{N}_r^T \widetilde{\mathbf{T}}_r, \quad \delta T_r \approx \mathbf{N}^T \delta\widetilde{\mathbf{T}}_r, \end{aligned} \tag{15.36}$$

where $N_{f,i}^T$ and $N_{r,i}^T$ denote the fluid temperature shape function and rock temperature shape function at node i, respectively. $\mathbf{T}_f$ and $\mathbf{T}_r$ are the vectors of the nodal values of the unknows.

A system of algebraic equations for the heat transport problem are obtained by substituting Eqs. (15.36) and boundary conditions into Eq. (15.34) and Eq. (15.35)

$$\mathbf{M}_f^T \dot{\widetilde{\mathbf{T}}}_f + \mathbf{K}_f^T \widetilde{\mathbf{T}}_f = 0, \tag{15.37}$$

$$\mathbf{M}_r^T \dot{\widetilde{\mathbf{T}}}_r + \mathbf{K}_r^T \widetilde{\mathbf{T}}_r = 0, \tag{15.38}$$

where

$$\mathbf{M}_f^T = \int_{S_c} \left(\mathbf{N}_f^T\right)^T \rho_f c_f \mathbf{N}_f^T dS,$$

$$\mathbf{K}_f^T = \int_{S_c} \left(\mathbf{N}_f^T\right)^T w_n \rho_f c_f \mathbf{v}_f \cdot \nabla\mathbf{N}_f^T dS \text{-} 2\int_{S_c} \left(\mathbf{N}_f^T\right)^T k_r \nabla\mathbf{N}_f^T dS,$$

$$\mathbf{M}_r^T = \int_{\Omega} \left(\mathbf{N}_r^T\right)^T \rho_r c_r \mathbf{N}_r^T d\Omega, \ \mathbf{K}_r^T = \int_{\Omega} \left(\nabla\mathbf{N}_r^T\right)^T k_r \nabla\mathbf{N}_r^T d\Omega.$$

For the coupled fluid-thermal solver, the element implemented is the same as that in the coupled fluid-fracture solver. The heat transfer analysis is performed by using ABAQUS.

15.3.3 Solution Strategy

The iterative solution strategy is presented below as flow chart shown in Figure 15.4. At a specific time $t+\Delta t$, the inputs of the coupled fluid-fracture solver at s^{th} iteration include the rock temperature $T_r^{(t)}$, rock displacement, and fluid pressure of the last iteration, i.e. $\mathbf{u}^{(t+\Delta t,\ s-1)}$ and $P_f^{(t+\Delta t,\ s-1)}$. Hence, the updated $\mathbf{u}^{(t+\Delta t,\ s)}$ and $P_f^{(t+\Delta t,\ s)}$ can be obtained. It should be noted that the initial rock displacement and fluid pressure are equal to those of previous time step (i.e. $\mathbf{u}^{(t+\Delta t,\ 0)} = \mathbf{u}^{(t)}$, $P_f^{(t+\Delta t,\ 0)} = P_f^{(t)}$). The iteration ends if normalized values of $\mathbf{u}^{(t+\Delta t,\ s)}$ and $P_f^{(t+\Delta t,\ s)}$ for two consecutive iterations meet the following conditions

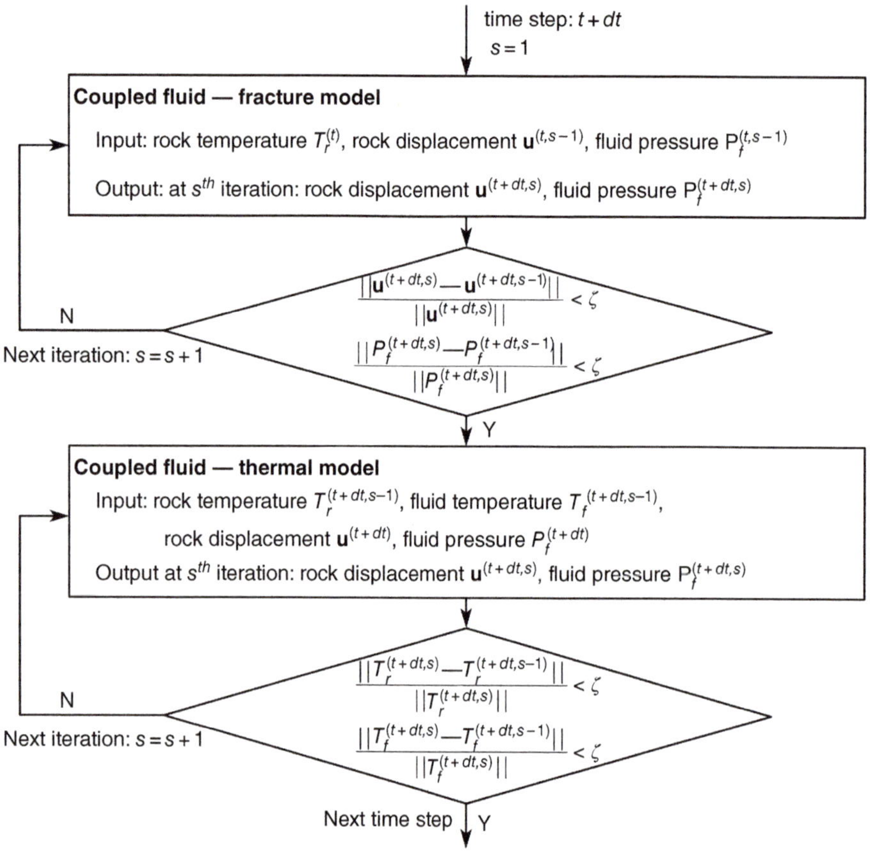

Figure 15.4 Solution scheme of the sequentially coupled thermoelastic model at time step $t+\Delta t$.

$$\frac{\left\|\mathbf{u}^{(t+\Delta t,\,s)}-\mathbf{u}^{(t+\Delta t,\,s-1)}\right\|}{\left\|\mathbf{u}^{(t+\Delta t,\,s)}\right\|}<\zeta,\quad \frac{\left\|P_f^{(t+\Delta t,\,s)}-P_f^{(t+\Delta t,\,s-1)}\right\|}{\left\|P_f^{(t+\Delta t,\,s)}\right\|}<\zeta, \tag{15.39}$$

where ζ is the specified tolerance, which is set to 0.01.

Next, the calculation proceeds to the coupled fluid-thermal solver with input $T_r^{(t+\Delta t,\,s-1)}$, $T_f^{(t+\Delta t,\,s-1)}$, $\mathbf{u}^{(t+\Delta t)}$, and $P_f^{(t+\Delta t)}$. The iteration solution scheme and the error tolerance are the same as those in the couple fluid-fracture model. The iterations are repeated until consecutive normalized values of $T_r^{(t+\Delta t,\,s)}$ and $T_f^{(t+\Delta t,\,s)}$ are within the specified tolerance η. Therefore, the rock temperature and fluid temperature at the time step $t+\Delta t$, i.e. $T_r^{(t+\Delta t)}$ and $T_f^{(t+\Delta t)}$, are obtained. The solution stream then proceeds to next time step.

15.4 Numerical Results

In this section, two cases, i.e. single- and double-fracture EGSs, are studied to show the fluid flow and thermal and mechanical behaviors of the EGS. In order to reveal the thermal effects attributed to long time fluid circulation on fracture propagation, two different total injection rates are selected, i.e. $Q_t = 0.1$ or $0.05\ m^3/s$. Besides, the injection rate is kept the same during the EGS operation period. In addition, since two fractures are far separated, the fluid injected is assumed to be distributed evenly to the multiple fractures. For example, if the number of fractures is 2, the individual injection rate for each fracture is $Q_t/2$. The mechanical and thermal properties of granite are chosen according to the literature [35, 36], and are listed in Table 15.1.

15.4.1 Fluid Flow and Production Temperature

Figure 15.5 shows the vector field of fluid velocity in the single-fracture case before the fracture propagates. As shown in Figure 15.6, the change of production temperature in two cases can be divided into two stages: a rapid decline stage during the first three years and a gradual decline after that. In the first year, a slightly higher production temperature is seen in the double-fracture case compared with single-fracture case if Q_t is the same. With the time increasing, the difference in the production temperature between the double-fracture case and single-fracture case becomes larger and larger. After 20 years of operation,

Table 15.1 Parameter used for the computation in the model.

Parameter	Value	Parameter	Value
Rock density, ρ_r (kg/m^3)	2640	Injection temperature, $T_{f,in}$ (°C)	40
Thermal conductivity of rock, k_r (W/m/°C)	2.76	Fluid density, ρ_f (kg/m^3)	1000
Specific heat capacity of rock, c_r (kJ/kg/°C)	1	Thermal conductivity of fluid, k_f (W/m/°C)	0.6
Young's Modulus of rock, E (GPa)	35	Specific heat capacity of fluid, c_f (kJ/kg/°C)	4.2
Poisson's ratio of rock, υ	0.12	Fluid dynamic viscosity, μ (Pa·s)	0.001
Thermal expansion coefficient of rock, α	8×10^{-6}	Reinjection well depth, H_i (m)	3505
Ground temperature, B_0 (°C)	25	Production well depth, H_p (m)	3905
Geothermal gradient, A_0 (°C/m)	0.057	Total injection rate, Q_t (m^3/s)	0.1, 0.05
Critical energy release rate, G_n^C (kN/m)	37	Fracture radius, R (m)	200
Vertical principal stress, σ_v (MPa)	110	Initial fracture aperture, w_0 (m)	0.005
Maximum horizontal principal stress, σ_H (MPa)	90	Number of fractures, N	1, 2
Minimum horizontal principal stress, σ_h (MPa)	70	Fracture spacing, d (m)	80
Depth of fracture center, H_0 (m)	3705	Time increment, Δt (s)	3600

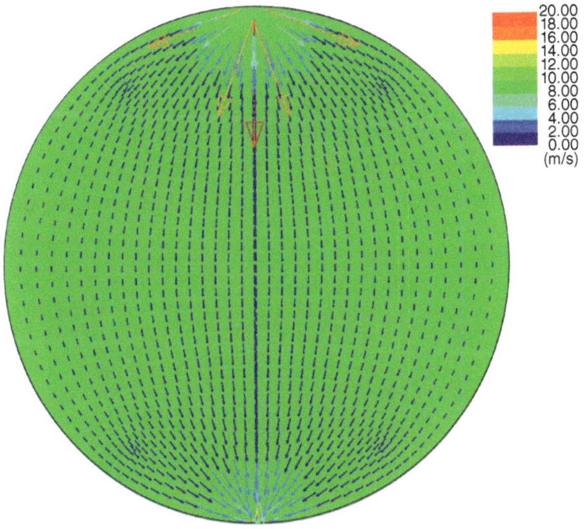

Figure 15.5 Vector of the fluid velocity in the single-fracture case.

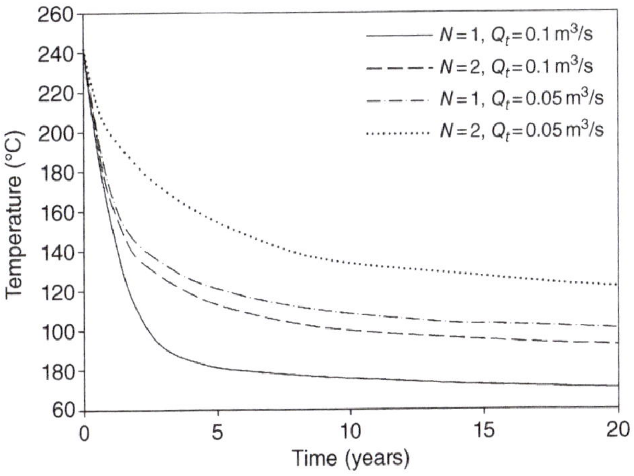

Figure 15.6 Production temperature of EGS for single- and double- fracture cases when Q_t = 0.1 or 0.05 m^3/s.

the production temperature of double-fracture case is 100 °C, roughly 30 °C larger than that of single-fracture case when Q_t is 0.1 m^3/s. Besides, the highest production temperature at 20-year is observed in the double-fracture case when Q_t is 0.05 m^3/s, at 122 °C, which means the production temperature decreases at least 120 °C during the 20-year operation period among the cases studied.

From Figure 15.6, the heat extraction in single-fracture case when Q_t is 0.1 m^3/s is the least efficient compared to other cases. This indicates that the number of fractures and the total injection rate play an important role in the production performance. As mentioned above, the injection temperature is kept constant in all cases. The injection rate is one of the main factors to determine how fast the heat can be absorbed by fluid from the reservoir during a given operational time. Therefore, when the fracture number is the same, the case with higher total injection rate will have worse production performance. Furthermore, when the total injection rate is the same, the injection rate for each fracture in double-fracture case is half of that in single-fracture case, so the cooling front of the fluid is slower, which allows the fluid to have more time to take heat from the rock formation. This results in a higher production temperature. In addition, in the double-fracture cases, the fracture spacing is large enough, so the thermal influence between neighboring fractures is weak.

15.4.2 Temperature Distribution in Rock Formation

To analyze the reservoir performance, the evolution of outlet temperature and reservoir temperature during the 20-year operational time are obtained. Figures 15.7 and 15.8 show the contour of the reservoir temperature in the vertical cross section ($x = 0$) for the single-fracture case after continuous injection of 2–20 years when Q_t is 0.05 and 0.1 m^3/s, respectively. The injection point is located at (0, 0, 200 m), which corresponds to a depth of 3505 m. It is observed that once the cold water is injected into fractures through the injection well, rock temperature near the injection well decreases sharply, forming thermal influence zones (THZs) with temperature below 90 °C. With the continuous injection of cold fluid, the THZ gradually expands. By comparing Figure 15.8 with Figure 15.7, it is found that the case with larger injection rate has stronger cooling effect and larger THZ. Hence, the single-fracture case with 0.1 m^3/s injection rate has a worse production performance, as shown in Figure 15.6.

Figure 15.9 shows the temperature distribution inside the surrounding rock formations for the double-fracture case in vertical cross section ($x = 0$) when Q_t is 0.1 m^3/s during 20 years. With time increasing, two THZs merge into one, resulting in the occurrence of interference zones between the fractures. In addition, compared with single-fracture case, the cooling rate in the rock temperature is much slower when the injection rate is kept the same.

Figure 15.10 shows the temperature distribution of the reservoir for the single-fracture case in sections of $z = 0$ m and $y = 0$ m, while Figure 15.11 for two-fracture case when Q_t is 0.1 m^3/s. It is also found that the single-fracture case has stronger cooling effect and larger THZ compared to the double-fracture case when the total injection rate is kept the same. Therefore, it indicates that the double-fracture case has a longer EGS operation life compared with the other.

15.4.3 Fracture Propagation

In this section, numerical calculations based on XFEM are carried out to show the thermal effects on fracture propagation behaviors during the 20-year operation period. In order to highlight the thermal effects, the in situ stresses are kept the same, while the total injection rate varies. The parameters used for the present calculations are listed in Table 15.1.

15.4.3.1 Single Fracture Case

Figure 15.12 shows the position of longest fracture front varying with time for single- fracture case under different total injection rates, Q_t. It is found that the larger the

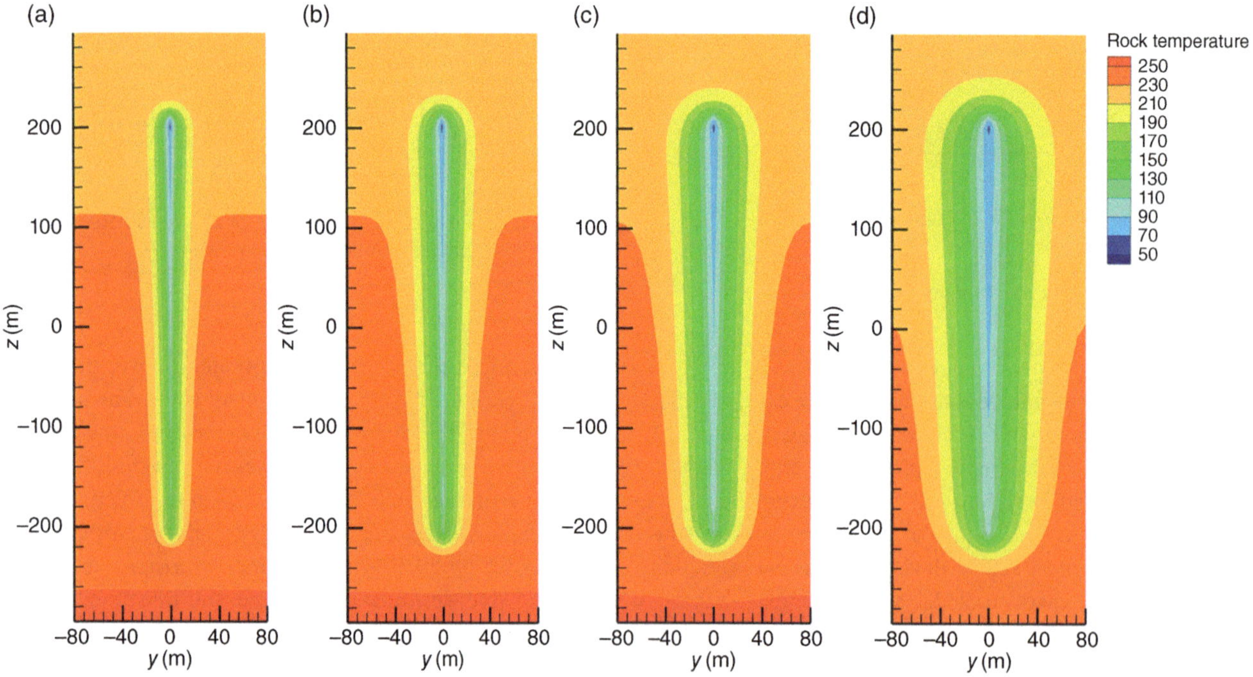

Figure 15.7 Contours of temperature distribution in the vertical cross section ($x = 0$) for single-fracture case when Q_t is 0.05 m^3/s: (a) $t = 2$ years; (b) $t = 5$ years; (c) $t = 10$ years; (d) $t = 20$ years.

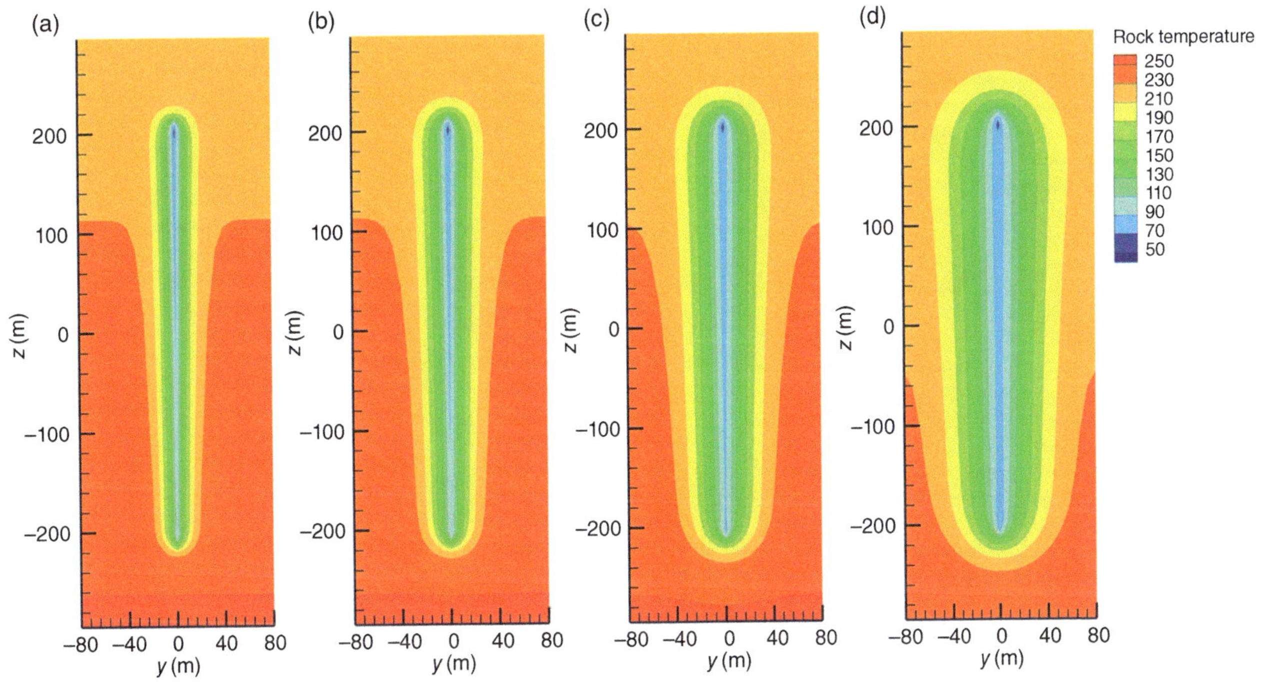

Figure 15.8 Contours of temperature distribution in the vertical cross section ($x = 0$) for single-fracture case when Q_t is 0.1 m^3/s: (a) $t = 2$ years; (b) $t = 5$ years; (c) $t = 10$ years; (d) $t = 20$ years.

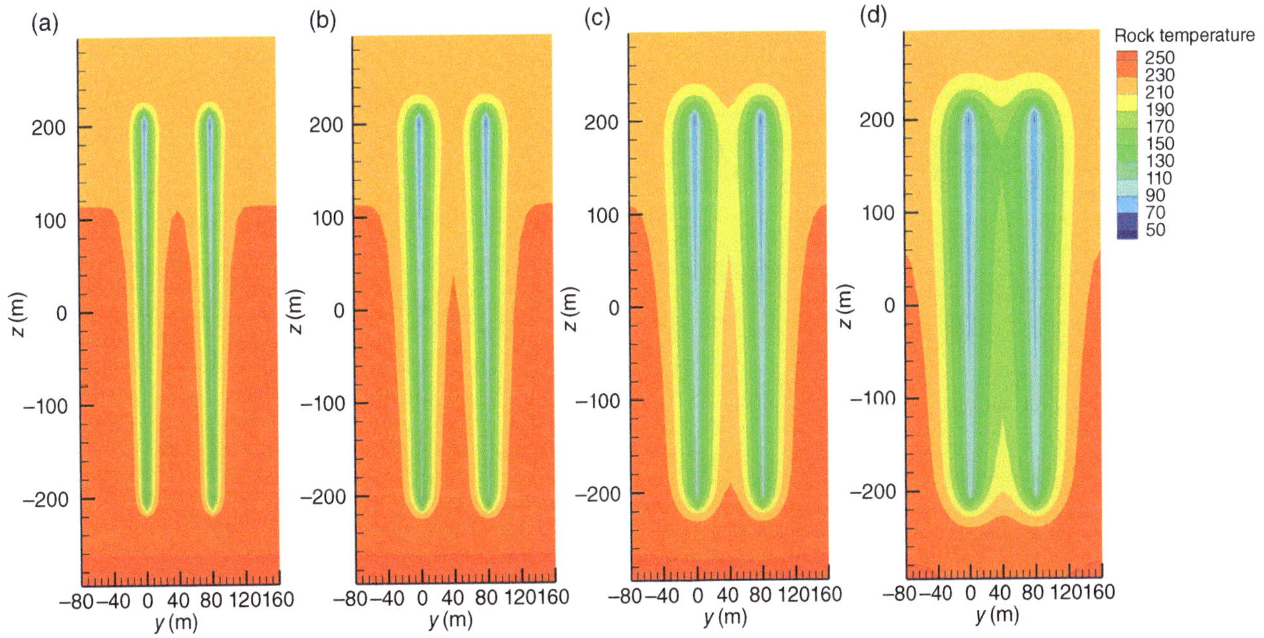

Figure 15.9 Contours of temperature change in the vertical cross section ($x = 0$) for double-fracture case when Q_t is 0.1 m^3/s: (a) $t = 2$ years; (b) $t = 5$ years; (c) $t = 10$ years; (d) $t = 20$ years.

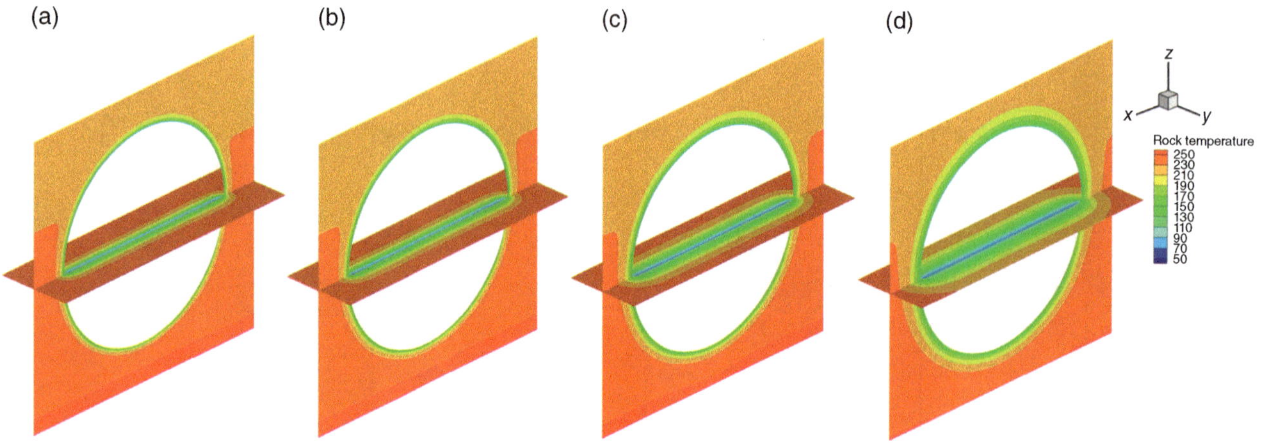

Figure 15.10 Contour of reservoir temperature in sections ($z = 0$, $y = 0$) for single-fracture case when Q_t is 0.1 m^3/s at different times: (a) $t = 2$ years; (b) $t = 5$ years; (c) $t = 10$ years; (d) $t = 20$ years.

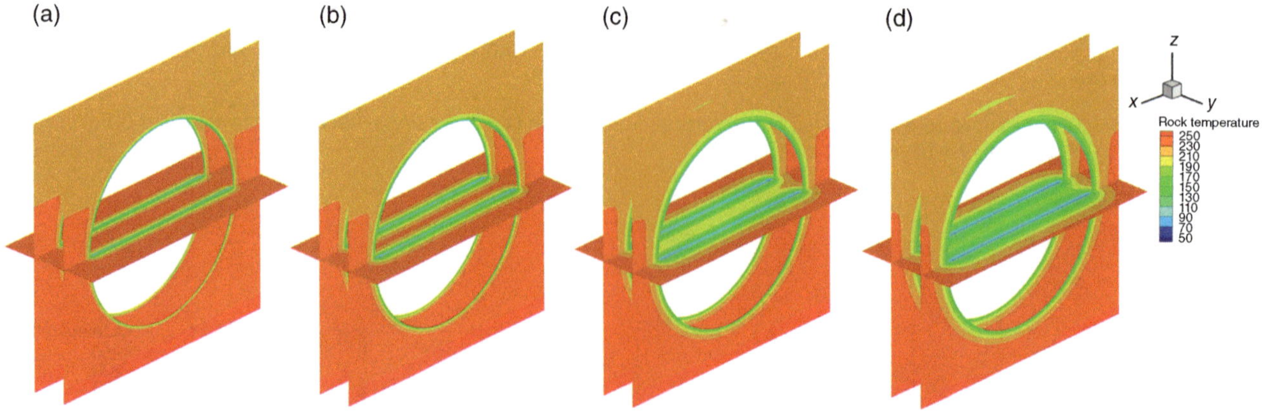

Figure 15.11 Reservoir temperature contour in sections ($z = 0$, $y = 0$, $y = 80$ m) for double-fracture case when Q_t is 0.1 m^3/s at different times: (a) $t = 2$ years; (b) $t = 5$ years; (c) $t = 10$ years; (d) $t = 20$ years.

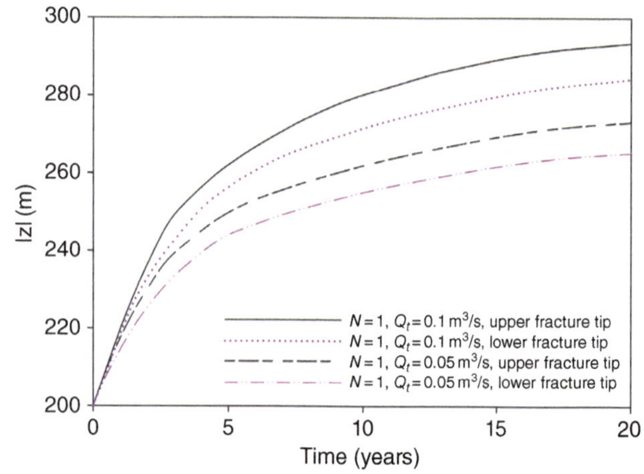

Figure 15.12 Evolution of fracture front for single-fracture case under different total injection rate.

injection rate, the longer the fracture front propagation. The upper fracture length reaches 293.52 m after 20 years when Q_t is 0.1 m^3/s, 20.35 m larger than that when Q_t is 0.05 m^3/s. This is mainly for two reasons. First, as shown in Figure 15.13, the fluid pressure along the z-axis in sections ($x = 0$, $y = 0$) is about 65.6 MPa at 20 years when Q_t is 0.1 m^3/s, roughly 12 MPa larger than when Q_t is 0.05 m^3/s. Second, larger injection rate, meaning stronger cooling effect, will induce larger thermal stress, leading to larger propagation speed. In addition, the lower front of the fracture has a smaller distance from the original location than the upper front of the fracture. For instance, when Q_t is 0.1 m^3/s, the lower front of the fracture propagates 84.43 m after 20 years, almost 9.09 m smaller than the upper front of the fracture. This is due to the larger thermal stress resulting from the stronger cooling at the injection point. As shown in Figure 15.7, the THZ around the upper fracture front is larger than that around the lower fracture front.

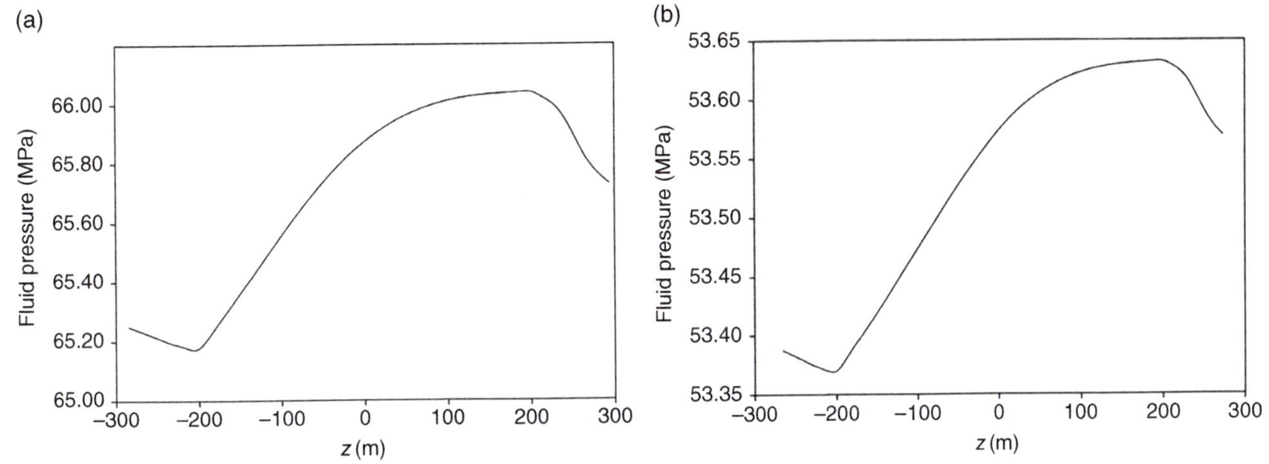

Figure 15.13 Fluid pressure along the z-axis for single-fracture case at 20 years under different total injection rates: (a) Q_t = 0.1 m^3/s and (b) Q_t = 0.05 m^3/s.

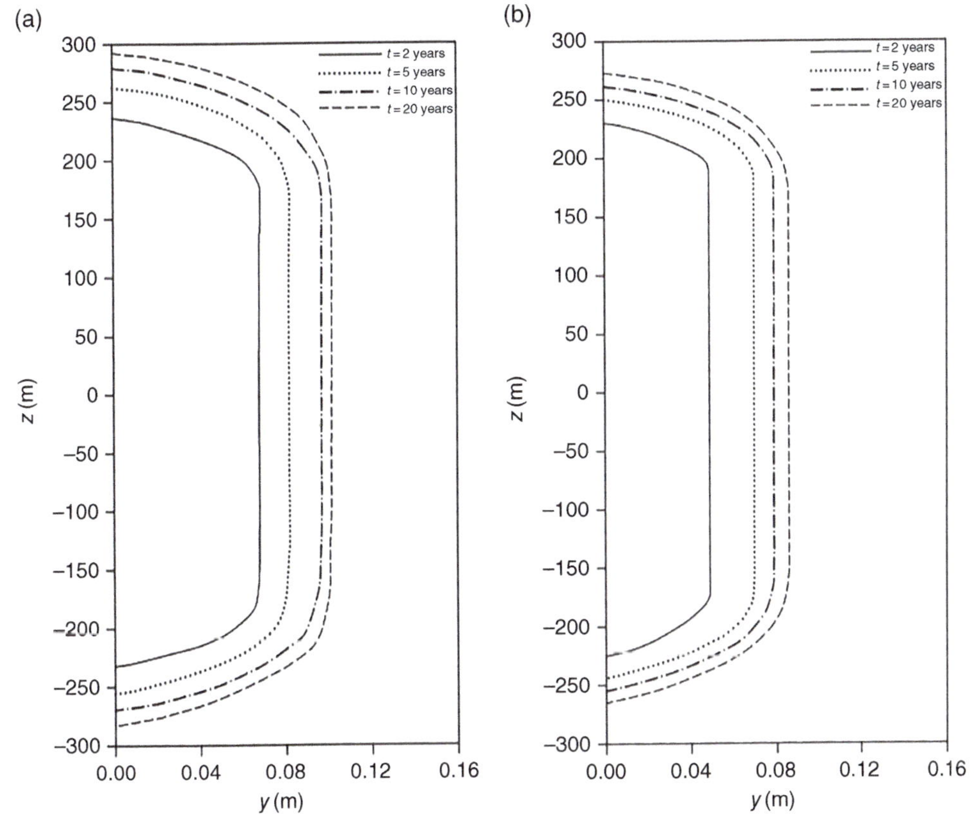

Figure 15.14 Half cross sectional (x = 0) shape of the fracture in single-fracture case when the total injection rate are: (a) 0.1 m^3/s and (b) 0.05 m^3/s.

Figure 15.14a, b display the variation of cross sectional ($x = 0$) shape of the fracture in single-fracture case when the total injection rates are 0.1 and 0.05 m^3/s, respectively. Due to geometrical symmetry in single-fracture case, only the right half of the fracture profile evolution in vertical section ($x = 0$) is provided. Firstly, it is observed that the fracture aperture increases with time. For instance, when Q_t is 0.1 m^3/s, the fracture opening at the right half amounts to a maximum of 10.19 cm after 20 years, while the maximum value is 6.85 cm after 2 years. It is due to the fact that the rock temperature around the fracture decreases with time and the THZ gradually expands, as shown in Figure 15.8, leading to increased tensile stress around the fracture and further fracture propagation. In addition, the larger injection rate, the larger mechanical opening induced. For example, the half fracture opening at 20 years reaches 8.60 cm when $Q_t = 0.05\ m^3/s$, around 1.59 cm smaller than that when $Q_t = 0.1\ m^3/s$.

15.4.3.2 Double Fracture Case

Figure 15.15 presents the position of fracture tip for double-fracture case under different total injection rates, Q_t. By comparing with Figure 15.12, when the total injection rate is the same, the fracture in single-fracture case propagates at a larger speed than that in the double-fracture case. For instance, the upper fracture front position in the double-fracture case is at $z = 269.03$ m when Q_t is $0.1\ m^3/s$, which is approximately 24.49 m closer to the injection well than that in the single-fracture case. This is mainly due to two reasons. First, from Figures 15.13 and 15.16, when Q_t is $0.1\ m^3/s$, the fluid pressure along the z axis in sections ($x = 0$, $y = 0$) is roughly 52.05 MPa at 20 years in the double-fracture case, approximately 13.55 MPa smaller than that in the single-fracture case. Second, the cooling effect in the single-fracture case is stronger than that in the double-fracture case due to larger induced thermal stress resulting from larger temperature decrease.

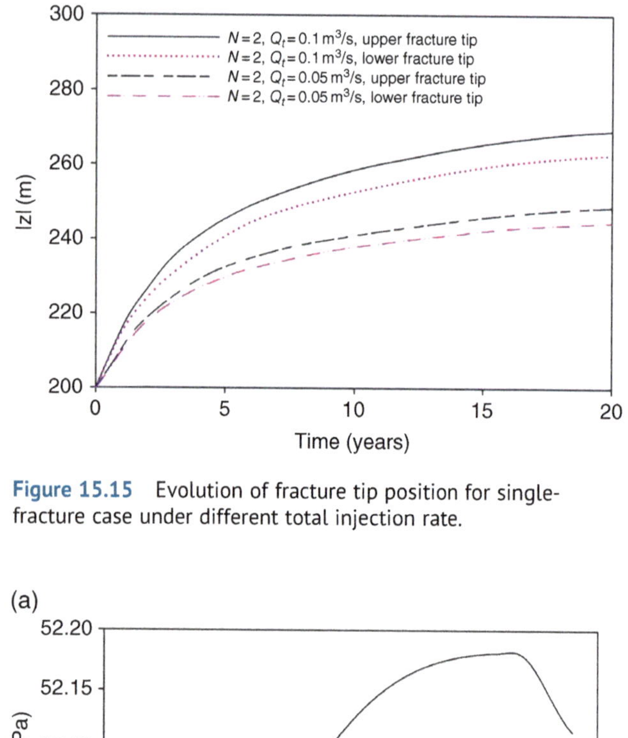

Figure 15.15 Evolution of fracture tip position for single-fracture case under different total injection rate.

In the two-fracture case, the behaviors of the two fractures are the same and symmetrical with respect to the plane of $y = 40$ m. Figure 15.17a, b show the shape of Fracture 2 in the vertical section ($y = 0$) when Q_t is $0.1\ m^3/s$ and Q_t is $0.05\ m^3/s$, respectively. It is observed that the fracture aperture is not symmetric with respect to the plane of $y = 80$ m. When Q_t is $0.1\ m^3/s$, the mechanical opening of the left half of the Fracture 2 at 20-year is 6.62 cm, nearly 1 cm smaller than that of the right half. The most significant reason is that the compression coming from the Fracture 1 results in the stress interference. In addition, by comparing Figure 15.17 with Figure 15.14, the fracture in the single-fracture case has a wider opening in the vertical section ($x = 0$) if the total injection rate is the same. This also demonstrates that the cooling rate plays an important role in fracturing propagation. Furthermore, due to the stress interference from the Fracture 1, it is found that the direction of fracture tips of the Fracture 2 points to the plane of $y = 40$ m in Figure 15.17. These further demonstrate that the thermal induced stress and fluid pressure significantly influences the fracturing behaviors. Therefore, thermal effects during the fluid circulation must be considered because the induced thermal stress may facilitate the fracture propagation even though the fluid pressure may not drive the crack propagation.

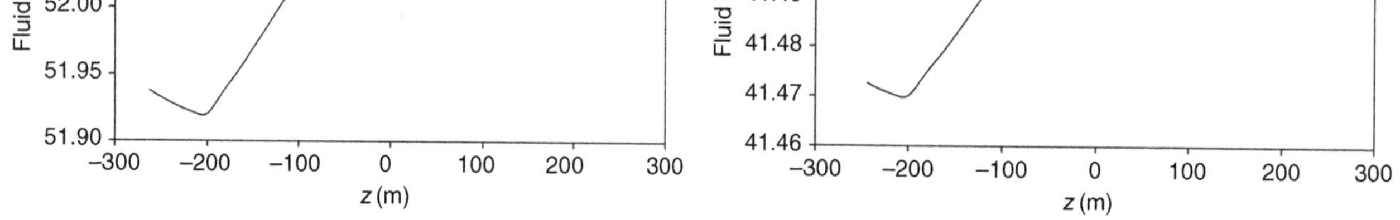

Figure 15.16 The fluid pressure along the z-axis for double-fracture case at 20 years under different total injection rate: (a) $0.1\ m^3/s$ and (b) $0.05\ m^3/s$.

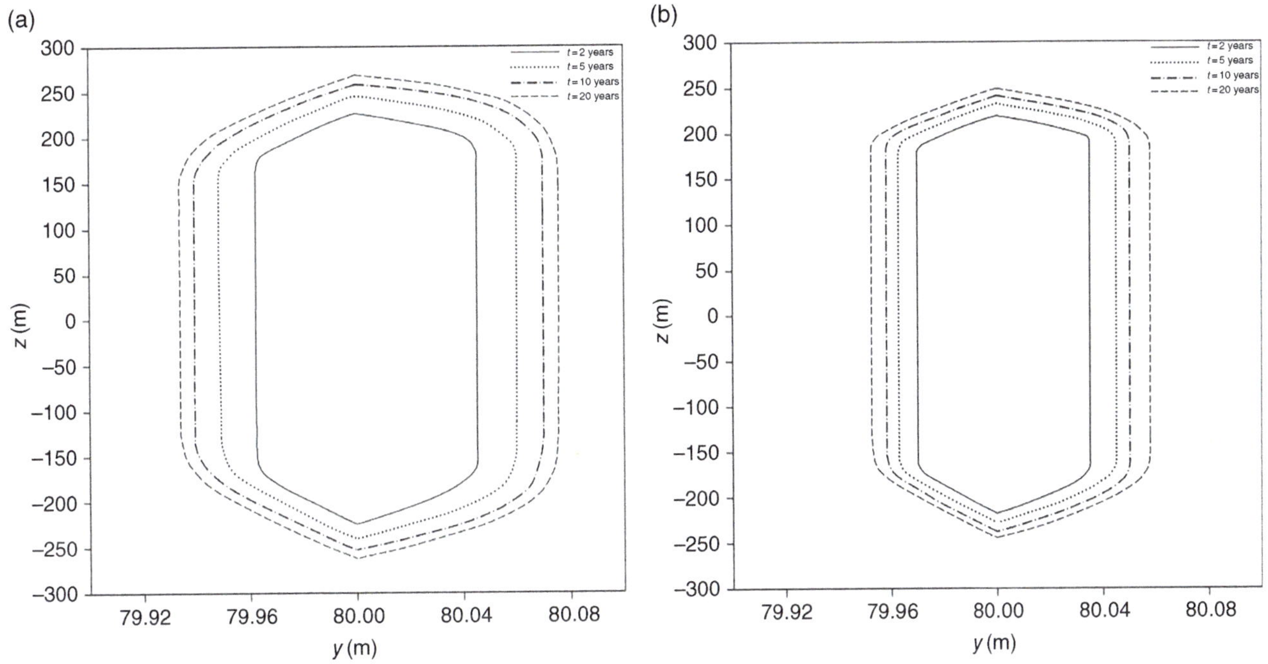

Figure 15.17 Cross sectional (x = 0) shape of the fracture in double-fracture case when the total injection rate are: (a) 0.1 m^3/s and (b) 0.05 m^3/s.

15.5 Conclusions

In this chapter, a numerical model for predicting heat extraction and fracture propagation for single- and double-fracture EGSs is developed. Numerical simulations using ABAQUS are carried out to investigate the thermal and mechanical behaviors of the reservoir after long time fluid circulation. The following conclusions are drawn:

1) The heat extraction can be separated into two stages: rapid decline stage in the first three years and gradual decline stage after that. The fluid injection rate plays a dominant role in determining how fast the heat is absorbed from the rock matrix, and thus affecting heat extraction. In the two cases, the two-fracture case has a better performance in production temperature (i.e. 122 °C after 30 years), while the single-fracture case is the least efficient system (i.e. merely 70 °C after 30 years).
2) Once the cold fluid is injected from the injection well, the reservoir temperature around the injection well decreases sharply, and the thermal influence zones expand along fractures with increasing time and merge into a thermal interference zone between fractures.
3) Intensive cooling and fluid pressure significantly influence fracture propagation, and the larger injection rate of one individual fracture means stronger thermal effect and fluid pressure. The single-fracture case has larger induced tensile stresses and fluid pressure, larger mechanical opening and longer extended fracture.
4) Thermal induced stresses cause fractures to propagate to a larger length (i.e. the upper fracture extends 93.52 m for the single-fracture case and 69.03 m for the double-fracture case when Q_t is 0.1 m^3/s after 20 years) and result in increased mechanical opening (i.e. 20.38 cm for the single-fracture case and 14.34 cm for the double-fracture case when Q_t is 0.1 m^3/s after 20 years). Therefore, thermal effects should be considered when modeling the heat extraction and mechanical behavior of enhanced geothermal systems during long time fluid circulation.

References

1 Wan, Z., Zhao, Y., and Kang, J. (2005). Forecast and evaluation of hot dry rock geothermal resource in China. *Renewable Energy* 30 (12): 1831–1846.

2 Liu, Y., Wang, G., Zhu, X., and Li, T. (2020). Occurrence of geothermal resources and prospects for exploration and development in China. *Energy Exploration & Exploitation* 39 (2): 536–552.

3 Lei, Z, Ranjith, P, Wu, B, and De Silva V, editors. (2021). Investigation of the fracture performance of soundless cracking demolition agents (SCDA) in different reservoir rocks. *55th US Rock Mechanics/Geomechanics Symposium*, Virtual, ARMA-2021-1749, OnePetro.

4 Guo, T., Gong, F., Wang, X. et al. (2019). Performance of enhanced geothermal system (EGS) in fractured geothermal reservoirs with CO2 as working fluid. *Applied Thermal Engineering*. 152: 215–230.

5 Zhang, F., Jiang, P., and Xu, R. (2013). System thermodynamic performance comparison of CO2-EGS and water-EGS systems. *Applied Thermal Engineering*. 61 (2): 236–244.

6 Wu, B, Zhang, X, and Jeffrey, R.G. (2016). Analysis of thermal effects on hydraulic fracturing near a horizontal well by using displacement discontinuity method. *SPE Asia Pacific Hydraulic Fracturing Conference*, Beijing, OnePetro.

7 Wu, B., Zhang, G., Zhang, X. et al. (2017). Semi-analytical model for a geothermal system considering the effect of areal flow between dipole wells on heat extraction. *Energy* 138: 290–305.

8 Wu, B., Lei, Z., Zhao, G. et al. (2022). An efficient model for predicting heat extraction from a multiple-well enhanced geothermal system in the presence of areal flow. *Applied Thermal Engineering* 210: 118399.

9 Wu, B., Zhang, X., Jeffrey, R.G. et al. (2016). A simplified model for heat extraction by circulating fluid through a closed-loop multiple-fracture enhanced geothermal system. *Applied Energy* 183: 1664–1681.

10 Xu, T., Yuan, Y., Jia, X. et al. (2018). Prospects of power generation from an enhanced geothermal system by water circulation through two horizontal wells: a case study in the Gonghe Basin, Qinghai Province, China. *Energy* 148: 196–207.

11 Lei, Z., Zhang, Y., Yu, Z. et al. (2019). Exploratory research into the enhanced geothermal system power generation project: the Qiabuqia geothermal field, Northwest China. *Renewable Energy* 139: 52–70.

12 Sun, Z., Xin, Y., Yao, J. et al. (2018). Numerical investigation on the heat extraction capacity of dual horizontal wells in enhanced geothermal systems based on the 3-D THM model. *Energies* 11 (2): 280.

13 Xu, C., Dowd, P.A., and Tian, Z. (2015). A simplified coupled hydro-thermal model for enhanced geothermal systems. *Applied Energy* 140: 135–145.

14 Jiang, F., Chen, J., Huang, W., and Luo, L. (2014). A three-dimensional transient model for EGS subsurface thermo-hydraulic process. *Energy* 72: 300–310.

15 Zhang, G., Liu, H., Zhang, J. et al. (2010). Three-dimensional finite element simulation and parametric study for horizontal well hydraulic fracture. *Journal of Petroleum Science and Engineering* 72 (3–4): 310–317.

16 Chen, Z. (2012). Finite element modeling of viscosity-dominated hydraulic fractures. *Journal of Petroleum Science and Engineering* 88: 136–144.

17 Phongthanapanich, S. and Dechaumphai, P. (2004). Adaptive Delaunay triangulation with object-oriented programming for crack propagation analysis. *Finite Elements in Analysis and Design* 40 (13–14): 1753–1771.

18 Crouch, S.L., Starfield, A.M., and Rizzo, F. (1983). *Boundary Element Methods in Solid Mechanics*. London: George Allen & Unwin.

19 Moës, N., Dolbow, J., and Belytschko, T. (1999). A finite element method for crack growth without remeshing. *International Journal for Numerical Methods in Engineering* 46 (1): 131–150.

20 Belytschko, T. and Black, T. (1999). Elastic crack growth in finite elements with minimal remeshing. *International Journal for Numerical Methods in Engineering* 45 (5): 601–620.

21 Chen, Z. (2013). An ABAQUS implementation of the XFEM for hydraulic fracture problems. *ISRM International Conference for Effective and Sustainable Hydraulic Fracturing, Brisbane*, International Society for Rock Mechanics and Rock Engineering.

22 Wang, X., Liu, C., Wang, H. et al. (2016). Comparison of consecutive and alternate hydraulic fracturing in horizontal wells using XFEM-based cohesive zone method. *Journal of Petroleum Science and Engineering* 143: 14–25.

23 Shi, J., Chopp, D., Lua, J. et al. (2010). Abaqus implementation of extended finite element method using a level set representation for three-dimensional fatigue crack growth and life predictions. *Engineering Fracture Mechanics* 77 (14): 2840–2863.

24 Batchelor, C.K. and Batchelor, G. (2000). *An Introduction to Fluid Dynamics*. Cambridge University Press.

25 Zienkiewicz, O.C., Taylor, R.L., Taylor, R.L., and Taylor, R. L. (2000). *The Finite Element Method: Solid Mechanics*. Butterworth-heinemann.

26 Chen, Z. (2015). A new enriched finite element method with application to static fracture problems with internal fluid pressure. *International Journal of Applied Mechanics* 7 (3): 1550037.

27 Abaqus V. 6.14 Documentation. *Dassault Systemes Simulia Corporation*. 2014;651:6.2.

28 Benzeggagh, M.L. and Kenane, M. (1996). Measurement of mixed-mode delamination fracture toughness of unidirectional glass/epoxy composites with mixed-mode bending apparatus. *Composites Science and Technology* 56 (4): 439–449.

29 Cheng, A.H., Ghassemi, A., and Detournay, E. (2001). Integral equation solution of heat extraction from a fracture in hot dry rock. *International Journal for Numerical and Analytical Methods in Geomechanics* 25 (13): 1327–1338.

30 Kreith, F. and Black, W.Z. (1980). *Basic Heat Transfer*. New York: Harper & Row.

31 Zhao, Z. (2014). On the heat transfer coefficient between rock fracture walls and flowing fluid. *Computers and Geotechnics* 59: 105–111.

32 Schulz, R. (1987). Analytical model calculations for heat exchange in a confined aquifer. *Journal of Geophysics* 61 (1): 12–20.

33 Loehnert, S., Mueller-Hoeppe, D., and Wriggers, P. (2011). 3D corrected XFEM approach and extension to finite deformation theory. *International Journal for Numerical Methods in Engineering* 86 (4–5): 431–452.

34 Holl, M., Rogge, T., Loehnert, S. et al. (2014). 3D multiscale crack propagation using the XFEM applied to a gas turbine blade. *Computational Mechanics* 53 (1): 173–188.

35 Weinert, S., Bär, K., and Sass, I. (2020). Thermophysical rock properties of the crystalline Gonghe Basin complex (Northeastern Qinghai–Tibet-Plateau, China) basement rocks. *Environmental Earth Sciences* 79 (3): 77.

36 Zhou, Z., Jin, Y., Zeng, Y. et al. (2020). Investigation on fracture creation in hot dry rock geothermal formations of China during hydraulic fracturing. *Renewable Energy* 153: 301–313.

16

Multiple Hydraulic Fractures Growth from a Highly Deviated Well: A XFEM Study

Yun Zhou[1] *and Diansen Yang*[2]

[1] *State Key Laboratory of Geomechanics and Geotechnical Engineering, Institute of Rock and Soil Mechanics, Chinese Academy of Sciences, Wuhan, Hubei, China*
[2] *School of Civil Engineering, Wuhan University, Wuhan, China*

16.1 Introduction

Hydraulic fracturing is an indispensable technology for permeability enhancement in the development of unconventional energy. The combination of horizontal well (HW) and multistage hydraulic fracturing has become an effective stimulation technology for low permeability reservoir [1, 2]. For example, this technology has been widely used in Erdos Basin, which is one of the major basins for oil and gas production in China, due to its low porosity and permeability. The reservoir lithology contains a very thick tight sandstone layer in between two thin layers of mudstone. However, the operators found that highly deviated wells (HAWs) can produce a larger gas production rate than HWs based on large amount of production data in Huaqing Oilfield. The well placement and multiple fracture clusters for HWs and HAWs are illustrated in Figure 16.1. The mechanisms for the production rate enhancement of HAWs are still not clear, and we hope to preliminarily tap them in this paper from the viewpoint of fracture mechanics.

A HAW refers to the wellbore trajectory in reservoir formations, whose inclination in most of its length has a high angle (60°–80°) with respect to the horizontal direction [3]. The main purpose of using both a HAW and a HW is to increase the contact area between the wellbore and pay zone, to eventually improve the oil recovery rate. Although the HWs have been widely and successfully used in the stimulation of unconventional gas reservoirs, the HWs are normally considered to be more expensive than the conventional deviated wells. In addition, it is inevitable to use HAWs in some specific cases, such as the oil and gas development in highly inclined strata. Since the application of HAWs in oil and gas industry is less than that of HWs, hydraulic fracture growth along the HAWs received less attention compared to vertical wells and HWs.

As the axis of a HAW is not necessarily aligned with the in-situ stress, El Rabaa [4] found that hydraulic fracture induced from this type of wells can be more complex than that from conventional vertical well. Following the early work of El Rabaa [4] and Weng [5], Chen et al. [6] developed a hydraulic fracturing criterion for highly deviated wells based on stress components around the borehole. By performing the hydraulic fracturing experiments in hydrostone with the wellbore cast in various orientations, Brumley and Abass [7] argued that the complex fracture geometry could occur along deviated wellbores. Recently, AL Isaee et al. [8] suggested that longitudinal planar fracture, transverse planar fracture and complex nonplanar fracture could be created from HAW, depending on the axis of wellbore relative to in situ stress. El Rabaa [4] further revealed that non-planar fracture geometries induced by highly deviated well could result in reduced fracture width and high treating pressure. The numerical simulations by Olson [9] also revealed the difficulty in HAW stimulation due to excess treating pressure and near wellbore screen-out. Some other researchers focused on the influence of completion parameters and wellbore orientations on the fracture initiation from HAW. For example, Van de Ketterij and de Pater [10] studied the relation between completion and treatment parameters and resulting fractures geometries. Zhu et al. [11] investigated the effect of perforation angle, well azimuth, and inclination angle on the fracture initiation and propagation from HAW by laboratory experiments. Later, Zhu et al. [12] adopted a coupled flow-stress-damage nonlinear finite element model to analyze the sensitivities of rock parameters, in situ stresses and deviated angle to the geometry of fracture initiated from HAW in

Mechanics of Hydraulic Fracturing: Experiment, Model, and Monitoring, First Edition.
Edited by Xi Zhang, Bisheng Wu, Diansen Yang, and Andrew Bunger.

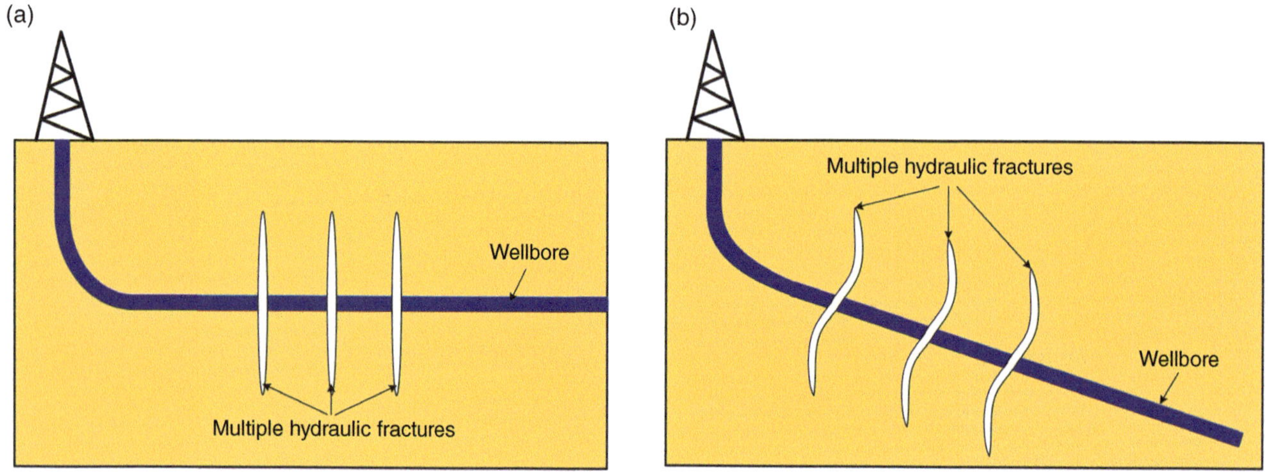

Figure 16.1 Stimulation techniques for tight sandstone reservoirs in Huaqing Oilfield: (a) Horizontal well, (b) Highly deviated well.

hard-rock. However, those studies mainly concentrated on the near wellbore fracture behaviors, while the subsequent fracture growth from the HAW in the presence of multiple fracture is not fully exploited.

Multi-cluster hydraulic fracturing technology becomes a prevalent approach to simulate unconventional gas reservoirs due to its high efficiency. Simultaneous fracture growth becomes a hot topic for maximizing the fracturing treatment benefits [13, 14]. Wu and Olson [15] demonstrated that the nonuniform development of multiple fractures was induced by the uneven partitioning of injection rate into each fracture. Chen et al. [16] concluded that perforation friction and interaction stress are two key factors that determine fluid partitioning among multiple fractures. Some researchers emphasized that an increase in perforation loss by decreasing number [17–19] or diameter [16, 17, 20] of perforations can help to uniformly divide the injection rate to counteract the stress shadow effect. However, large pressure drop at perforations will increase the total injection pressure and energy consumption, and eventually produce a pronounced adverse effect on stimulation operation [21]. Increasing the cluster spacing is an effective way to reduce the stress shadow, but large spacing will reduce the number of fractures in a certain fracturing stage [19, 22, 23]. Hence, too small perforation number or too large cluster spacing is not recommended to mitigate the stress shadow effect [19]. According to this, some researchers adopted nonuniform cluster spacing [17, 24, 25] or perforation distribution [18, 19] to optimize the fracturing treatment. Apart from these typical methods, the stress interference induced by multi-well fracturing (such as zipper fracturing) can be utilized to promote the uniform growth of multiple fractures [20, 22, 26]. Based on two-fracture interaction, Chen et al. [27, 28] found that higher injection pressure levels can promote simultaneous growth of multiple fractures if there is a different size. Li et al. [29] and Zhou et al. [30] emphasized the use of temporarily plugging technology in generating a more uniform fracture network. In addition, multi-field coupling effect, such as hydro-mechanical (HM) [23, 31] or thermal-mechanical [25] coupling, also can decrease the stress interference between propagating fractures. However, these studies are all targeted to improve the simultaneous growth of multiple fractures along the HWs. Whether these measures can be applicable to HAWs is still a question.

Since the XFEM can get rid of remeshing in handling fracture propagation problems [32], it is adopted in this paper to simulate the multiple fractures propagation along the HAW. A large number of hydraulic fracturing models have been established based on XFEM, and the major differences between these models are how to solve and couple the fluid flow in fractures and solid matrix. For example, Khoei et al. [33] adopted finite difference method, while Shi et al. [34] used finite element method to discretize the fluid flow along the fracture without considering the fluid seepage in the porous matrix. To take the flow exchange between matrix and fractures into account, Salimzadeh and Khalili [35] added additional pressure nodes to the fracture, whereas Paul et al. [36] imposed additional flux degrees of freedom (DOFs) on the vertex nodes of vital edge. Remij et al. [37] combined additional pressure nodes and Heaviside enrichment function to make the pressure field discontinuous over the fracture, whereas Paul et al. [36] adopted Lagrange multipliers to enforce the pressure continuity condition across the fracture. In this paper, the fully HM-coupled model proposed by Zhou et al. [38] is adopted to obtain the mass transfer between matrix and fractures. The Ridge enrichment function [39] is used to describe the continuous pressure field across the fracture, which is similar to the model proposed by Khoei et al. [40]. However, our model adopted linear elastic fracture mechanics (LEFM) to calculate the stress around the fracture tip, while Khoei et al. [40] used cohesive zone model to simulate the inelastic deformation of the fracture. To study the multizone hydraulic fracturing treatments under XFEM framework, Vahab and Khalili [41] proposed a consistent computational algorithm to determine the distribution of fracturing fluid flow. Zeng et al. [42] adopted Newton iteration method to solve the fluid partitioning among multiple fractures to avoid imposing constraint conditions at fracture-well intersection. In this paper, the secant method is used to satisfy the pressure continuity and mass conservation conditions among the multi-cluster perforations. Hence, the first innovation of this work is the established model, which can simulate the multi-cluster fracturing process in porous media. Employing this model to investigate the propagation of multiple competitive fractures along the HAW is another innovation.

In this paper, we will develop a XFEM-based hydraulic fracture simulator for arbitrary 2D fracture growth. To model the stress interference between multiple fractures, the governing equations are first presented. In addition, the weak forms of both stress equilibrium and fluid mass continuity equations are turned into a system of linear equations through XFEM-type enrichments in space and Newmark scheme in time. The resulting HM-coupled equations are then solved by Newton–Raphson iteration method. Following numerical validation, the model is applied to the HAW stimulation cases with various fracture placements. The numerical results are presented for parametric studies and some preliminary conclusions are then drawn.

16.2 Problem Formulation

Due to the high complexity and time-consumption of 3D hydraulic fracturing problems, the 3D problems are often transformed to 2D problems to speed up the calculation. In this paper, the HW and HAW are assumed to be drilled in the σ_v–σ_h plane, which is an ideal but pursued situation in oil and gas production, where σ_v and σ_h is vertical confining stress and minimum horizontal confining stress, respectively. Then the actual 3D problems can be simplified. The well-known KGD model [43, 44] is a classical simplified 2D model on horizontal plane (σ_H–σ_h plane) with a constant fracture height, as shown in Figure 16.2a. This model is usually adopted to investigate the variation of fracture length, pressure, and width with injection time or other factors. To investigate the variation of fracture height of multiple fractures in HAW and HW, the actual 3D problem is simplified to 2D problem on vertical plane (σ_v–σ_h plane) in this paper, as shown in Figure 16.2b. The fracture length is taken as fixed and the simplified 2D model only considers the vertical growth of multiple fractures. Hence, the major differences between these two models are the model assumption and research objective. In addition, the thickness of the reservoir is assumed as very large in this paper, which is the actual condition of the Huaqing Oilfield. The mechanical properties of the rock are homogeneous and isotropic, regardless of the change of the in-situ stress along the depth. The boundary conditions of the simplified multi-cluster hydraulic fracturing model are stated as follows.

Figure 16.3a provides the schematic of multi-cluster hydraulic fracturing in a 2D domain (Ω). Several hydraulic fractures are perforated perpendicular to the wellbore axis

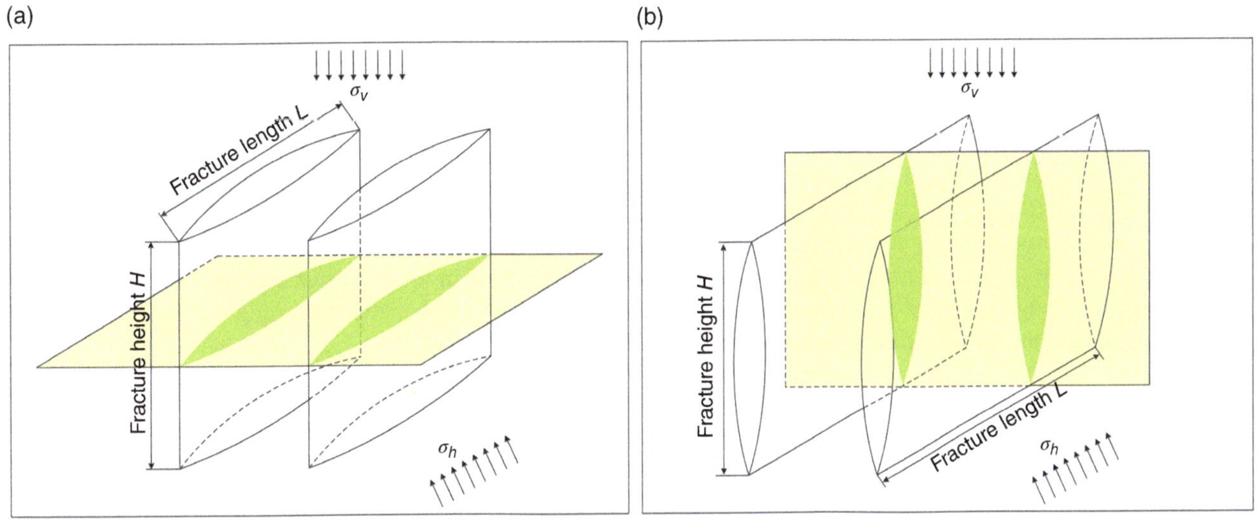

Figure 16.2 Simplified 2D model on different plane: (a) Classical KGD model (on σ_H–σ_h plane), (b) 2D model in this paper (on σ_v–σ_h plane).

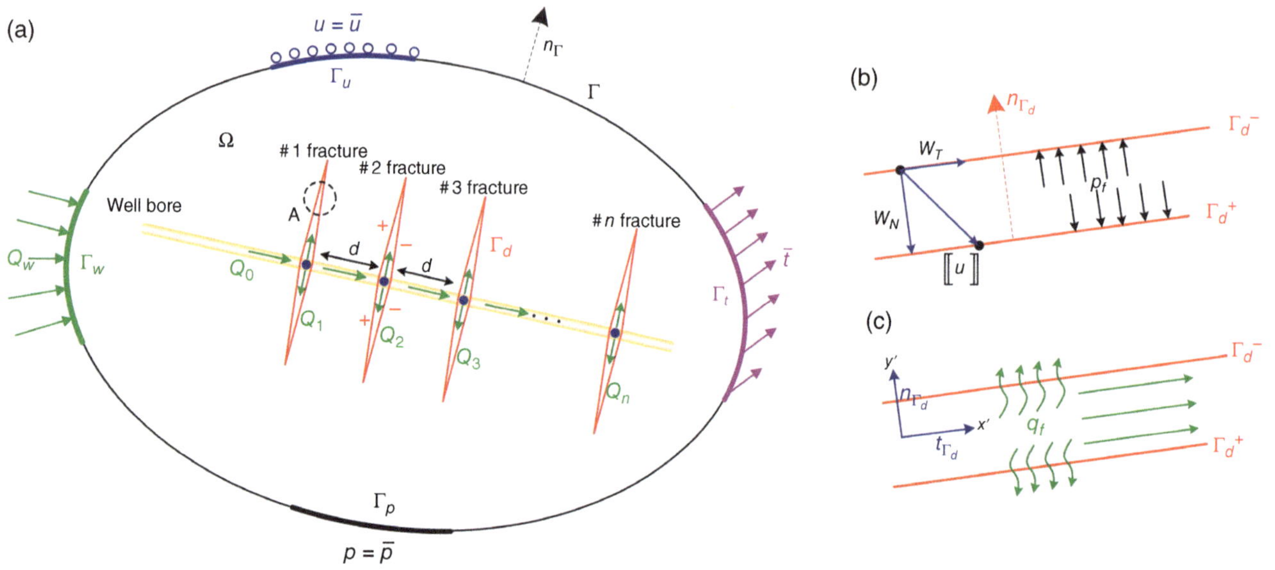

Figure 16.3 Problem configuration and boundary conditions for the simplified 2D model.

in one fracturing stage. The total fracture number is denoted as n, and the fracture spacing as d. The exterior boundary of the domain is Γ, and the unit outward vector of Γ is $\boldsymbol{n}_\Gamma$. The displacement and traction boundary conditions imposed on Γ are represented as

$$\begin{aligned} \boldsymbol{u} &= \bar{\boldsymbol{u}} \quad on\,\Gamma u, \\ \boldsymbol{\sigma}\cdot\boldsymbol{n}_\Gamma &= \bar{\boldsymbol{t}} \quad on\,\Gamma u, \end{aligned} \tag{16.1}$$

where $\bar{\boldsymbol{u}}$ is prescribed displacement imposed on boundary Γ_u, $\bar{\boldsymbol{t}}$ is prescribed traction applied on boundary Γ_t, and $\boldsymbol{\sigma}$ is stress tensor. Note that $\Gamma_u \cup \Gamma_t = \Gamma$ and $\Gamma_u \cap \Gamma_t = \emptyset$ hold for the solid phase.

In addition, for the fluid phase, the pore pressure and flux boundary conditions imposed on Γ are

$$\begin{aligned} p &= \bar{p} \quad on\,\Gamma_p, \\ Q_w &= \boldsymbol{v}_w\cdot\boldsymbol{n}_\Gamma \quad on\,\Gamma_w \end{aligned} \tag{16.2}$$

where $\bar{p}$ is prescribed pore pressure imposed on Γ_p, Q_w is prescribed injection rate applied on Γ_w, and $\mathbf{v}_w$ is outflow rate of fluid phase. Note that $\Gamma_p \cup \Gamma_w = \Gamma$ and $\Gamma_p \cap \Gamma_w = \emptyset$ hold for the fluid phase.

In addition to the flux boundary conditions applied on Γ, the constant flux rate Q_0 is used to inject fracturing fluid into the domain through the wellbore. The total injection rate Q_0 is divided into multiple fractures, which can be expressed as

$$Q_0 = \sum_{i=1}^{n} Q_i \tag{16.3}$$

where Q_i is the flux of fluid injected into the ith fracture counted from the uppermost top fracture, as is shown in Figure 16.3a, representing the flux boundary condition for the ith fracture domain. Each fracture domain Ω' is defined by fracture surface $\Gamma_d^+ \cup \Gamma_d^-$, as is shown in Figures 16.3b and c. And the traction boundary condition imposed on internal fracture surface Γ_d ($\Gamma_d = \Gamma_d^+$) is

$$\boldsymbol{\sigma}\cdot\boldsymbol{n}_{\Gamma_d} = -p_f\boldsymbol{n}_{\Gamma_d} \tag{16.4}$$

where p_f is fluid pressure applied on fracture surface, $\boldsymbol{n}_{\Gamma_d}$ is unit normal vector to the positive face of fracture Γ_d^+ ($\boldsymbol{n}_{\Gamma_d} = \boldsymbol{n}_{\Gamma_d^+} = -\boldsymbol{n}_{\Gamma_d^-}$). Since the permeable matrix is considered, mass transfer between fractures and matrix can be described as flux boundary condition imposed on the fracture surface, as is shown in Figure 16.3c, that is

$$[[\mathbf{v}_w]]\cdot\boldsymbol{n}_{\Gamma_d} = q_f \tag{16.5}$$

where $[[\mathbf{v}_w]] = \mathbf{v}_w^+ - \mathbf{v}_w^-$ denotes the difference of flow rate between Γ_d^+ and Γ_d^-, and q_f represents the leakage rate of fluid phase per unit length of fracture.

16.2.1 Governing Equations

To describe the coupling between solid deformation and fluid flow in multi-cluster hydraulic fractures in saturated porous media, the governing equations are stated as blow.

16.2.1.1 Solid Deformation

To consider the solid deformation in saturated porous media, the mixture theory presented by Biot [45] is adopted. For a permeable matrix containing single fluid phase, the momentum balance equation can be written as [46]

$$\nabla \cdot (\boldsymbol{\sigma}' - \alpha p\mathbf{I}) + \rho(\boldsymbol{b} - \ddot{\boldsymbol{u}}) = \boldsymbol{0} \tag{16.6}$$

where $\boldsymbol{\sigma}'$ is effective stress tensor, which can be linked to stress tensor $\boldsymbol{\sigma}$ by $\boldsymbol{\sigma} = \boldsymbol{\sigma}' - \alpha p\mathbf{I}$. The parameter α is Biot's coefficient, p is pore pressure of fluid, and $\mathbf{I}$ is the identity tensor. $\rho = (1 - \phi)\rho_s + \phi\rho_f$ is the average density of the mixture, where ϕ is porosity. ρ_s and ρ_f are mass densities of solid and fluid phase, respectively. $\boldsymbol{b}$ is acceleration vector associated with body force, and $\ddot{\boldsymbol{u}}$ is acceleration vector of solid phase.

In addition, the constitutive relation between the stress and strain can be expressed by Hooke's Law,

$$d\boldsymbol{\sigma}' = \boldsymbol{D} : d\boldsymbol{\varepsilon} \tag{16.7}$$

where $\boldsymbol{D}$ is tangential modulus tensor and $\boldsymbol{\varepsilon}$ is strain tensor of the solid matrix.

16.2.1.2 Fluid Flow in Matrix

To describe the fluid flow in permeable matrix, the fluid velocity $\mathbf{v}_w$ can be obtained according to the Darcy seepage law [47]

$$\boldsymbol{v}_w = \boldsymbol{k}_w\left(- \nabla p + \rho_f(\boldsymbol{b} - \ddot{\boldsymbol{u}})\right) \tag{16.8}$$

where $\boldsymbol{k}_w = \boldsymbol{k}_m/\mu$, where $\boldsymbol{k}_m$ is intrinsic permeability of porous matrix and μ is the dynamic fluid viscosity. According to mass conservation, the continuity equation of pore fluid can be expressed as [48]

$$\nabla \cdot \boldsymbol{v}_w + \alpha\nabla \cdot \dot{\boldsymbol{u}} + \frac{\dot{p}}{C_c} = \boldsymbol{0} \tag{16.9}$$

where $\dot{\boldsymbol{u}}$ is velocity vector of solid phase, $\dot{p}$ is the derivative of pore pressure with respect to time, and C_c is the bulk modulus of total mixture defined as [40]

$$\frac{1}{C_c} = \frac{\alpha - \phi}{C_s} + \frac{\phi}{C_f} \tag{16.10}$$

where C_s and C_f are bulk moduli of the solid phase and fluid phase, respectively.

16.2.1.3 Fluid Flow in Fractures

To describe the fluid flow in hydraulic fractures, the continuity equation of fracturing fluid along the fractures can be derived in the same manner as Eq. (16.9). We have [48]

$$\nabla \cdot \boldsymbol{v}_f + \nabla \cdot \dot{\boldsymbol{u}} + \frac{\dot{p}}{C_f} = \boldsymbol{0} \tag{16.11}$$

where $\mathbf{v}_f$ is the Darcy velocity vector of fluid flow along the fractures, which can be written as [47]

$$\boldsymbol{v}_f = k_f\left(- \nabla p + \rho_f(\boldsymbol{b} - \ddot{\boldsymbol{u}})\right) \tag{16.12}$$

where k_f is the permeability of fractures dependent on the normal opening of fractures w_N[34]

$$k_f = \frac{w_N^2}{12\mu} \tag{16.13}$$

16.2.1.4 Flow Rate Division to Multiple Fractures

The flux redistribution among multi-cluster hydraulic fractures is analogous to the flow of electric current, which can be obtained by applying Kirchoff's first and second laws [49, 50]. The distribution of fluid into multiple perforation clusters is a dynamic process depended on wellbore friction, perforation friction, and fluid pressure at entry points of fractures [51], as is shown in Figure 16.3a. According to the mass conservation, total injection rate should be the sum of the injection rates of all intersected fractures by the wellbore, which is governed by Eq. (16.3). Since the wellbore friction is negligible for closely spaced fractures (small fracture spacing), the pressure continuity condition can be obtained as [49]

$$p_{m,1} + p_{p,1} = \cdots = p_{m,i} + p_{p,i} = \cdots = p_{m,n} + p_{p,n} \quad i = 1\ldots n \tag{16.14}$$

where $p_{m,i}$ is fluid pressure at entry point of the ith fracture, $p_{p,i}$ is pressure loss generated by perforation frictional, which is expressed as [52]

$$p_{p,i} = \frac{8\rho_f Q_i^2}{\pi^2 n_{p,i}^2 d_{p,i}^4 C^2} \tag{16.15}$$

where Q_i is the injection rate of the ith fracture, $n_{p,\,i}$ is the number of perforations for the ith fracture, $d_{p,\,i}$ is the diameter of perforation for the ith fracture, and C is dimensionless discharge coefficient that is taken as 0.6 in this paper. Hence, the simultaneous satisfaction of Eq. (16.14) and Eq. (16.3) has been made for the coupling process of rock deformation and fluid flow.

16.2.1.5 Fracture Propagation

The stress solutions of LEFM are adopted to calculate the stress intensity factor so as to check if the propagation criterion is met. The propagation direction θ_p is determined by the maximum hoop tensile stress criterion [53],

$$\theta_p = 2\arctan\frac{1}{4}\left[\frac{K_{\mathrm{I}}}{K_{\mathrm{II}}} - sign(K_{\mathrm{II}})\sqrt{\left(\frac{K_{\mathrm{I}}}{K_{\mathrm{II}}}\right)^2 + 8}\right] \tag{16.16}$$

where K_{I} and K_{II} are mode I and II stress intensity factor, respectively. Note that the interaction integral method presented by Moës and Belytschko [54] is adopted to calculate the stress intensity factors K_{I} and K_{II}.

16.2.2 Weak Forms

The weak form of momentum balance equation for solid phase can be derived by multiplying the Eq. (16.6) by test function $\delta \boldsymbol{u}(x, t)$, and then integrating their product over the domain Ω. Upon applying the divergence theorem along the Γ_d and Γ_t, and imposing the internal and external traction conditions, Eq. (16.6) can be rewritten as

$$\int_{\Omega} \delta \boldsymbol{u} \rho(\boldsymbol{b} - \ddot{\boldsymbol{u}}) d\Omega + \int_{\Gamma_t} \delta \boldsymbol{u} \cdot \bar{\boldsymbol{t}} d\Gamma - \int_{\Gamma_d} [[\delta \boldsymbol{u}]] p_f \boldsymbol{n}_{\Gamma_d} d\Gamma - \int_{\Omega} \nabla \delta \boldsymbol{u} : \boldsymbol{\sigma} d\Omega = \boldsymbol{0} \tag{16.17}$$

where $[[\delta \boldsymbol{u}]] = \delta \boldsymbol{u}^+ - \delta \boldsymbol{u}^-$ denotes the step of test displacement function across the fracture surface.

The weak form of continuity equation for fluid flow in matrix can be derived by multiplying the Eq. (16.9) by test function $\delta p(x, t)$ and integrating the product over the domain Ω. By applying the divergence theorem, and imposing the external flux conditions on Γ_w, Eq. (16.9) can be rewritten as

$$\int_{\Omega} \delta \boldsymbol{u} \rho(\boldsymbol{b} - \ddot{\boldsymbol{u}}) d\Omega + \int_{\Gamma_t} \delta \boldsymbol{u} \cdot \bar{\boldsymbol{t}} d\Gamma - \int_{\Gamma_d} [[\delta \boldsymbol{u}]] p_f \boldsymbol{n}_{\Gamma_d} d\Gamma - \int_{\Omega} \nabla \delta \boldsymbol{u} : \boldsymbol{\sigma} d\Omega = \boldsymbol{0} \tag{16.18}$$

where $[[\boldsymbol{v}_w]] = \boldsymbol{v}_w^+ - \boldsymbol{v}_w^-$ denotes the step of fluid velocity across the fracture, implying the mass transfer between the matrix and fractures. In addition, the pressure field is continuous across the fracture, that is, $p^+ = p^-$ is assumed.

Similarly, the weak form of fluid mass continuity equation for fluid flow in fractures can be obtained by multiplying the test pressure function $\delta p(x, t)$ and integrating the product over the fracture domain Ω'. This leads to

$$\int_{\Gamma_d} \delta p q_f d\Gamma = \int_{\Omega'} \delta p \nabla \cdot \dot{\boldsymbol{u}} d\Omega' + \int_{\Omega'} \delta p \frac{\dot{p}}{C_f} d\Omega' - \int_{\Omega'} \nabla \delta p \cdot \boldsymbol{v}_f d\Omega' \tag{16.19}$$

Note that the first term of Eq. (16.19) can be simplified by integrating the internal flux boundary condition Eq. (16.5) over fracture surface Γ_d, as follows

$$\int_{\Gamma_d} \delta p q_f d\Gamma = \int_{\Gamma_d} \delta p [[\boldsymbol{v}_w]] \cdot \boldsymbol{n}_{\Gamma_d} d\Gamma \tag{16.20}$$

Hence, the weak form of fluid continuity equation in permeable matrix, Eq. (16.18), and the weak form of fluid mass continuity equation in fracture, Eq. (16.19), can be combined to one equation by inserting Eq. (16.19) and Eq. (16.20) into Eq. (16.18).

16.3 Numerical Method

In this section, the approximation methods of the displacement and pressure fields based on XFEM are first presented. The discretized HM-coupled equations are obtained by substituting the test functions into the weak forms of governing equations. The generalized Newmark scheme is employed to discretize the HM-coupled equations in time domain. The Newton–Raphson method is adopted to iteratively solve the HM-coupled equations at each time step. Meanwhile, the secant method is applied so that the pressure continuity and mass conservation conditions are satisfied at the fracture-well intersection points to find the flow rate division for the multiple fracture system.

16.3.1 XFEM Approximation of *u*(*x*, *t*) and *p*(*x*, *t*)

The XFEM approximation of the displacement can be obtained by incorporating appropriate enrichment functions into standard shape functions based on the concept of partition of unity [55]. For displacement fields, the Heaviside enrichment function $H(\boldsymbol{x})$ is adopted to describe the displacement discontinuity across the fracture surfaces, while the asymptotic crack tip functions $B_\alpha(\boldsymbol{x})$ are adopted to describe the stress singularity around the fracture tip. Hence, the approximation of displacement field can be written as

$$\boldsymbol{u}(\boldsymbol{x}, t) = \boldsymbol{N}_u^{std} \overline{\boldsymbol{u}}(t) + \boldsymbol{N}_u^{Hev} \overline{\boldsymbol{a}}(t) + \boldsymbol{N}_u^{Tip} \overline{\boldsymbol{b}}(t) \tag{16.21}$$

where $\boldsymbol{N}_u^{std} \overline{\boldsymbol{u}}(t)$ is standard part of displacement approximation, $\boldsymbol{N}_u^{Hev} \overline{\boldsymbol{a}}(t)$ is enriched part associated with Heaviside function, and $\boldsymbol{N}_u^{Tip} \overline{\boldsymbol{b}}(t)$ is enriched part due to asymptotic tip functions. $\boldsymbol{N}_u^{std} = \{N_i(\boldsymbol{x})\}$, $\boldsymbol{N}_u^{Hev} = \{N_i(\boldsymbol{x})(H(\boldsymbol{x}) - H(\boldsymbol{x}_i))\}$, and $\boldsymbol{N}_u^{Tip} = \{N_i(\boldsymbol{x})(B_\alpha(\boldsymbol{x}) - B_\alpha(\boldsymbol{x}_i))\}$ are matrices of standard and enriched shape functions. $\overline{\boldsymbol{u}}$, $\overline{\boldsymbol{a}}$, and $\overline{\boldsymbol{b}}$ are the displacement and enriched nodal variables, and their DOFs depend on the choices of shape functions. The expression of $H(\boldsymbol{x})$ and $B_\alpha(\boldsymbol{x})$ can be referred to Khoei [47].

For the pressure field, the Ridge function $\psi(\boldsymbol{x})$, known as modified level-set enrichment function [39], is adopted to describe the discontinuity of pressure gradient across the fracture [56]. And the approximation of pressure field can be written as

$$p(\boldsymbol{x}, t) = \boldsymbol{N}_p^{std}\overline{\boldsymbol{p}}(t) + \boldsymbol{N}_p^{Rid}\overline{\boldsymbol{d}}(t) \tag{16.22}$$

where $\boldsymbol{N}_p^{std}\overline{\boldsymbol{p}}(t)$ is standard part of pressure approximation, and $\boldsymbol{N}_p^{Rid}\overline{\boldsymbol{d}}(t)$ is enriched part. $\boldsymbol{N}_p^{std} = \{N_i(\boldsymbol{x})\}$ and $\boldsymbol{N}_p^{Rid} = \{N_i(\boldsymbol{x})\psi(\boldsymbol{x})\}$ are matrices of standard and enriched shape functions, respectively. $\boldsymbol{p}$ and $\boldsymbol{d}$ are pressure and enriched nodal variables for the pressure field, respectively. The formulation of Ridge function $\psi(\boldsymbol{x})$ can be referred to Moës et al. [39].

16.3.2 Spatial and Time Discretization

When $\delta\boldsymbol{u}(\boldsymbol{x}, t)$ and $\delta p(\boldsymbol{x}, t)$ is considered as the variations of $\boldsymbol{u}(\boldsymbol{x}, t)$ and $p(\boldsymbol{x}, t)$, the discretized HM-coupled equations can be obtained by substituting Eq. (16.21) and Eq. (16.22) into the weak form.

$$\begin{cases} \boldsymbol{K}\overline{\boldsymbol{U}} - \boldsymbol{Q}\overline{\boldsymbol{P}} + \boldsymbol{M}\ddot{\overline{\boldsymbol{U}}} + \boldsymbol{f}^{\text{int}} - \boldsymbol{f}^{ext} = \boldsymbol{0} \\ \boldsymbol{H}\overline{\boldsymbol{P}} + \boldsymbol{C}\ddot{\overline{\boldsymbol{U}}} + \boldsymbol{Q}^T\dot{\overline{\boldsymbol{U}}} + \boldsymbol{S}\dot{\overline{\boldsymbol{P}}} + \boldsymbol{q}^{int} - \boldsymbol{q}^{ext} = \boldsymbol{0} \end{cases} \tag{16.23}$$

The readers can refer to our early work [38] for the detailed expressions of stiffness matrix $\boldsymbol{K}$, coupling matrix $\boldsymbol{Q}$, solid mass matrix $\boldsymbol{M}$, internal force vector $\boldsymbol{f}^{int}$, external force vector $\boldsymbol{f}^{ext}$, permeability matrix $\boldsymbol{H}$, fluid mass matrix $\boldsymbol{C}$, compressibility matrix $\boldsymbol{S}$, internal flux vector $\boldsymbol{q}^{int}$, and external flux vector $\boldsymbol{q}^{ext}$. $\overline{\boldsymbol{U}} = [\overline{\boldsymbol{u}}\ \overline{\boldsymbol{a}}\ \overline{\boldsymbol{b}}]$ and $\overline{\boldsymbol{P}} = [\overline{\boldsymbol{p}}\ \ \overline{\boldsymbol{d}}]$ are the vectors for assembled nodal displacement and pressure variables, respectively.

In Eq. (16.23), the terms including time derivatives, like the velocity, acceleration and pressure changing rate are discretized using the generalized Newmark scheme [57].

16.3.3 Solution Strategy

16.3.3.1 Solution of HM-Coupled Equations

The Newton–Raphson method is used to solve the system of HM-coupled equations, Eq. (16.23). In particular, we use the iteration method to minimize the residuals of the two equations, denoted by $\boldsymbol{\psi}_{n+1}^U$ and $\boldsymbol{\psi}_{n+1}^P$ at time t_{n+1}, associated with stress equilibrium and fluid flow, which are defined as

$$\begin{cases} \boldsymbol{\psi}_{n+1}^U = \boldsymbol{K}\overline{\boldsymbol{U}}_{n+1} - \boldsymbol{Q}\overline{\boldsymbol{P}}_{n+1} + \dfrac{\boldsymbol{M}}{\beta\Delta t^2}\overline{\boldsymbol{U}}_{n+1} + \boldsymbol{f}_{n+1}^{int} - \boldsymbol{G}_{n+1}^U \\ \boldsymbol{\psi}_{n+1}^P = \boldsymbol{H}\overline{\boldsymbol{P}}_{n+1} + \dfrac{\boldsymbol{C}}{\beta\Delta t^2}\overline{\boldsymbol{U}}_{n+1} + \dfrac{\gamma\boldsymbol{Q}^T}{\beta\Delta t}\overline{\boldsymbol{U}}_{n+1} + \dfrac{\boldsymbol{S}}{\theta\Delta t}\overline{\boldsymbol{P}}_{n+1} + \boldsymbol{q}_{n+1}^{int} - \boldsymbol{G}_{n+1}^P \end{cases} \tag{16.24}$$

where Δt is the time step, β, θ, and γ are the parameters used in Newmark algorithm, $\boldsymbol{G}_{n+1}^U$ and $\boldsymbol{G}_{n+1}^P$ are the residuals calculated based on the displacement and pressure fields at time t_n. Following the Newton–Raphson iteration method, the incremental displacement and pressure components at the ith iteration step during the time period $[t_n, t_{n+1}]$ can be obtained as

$$\begin{bmatrix} d\boldsymbol{U}_{n+1} \\ d\boldsymbol{P}_{n+1} \end{bmatrix}^{i+1} = -\begin{bmatrix} \boldsymbol{\psi}_{n+1}^U \\ \boldsymbol{\psi}_{n+1}^P \end{bmatrix}^i \cdot \left(\begin{bmatrix} \dfrac{\partial\boldsymbol{\psi}^U}{\partial\boldsymbol{U}} & \dfrac{\partial\boldsymbol{\psi}^U}{\partial\boldsymbol{P}} \\ \dfrac{\partial\boldsymbol{\psi}^P}{\partial\boldsymbol{U}} & \dfrac{\partial\boldsymbol{\psi}^P}{\partial P} \end{bmatrix}_{n+1}^i \right)^{-1} \tag{16.25}$$

where i is Newton–Raphson iteration step and $n+1$ is the time step. The detailed expression for each term in the above Jacobian matrix is referred to Zhou et al. [38].

16.3.3.2 Solution of Flow Rate Division

Since the influx for a fracture is a part of the solution, the secant method is employed to determine the influxes for multiple fractures. Figure 16.4 shows the iteration method. Firstly, injection rate of each fracture is assumed as $Q_i = Q_0/n$. The pressure drop $p_{p,i}$ induced by perforation friction of each fracture can be obtained by Eq. (16.15). In addition, the fluid pressure at the entry point of the fracture $p_{m,i}$ can be calculated by solving the HM-coupled equations according to Eq. (16.23). To ensure the pressure continuity condition Eq. (16.14), the iteration target pressure p_{avg} is defined as

$$p_{avg} = \left(\sum_{i=1}^{n} p_i\right)/n \tag{16.26}$$

where $p_i = p_{p,i} + p_{m,i}$ is called the total pressure for ith fracture. The error ε_p defined to evaluate the variation in total pressure among multiple fractures, is

$$\varepsilon_p = \sum_{i=1}^{n} \left(\frac{p_i - p_{avg}}{p_{avg}}\right)^2 \tag{16.27}$$

Then compare the value of ε_p with the tolerance (10^{-4} in this paper) to determine whether the pressure continuity condition is satisfied. If so, the accurate flux for each fracture is obtained. If not, the secant method is employed to update the influxes for multiple fractures based on the values of p_i and Q_i at previous iteration step, which are expressed as

$$Q_i^{(k+1)'} = Q_i^{(k)} + \frac{\left(Q_i^{(k)} - Q_i^{(k-1)}\right)\left(p_{avg}^{(k)} - p_i^{(k)}\right)}{p_i^{(k)} - p_i^{(k-1)}} \tag{16.28}$$

where the superscript k denotes the number of iteration steps of secant method. However, the obtained $Q_i^{(k+1)'}$

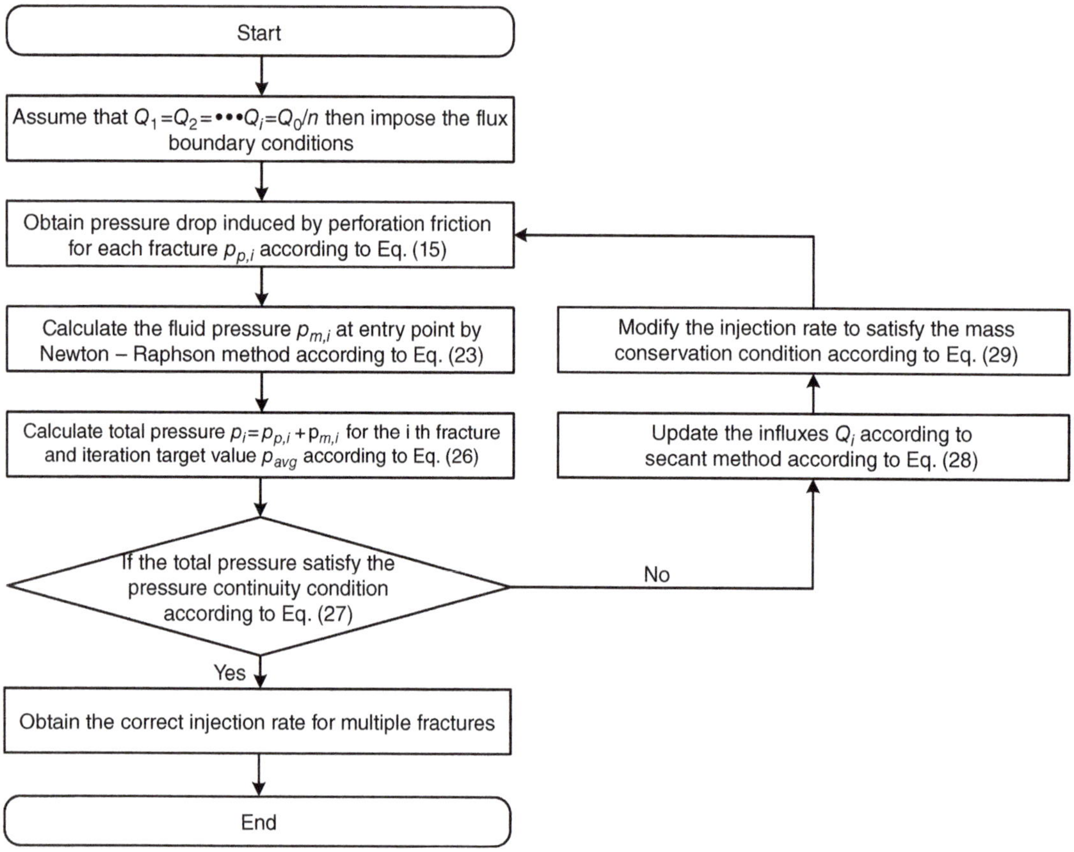

Figure 16.4 Flowchart of solving the problem of flow rate division for multiple fractures.

may not always satisfy the mass conservation condition. The following formulation is employed to adjust the influx for each fracture.

$$Q_i^{(k+1)} = Q_0 Q_i^{(k+1)'} / \sum_{i=1}^{n} Q_i^{(k+1)'} \tag{16.29}$$

Based on the modified influxes of multiple fractures, the total pressure of each fracture p_i can be updated by solving the coupled problem. And the above iteration process is repeated until ε_p is less than 10^{-4}.

16.4 Numerical Results

16.4.1 Verification of the Model

The verification of the proposed XFEM model against KGD problem can be referred to Zhou et al. [38]. To verify the correctness of the numerical model in dealing with the flow rate division and the stress interference between multiple fractures, the problem, considering the simultaneous growth of four fractures, proposed by Wu [58] is adopted in this section, as shown in Figure 16.5a. Note that this verification case is simplified to 2D on the horizontal (σ_H–σ_h) plane with the constant fracture height $h = 30.48$ m. Due to the symmetry of the computational zone, the influx for #1 fracture is equal to that for #4 fracture, and the injection rate of #2 fracture is equal to that of #3 fracture. Hence, only a quarter of the computational zone (80×120 m) is considered to reduce the computational cost, as is shown in Figure 16.5b. The vertical displacement is confined along the bottom boundary ($u_y = 0$) and the horizontal displacement is confined along the right boundary ($u_x = 0$). The pore pressure boundary $p = 0$ is also applied on the left and top boundary. The minimum horizontal confining stress $\sigma_h = 30.69$ MPa is imposed along the left boundary and maximum horizontal confining stress $\sigma_H = 31.38$ MPa is applied on the top boundary. Since only half of the #1 and #2 fracture are considered, the entry point of fractures are located on the bottom line with the fracture spacing equal to 100 ft. ($d = 30.48$ m). The distance between the entry point of #2 fracture and center of symmetry is $d/2$. The initial length of two fractures are 1.25 m and other input parameters are listed in Table 16.1.

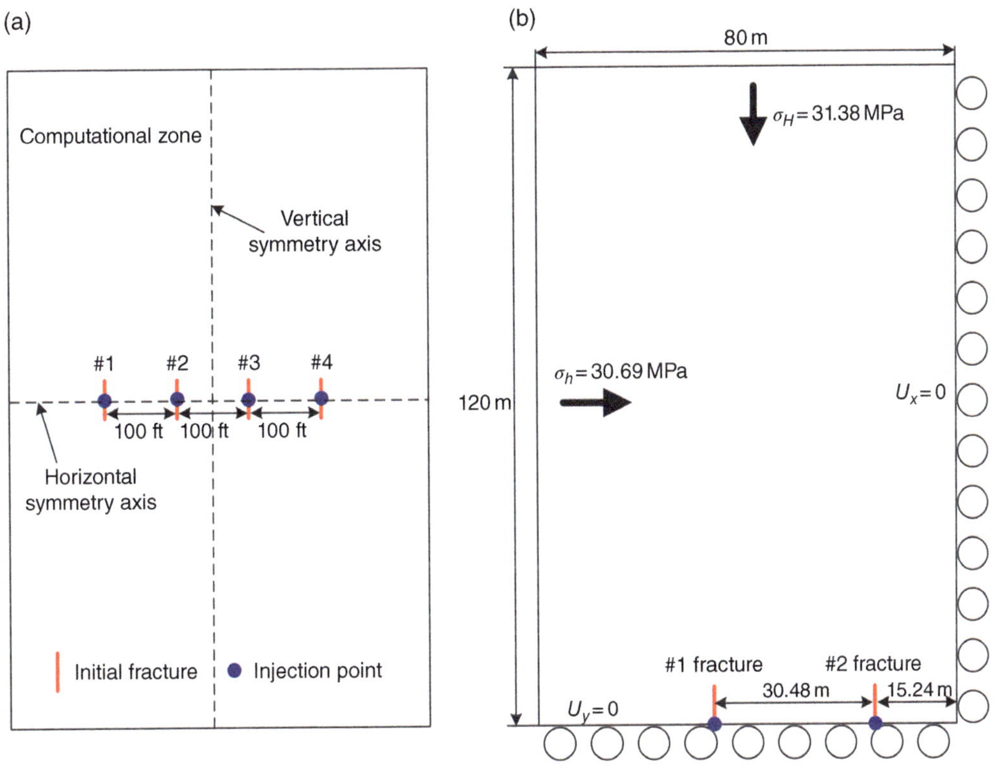

Figure 16.5 Computational model for verification: (a) Problem proposed by Wu [58] (*Source:* Adapted from Wu [58]), (b) Numerical model in this section.

Table 16.1 Input parameters of verification case.

Parameter	Value	Parameter	Value
Elastic modulus E	45 GPa	Fracture height h	30.48 m
Poisson's ratio ν	0.2	Porosity ϕ	0.08
Solid density ρ_s	2700 kg/m^3	Matrix permeability k_m	0.2 mD
Fluid density ρ_f	1200 kg/m^3	Biot coefficient α	0.48
Injection rate Q_0	9.54 m^3/min	Bulk moduli of solid phase C_s	36 GPa
Fluid viscosity μ	1 mPa s	Bulk moduli of fluid phase C_f	3 GPa
Perforations number n_p	15	Critical stress intensity factor K_c	1 MPa·m$^{0.5}$
Perforation diameter d_p	16 mm		

Figure 16.6a plots the propagation paths of two fractures when $T = 180$ s, which shows good agreements with those obtained by Wu [58]. Due to the stress interference from other fractures, the path of #1 and #2 fracture becomes curved. Figure 16.6b is the division of flow rate for two fractures, which suggests that the variation of flow rate obtained by the proposed numerical model shows little difference between the results obtained by Wu [58]. Due to the influence of stress shadow, flow rate for #2 fracture is less than that of #1 fracture, resulting the preferential propagation of #1 fracture, which qualitatively validates our numerical model.

16.4.2 Multi-Cluster Hydraulic Fracturing in High-Angle Well

In order to reduce the complexity in analyzing numerical results for parametric study, people always like to use dimensionless parameters [59]. We will introduce a few dimensionless parameters to facilitate discussion on

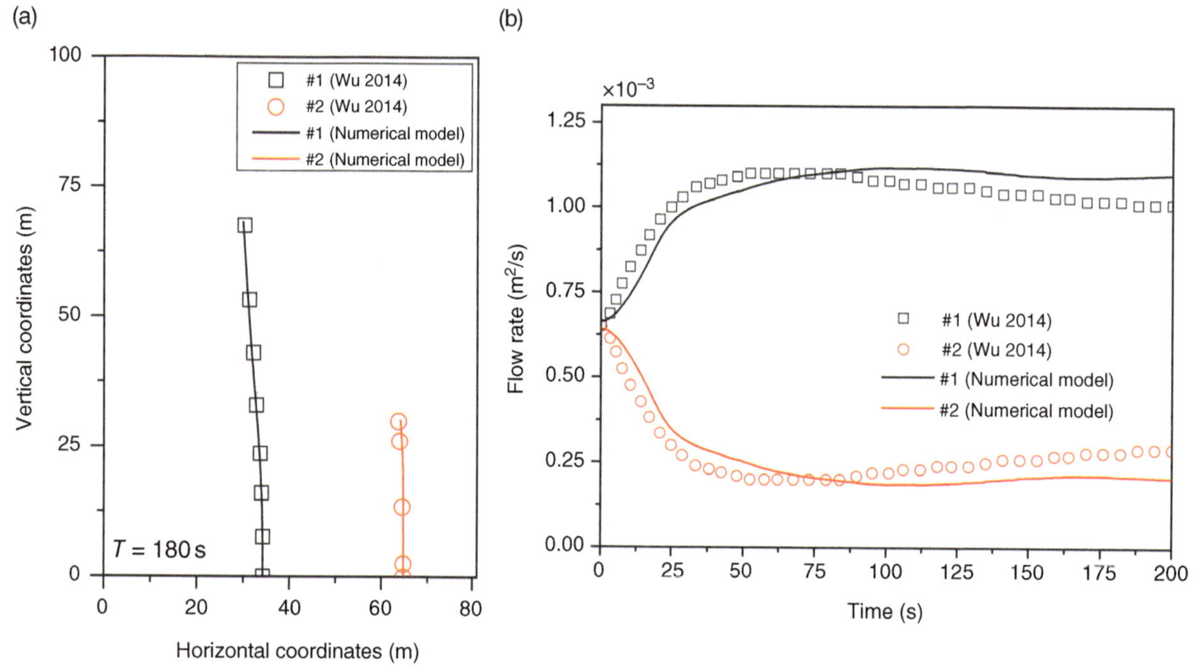

Figure 16.6 Comparison of numerical results: (a) fracture path, (b) flow rate for each fracture.

numerical results. In particular, a dimensionless toughness κ represents the energy dissipation ratio of fracture propagation to the viscous fluid flow. A dimensionless parameter Γ_m denotes the competition between the perforation loss and stress interaction when the fracture propagation is in the viscosity dominated regime. A dimensionless parameter C_v is introduced to characterize the extent of uniform fracture growth in the presence of multiple fractures. The definition of these dimensionless parameters can be referred to Appendices.

16.4.2.1 Model Set-up

Figure 16.7 displays the 2D fracture and well placements for both HAW and HW stimulations of a gas reservoir that is of a size of 100 m by 100 m. Since the symmetry does not hold for HAW, a full domain is being used in this section. Note that the simulation case is simplified to 2D on the vertical (σ_v–σ_h) plane with the constant fracture length 30 m. The long solid line represents the well bore and it is inclined relative to the direction of σ_h in Figure 16.7a, while it is horizontal in Figure 16.7b. To compare the numerical results,

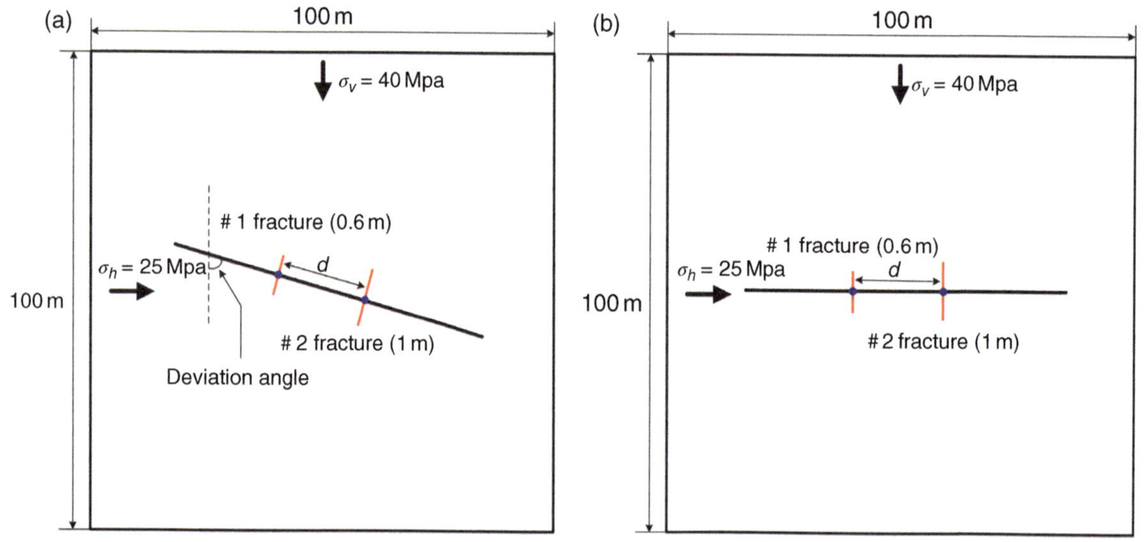

Figure 16.7 Computational model: (a) HAW, (b) HW.

the fracturing spacing is assumed to be the same, denoted by d. The red lines represent the initial fracture that are placed as a straight line normal to the well. For these two fracture systems the initial fracture height is either 0.6 m (#1 fracture) or 1 m (#2 fracture). Then the initial fracture height ratio η is equal to 0.6. The effects of initial fracture length on the simultaneous growth of multiple fractures have been investigated by Chen et al. [27]. Their numerical results show that a larger difference in initial fracture lengths will act against a simultaneous growth. The fracture-well intersection points are illustrated by the small dots in Figure 16.7. Apart from these classical bi-wing fractures with two identical branches, we also consider the placement where the fracture has only the up growing branch for a fracture. This is the so-called single wing fractures [28].

The perforation diameter of each fracture takes seven different values from 10 to 25 mm (at an even increment of 2.5 mm), and it is the same for all fractures considered in one case. The horizontal and vertical tractions are imposed on the reservoir boundary. Particularly, these confining stress are fixed, and zero pore pressure is applied to the boundary. The large confining stress difference is employed to mimic the real stress state in Huaqing Oilfield, although the values may be different than those used here. Obviously, as the operational parameters, the deviation angle, the fracture spacing, placement and number can be varied to find the best efficiency of multi-cluster hydraulic fracturing treatments.

In addition, the material constants and the injection rate used in all calculations are listed in Table 16.2 unless otherwise specified. Based on these inputs, we can list the changeable parameters in Table 16.3.

16.4.2.2 Operational Parameters

To investigate the effect of operational parameters on the growth of multiple fractures in HAW, three different fluid viscosities $\mu = 1$, 10 and 100 mPa s are adopted. Hence, the corresponding dimensionless toughness can be obtained as $\kappa = 1.6795$, 0.9444, 0.5311. Note that the bi-wing fracture placement is used in Figure 16.8 and the calculation stops when the height of a fracture reaches 40 m. Figure 16.8a compares the values of C_v at the end of computation between HAW and HW when $\kappa = 1.6795$, 0.9444 and 0.5311. It illustrates that the value of C_v for HAWs is always lower than that for HWs for various conditions of dimensionless toughness and perforation diameter. This means that HAW can promote the simultaneous growth of

Table 16.2 Material constants.

Parameter	Value	Parameter	Value
Elastic modulus E	18.6 GPa	Matrix permeability k_m	0.2 mD
Poisson's ratio ν	0.16	Biot coefficient α	0.48
Solid density ρ_s	2700 kg/m^3	Bulk moduli of solid phase C_s	36 GPa
Fluid density ρ_f	1000 kg/m^3	Bulk moduli of fluid phase C_f	3 GPa
Injection rate per fracture Q/n	0.025 m^3/s	Critical stress intensity factor K_c	2 MPa·m$^{0.5}$
Perforations number n_p	30	Horizontal confining stress σ_x	25 MPa
Fracture length h	30 m	Vertical confining stress σ_y	40 MPa
Porosity ϕ	0.08		

Table 16.3 Variable parameters.

Influencing factor	Dimensionless toughness	Deviation angle (°)	Fracture spacing (m)	Fracture placement	Fracture number
Operational parameters	1.68, 0.94, 0.53	75	10	bi-wing	2
Deviation angle	0.94	82.5, 75, 67.5	5	bi-wing	2
Fracture spacing	0.94	75	2, 5, 10	bi-wing	2
Fracture placement	0.94	75	2,5	single-wing	2
Fracture number	0.94	75	5	bi-wing	3, 4

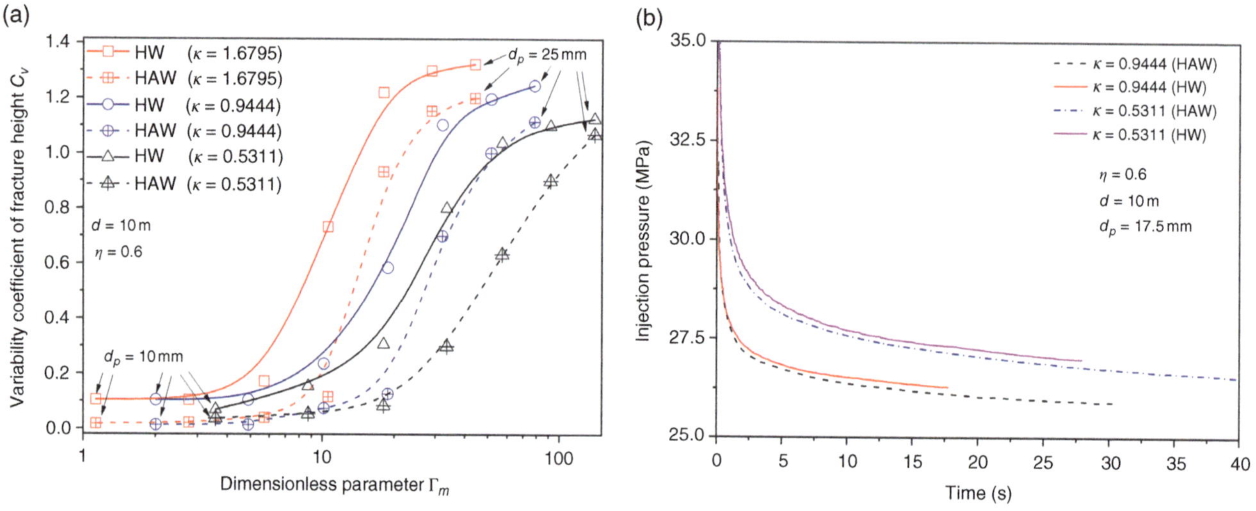

Figure 16.8 Comparison of numerical results between HW and HAW: (a) Variability coefficient of fracture height, (b) Injection pressure.

multiple fractures when all conditions are the same. In addition, the values of C_v decreases with the reduction of Γ_m and dimensionless toughness κ. This suggests that low perforation diameter or high fluid viscosity can promote the uniform propagation of multiple fractures under the same well type, which can be qualitatively validated by the numerical results by Lecampion and Desroches [14] and Chen et al. [27]. Figure 16.8b compares the injection pressure between HAW and HW when $\kappa = 1.6795$, 0.9444. Numerical results of Figure 16.8b illustrate that the injection pressure for HAW is slightly lower than that for HW, which suggests that lower energy is consumed by HAW. Figure 16.8 illustrates that HAW can promote the uniform growth of multiple fractures with consuming lower energy than HW.

To figure out the reason, the final geometry of two fractures for both HAW and HW are plotted in Figure 16.9 using $d_p = 17.5$ mm, $\kappa = 0.9444$ and 0.5311. Since the initial height of #1 fracture is shorter than that of #2 fracture, propagation of #1 fracture is restrained by the stress shadow induced by #2 fracture in HW. However, for HAW cases, the injection points of two fractures have a difference in height and this breaks the symmetry of fracture growth. Although the classical bi-wing fractures are expected, upward propagation of #1 fracture and downward propagation of #2 fracture become dominant as shown in Figure. 16.9a and c. Because two fractures propagate in the opposite direction, the fracture opening of #1 and #2 is maintained large and this will assist in proppant transport. On the other hand, the fracture symmetry is maintained for HW cases. The fracture opening at the fracture-well intersection points of #1 fracture is limited as shown in Figure 16.9b and d. Although the perforation loss can help the propagation of shorter fracture, the reduced fracture opening of the shorter fracture can affect the subsequent proppant transport.

16.4.2.3 Deviation Angle

In this subsection, three different deviation angles (82.5°, 75°, and 67.5°) are used to investigate their influence on the propagation of multiple fractures in HAW. As shown in Table 16.3, dimensionless toughness $\kappa = 0.9444$ and fracture spacing $d = 5$ m are adopted. Note that two bi-wing hydraulic fractures are simultaneously initiated from the wellbore and the calculation stops when the height of a fracture reaches 40 m. Figure 16.10a compares the values of C_v at the end of computation between different deviation angles. In addition, numerical results of HW (deviation angle 90°) with $\kappa = 0.9444$ and $d = 5$ m are also plotted in the Figure 16.10a. It shows that the values of C_v decrease with the reduction of deviation angle. This illustrates that the increase of the deviation of HAW with respect to minimum in-situ stress on σ_v–σ_h plane (decrease deviation angle) can promote the uniform growth of multiple fractures. Figure 16.10b compares the injection pressure of HAW with different deviation angles, it shows that injection pressure decreases with the reduction of deviation angle. Numerical results of Figure 16.10 suggest that the reduction of the deviation angle of HAW in a certain range (>67.5°) can enhance the efficiency of hydraulic fracturing stimulation.

Figure 16.11 compares the geometry of two fractures with different deviation angles when $d = 5$ m, $d_p = 15$ mm and $T = 16$ s. With the decrease of deviation angle, the deviation of two fractures in height direction increases. As mentioned before, the height difference of two injection points leads to

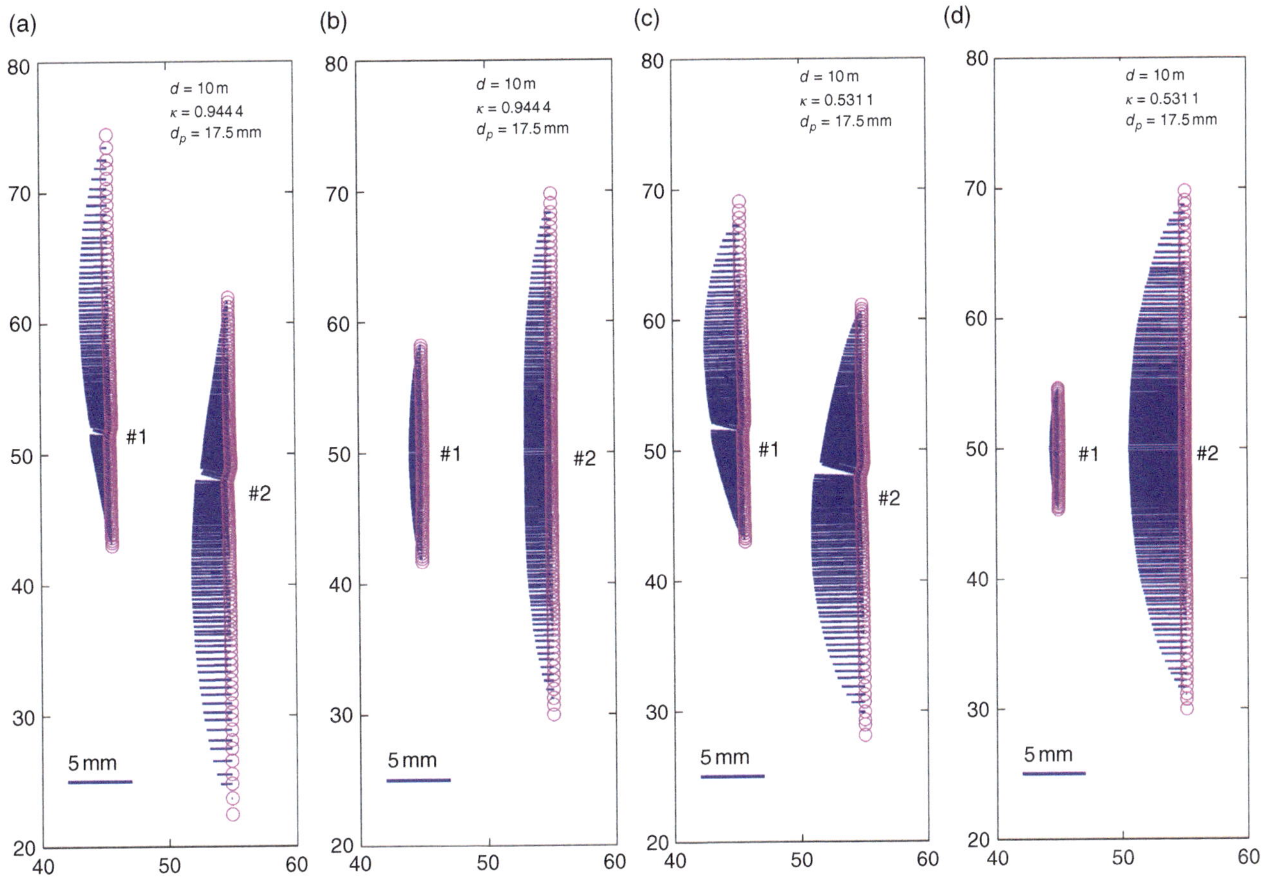

Figure 16.9 Fracture geometry of HW and HAW: (a) HAW (T = 30s), (b) HW (T = 18 s), (c) HAW (T = 30s), and (d) HW (T = 28 s).

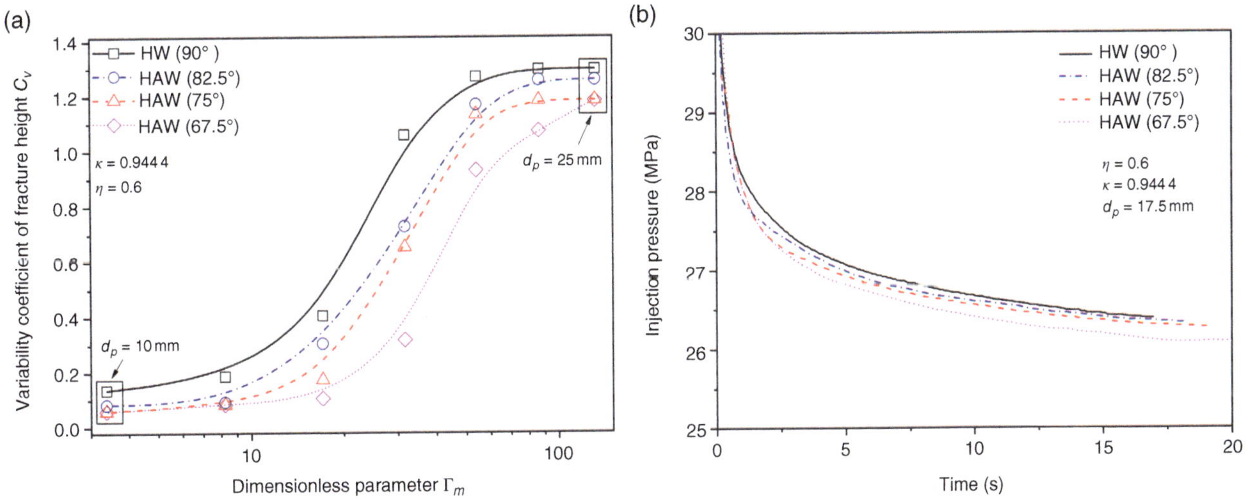

Figure 16.10 Comparison of numerical results between HW and HAW: (a) Variability coefficient of fracture height, (b) Injection pressure.

the asymmetry propagation of two fractures and thus promotes the uniform growth of multiple fractures. Figure 16.11 further illustrates this mechanism and reveals that the larger deviation in height can better promote the uniform propagation of two fractures. When the deviation angle is larger than 82.5°, the fracture geometry shows a preferential growth pattern with #1 fracture being restrained by #2 fracture. However, two fractures almost

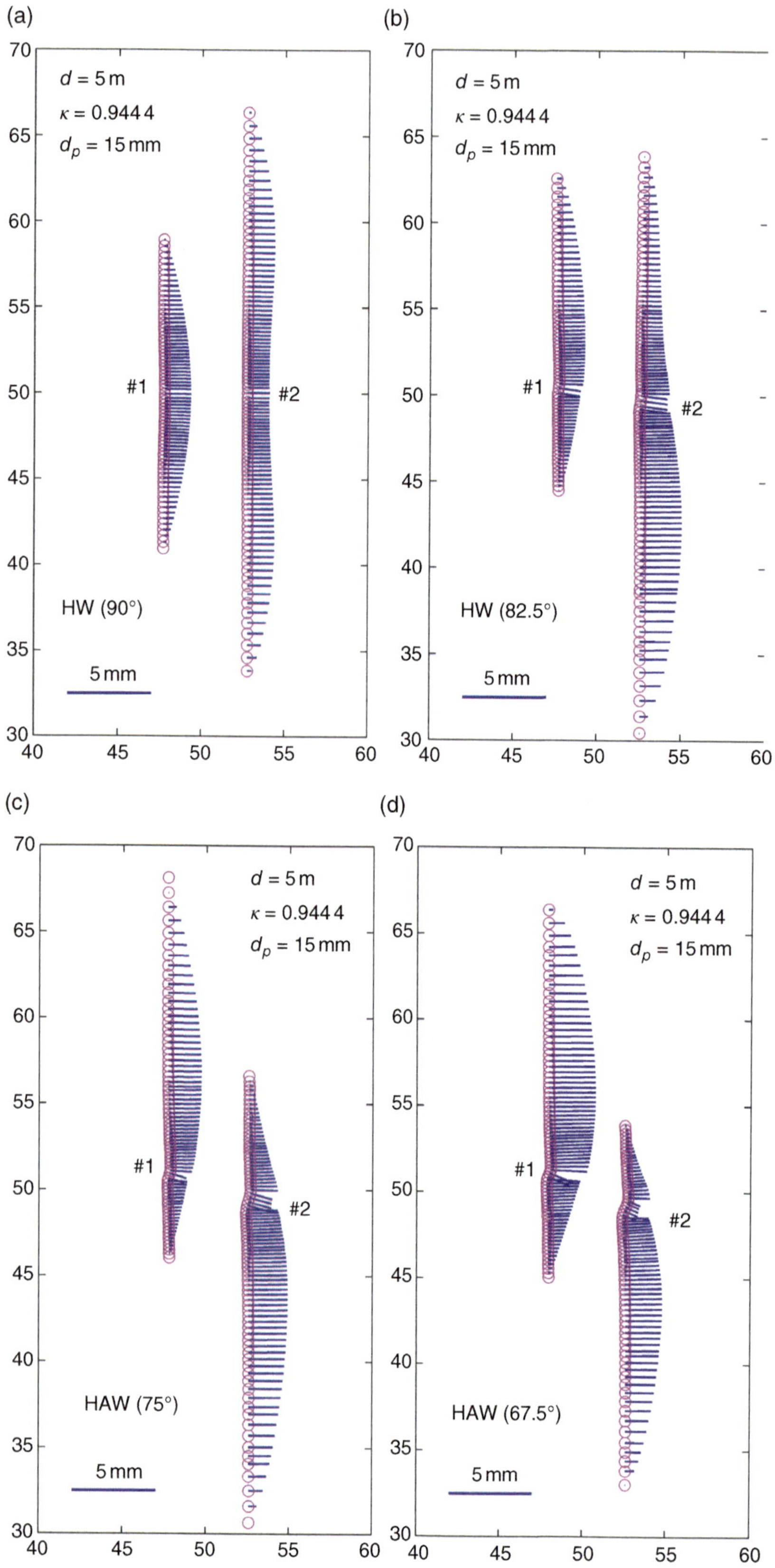

Figure 16.11 Fracture geometry of HW and HAW: (a) HW ($T = 16$ s), (b) 82.5° HAW ($T = 16$ s), (c) 75° HAW ($T = 16$ s), (d) 67.5° HAW ($T = 16$ s).

have the same fracture height when deviation angle is less than 82.5°. Although our study suggests the advantage of the reduction of deviation angle in HAW to promote the uniform growth of multiple fractures, the numerical results should be carefully applied to engineering practice when the deviation angle is less than 67.5°. Because the plane assumption of the numerical model could be invalid when the deviation angle becomes too low, the competitive growth of multiple fractures in this type of well needs further study.

16.4.2.4 Fracture Spacing

As shown in Table 16.3, three different fracture spacing d = 2, 5, and 10 m are adopted. The condition $\kappa = 0.9444$ indicates that the considered cases are all in the viscosity dominated regime. Note that bi-wing fracture is used in this subsection and the calculation stops when the height of a fracture reaches 40 m. Figure 16.12a and b compare the values of C_v at the end of computation between HAW and HW when d = 10, 5 and 2 m. It illustrates that the value of C_v for HAWs is always lower than that for HWs for different fracture spacing and perforation diameter. This means that HAW can promote the simultaneous propagation of multiple fractures. Besides, with the reduction of fracture spacing, the value of C_v increases for both HAW and HW. Since the stress interference increases with the reduction of fracture spacing, preferential fracture growth of longer fracture is facilitated.

Figure 16.12c and d compare the injection pressure between HAW and HW when d = 10, 5, and 2 m. Obviously, the injection pressure of HAWs is always lower than that of HWs. Due to stress interference in HW, the injection pressure significantly increases as the decreasing of fracture spacing. However, the injection pressure appears to be insensitive to fracture spacing for HAWs. This means that the stress interference between neighboring fractures in

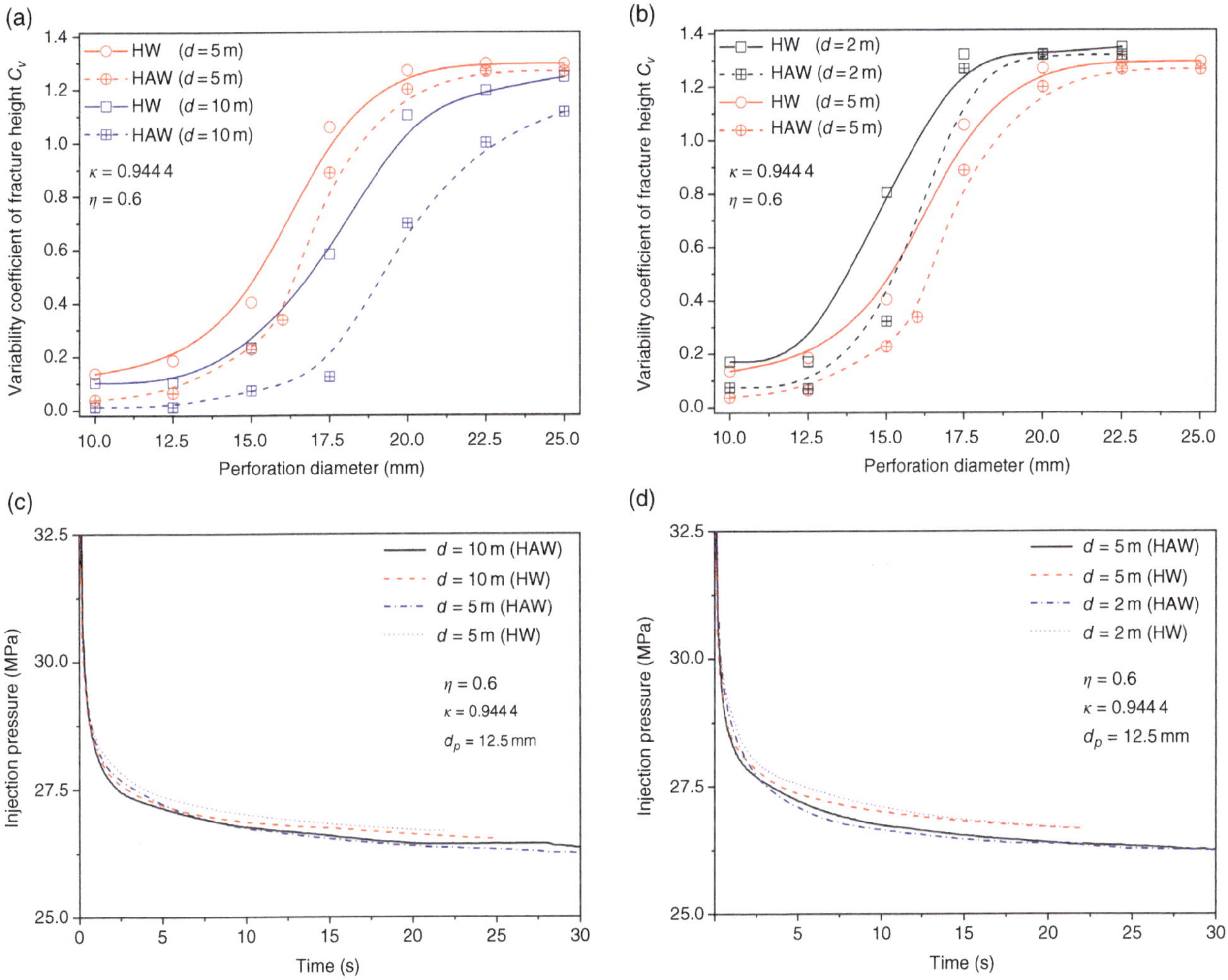

Figure 16.12 Comparison of numerical results between HW and HAW: (a) Variability coefficient when d = 5 and 10 m, (b) Variability coefficient when d = 2 and 5 m, (c) Injection pressure when d = 5 and 10 m, (d) Injection pressure when d = 2 and 5 m.

HAW is not significant, even though closely spaced fractures are created. This also suggests that closely spaced fractures can be created in HAW with less energy consumption. To find out the reason, the final geometry of two fractures for both HAW and HW cases are plotted in Figure 16.13 using $d_p = 12.5$ mm, $d = 2$ m and 5 m. In HW, the closer the two fracture, the shorter the #1 fracture, thus facilitating preferential fracture growth of the #2 fracture. In HAW, the downward propagation of #1 fracture and upward propagation of #2 fracture are further restrained when the fracture spacing reduces to 2 m. This reduces the injection pressure and promotes growth of upward propagation of #1 fracture and downward propagation of #2 fracture. Based on the numerical results, the HAWs will reduce the stress shadow effect and promote uniform propagation of multiple fractures.

16.4.2.5 Fracture Placement

Since the confining stress increases as the depth increases, the downward propagation of hydraulic fractures would be more difficult than the upward propagation. In a limit, the downward propagation of fractures is not allowed in some conditions, resulting in the so called single-wing hydraulic fractures. Therefore, as emphasized above, the real fracturing process may have three potential types of fracture placements. For comparing the effect of fracture placement, the fracture spacing is taken as 2 and 5 m. Note that the calculation is stopped when the height of a fracture is more than 25 m in this subsection, although different fracture placements (bi-wing HAW, bi-wing HW, and single-wing HAW) are used.

Figure 16.14a and b show C_v for three different fracture placements when $d = 2$ and 5 m. It demonstrates that C_v for single-wing fracture from a HAW is significantly lower than that for bi-wing fracture from a HAW and HW under the condition of a large value of Γ_m (or small perforation loss). However, with the reduction of Γ_m, the difference in C_v between different fracture placements disappears. Therefore, the perforation loss becomes dominated in controlling the final fracture patterns when Γ_m is less than 20.

Similar to the numerical results by Chen et al. [28], the limited width at the entry of single-wing fractures increases the flow resistance. This directly results in a higher injection pressure as shown in Figure 16.14c and d. Because of the same injection rate used, more energy is thus consumed for the single-wing fracture growth. However, if a more viscous fracturing fluid is used, the required pressure level for simultaneous single-wing fracture growth can be

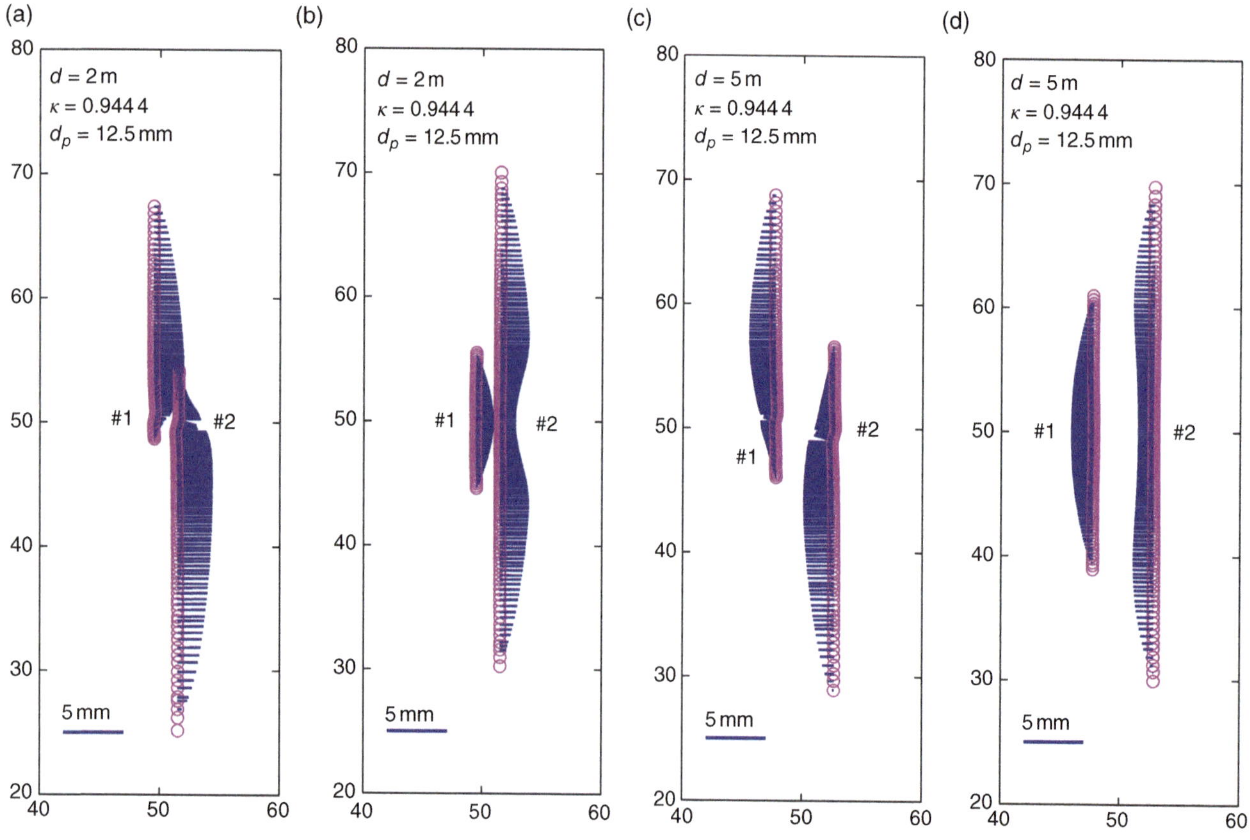

Figure 16.13 Fracture geometry of HW and HAW: (a) HAW (T = 18 s), (b) HW (T = 20s), (c) HAW (T = 18 s), (d) HW (T = 18 s).

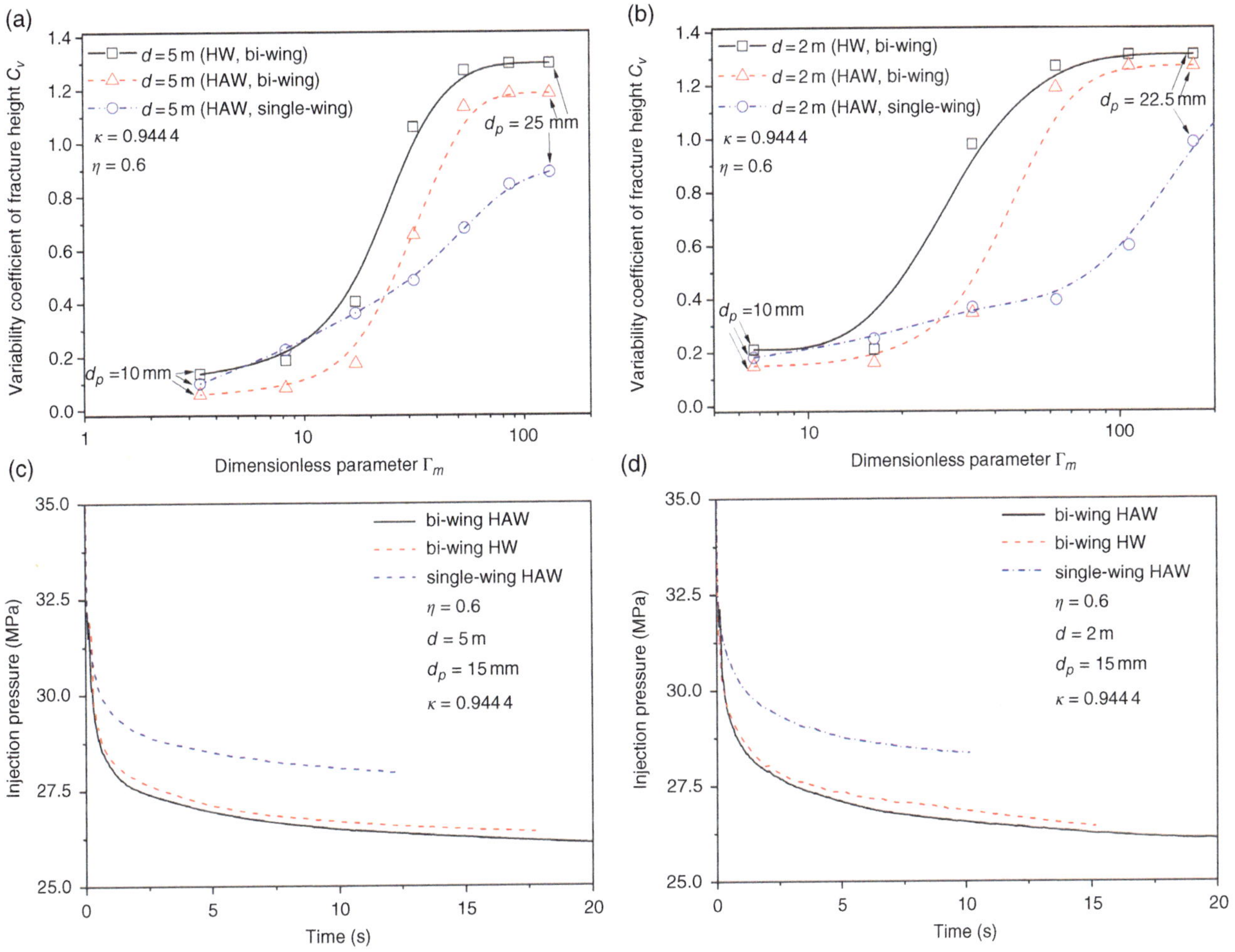

Figure 16.14 Comparison of numerical results between HW and HAW: (a) Variability coefficient when d = 5 m, (b) Variability coefficient when d = 2 m, (c) Injection pressure when d = 5 m, (d) Injection pressure when d = 2 m.

readily reached. In addition, the perforation friction effect will become weak as fracture height increases and the fracture tortuosity is minimized by increasing fluid viscosity and reducing injection rates [5]. More importantly, we can create a more closely spaced fracture pattern if the pressure is maintained at a relatively high level, which also implies a significant increase in gas production rate as measured on site. Of course, we need to constrain the fracture growth direction in practice. Fortunately, the increasing trend of the confining stress with increasing depth may help us to constrain the significant downward fracture growth from a HAW. The other factors to facilitate single-wing fracture growth also need further investigation.

The mechanism for promoting simultaneous single-wing fracture growth discussed by Zhang et al. [60] and Chen et al. [28] is recovered in our numerical results. The fracture width profiles for different fracture placements are plotted in Figure 16.15 when T = 11.5 s and d_p = 15 mm. In Figure 16.15c and f, there is a section along the longer fracture with a valley width in between two peaks. This section is located at the height right close to the tip of the shorter fracture. As discussed by Zhang et al. [60], there is a relatively low stress zone to assist in the upward growth of the shorter fracture.

16.4.2.6 Fracture Number

In previous sections, numerical results show that the use of HAWs can promote the simultaneous growth of two hydraulic fractures. Actually, this conclusion can be extended to the cases with more than two hydraulic fractures. Figure 16.16c and d provides the comparisons of numerical results when the fracture number is increased to n = 3 and n = 4. Note that the initial height of each fracture is equal to 0.5 m (η = 1) and the calculation is ended when the height of a fracture is over 30 m. Figure 16.16a shows the change of variability coefficient C_v with respect to dimension parameter Γ_m. The C_v value for HAWs is obviously less than that for HWs. As is shown in Figure 16.16a, three fractures along

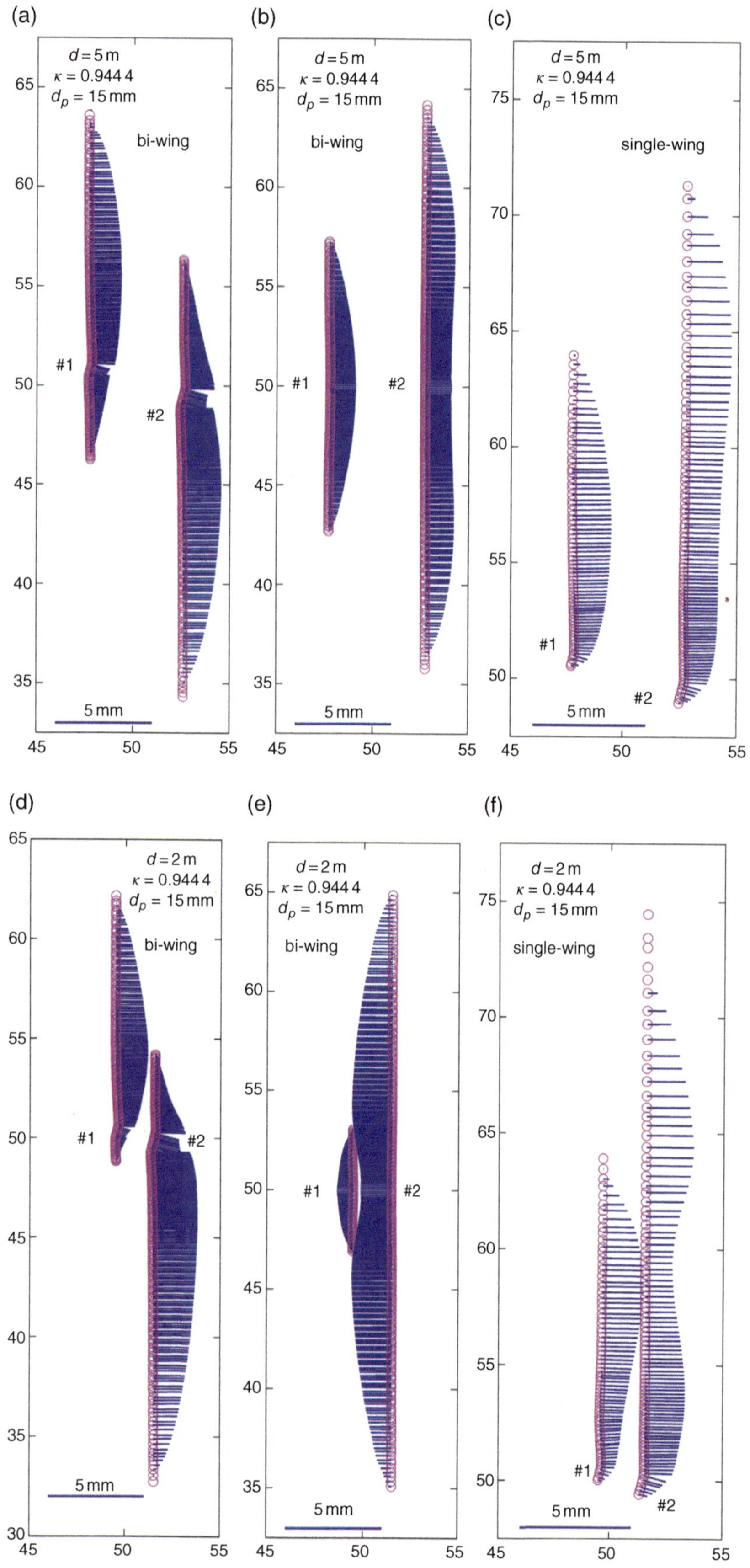

Figure 16.15 Geometry of bi/single-wing fracture in HW and HAW (T = 11.5 s): (a) bi-wing fracture of HAW when d = 5 m, (b) bi-wing fracture of HW when d = 5 m, (c) single-wing fracture of HAW when d = 5 m, (d) bi-wing fracture of HAW when d = 2 m, (e) bi-wing fracture of HW when d = 2 m, (f) single-wing fracture of HAW when d = 2 m.

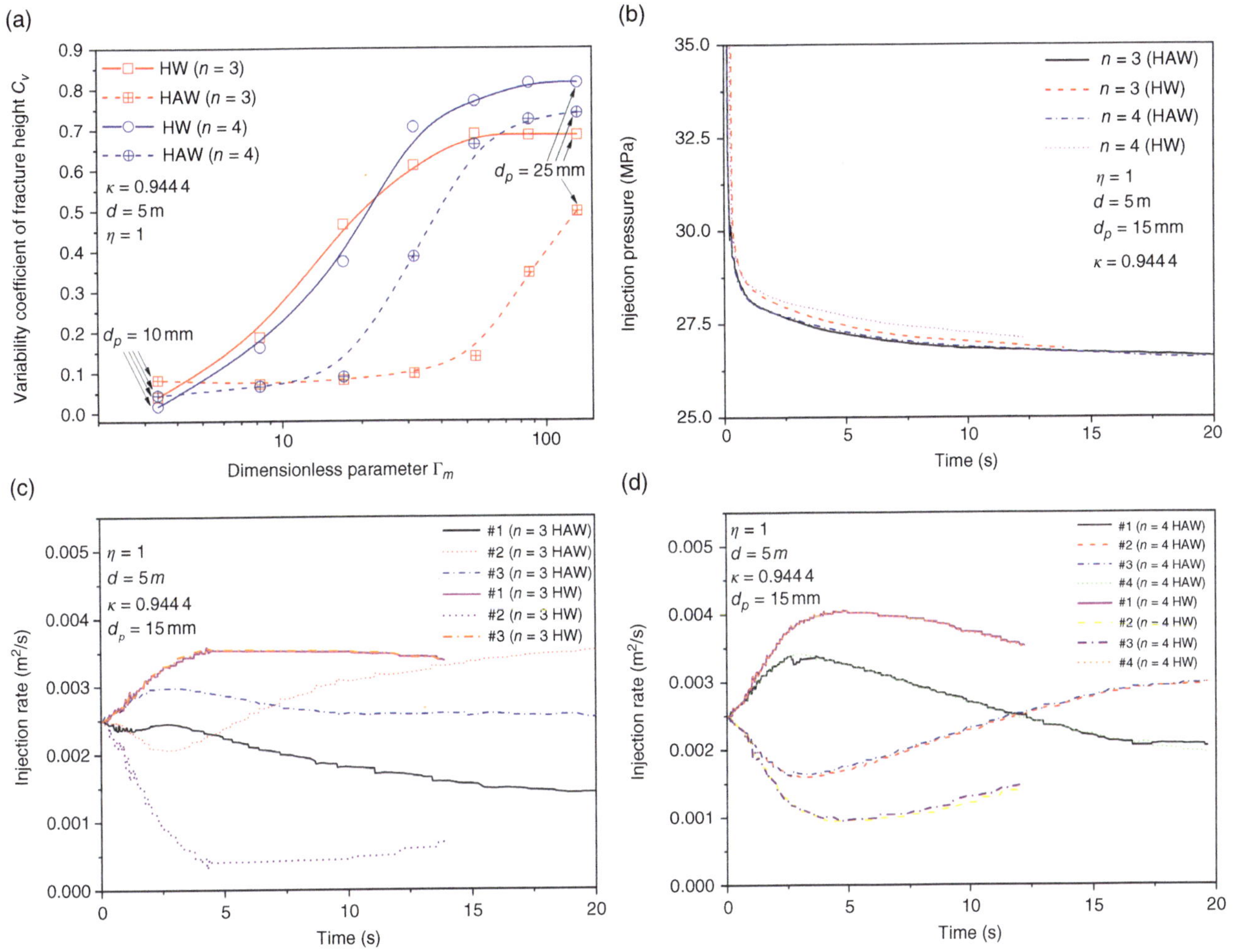

Figure 16.16 Comparison of numerical results between HW and HAW: (a) Variability coefficient of fracture height, (b) Injection pressure, (c) Flow rate division when n = 3, (d) Flow rate division when n = 4.

the HAW can simultaneously propagate when $\Gamma_m < 20$, whereas three fractures along the HW cannot simultaneously propagate until $\Gamma_m < 3$. Figure 16.16a also shows that the ability to promote the uniform growth of multiple fractures decreases with the increasing of n. In addition, the curve of injection pressure, as shown in Figure 16.16b, for HAWs is slightly lower than that for HWs when $n > 2$. On the other hand, the injection pressure for HWs increases with the increase of n, while the energy consumed for HAWs cannot be readily distinguished when $n > 2$. This means that HAWs can promote uniform growth of more than two fractures with less energy consumption.

Figure 16.16c and d show the flow division of multiple fractures for both HAW and HW when d_p = 15 mm. In Figure 16.16c, the fluid injection rate of #2 fracture in HAW decreases first and then increases to about 0.0035 m^2/s, while the flux of #2 fracture in HW continuously decreases to about 0.0005 m^2/s. This discrepancy shows the ability of HAW to adjust the flux of multi-cluster fractures. Figure 16.16c shows the flow rate of multiple fractures along HAW are more uniform than those along HW. Figure 16.17 displays the fracture geometry for both HAWs and HWs when n = 3 and 4. Due to stress interference and height difference, all fracture geometries become nonsymmetry for HAW cases, as shown in Figure 16.17a and c. The nonsymmetric propagation patches would be one reason that C_v is smaller for HAWs. In particular, the fracture interaction among four fractures would be reduced when the fractures are aligned in a step pattern as shown in Figure 16.17c. However, when the fracture height is long enough, the fracture interaction would play an important role, of which the effect on the simultaneous propagation of multiple fractures along HAW needs further investigation.

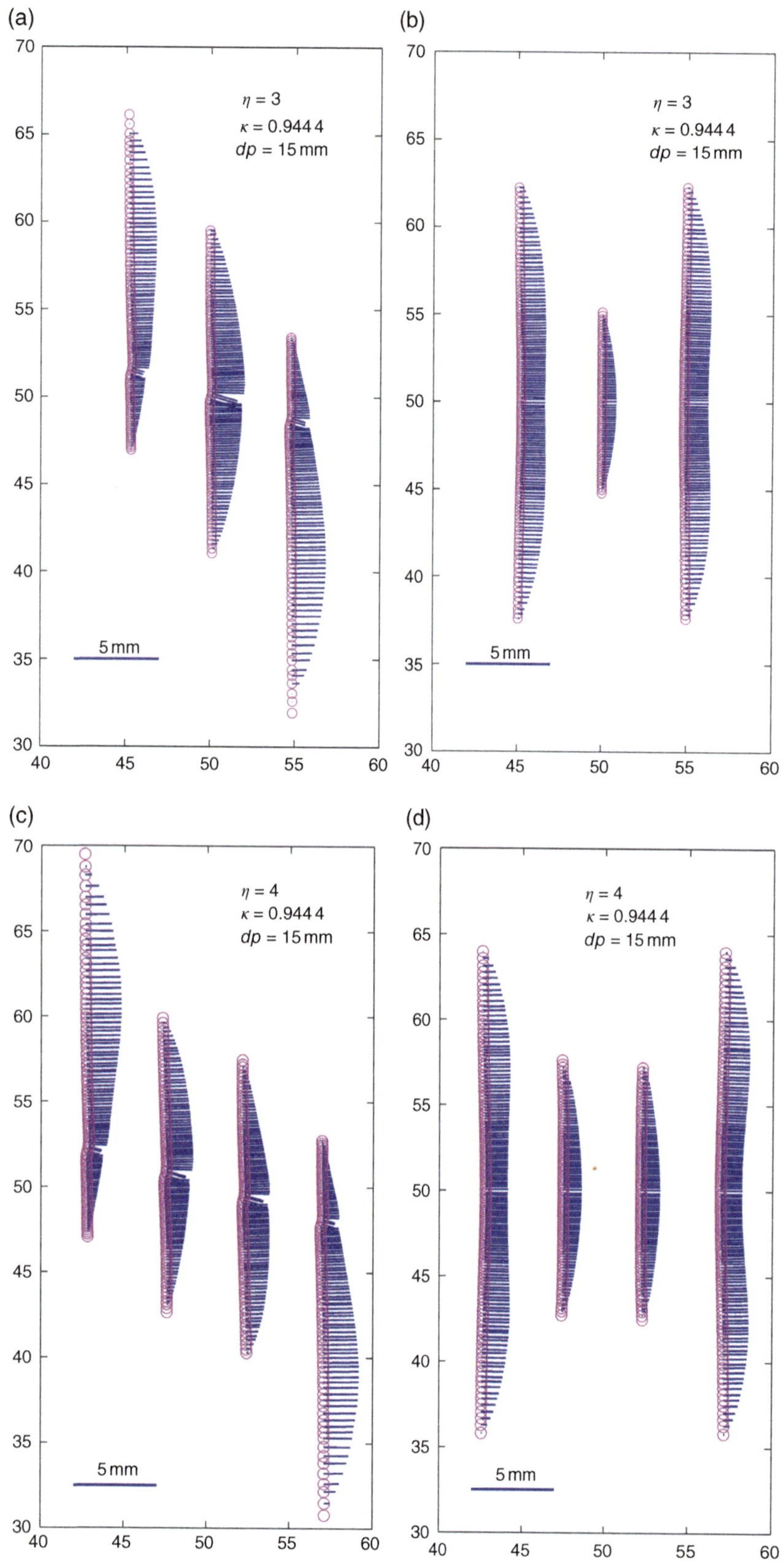

Figure 16.17 Fracture geometry of HW and HAW (T = 11 s): (a) HAW with n = 3, (b) HW with n = 3, (c) HAW with n = 4, (d) HW with n = 4.

16.5 Discussion

The influence of operational parameters, deviation angle, fracture spacing, placement and number on the simultaneous propagation of multiple fractures are investigated. According to the numerical results of parameters analysis, low perforation diameter, high fluid viscosity and single-wing fracture can promote the uniform propagation of multiple fractures in both HW and HAW. However, these methods will increase the injection pressure and energy consumption during stimulation treatments. Based on numerical cases of different deviation angle, the values of C_v and injection pressure decrease with the reduction of deviation angle. This means that the deviation of wellbore relative to minimum horizontal in situ stress can promote the uniform growth of multiple fractures without increasing energy consumption. Numerical results of fracture spacing and number further illustrates this effect. On the one hand, the injection pressure appears to be insensitive to the fracture spacing for HAWs. On the other hand, the injection pressure for HAWs does not increase with the increase of fracture number. These results indicate that HAWs can generate more closely spaced fractures than HWs. Hence, the potential engineering application of this work is the adoption of HAW to mitigate the stress shadow effect and improve the hydraulic fracturing efficiency.

Since the real 3D hydraulic fracturing problems are simplified to 2D problem on vertical plane (σ_v–σ_h plane) without considering the influence of fracture length, the assumption is not completely consistent with the actual condition. In addition, Fracture interaction may dominate when the fracture height is long enough. To further maintain simultaneous fracture growth, people must resort to some devices in promoting perforation resistance. However, the numerical model proposed in this paper can preliminarily reveal the mechanisms that the production rate can be enhanced after HAW fracturing stimulation in Huaqing Oilfield. Although the mechanisms for enhancing production rates are still not fully exploited due to the problem complexity and the difficulty in measuring fracture patterns. Many factors associated with fracture patterns need further investigations, such as fracture placement and device functionality. Obviously, the extension of multi-cluster fracturing from HW to HAW will pose some new challenges to hydraulic fracturing study, especially on fracture growth pattern selections for reservoir stimulation.

16.6 Conclusions

In this study, we have investigated the effect of well types (HW and HAW) on the simultaneous growth of multiple fractures, with an aim to explain why the production rate can be enhanced after HAW fracturing stimulation. To this end, a fully coupled XFEM model is proposed to handle the nonlinear process of multi-cluster hydraulic fracturing in a reservoir. The resulting HM-coupled equations are solved by Newton–Raphson iteration method, together with the secant method for satisfying the pressure continuity and mass conservation conditions at the fracture-well intersections. Then, the model is validated by the problem proposed by Wu [58] while considering the simultaneous growth of four hydraulic fractures.

The comparisons of numerical results for HAWs and HWs are made to investigate the influence of several factors on the simultaneous propagation of multiple fractures. The major conclusion can be drawn as follows:

1) Reducing perforation diameter, increasing fluid viscosity, and enhancing fracture spacing can promote the uniform growth of multiple fractures in both HWs and HAWs.
2) The variability coefficient of fracture height and injection pressure is lower for HAW than those for HW. This implies that more closely spaced fractures can be generated without increasing energy consumption for HAWs.
3) Fracture placement is another factor in creating more closely spaced fractures. The single-wing fracture has an advantage over the bi-wing fractures in this aspect without assistance of the devices.
4) Numerical results of fracture geometry demonstrate that the preferential but opposite fracture growth direction may contribute to the weakening of the stress shadow effects, no matter how many fractures are generated.

Appendix 16.A Dimensionless Toughness κ

According to Detournay [59], dimensionless toughness κ represents the energy dissipation ratio of fracture propagation to the viscous fluid flow, which can be written as

$$\kappa = \frac{K_c{}'}{(E'\mu' Q/n)^{1/4}} \tag{16.A.1}$$

where $K_c{}' = 4K_c\sqrt{2/\pi}$, $E' = E/(1\text{-}\nu^2)$, $\mu' = 12\,\mu$. $\kappa > 4$, called toughness dominated regime, means that most energy are consumed for creating new fracture face, while $\kappa < 1$, called viscosity dominated regime, denotes that most energy are used for viscous flow. Note that the defined κ is a nominal dimensionless toughness, which is only strictly correct if all fractures receive identically the same inflow of fluid.

Appendix 16.B Dimensionless Parameter Γ_m

This parameter measures the perforation loss. In reality, the perforation diameter for multi-cluster fracturing varies from 10 to 25 mm. The dimensionless parameter Γ_m proposed by Lecampion and Desroches [14] is defined as

$$\Gamma_m = \frac{E'^{3/4}\mu'^{1/4}}{\varphi_p d^{3/4}(Q/n)^{7/4}} \tag{16.A.2}$$

where $\varphi_p = \frac{8\rho_f}{\pi^2 n_p^2 d_p^4 C^2}$ denotes the perforation friction coefficient. Dimensionless parameter Γ_m denotes the competition between the perforation loss and stress interaction when the fracture propagation is in the viscosity dominated regime. Γ_m increases with the increase of perforation diameter d_p. Lecampion and Desroches [14] concluded that a small value of Γ_m promotes simultaneous growth of multi-cluster fractures, while a large value promotes preferential growth. It is found by Lecampion and Desroches [14] that $\Gamma_m < 1$ appears more than sufficient to ensure simultaneous growth.

Appendix 16.C Dimensionless Variability Coefficient C_v

In order to evaluate the extent of uniform multi-cluster fracture growth, the dimensionless variability coefficient C_v proposed by Chen et al. [27], is given as

$$C_v = \frac{\text{std}(h)}{\text{mean}(h)} \tag{16.A.3}$$

where std(h) is the standard deviation of all fracture heights and mean(h) is average height of all fractures. It is easy to know that C_v varies from 0 to $\sqrt{n}$. Large C_v suggests that the great differences between fracture heights and a preferential growth pattern for multiple fractures, while small C_v implies the more uniform fracture growth patterns and a potentially expected fracturing treatment.

References

1 Li, Y., Deng, J., Liu, W. et al. (2017). Numerical simulation of limited-entry multi-cluster fracturing in horizontal well. *Journal of Petroleum Science and Engineering* 152: 443–455. https://doi.org/10.1016/j.petrol.2017.03.023.

2 Yang, Z., Yi, L., Li, X., and He, W. (2018). Pseudo-three-dimensional numerical model and investigation of multi-cluster fracturing within a stage in a horizontal well. *Journal of Petroleum Science and Engineering* 162: 190–213. https://doi.org/10.1016/j.petrol.2017.12.034.

3 Passey, Q.R., Yin, H., Rendeiro, C.M., and Fitz, D.E., 2005. Overview of high-angle and horizontal well formation evaluation: issues, learnings, and future directions. *46th Annual Logging Symposium*, New Orleans, LO.

4 El Rabaa, W. (1989). Experimental study of hydraulic fracture geometry initiated from horizontal wells. *64th Annual Technical Conference and Exhibition*. San Antonio, TX. https://doi.org/10.2118/19720-MS

5 Weng, X. (1993). Fracture initiation and propagation from deviated wellbores. *SPE Annual Technical Conference and Exhibition*. Houston, TX. https://doi.org/10.2118/26597-MS

6 Chen, M., Chen, Z., and Huang, R.Z. (1995). Hydraulic fracturing of highly deviated wells. *Journal of China University of Petroleum* 19 (2): 30–35.

7 Brumley, J.L. and Abass, H.H. (1996). Hydraulic fracturing of deviated wells: interpretation of breakdown and initial fracture opening pressure. *SPE Eastern Regional Meeting*, Columbus, OH. https://doi.org/10.2118/37363-MS

8 AL Isaee, O., Chavez Florez, J., Ali, N. et al. (2019). 3D Petrophysical and geomechanical view for enhancing hydraulic fracturing of horizontals and highly deviated wells. *Abu Dhabi International Petroleum Exhibition & Conference*. Abu Dhabi, UAE. https://doi.org/10.2118/197901-MS

9 Olson, J.E. (1995). Fracturing from highly deviated and horizontal wells: numerical analysis of non-planar fracture propagation. *Low Permeability Reservoirs Symposium*. Denver, CO. https://doi.org/10.2118/29573-MS

10 Van de Ketterij, R.G. and de Pater, C.J. (1997). Experimental study on the impact of perforations on hydraulic fracture tortuosity. *SPE European Formation Damage Conference*. Hague, Netherlands. https://doi.org/10.2118/38149-MS

11 Zhu, H., Deng, J., Liu, S. et al. (2014). Hydraulic fracturing experiments of highly deviated well with oriented perforation technique. *Geomechanics and Engineering* 6 (2): 153–172. https://doi.org/10.12989/gae.2014.6.2.153.

12 Zhu, H., Zhao, X., Guo, J. et al. (2015). Coupled flow-stress-damage simulation of deviated-wellbore fracturing in hard-rock. *Journal of Natural Gas Science and Engineering* 26: 711–724. https://doi.org/10.1016/j.jngse.2015.07.007.

13 Bunger, A.P. (2013). Analysis of the power input needed to propagate multiple hydraulic fractures. *International Journal of Solids and Structures* 50 (10): 1538–1549. https://doi.org/10.1016/j.ijsolstr.2013.01.004.

14 Lecampion, B. and Desroches, J. (2015). Simultaneous initiation and growth of multiple radial hydraulic fractures from a horizontal wellbore. *Journal of the Mechanics and Physics of Solids* 82: 235–258. https://doi.org/10.1016/j.jmps.2015.05.010.

15 Wu, K. and Olson, J.E. (2016). Mechanisms of simultaneous hydraulic-fracture propagation from multiple perforation clusters in horizontal wells. *SPE Journal* 21 (3): 1000–1008. https://doi.org/10.2118/178925-PA.

16 Chen, M., Zhang, S., Ma, X. et al. (2018). A semi-analytical model for predicting fluid partitioning among multiple hydraulic fractures from a horizontal well. *Journal of Petroleum Science and Engineering* 171: 1041–1051. https://doi.org/10.1016/j.petrol.2018.07.031.

17 Lu, Y., Li, H., Lu, C. et al. (2019). The effect of completion strategy on fracture propagation from multiple cluster perforations in fossil hydrogen energy development. *International Journal of Hydrogen Energy* 44 (14): 7168–7180. https://doi.org/10.1016/j.ijhydene.2019.01.121.

18 Yi, S., Wu, C.H., and Sharma, M.M. (2020). Optimization of plug-and-perforate completions for balanced treatment distribution and improved reservoir contact. *SPE Journal* 25 (2): 558–572. https://doi.org/10.2118/194360-PA.

19 Zhao, Z., Zheng, Y., Kang, Y. et al. (2021). Investigation on nonuniform extension of hydraulic fracture in shale gas formation. *Geofluids* 5516040. https://doi.org/10.1155/2021/5516040.

20 Zhao, J., Chen, X., Li, Y., and Fu, B. (2016). Simulation of simultaneous propagation of multiple hydraulic fractures in horizontal wells. *Journal of Petroleum Science and Engineering* 147: 788–800. https://doi.org/10.1016/j.petrol.2016.09.021.

21 Nikolskiy, D. and Lecampion, B. (2019). Modeling of simultaneous propagation of multiple blade-like hydraulic fractures from a horizontal well. *Rock Mechanics and Rock Engineering* 1–18. https://doi.org/10.1007/s00603-019-02002-4.

22 Liu, X., Rasouli, V., Guo, T. et al. (2020). Numerical simulation of stress shadow in multiple cluster hydraulic fracturing in horizontal wells based on lattice modelling. *Engineering Fracture Mechanics* 238: 107278. https://doi.org/10.1016/j.engfracmech.2020.107278.

23 Manchanda, R., Bryant, E.C., Bhardwaj, P. et al. (2018). Strategies for effective stimulation of multiple perforation clusters in horizontal wells. *SPE Production & Operations* 33 (3): 539–556. https://doi.org/10.2118/179126-PA.

24 Peirce, A. and Bunger, A. (2015). Interference fracturing: nonuniform distributions of perforation clusters that promote simultaneous growth of multiple hydraulic fractures. *SPE Journal* 20 (2): 384–395. https://doi.org/10.2118/172500-PA.

25 Sun, T., Zeng, Q., and Xing, H. (2021). A model for multiple hydraulic fracture propagation with thermo-hydro-mechanical coupling effects. *Energies* 14 (4): 894. https://doi.org/10.3390/en14040894.

26 Saberhosseini, S.E., Chen, Z., and Sarmadivaleh, M. (2021). Multiple fracture growth in modified zipper fracturing. *International Journal of Geomechanics* 21 (7): 04021102. https://doi.org/10.1061/(ASCE)GM.1943-5622.0002040.

27 Chen, X., Zhao, J., Yan, W., and Zhang, X. (2019a). Numerical investigation into the simultaneous growth of two closely spaced fluid-driven fractures. *SPE Journal* 24 (1): 274–289. https://doi.org/10.2118/194188-PA.

28 Chen, X., Zhao, J., Li, Y. et al. (2019b). Numerical simulation of simultaneous hydraulic fracture growth within a rock layer: implications for stimulation of low-permeability reservoirs. *Journal of Geophysical Research: Solid Earth* 124 (12): 13227–13249. https://doi.org/10.1029/2019JB017942.

29 Li, J., Dong, S., Hua, W. et al. (2020). Numerical simulation of temporarily plugging staged fracturing (TPSF) based on cohesive zone method. *Computers and Geotechnics* 121: 103453. https://doi.org/10.1016/j.compgeo.2020.103453.

30 Zhou, F., Su, H., Liang, X. et al. (2019). Integrated hydraulic fracturing techniques to enhance oil recovery from tight rocks. *Petroleum Exploration and Development* 46 (5): 1065–1072. https://doi.org/10.1016/S1876-3804(19)60263-6.

31 Zeng, Q., Liu, W., and Yao, J. (2018b). Hydro-mechanical modeling of hydraulic fracture propagation based on embedded discrete fracture model and extended finite element method. *Journal of Petroleum Science and Engineering* 167: 64–77. https://doi.org/10.1016/j.petrol.2018.03.086.

32 Belytschko, T. and Black, T. (1999). Elastic crack growth in finite elements with minimal remeshing. *International Journal for Numerical Methods in Engineering* 45 (5): 601–620. https://doi.org/10.1002/(SICI)1097-0207(19990620)45:5<601::AID-NME598>3.0.CO;2-S.

33 Khoei, A.R., Hirmand, M., Vahab, M., and Bazargan, M. (2015). An enriched FEM technique for modeling hydraulically driven cohesive fracture propagation in impermeable media with frictional natural faults: numerical and experimental investigations. *International Journal for Numerical Methods in Engineering* 104 (6): 439–468. https://doi.org/10.1002/nme.4944.

34 Shi, F., Wang, X., Liu, C. et al. (2017). An XFEM-based method with reduction technique for modeling hydraulic fracture propagation in formations containing frictional natural fractures. *Engineering Fracture Mechanics* 173: 64–90. https://doi.org/10.1016/j.engfracmech.2017.01.025.

35 Salimzadeh, S. and Khalili, N. (2016). Fully coupled XFEM model for flow and deformation in fractured porous media with explicit fracture flow. *International Journal of Geomechanics* 16 (4): 04015091. https://doi.org/10.1061/(ASCE)GM.1943-5622.0000623.

36 Paul, B., Faivre, M., Massin, P. et al. (2018). 3D coupled HM–XFEM modeling with cohesive zone model and applications to non planar hydraulic fracture propagation and multiple hydraulic fractures interference. *Computer Methods in Applied Mechanics and Engineering* 342: 321–353. https://doi.org/10.1016/j.cma.2018.08.009.

37 Remij, E.W., Remmers, J.J., Huyghe, J.J., and Smeulders, D.D. (2015). The enhanced local pressure model for the accurate analysis of fluid pressure driven fracture in porous materials. *Computer Methods in Applied Mechanics and Engineering* 286: 293–312. https://doi.org/10.1016/j.cma.2014.12.025.

38 Zhou, Y., Yang, D., Zhang, X. et al. (2020). Numerical investigation of the interaction between hydraulic fractures and natural fractures in porous media based on an enriched FEM. *Engineering Fracture Mechanics* 235: 107175. https://doi.org/10.1016/j.engfracmech.2020.107175.

39 Moës, N., Cloirec, M., Cartraud, P., and Remacle, J.F. (2003). A computational approach to handle complex microstructure geometries. *Computer Methods in Applied Mechanics and Engineering* 192 (28–30): 3163–3177. https://doi.org/10.1016/S0045-7825(03)00346-3.

40 Khoei, A.R., Vahab, M., and Hirmand, M. (2018). An enriched–FEM technique for numerical simulation of interacting discontinuities in naturally fractured porous media. *Computer Methods in Applied Mechanics and Engineering* 331: 197–231. https://doi.org/10.1016/j.cma.2017.11.016.

41 Vahab, M. and Khalili, N. (2018). X-FEM modeling of multizone hydraulic fracturing treatments within saturated porous media. *Rock Mechanics and Rock Engineering* 51: 3219–3239. https://doi.org/10.1007/s00603-018-1419-z.

42 Zeng, Q., Liu, Z., Wang, T. et al. (2018a). Fully coupled simulation of multiple hydraulic fractures to propagate simultaneously from a perforated horizontal wellbore. *Computational Mechanics* 61 (1): 137–155. https://doi.org/10.1007/s00466-017-1412-5.

43 Geertsma, J. and de Klerk, F. (1969). A rapid method of predicting width and extent of hydraulically induced fractures. *Journal of Petroleum Technology* 21: 1571–1581. https://doi.org/10.2118/2458-PA.

44 Khristianovic, S.A. and Zheltov, Y.P. (1955). Formation of vertical fractures by means of highly viscous liquid. *4th World Petroleum Congress*. Rome, Italy.

45 Biot, M.A. (1956). General solutions of the equations of elasticity and consolidation for a porous material. *Journal of Applied Mechanics* 23 (1): 91–96.

46 Zienkiewicz, O.C. and Shiomi, T. (1984). Dynamic behaviour of saturated porous media; the generalized Biot formulation and its numerical solution. *International Journal for Numerical and Analytical Methods in Geomechanics* 8 (1): 71–96. https://doi.org/10.1002/nag.1610080106.

47 Khoei, A.R. (2014). *Extended Finite Element Method: Theory and Applications*. Wiley https://doi.org/10.1002/9781118869673.

48 Mohammadnejad, T. and Khoei, A.R. (2013). Hydro-mechanical modeling of cohesive crack propagation in multiphase porous media using the extended finite element method. *International Journal for Numerical and Analytical Methods in Geomechanics* 37 (10): 1247–1279. https://doi.org/10.1002/nag.2079.

49 Elbel, J.L., Piggott, A.R., and Mack, M.G. (1992). Numerical modeling of multilayer fracture treatments. *Permian Basin Oil and Gas Recovery Conference*. Midland, TX. https://doi.org/10.2118/23982-MS

50 Siriwardane, H.J. and Layne, A.W. (1991). Improved model for predicting multiple hydraulic fracture propagation from a horizontal well. *SPE Eastern Regional Meeting*. Lexington, KY. https://doi.org/10.2118/23448-MS

51 Wu, K. and Olson, J.E. (2015). Simultaneous multifracture treatments: fully coupled fluid flow and fracture mechanics for horizontal wells. *SPE Journal* 20 (2): 337–346. https://doi.org/10.2118/167626-PA.

52 Crump, J.B. and Conway, M.W. (1988). Effects of perforation-entry friction on bottomhole treating analysis. *Journal of Petroleum Technology* 40 (8): 1–041. https://doi.org/10.2118/15474-PA.

53 Erdogan, F. and Sih, G.C. (1963). On the crack extension in plates under plane loading and transverse shear. *Journal of Basic Engineering* 85 (4): 519–525. https://doi.org/10.1115/1.3656897.

54 Moës, N. and Belytschko, T. (2002). Extended finite element method for cohesive crack growth. *Engineering Fracture Mechanics* 69 (7): 813–833. https://doi.org/10.1016/S0013-7944(01)00128-X.

55 Melenk, J.M. and Babuska, I. (1996). The partition of unity finite element method: basic theory and applications. *Computer Methods in Applied Mechanics and Engineering* 139 (1–4): 289–314. https://doi.org/10.1016/S0045-7825(96)01087-0.

56 Réthoré, J., Borst, R.D., and Abellan, M.A. (2007). A two-scale approach for fluid flow in fractured porous media. *International Journal for Numerical Methods in Engineering* 71 (7): 780–800. https://doi.org/10.1002/nme.1962.

57 Zienkiewicz, O.C., Chan, A.H.C., Pastor, M. et al. (1999). *Computational Geomechanics*. Chichester: Wiley.

58 Wu, K. (2014). Numerical modeling of complex hydraulic fracture development in unconventional reservoirs. Doctoral dissertation. The University of Texas at Austin. Austin, TX.

59 Detournay, E. (2004). Propagation regimes of fluid-driven fractures in impermeable rocks. *International Journal of Geomechanics* 4 (1): 35–45. https://doi.org/10.1061/(ASCE)1532-3641(2004)4:1(35).

60 Zhang, X., Bunger, A.P., and Jeffrey, R.G. (2014). Mechanics of two interacting magma-driven fractures: a numerical study. *Journal of Geophysical Research: Solid Earth* 119: 8047–8063. https://doi.org/10.1002/2014JB011273.

17

Hydraulic Fracturing-Induced Slip on a Permeable Fault

Xi Zhang[1,2], R. G. Jeffrey[2], and J. Yang[3]

[1] Faculty of Engineering, China University of Geosciences, Wuhan, China
[2] SCT Operations Pty. Ltd., Wollongong, New South Wales, Australia
[3] Institute of Rock and Soil Mechanics, Chinese Academy of Sciences, Wuhan, China

17.1 Introduction

In the past two decades, there has been a surge of research activities on characterizing the fluid injection-associated seismic events. As we know, hydraulic fracturing is generally accompanied by low-magnitude, long-duration microseismic events that are visualized as a diffusive cloud from the injection point [1, 2], although there is unlikely the emergence of the damaging earthquakes [3]. Close to the injection well, the injection-induced microseismic events, which nucleate over relatively short injection time scales [4], are important to the determination of fracture growth extent and stimulated reservoir volume. For some injections, the seismic signals have been determined to come from slipping zones along an existing fault that is subject to a pore pressure change that results in aseismic slip [5]. The induced low-magnitude seismic events can be thought of as the product of the process of fault slip reactivation as a result of the increased fluid pressure. Prior to the induced events, the fault may experience a relatively long period of slow slip [6], which is so slow that it can be treated as aseismic. Although the mechanisms that occur as aseismic slip nucleates induced seismic events are still open, the spatial extent of hydraulic-driven aseismic slip should play a role and has been extensively studied [7–9]. The physical relationships between fluid flow, fault aseismic slip, and stress changes leading to induced microseismic events must be fundamental to understand how faults become unstable in slip under fluid injection.

Studies of injection-induced microseismicity often model a preexisting fault as a slipping patch with a finite permeability. As a limiting force to resisting slip, the shear strength τ_s along the closed portion of the fault can be given in the presence of fluid pressure p_f as

$$\tau_s = f\left(\sigma_n - p_f\right) \tag{17.1}$$

where f is the coefficient of friction and σ_n is the normal stress acting to close the fault surfaces.

The fault is regarded as a shear weakening surface in resisting the shear movement with some limited and reducing frictional strength. Only with the presence of weakening can the rapid slip occur and arrest. Two source mechanisms can contribute to the occurrence of microseismicity, provided that the fault slip zones are extremely thin. One is the increasing fluid pressure and the other is a lower coefficient of friction which decreases with slip. This slip weakening process is motivated by experimental measurements and by the observations that major faults are statically strong, but dynamically weak. A widely accepted hypothesis is that the lowest residual strength is an important parameter to distinguish stable from unstable faults [8]. It is argued that the coefficient of friction experiences a slip weakening process in the presence of many factors such as gouge granulation, contact loss on an uneven surface due to flash heating, and larger difference in compressibility between water and gouge. But the real situation is more complex.

Pressure diffusion inside the fault is dependent on the fault permeability and its evolution in responses to stress and fault geometry changes. The general fault architectures contain the mature, well-developed fault zone that consists of an inner thin low-permeability core and two outside broader high-permeability damage zones, marked by thick gouge layers [10]. Shallow, poorly developed faults are found to often have a wider aperture right across the thin fault surfaces and are typically modeled as fluid conduits with parallel walls [11]. Actually, the change in the along-fault architecture plays a role in the conduction of

Mechanics of Hydraulic Fracturing: Experiment, Model, and Monitoring, First Edition.
Edited by Xi Zhang, Bisheng Wu, Diansen Yang, and Andrew Bunger.

fluid during the slipping process, possibly generating more continuous seismic signals in space. The individual seismogenic zones may be associated with low-permeability but slip-weakening materials, while high-permeability and less slip-weakening materials would be a major contribution of the arrest of rapid slip.

The use of microseismic monitoring of hydraulic fracturing has also provided great value for measurement of fracture geometry and optimization of well stimulations, completions, and field development, through understanding how a formation responds to fluid injection [12]. Although there are many uncertainties, the mechanics for the generation of microseismicity at different length scales is fundamentally important to fracturing stimulation optimization because the understanding of conditions that lead to seismic versus aseismic and localized versus widespread fault reactivation is central to hydraulic fracturing treatments. Several questions may arise from the observations, such as (i) how does the seismic event nucleate from the slow slip, both with and without the presence of pressurized fluids? (ii) Can the rupture be stopped by a fluid-transport barrier such as geometrical fault complexities and is the manifestation of slip rate heterogeneities common? (iii) Can we use the radiated energy of a microseismic event in order to identify the slip event sources on a fault?

17.2 Model Setup

Let us denote by V, the elastic body bounded by a closed boundary S, enclosing the fault Σ as shown in Figure 17.1a. As the elastic deformation occurs in the body, the internal stress σ_{ij} and strain ε_{ij} are generated. On S, there is a prescribed traction vector $\sigma \cdot n$ in which n stands for the unit vector normal. The preexisting fault in V is pressurized by the injected fluid. On the fault surface Σ, there are a slip displacement vector u and the normal unit vector ν pointing from the negative to positive side of the fault surface. If the surface is opened by the fluid pressure, the generated fracture opening is denoted by w. As many models, the injection rate is assumed to be constant and the fluid dynamic viscosity is denoted by μ.

As shown in Figure 17.1b, there are four different portions of a fault surface around the injection well according to their deformation and flow modes. There are no predetermined orders for the moving boundaries of these zones in the presence of varying fluid pressure and preexisting hydraulic aperture. For instance, if the fluid front lags behind the slip zone tip, a fluid lag zone exists. In general, the lag zone is defined as the gap area between the fluid front and the slip or opening front existing ahead of the fluid front. Without loss of generality, we assume that the fault surface is large compared to these front-defined areas, recognizing that in reality there are many small faults around the injection well. Therefore, we can define the moving opening, fluid and slip fronts as C_o, C_f, and C_s, respectively, and their length counterparts for a two-dimensional (2-D) problem are denoted as l_o, l_f, and l_s, while the length of the entire fault is denoted by L.

As shown in Figure 17.1b, the inner part is the open area where the fracture surfaces are separated by the elevated pressure overcoming the confining stress σ generated by the traction on S. In the intermediate-inner portion, a fluid occupied area subject to an internal pressure p may compete in size with a frictional slipping zone where the surface offers the resistance to slipping characterized by the shear strength τ_s. Inside the open zone, τ_s is zero as the contact is lost. The resultant shear stress τ generated by the traction on S tends to drive the slipping zone to grow.

The 2-D model for slip development is commonly based on the condition that the fault surface is entirely in contact. This 2-D model is extended to hydraulic fracturing cases by including the presence of both fracture shearing and

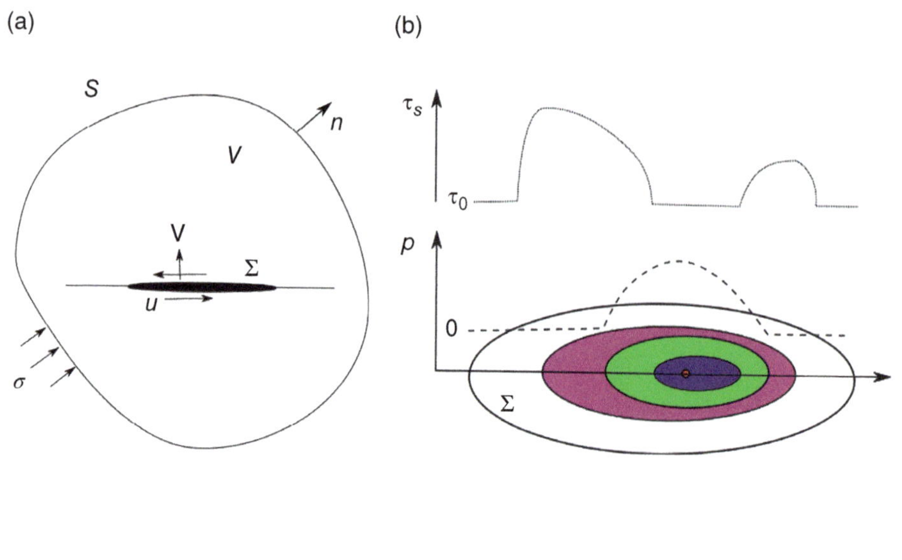

Figure 17.1 (a) A fault embedded in an elastic formation that is subject to external loading and internal fluid pressure caused by fluid injection; (b) schematic diagram of shear stress (solid line) and pressure (dotted line) distributions on a fault plane, where the injection well is identified by a circle on the horizontal axis. Due to the pressure variations, the fault plane can be characterized from the outer to the center (injection well) as sticking, slipping, fluid-filled, and opening zones.

opening. Assume that the fault plane is in quasi-static elastic deformation before the dynamic rupture is nucleated. For a patch $[x_1,x_2]$ on the planar fault up to the sticking zone, the equations of momentum balance are given by

$$\sigma(x_0) - p_f(x_0) = \frac{E'}{4\pi}\int_{x_1}^{x_2} K(x_0, x)w(x)dx \quad (17.2)$$

$$\tau(x_0) - \tau_s(x_0) = \frac{E'}{4\pi}\int_{x_1}^{x_2} K(x_0, x)u(x)dx \quad (17.3)$$

where E' is the plane-strain modulus of the rock, and the kernel $K(x_0, x)$ is the Green's function. These equations can be extended to more general cases. The governing equations for nonplanar fault planes are provided in Zhang et al. [13].

A constitutive friction law is applied in most of the numerical models discussed below to the evolution of τ_s given in Eq. (17.1). As utilized by many studies, the coefficient of friction is considered as slip weakening, that is, it decreases with the slip in a functional form $f(u)$. The truncated linear slip-weakening friction law is commonly used to describe the decrease of τ_s with slip for the transition from aseismic to seismic slip [8, 14]. In particular, the truncated linear slip-weakening of $f(u)$ is given by

$$f(u) = \begin{cases} f_s & \text{if } u = 0 \\ f_s - \dfrac{f_s - f_d}{u_c}|u| & \text{if } 0 < |u| \le u_c \\ f_d & \text{if } |u| > u_c \end{cases} \quad (17.4)$$

where f_s and f_d are the static and the sliding or residual coefficients of friction, respectively, and u_c is the critical slip weakening distance at which the coefficient of friction is reduced to the residual value. The range of the value of u_c is discussed by Uenishi and Rice [14] and Cocco et al. [15]. If σ_n is uniform along the fault, the fault strength calculation used here is the same as that employed by Uenishi and Rice [14]. In many calculations, before slip initiation, $\tau_s = f_s\sigma_n$ for a point on the fault with the varying σ_n. Evolution of f based on a rate- and state-dependent form is also widely used for nucleation of earthquake slip events [4].

Along the fault, the cracks constitute the closed and open portions for pressure diffusion depending on the opening magnitude or the fault conductivity. The governing equations for fluid flow in an open section, or a hydraulic fracture, are given below. If the hydraulic aperture offering fault conductivity is defined using an averaging scheme, the well-known cubic law can also provide a good estimate. Therefore, the fluid volumetric flux q_P is given by Poiseuille law,

$$q_P = -\frac{(w+b)^3}{\mu'}\frac{\partial p_f}{\partial x} \quad (17.5)$$

where b is the deformation-independent hydraulic aperture that contributes to fluid flow and $\mu' = 12\mu$.

For viscous fluid flow, pressure gradient and permeability will change along the fluid-occupied portion, in contrast to some simpler models using inviscid flow assumptions. If the fluid leak-off in the transverse direction of the fault is neglected, the mass continuity in a fault portion filled with fluids requires

$$\frac{\partial(w+b)}{\partial t} + \frac{\partial q_P}{\partial x} = 0 \quad (17.6)$$

The fluid diffusion in the closed fault portion depends on the fault permeability or the hydraulic aperture b. In the previous work [8], this longitudinal permeability k is assumed to be constant along the fault. However, the inelastic change of the porosity with slip should be accounted for and the permeability varies from its initial value. The evolution equation of fault permeability in terms of b may be associated with fault slip, for instance, the linear shear-induced dilation law [9]. A more complete approach requires connecting the change in permeability with gouge porosity to find a hydraulic diffusivity for pressure diffusion. For example, in terms of the hydraulic aperture, an empirical equation for permeability, such as Kozeny–Carman equation, could be used as follows

$$\frac{\phi}{\phi_0} = \frac{b}{b_0} \text{ and } \frac{k}{k_0} = \left(\frac{b}{b_0}\right)^3 \quad (17.7)$$

where $k_0 = b_0^2/12$.These assumptions are valid if $\phi - \phi_0$ is small in most situations. For a fracture channel, we have $\phi_0 = 1$.

The continuity of fluid mass in the porous medium leads to

$$\frac{\partial q_D}{\partial x} + \dot{m} = 0 \quad (17.8)$$

where the Darcy's flux q_D and the rate of the change of fluid mas $\dot{m}$ are, respectively,

$$q_D = -\rho\frac{k}{\mu}\frac{\partial p_f}{\partial x} \text{ and } \dot{m} = \rho\dot{\phi} \quad (17.9)$$

If the fluid is incompressible, substituting Eq. (17.9) in Eq. (17.8) leads to

$$\phi_0\frac{\partial b}{\partial t} + \frac{\partial}{\partial x}\left(\frac{b^3}{\mu'}\frac{\partial p_f}{\partial x}\right) = 0 \quad (17.10)$$

The boundary conditions for the proposed model are given as follows. At the inlet, the fluid flux into either side of the fracture is assumed to be the same when the problem considered is of geometric symmetry. For the 2-D problems considered the flux into each side is equal to half of the injection rate per unit fracture thickness Q_0, that is,

$$q_P(x=0) = Q_0/2 \quad (17.11)$$

and at the opening and sliding fronts, we have, respectively,

$$w(x = l_o) = 0 \text{ and } u(x = l_s) = 0 \tag{17.12}$$

and at the fluid front, we have

$$p(x = l_f) = 0 \tag{17.13}$$

The fluid front can be characterized by its moving velocity, which is equal to the volumetric flux divided by the hydraulic fracture width at the fluid front, that is,

$$\dot{l}_f = \lim_{x \to l_f} \frac{q_D(x, t)}{w(x, t) + b(x, t)} \tag{17.14}$$

indicating that the flux is finite at the fluid front. In addition, the opening and slip fronts can be tracked based on the opening and slip magnitudes. It is assumed that their velocities are so small that the inertia effect of rock deformation is not important.

Initially, the rock mass is stationary, and the frictional strength is sufficient to resist the applied shear stress. The fault portion ahead of the fluid front is empty of fluid. In the model, the fault is assigned a constant initial hydraulic aperture b_0, either constant or having a known specified functional behavior.

It must be noted that the convergence of numerical solutions must be tested if only one grid scheme is used for fracture deformation, fluid flow, and frictional slip, because the results show strong mesh sensitivity. For the methods with single-grid discretization, to capture the slip front in addition to the fluid front, a relatively small element size must be used, for example, in the boundary element method framework for solving the above equations. A coarse mesh can create artificial slip zone size oscillations [9]. The other solution methods include semi-analytical solutions based on a moving grid scheme [8], which can provide benchmark solutions to other numerical methods.

17.3 Summary of Modeling Results

In this section, we summarize the present knowledge of the modeling results on quasi-static aseismic slip development along a fault and the potential occurrence of dynamic slip. In the following, the results are categorized based on fault conductivity and fault geometrical and structural complexities.

17.3.1 Fully Closed Fractures

17.3.1.1 Constant Fault Permeability

The constant-permeability fluid-saturated fault simply provides a conduit for pressurized fluid movement that may lead to the nucleation of dynamic slip. If the fault permeability is given as constant, the pressure change along the infinite fluid-infiltrated fault under a constant injection rate can be expressed as [8]

$$p_f(x, t) - p_0 = \Delta p(t)\Pi(x) \tag{17.15}$$

where p_0 is reservoir pressure, $\Delta p(t) = \frac{2}{\sqrt{\pi}} \frac{Q_0}{\alpha} \sqrt{\alpha t}$, in which α is the fault diffusivity, proportional to fault permeability, and $\Pi(\xi) = exp\left(-\frac{\xi^2}{4}\right) - \frac{\sqrt{\pi}}{2}|\xi|\mathrm{Erfc}\left(\frac{|\xi|}{2}\right)$, in which $\xi = \frac{x}{\sqrt{\alpha t}}$. The pressure solution under the conditions of constant pressure injection for an infinite fluid-infiltrated fault can also be found in Garagash and Germanovich [8].

Even for the simple case with a mode II fracture in homogeneous solids, an increasing fluid pressure can eventually nucleate a dynamic slip process [16]. On the other hand, it must be noted that the aseismic slip initiates within the pressurized region and then continues to develop in a sustained manner beyond the pressure front before the unstable slip occurs. Therefore, the aseismic slip stage acts as a precursor for earthquake nucleation in the presence of hydraulic diffusion. As discussed by Uenishi and Rice [14], the nucleation length for dynamic slip based on the slip-weakening friction law is independent of the loading conditions in the absence of fluid flow. Garagash and Germanovich [8] have extended this consideration to the hydraulic shearing with the friction law of limited slip weakening with a residual strength.

In addition, if the fluid pressure level is moderate, their snap-through numerical results showed that the quasistatic solution under the constant pressure injection conditions can be recovered after a run-out distance of dynamic slip. This implies the transient nature of some dynamic slip processes.

As an important outcome in their study, Garagash and Germanovich [8] show that the slip stability depends on the relative values of the residual strength in the friction law and the background shear stress, as presented in a phase diagram. If the residual strength is larger than the background shear stress, the fluid pressure can reach a much larger value, but the fault is still ultimately stable. For a spatially varying pressure profile along the fault patch, this condition is helpful in predicting the arrest of fault growth, which may take place in the field.

17.3.1.2 Enhanced Fault Permeability

For most cases, fault permeability is unlikely to be constant and the fault may be unsaturated at the beginning. People have long recognized the correlation between slip development and permeability [17–19]. When a fault is in slipping, the surface roughness can create a dilatancy effect to enhance the closed fault permeability [20, 21]. In a field test, Guglielmi et al. [5] observed a 20-fold increase in permeability over the original value during the slow (~4 μ m/s) aseismic slip process after fluid injection into a natural fault. Baisch et al. [22] also measured a 20-fold permeability increase after fluid injection at a geothermal site. The basic relation for slip-induced dilation can be given as

$$\dot{b} = \beta b \dot{p}_f + \lambda \dot{u} \tag{17.16}$$

where the first term arises from the fluid and pore compressibility in the gouge layer. A constant slope λ with a saturated value for dilatancy is used in Zhang et al. [9] and a decreasing slope with slip is also used [23].

Substituting Eq. (17.16) into Eq. (17.10), we have a diffusion equation for fluid pressure along the closed fracture portion

$$\frac{\partial p_f}{\partial t} + \frac{\lambda}{\phi_0 \beta b}\frac{\partial u}{\partial t} + \frac{\partial}{\phi_0 \beta \mu' b \partial x}\left(b^3 \frac{\partial p_f}{\partial x}\right) = 0 \tag{17.17}$$

If b is constant, we can find the fault diffusivity $\alpha = k/\phi_0 \beta \mu$. If the dilatant hardening disappears, the second term on the left-hand side is eliminated [9]. It is well recognized that the dilatancy is strongly constrained and the dilation can be suppressed by the smoothing of small-scale fracture surface roughness [23].

The dilatant hardening can significantly increase the aseismic slip zone size and extend the injection time for quasistatic stable fault growth under a constant injection rate. This hardening effect relies on both λ and b_0. If they are larger and the injection rate is not too large, the dynamic slip can be avoided [9, 23].

It must be noted that the fault instability analysis of the coupled fracture growth and fluid flow problem addressed numerically is more complex in the presence of dilatancy. Numerical results must be treated with care, especially by a check for any mesh sensitivity and using a reasonable dilatancy hardening factor. The fluid front can lag behind the aseismic slip front (figure 15 in [9]) and the rapid slip occurs on the non-pressurized fault path, which is adjacent to the fluid-occupied region. This is consistent with the field experimental measurements [5] and the analytical results [4, 17, 24].

For a larger dilatancy effect, the slip zone extension experiences a faster, but not drastically higher rate, then the slip rate decreases progressively to a lower level [9], in contrast to the abrupt arrest of slip for earthquake slip in the absence of pressurized fluids. The fast slip events manifest themselves obviously transient natures. These slip and fluid flow behaviors would be used as the characteristics to find the boundaries of hydraulic fracturing treatment areas.

Numerical results demonstrated that the transition from aseismic to dynamic slip is significantly delayed by the dilatant hardening, using a cumulative 20-fold increase in permeability during a field test [5]. The permeability increase can facilitate fluid flow, but the injection pressure and slip rates are reduced accordingly to stabilize fault slip. However, it must be noted that the pressure drop is limited and the stress state at the points on the fault will be still in marginally pressurized regime in terms of the static friction law, the so-called conditionally stable regime. If the fault goes into a region with strong reduction in permeability, unstable rupture may be produced. The sudden loss and gain of the slip stability in this manner pose a difficulty in maintaining numerical convergency.

This conditionally stable regime is not vanishing for a point on the highly permeable fault. By using a perturbation of the fault conductivity, numerical results show that the pressure can still fall into the vicinity of the marginally pressurized regime [9]. The slip zone size can reach tens of meters if the critical slip weakening distance is centimeters or larger in size [14]. This indicates that the dilatant hardening is not short-lived, but evolves with the fault size and stress conditions unless the dilatancy effect is fully suppressed.

17.3.1.3 Fault Permeability Reduction

A permeability reduction can also take place because of fluid circulation in the fault. The mass convection carried by fluids from remote locations with subsequent mineral precipitation can partially seal the fault conduit. The formation permeability can also be altered by chemical reactions or fine particle movement if not well prepared for a fracturing treatment. For the fault healing due to fluid transport, at a higher flow rate, permeability can suffer a decrease of two orders of magnitude [25]. The small permeability can slow down fluid flow and lead to an increase in fault strength. This will give rise to slip deceleration and eventual healing of the slipping patch. Of course, the initiation site of slip healing is also important, as fluid pressure varies spatially on the fault. In general, the fault strengthening can reduce the slip zone size and can cause a growing slip zone to be arrested, but may provoke a strong earthquake.

Fault healing rates depend on pressure, temperature, and mineralogy. This process may promote a more heterogeneous spatial distribution of fault strength, thus generating high-frequency seismic events, in contrast to what we know about dilatancy effect. In addition, the healing effect can prolong the duration of slow slip until nucleation of seismic events. Due to the site heterogeneity, as discussed by [26], the introduction of fault sections that heal rapidly, will produce earthquake swarms. But how rock and fluid interact with each other in the presence of fault healing is worth further investigation.

It was long recognized that the dilatancy and healing effects can coexist in one fault. Their combination, behaving as the well-known fault valving effect [27], may control the evolution of fault permeability and microseismicity.

17.3.2 Partially Opened Fracture

When the fault is not subject to a large shear stress, it may be at a small angle with respect to the maximum principal stress. The elevated pressure under constant injection rate can open some parts of the fault, which become hydraulic fractures with mechanical opening. Fracture opening in the opening portion can generate a much larger permeability than shear-induced dilation in the closed portion to favor fluid invasion. In the opened patch, the frictional strength is zero and the normal stress redistribution can extend the slip zone size. In addition, the opening hydraulic fractures can change the stress of the closed parts if the fault is not planar [28].

17.3.2.1 Planar Fault

When the fault is planar and the inelastic permeability change is not considered, the governing equations for rock deformation are given by Eqs. (17.2) and (17.3) and the fluid flow is controlled by Eqs. (17.6). Slip does not affect the normal stress distribution along the fault and the hydraulic fracture problem can be solved independently [29]. The numerical results by [29] indicated that at the nucleation of dynamic slip, the slipping patch size shows an exponential growth law with the hydraulic fracture zone size. It is inferred that the nucleation of the dynamic slip occurs when this critical hydraulic fracture size is reached. Based on the results, the injected fluid volume becomes an important parameter in avoiding the occurrence of dynamic slip.

If the dilatant hardening is included in permeability calculation, it can change the fluid pressure distribution through the variation of fault permeability and the fluid flow relies on Eq. (17.17) too. In turn the fracture opening change is also affected due to their coupling. When the dilatancy effect is strong, the open zone size is constrained and even disappears (figure 15 in [9]). Generally, with the help of permeability enhancement, the pressure level is limited, as well as the opening zone size. Sometimes, no dynamic slip may occur when the injection rate is controlled.

17.3.2.2 Nonplanar Fault

Injection-induced seismic events are associated with specific geological formations, especially when a fault has some no-planar segments. In particular, the nucleated dynamic slip can be arrested by the change of slip planes and thus the slip rates are constrained by the section with higher frictional strength. The slip patterns may be separated by these special geological structures resembling earthquake swarms in volcanic environments [30] and the measured slip rates may vary significantly when the slip is reactivated at different parts and become unsteady [31]. In addition, these fault geometrical complexities can act as the barrier for dynamic slip, in a similar way as the additional rate-hardening effect in the friction law to arrest slip.

For a fault with dilational jogs, the slip rate heterogeneity can be measured in the field [32]. The dilational jogs can slow down fluid penetration, but the opened jogs can reduce the normal stress on the fault steps, both creating an increase in fluid pressure and inducing a rapid slip movement after the flow barrier is overcome [13]. The numerical results also indicated that the slip rate can reach values typical of microearthquake events caused by the jogs, accompanied by rising and then dropping pressure. It is found in their numerical results that the drop in injection pressure is finite and progressive. The entire microearthquake process is transient.

Additionally, the fracturing fluid can be trapped in high permeability portions on a fault or diverted into fault branches. Fault geometrical variations can produce much richer seismic slip patterns in the stimulated reservoirs, and on the other hand, the spatiotemporally varying slip signals may provide an alternative interpretation for fault structural complexities and local slip estimates. Along the long or short slipping patches, there are different deformation patterns for shear slip and tensile opening. The elastic stress change and pore pressure diffusion can yield much richer complexities in slip patterns on the fault, such as discrete slipping patches, for us to decipher the locations of dilational jogs.

17.4 Radiated Energy

Apart from the source mechanisms and manifested features of injection induced microseismicity, the amount of energy radiated from an earthquake or more specifically a fast slip transient caused by fluid injection can be measured using acoustic emission sensors based on earthquake signal treatments and spectral ratios. It is reported that the seismic energy is only a few percent of injection energy in both laboratory and in situ experiments [33, 34]. As we know, microseismic monitoring of hydraulic fractures is considered as one of the very few methods available to determine how reservoir rocks and bounding formations respond to fracture stimulation treatments. Although it is only a small portion of input energy, radiated energy carried by seismic waves would provide an important means to estimate the fracture shape and stress drop on the fault because its distribution along the fault can be measured. The quantification of radiated energy is still a challenge because the conventional method based on the similarity between the moment to radiated energy does not provide the fluid flow and fracture extents for injection induced fracture growth as there is no unified fracture model available to connect them.

There are many approaches to deriving the energy conservation laws for an earthquake event [35–39]. We will attempt to extend their energy budget consideration to an arrested fast slip event during hydraulic fracturing (e.g. an event that undergoes fast slip but then is arrested because of increased fault strength) when fault slip is stabilized by coupling with fluid flow, as we undertook to do in our previous study for a dry fault under heterogeneous stress states [40]. For a constant injection rate case, we do not compute the dynamic energy changes during a rapid slip event, but calculate the radiated energy based on the start and end stress and deformation states of the arrested event. For simplicity, the energy consumption in generating new fault surface and associated with microstructure changes is not considered here.

During an arrested event, the sudden fault slip jump can generate additional displacement and stress in the area V. Using the notations in Figure 17.1, the total stress and displacement at a point can be written as

$$\begin{aligned}\bar{\sigma}_{ij} &= \sigma_{ij} + \Delta\sigma_{ij}\\ \bar{v}_i &= v_i + \Delta v_i\end{aligned} \tag{17.18}$$

where $\Delta\sigma_{ij}$ and Δv_i are the corresponding increments of stress and displacement after the event. It is noted that they are both time-dependent for the actual dynamic slip process. In the following, instead of the rates of stress $\dot{\sigma}_{ij}$ and displacement $\dot{v}_i$, we use their increments $\Delta\sigma_{ij}(t)$ and $\Delta v_i(t)$, which are therefore a function of real time t.

Following the method used by Rudnicki and Freund [38] and Okubo et al. [36], a general energy budget for an earthquake can be written in the incremental form, equivalent to Eq. (37) in Rudnicki and Freund [38]. For an injection-induced event, additional energy parts associated with viscous flow must be added [41]. Therefore, we have the energy budget equation

$$\Delta E_P - E_R - \Delta K - \Delta U = \Delta W_f + \Delta D_f \tag{17.19}$$

where each energy component is given as:

1) Radiated energy

$$E_R = -\int_{t_0}^{t_1} dt \int_S \Delta\sigma_{ij}(t) n_j \dot{v}_i dS \tag{17.20}$$

where t_0 and t_1 are the starting and end time instants for an event, the difference of which is assumed to be large enough that the seismic waves created by the arrested event totally pass the boundary S. If S is limited in size, the time difference can be small.

2) Input energy change ΔE_P includes the first part which is a potential upon using a fixed stress σ_{ij} on S between t_0 and t_1

$$\Delta E_P^1 = \int_S \sigma_{ij} n_j \Delta v_i dS \tag{17.21}$$

and the second part is associated with the input power by injection

$$\Delta E_P^2 = \int_{t_0}^{t_1} Q_0 p_{f,in} dt \tag{17.22}$$

where $p_{f,\,in}$ is the injection pressure.

3) Kinetic energy change between t_0 and t_1

$$\Delta K = K(t_1) \tag{17.23}$$

in which the kinetic energy is defined as $K = 1/2\int_V \rho \dot{v}^2 dV$;

4) Strain energy change between t_0 and t_1

$$\Delta U = U(t_1) - U(t_0) \tag{17.24}$$

in which the reversible strain energy is defined as $U = 1/2\int_V \sigma_{ij}\varepsilon_{ij} dV$;

5) Dissipation due to the work done in overcoming the fault strength

$$\Delta W_f = -\int_{t_0}^{t_1} dt \int_L \bar{\sigma}_{ij} n_j \dot{v}_i dL = \int_L \left(\int_{D_1}^{D_2} \tau_s du \right) dL \tag{17.25}$$

6) Dissipation in viscous flow in the fracture

$$\Delta D_f = -\int_{t_0}^{t_1} dt \int_L q_P \frac{\partial p_f}{\partial L} dL \tag{17.26}$$

It must be noted that we use the full fault length L in above definitions, instead of using the individual moving fronts, for simplicity.

In most cases, we can assume that the reversible kinetic energy vanishes at the start and end of the arrested event and therefore ΔK disappears in Eq. (17.19). On the other hand, during an event, the items associated with fluid flow still change slowly, but the slip zone size may experience a rapid increase. Therefore, ΔE_P^2 and ΔD_f can be calculated with a good accuracy. For the reversible ΔE_P^1 and ΔU, we can obtain their values through their quasi-static solutions at t_0 and t_1 as Rudnicki and Freund [38] did. Only ΔW_f involves the complex process of slip development on the fault, depending on the selected friction law. For the linear

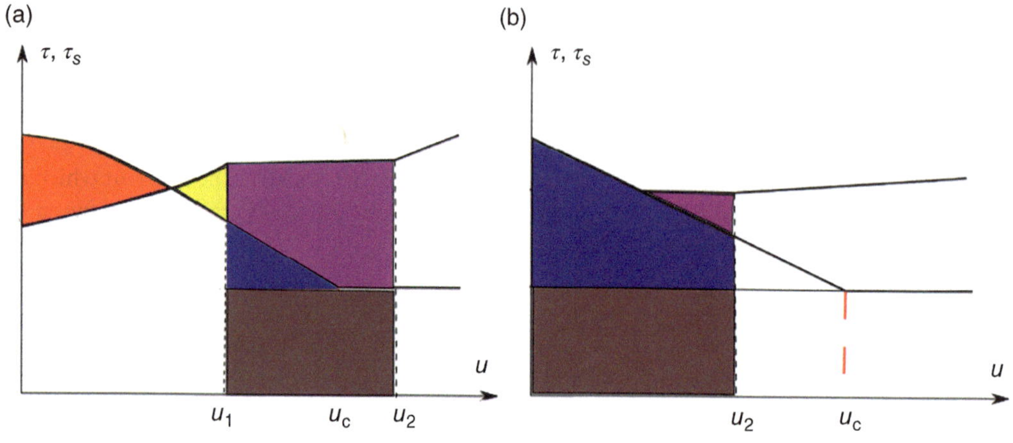

Figure 17.2 Variations of resultant shear stress and frictional strength during a microseismic event for the points (a) in the slipping patch prior to the event and (b) entering the slipping patch during the event.

slip weakening friction law, the numerical results for the points both in the slipping patch prior to the event and entering the slipping patch during the event are provided in Figure 17.2. Clearly, the linear decreasing frictional strength with increasing slip can be obeyed. This can simplify the calculation of ΔW_f.

To calculate E_R, it is better to transfer all items to their counterparts on the fault if the boundary integral formulation of the governing equations is used. In particular, the problem for a constant remote stress state, can be reduced to two subproblems based on the superposition method. One subproblem is the infinite medium containing no fault, and the fault is subject to $-\sigma_n$ and $-\tau$ in Figure 17.3a. One can find that this subproblem does not give rise to any change in either ΔU or ΔE_P^1 between the two time instants because the stress in the body remains the same, as well as the deformation. The other subproblem is shown in Figure 17.3b with known τ and τ_s on the fault plane at the two-time instants. This subproblem with no stress on remote boundary S can be solved through the energy method in fracture mechanics [42]. Therefore, we can calculate ΔE_P^1 and ΔU separately based on the subproblems in Figure 17.3b.

Following fracture energy analysis, ΔE_P^1 should be equal to the work done on the fault plane by the resultant force from the applied stresses. Hence, we can calculate ΔE_P^1 as

$$\Delta E_P^1 = \int_L \tau(u_2 - u_1)dL \tag{17.27}$$

where the slip at a point on L changes from u_1 to u_2 during the event.

Additionally, the strain energy at the start and end of the arrested event can be computed based on the following identity for a static body [37]

$$U = \frac{1}{2}\int_L \sigma_{ij} v_i n_j dL \tag{17.28}$$

Therefore, the strain energy at the start and end of the arrested event can be evaluated by the integration of unbalanced stresses (e.g. $\tau - \tau_s$ and σ_n in Figure 17.3b) and their induced displacement components over the slipping zone. Up to now, the calculation of E_R only involves the stress and displacement on the fault for the given configuration.

Also, we can combine Eq. (17.27) and Eq. (17.25) as they are related to the traction and displacement on the fault plane. Then we can write their sum as

$$\Delta W = \Delta E_P^1 - \Delta W_f = \int_L \left(\int_{D_1}^{D_2} (\tau - \tau_s) du \right) dL \tag{17.29}$$

Therefore, E_R is given by

$$E_R = \Delta W - \Delta U + \left(\Delta E_P^2 - \Delta D_f\right) \tag{17.30}$$

where the last two items on the right-hand side are associated with fluid flow in the fracture. According to the requirement for fluid flow along the fracture, the fluid volumetric flux during the event is finite, and so is the fluid pressure change, as confirmed by numerical results [13]. If the duration for the event is extremely short, these two terms vanish and Eq. (17.30) therefore is the same as the formula presented by Yang et al. [40].

17.5 Conclusions and Future Work

In this chapter, we attempt to review some analytical and numerical approaches to studying hydraulic-fracturing induced slip along a fault. With simplifying assumptions such as assigning a friction law and assuming an infinite medium, we have formulated the hydraulic driven faulting problems based on the coupled multiple physics, considering rock deformation, fluid flow in the fracture and friction on the fault. The salient features of these problems are the

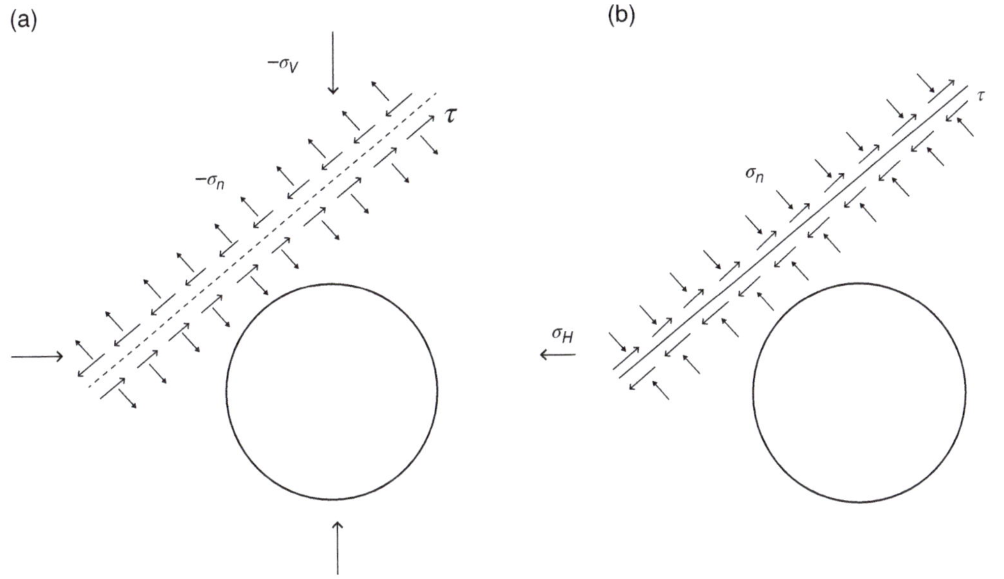

Figure 17.3 Decomposition of the 2-D problem for a fault near a tunnel illustrated in Figure 17.1b during the seismic event into two subproblems: (a) unfaulted body subjected to two principal stresses (σ_H and σ_V) and (b) faulted body. Note that the normal (σ_n) and shear (τ) stresses at a point on the fault in (b) have the same value as those in (a), but with opposite signs, and τ_s that varies with slip is not shown in (b). The numerical solution process determines τ_s for portions of the fault that undergo slip.

moving boundaries associated with different mechanisms. The nucleation of dynamic slip requires care in validating the numerical results. Particularly, we summarized the numerical results for nucleation of dynamic slip for both closed faults and partially opened hydraulic fractures. The dilatant strengthening effect of the fault surfaces can delay or eliminate the nucleation of dynamic slip.

The nucleation of dynamic slip is asserted to depend on the inherent physical and geometrical properties of the fault. Although, the slip rate is constrained in the numerical results, some important features associated with dynamic slip have been obtained. For instance, the slip front with strong movement can occur ahead of the fluid zone and with increasing fault permeability, a progressive slip rate reduction may imply fluid filling. The fault geometrical heterogeneities can result in perturbations to the local stress field and generate and then arrest a rapid slipping process. A discontinuous slip patterns may be produced, exhibiting many features of high-frequency microseismic events.

Quantifying the radiated energy for a microseismic event is useful in estimating the slip zone size and the stress drop associated with earthquake events. Based on the quasistatic results, we have presented a method to calculate the radiated energy for a fast slip transient.

Many important issues are not included in this chapter. Most models neglect the fluid leak-off from the hydraulic fracture into the surrounding rock. The fault plane used for the models is rather simple compared to the actual situation and a three-dimensional fault geometry considering the creation of new fault planes and off-fault damage is required [43]. Also, the friction law used so far is too simple to describe the comprehensive behaviors as the rate- and state-dependent friction laws do [44, 45]. A more accurate energy calculation may require elastodynamic models [46]. In addition, the studies on slip development in the field and experiments are not summarized here. Validation of the physical principles presented against what we observe is urgently needed.

Acknowledgment

The authors thank Brice Lecampion for carefully reviewing the manuscript.

References

1 Bao, X. and Eaton, D.W. (2016). Fault activation by hydraulic fracturing in western Canada. *Science* 354 (6318): 1406–1409. https://doi.org/10.1126/science.aag2583.

2 Ellsworth, W.L. (2013). Injection-induced earthquakes. *Science* 341 (6142): 1225942. https://doi.org/10.1126/science.1225942.

3 National Research Council (2013). *Induced Seismicity Potential in Energy Technologies*. Washington, DC: The National Academies Press https://doi.org/10.17226/13355.

4 Eyre, T.S., Eaton, D.W., Garagash, D.I. et al. (2019). The role of aseismic slip in hydraulic fracturing–induced seismicity. *Science Advances* 5 (8): 1–11. https://doi.org/10.1126/sciadv.aav7172.

5 Guglielmi, Y., Cappa, F., Avouac, J. et al. (2015). Seismicity triggered by fluid injection – induced aseismic slip. *Science* 348 (6240): 1224–1227. https://doi.org/10.1126/science.aab0476.

6 Das, I. and Zoback, M.D. (2013). Long-period, long-duration seismic events during hydraulic fracture stimulation of a shale gas reservoir – part 2: location and mechanisms. *Geophysics* 78 (6): KS109–KS117. https://doi.org/10.1190/1.3609093.

7 Cappa, F., Guglielmi, Y., Nussbaum, C., and Birkholzer, J. (2018). On the relationship between fault permeability increases, induced stress perturbation, and the growth of aseismic slip during fluid injection. *Geophysical Research Letters* 45 (20): 11012–11020. https://doi.org/10.1029/2018GL080233.

8 Garagash, D.I. and Germanovich, L.N. (2012). Nucleation and arrest of dynamic slip on a pressurized fault. *Journal of Geophysical Research: Solid Earth* 117 (February): 1–27. https://doi.org/10.1029/2012JB009209.

9 Zhang, X., Jeffrey, R.G., and He, C. (2016). A numerical model for the effect of permeability change on faulting during fluid injection. *Journal of Geophysical Research: Solid Earth* 124 (2): 2080–2101. https://doi.org/10.1029/2018JB016073.

10 Wibberley, C.A.J. and Shimamoto, T. (2002). Internal structure and permeability of major strike-slip fault zones: the median tectonic line in Mie prefecture, Southwest Japan. *Journal of Structural Geology* 25 (1): 59–78. https://doi.org/10.1016/S0191-8141(02)00014-7.

11 Caine, J.S., Evans, J.P., and Forster, C.B. (1996). Fault zone architechture and permeability structure. *Geology* 24 (11): 1025–1028. https://doi.org/10.1130/0091-7613(1996)024<1025.

12 Warpinski, N.R. (2014). A review of hydraulic-fracture induced microseismicity. *48th US Rock Mechanics/Geomechanics Symposium*, ARMA 14-7774. 1–4 June 2014, Minneapolis, Minnesota, American Rock Mechanics Association.

13 Zhang, X., Wu, B., Jeffrey, R.G. et al. (2019). Changes of slip rate and slip-plane orientation by fault geometrical complexities during fluid injection. *Journal of Geophysical Research: Solid Earth* 124 (8): 9226–9246. https://doi.org/10.1029/2019JB017856.

14 Uenishi, K. and Rice, J.R. (2003). Universal nucleation length for slip-weakening rupture instability under nonuniform fault loading. *Journal of Geophysical Research* 108 (B1): https://doi.org/10.1029/2001JB001681.

15 Cocco, M., Tinti, E., Marone, C., and Piatanesi, A. (2009). Scaling of slip weakening distance with final slip during dynamic earthquake rupture. In: *International Geophysics*, 1ste, vol. 94. Elsevier Inc. https://doi.org/10.1016/S0074-6142(08)00007-7.

16 Viesca, R. (2021). Self-similar fault slip in response to fluid injection. *Journal of Fluid Mechanics* https://arxiv.org/abs/2102.

17 McClure, M.W. and Horne, R.N. (2011). Investigation of injection-induced seismicity using a coupled fluid flow and rate/state friction model. *Geophysics* 76 (6): WC181–WC198. https://doi.org/10.1190/geo2011-0064.1.

18 Pearson, C. (1981). The relationship between microseismicity and high pore pressures during hydraulic stimulation experiments in low permeability granitic rocks. *Journal of Geophysical Research* 86: 7855–7864.

19 Shapiro, S.A., Huenges, E., and Borm, G. (1997). Estimating the crust permeability from fluid-injection-induced seismic emission at the KTB site. *Geophysical Journal International* 131 (2): F15–F18. https://doi.org/10.1111/j.1365-246X.1997.tb01215.x.

20 Power, W.L., Tullis, T.E., Brown, S.R. et al. (1987). Roughness of natural fault surfaces. *Geophysical Research Letters* 14 (1): 29–32.

21 Zhang, X., Jeffrey, R.G., and María Llanos, E. (2005). On plane-strain fluid-driven shear fracture propagation in elastic solids. *Geophysical Journal International* 163 (1): 419–430. https://doi.org/10.1111/j.1365-246X.2005.02755.x.

22 Baisch, S., Vörös, R., Rothert, E. et al. (2010). A numerical model for fluid injection induced seismicity at Soultz-sous-Forets. *International Journal of Rock Mechanics and Mining Sciences* 47: 405–413.

23 Ciardo, F. and Lecampion, B. (2019). Effect of dilatancy on the transition from aseismic to seismic slip due to fluid injection in a fault. *Journal of Geophysical Research: Solid Earth* 124 (4): 3724–3743. https://doi.org/10.1029/2018JB016636.

24 Bhattacharya, P. and Viesca, R.C. (2019). Fluid-induced aseismic fault slip outpaces pore-fluid migration. *Science* 364 (6439): 464–468. https://doi.org/10.1126/science.aaw7354.

25 Elkhoury, J.E., Detwiler, R.L., and Ameli, P. (2015). Can a fractured caprock self-heal? *Earth and Planetary Science Letters* 417: 99–106. https://doi.org/10.1016/j.epsl.2015.02.010.

26 McLaskey, G.C., Thomas, A.M., Glaser, S.D., and Nadeau, R.M. (2012). Fault healing promotes high-frequency earthquakes in laboratory experiments and on natural faults. *Nature* 491 (7422): 101–104. https://doi.org/10.1038/nature11512.

27 Sibson, R.H. (1992). Implications of fault-valve behaviour for rupture nucleation and recurrence. *Tectonophysics* 211 (1–4): 283–293. https://doi.org/10.1016/0040-1951(92) 90065-E.

28 Jeffrey, R.G., Zhang, X., and Thiercelin, M. (2009). Hydraulic fracture offsetting in naturally fractured reservoirs: quantifying a long-recognized process. *Proceedings of SPE Hydraulic Fracturing Technology Conference*. https://doi.org/10.2118/119351-MS

29 Azad, M., Garagash, D.I., and Satish, M. (2017). Nucleation of dynamic slip on a hydraulically fractured fault. *Journal of Geophysical Research: Solid Earth* 1–19: https://doi.org/10.1002/2016JB013835.

30 Shelly, D.R., Ellsworth, W.L., and Hill, D.P. (2016). Fluid-faulting evolution in high definition: connecting fault structure and frequency-magnitude variations during the 2014 Long Valley Caldera, California, earthquake swarm. *Journal of Geophysical Research: Solid Earth* 121: 1776–1795. https://doi.org/10.1002/2015JB012719.Received.

31 Hatch, R.L., Abercrombie, R.E., Ruhl, C.J., and Smith, K.D. (2020). Evidence of aseismic and fluid-diven processes in a small complex seismic swarm near Virginia City, Nevada. *Geophysical Research Letters* 47 (4): 1–11. https://doi.org/10.1029/2019GL085477.

32 Rutledge, J.T., Phillips, W.S., and Mayerhofer, M.J. (2004). Faulting induced by forced fluid injection and fluid flow forced by faulting: an interpretation of hydraulic-fracture microseismicity, Carthage Cotton Valley gas field, Texas. *Bulletin of the Seismological Society of America* 94 (5): 1817–1830. https://doi.org/10.1785/012003257.

33 De Barros, L., Cappa, F., Guglielmi, Y. et al. (2019). Energy of injection-induced seismicity predicted from in-situ experiments. *Scientific Reports* 9 (1): 1–12. https://doi.org/10.1038/s41598-019-41306-x.

34 Goodfellow, S.D., Nasseri, M.H.B., Maxwell, S.C., and Young, R.P. (2015). Hydraulic fracture energy budget: insights from the laboratory. *Geophysical Research Letters* 42 (9): 3179–3187. https://doi.org/10.1002/2015GL063093.

35 Kaneko, Y., Fukuyama, E., and Hamling, I.J. (2017). Slip-weakening distance and energy budget inferred from near-fault ground deformation during the 2016 Mw7.8 Kaikōura earthquake. *Geophysical Research Letters* 44 (10): 4765–4773. https://doi.org/10.1002/2017GL073681.

36 Okubo, K., Bhat, H.S., Rougier, E. et al. (2019). Dynamics, radiation and overall energy budget of earthquake rupture with coseismic off-fault damage. *Journal of Geophysical Research: Solid Earth* 124 (11): 11771–11801. https://doi.org/10.1029/2019JB017304.

37 Rivera, L. and Kanamori, H. (2005). Representations of the radiated energy in earthquakes. *Geophysical Journal International* 162 (1): 148–155. https://doi.org/10.1111/j.1365-246X.2005.02648.x.

38 Rudnicki, J.W. and Freund, L.B. (1981). On energy radiation from seismic sources. *Bulletin of the Seismological Society of America* 71 (3): 585–595. https://doi.org/10.2307/1779766.

39 Tinti, E., Spudich, P., and Cocco, M. (2005). Earthquake fracture energy inferred from kinematic rupture models on extended faults. *Journal of Geophysical Research: Solid Earth* 110 (12): 1–25. https://doi.org/10.1029/2005JB003644.

40 Yang, J., Yang, D., Zhang, X. et al. (2020). Energy budget and fast rupture on a near-excavation fault: implications for mitigating induced seismicity. *Journal of Geophysical Research: Solid Earth* 1–23: https://doi.org/10.1029/2020jb019360.

41 Lecampion, B. and Detournay, E. (2007). An implicit algorithm for the propagation of a hydraulic fracture with a fluid lag. *Computer Methods in Applied Mechanics and Engineering* 196 (November): 4863–4880. https://doi.org/10.1016/j.cma.2007.06.011.

42 Freund, L.B. (1990). *Dynamic Fracture Mechanics*. Cambridge University Press.

43 Sáez, A., Lecampion, B., Bhattacharya, P., and Viesca, R.C. (2021). Three-dimensional fluid-driven stable frictional ruptures. *Journal of the Mechanics and Physics of Soilds* https://eartharxiv.org/repository/view/2686.

44 Dublanchet, P. (2019). Fluid driven shear cracks on a strengthening rate-and-state frictional fault. *Journal of the Mechanics and Physics of Solids* 132: 103672. https://doi.org/10.1016/j.jmps.2019.07.015.

45 Garagash, D.I. (2021). Fracture mechanics of rate-and-state faults and fluid injection induced slip. *Philosophical Transactions Of the Royal Society A: Mathematical, Physical and Engineering Sciences* https://royalsocietypublishing.org/doi/abs/10.1098/rsta.2020.0129.

46 Lapusta, N., Rice, J.R., Ben-Zion, Y., and Zheng, G. (2000). Elastodynamic analysis for slow tectonic loading with spontaneous rupture episodes on faults with rate- and state-dependent friction. *Journal of Geophysical Research* 105 (B10): 23765–23789. https://doi.org/10.1029/2000JB900250.

Index

Note: Page numbers in *italics* refer to figures; page numbers in **bold** refer to tables.

Mechanics of Hydraulic Fracturing: Experiment, Model, and Monitoring, First Edition.
Edited by Xi Zhang, Bisheng Wu, Diansen Yang, and Andrew Bunger.

g

k

l

m

Printed in the USA/Agawam, MA
September 16, 2026

910661.198